PLANT OMICS AND CROP BREEDING

PLANT OMICS AND CROP BREEDING

Edited by
Sajad Majeed Zargar, PhD
Vandna Rai, PhD

Apple Academic Press Inc.
3333 Mistwell Crescent
Oakville, ON L6L 0A2 Canada

Apple Academic Press Inc.
9 Spinnaker Way
Waretown, NJ 08758 USA

©2017 by Apple Academic Press, Inc.

First issued in paperback 2021

No claim to original U.S. Government works

ISBN-13: 978-1-77463-047-1 (pbk)
ISBN-13: 978-1-77188-455-6 (hbk)

Library and Archives Canada Cataloguing in Publication

Plant omics and crop breeding / edited by Sajad Majeed Zargar, PhD, Vandna Rai, PhD.

Includes bibliographical references and index.
Issued in print and electronic formats.

ISBN 978-1-77188-455-6 (hardcover).--ISBN 978-1-31536593-0 (PDF)

1. Agricultural biotechnology. 2. Crop improvement. 3. Plant breeding. I. Zargar, Sajad Majeed, author, editor II. Rai, Vandna, author, editor

| SB106.B56P63 2017 | 630 | C2017-900920-6 | C2017-900921-4 |

Library of Congress Cataloging-in-Publication Data

Names: Zargar, Sajad Majeed, editor. | Rai, Vandna, editor.
Title: Plant omics and crop breeding / editors: Sajad Majeed Zargar, Vandna Rai.
Description: Waretown, NJ : Apple Academic Press, 2017. | Includes bibliographical references and index.
Identifiers: LCCN 2017003498 (print) | LCCN 2017005254 (ebook) | ISBN 9781771884556 (hardcover : alk. paper) | ISBN 9781315365930 (ebook)
Subjects: LCSH: Plant biotechnology--Research. | Plant breeding--Research.
Classification: LCC SB106.B56 P595 2017 (print) | LCC SB106.B56 (ebook) | DDC 631.5/3--dc23
LC record available at https://lccn.loc.gov/2017003498

Apple Academic Press also publishes its books in a variety of electronic formats. Some content that appears in print may not be available in electronic format. For information about Apple Academic Press products, visit our website at **www.appleacademicpress.com** and the CRC Press website at **www.crcpress.com**

ABOUT THE EDITORS

Sajad Majeed Zargar, PhD

Sajad Majeed Zargar, Ph.D, is currently an Assistant Professor at Sher-e-Kashmir University of Agricultural Sciences and Technology of Kashmir (SKUAST-Kashmir) in India. He was previously a Visiting Professor at the Nara Institute of Science and Technology, Japan. He has worked as an Assistant Professor at SKUAST-Jammu, Baba Ghulam Shah Badshah University (BGSB) in India. He has also worked as a Scientist at Advanta India Limited, Hyderabad, India, and TERI (The Energy and Resources Institute), New Delhi, India. Dr. Zargar has received several awards for his work and research. His editorial activities and scientific membership include publishing research and review articles in international journals and as a reviewer. He has been affiliated with the *Journal of Phytology,* the *Australian Journal of Crop Sciences,* the *African Journal of Biotechnology, Scientia Horticulture, Methods in Ecology and Evolution, Journal of Soil Science and Plant Nutrition,* and many others. Dr. Zargar has been invited to give many lectures at professional meetings and workshops and has received grants for research projects under his supervision.

Vandna Rai, PhD

Vandna Rai, Ph.D, is currently a Senior Scientist at the National Research Centre on Plant Biotechnology, IARI, in New Delhi, India. She was previously a Senior Scientist at the Directorate of Rice Research in Hyderabad, India. She has received awards, such as from the International Union for Biochemistry and Molecular Biology (IUBMB) at Kyoto, Japan, and has received travel grants for attendance at American Society for Plant Biology (ASPB) meetings. Dr. Rai is involved in teaching courses at postgraduate levels. With over 16 years of active research experience working the area of plant biochemistry, molecular biology, and genomics, she has published research articles in peer-reviewed journals and has written several book chapters. Dr. Rai has mapped several important genes and QTLs for abiotic stress tolerance in rice and has been involved in proteome profiling of rice and pigeonpea to find salt stress tolerant genes and proteins.

CONTENTS

LIST OF CONTRIBUTORS

Abha Agnihotri
Amity Institute of Microbial Technology, Amity University, Noida 201313, Uttar Pradesh, India. E-mail: agnihotri.abha@gmail.com

Ganesh Kumar Agrawal
Research Laboratory for Biotechnology and Biochemistry (RLABB), GPO Box 13265, Kathmandu, Nepal; GRADE (Global Research Arch for Developing Education) Academy Private Limited, Adarsh Nagar 13, Main Road, Birgunj, Nepal

Saurabh Anand
Department of Botany, University of Delhi, New Delhi 110007, India

Jasroop K. Aneja
Amity Institute of Microbial Technology, Amity University, Noida 201313, Uttar Pradesh, India

Sandip Das
Department of Botany, University of Delhi, New Delhi 110007, India. E-mail: sdas@botany.du.ac.in; sandipdas04@gmail.com

Renu Deswal
Department of Botany, University of Delhi, New Delhi 110007, India

Madhura Bhat G.
Department of Crop Physiology, University of Agricultural Sciences, GKVK, Bangalore, India

B. B. Gupta
Division of Plant Breeding and Genetics, Sher-e-Kashmir University of Agricultural Sciences and Technology of Jammu, Chatha, Jammu 180 009, Jammu and Kashmir, India

Ajai Prakash Gupta
Quality Control and Assurance Division, Indian Institute of Integrative Medicine, Jammu180001, Jammu and Kashmir, India

Ravi Gupta
Department of Plant Bioscience, Life and Industry Convergence Research Institute, Pusan National University, Miryang 627707, South Korea. E-mail: ravigupta@pusan.ac.kr

Suphla Gupta
Plant Biotechnology Division, Indian Institute of Integrative Medicine, Jammu180001, Jammu and Kashmir, India; Faculty at Academic of Scientific and Innovative Science, Indian Institute of Integrative Medicine, Jammu180001, Jammu and Kashmir, India. E-mail: suphlabg@gmail.com

M. Tofazzal Islam
Department of Biotechnology, Bangabandhu Sheikh Mujibur Rahman Agricultural University, Gazipur 1706, Bangladesh

Ajay Jain
National Research Centre on Plant Biotechnology, Indian Agriculture Research Institute, New Delhi 110012, India

Neha Jain
National Research Centre on Plant Biotechnology, Indian Agriculture, Research Institute, New Delhi 110012, India

Nelofer Jan
Plant Molecular Biology Laboratory, Department of Botany, University of Kashmir, Hazratbal, Srinagar 190006, Jammu and Kashmir, India

Riffat John
Plant Molecular Biology Laboratory, Department of Botany, University of Kashmir, Hazratbal, Srinagar 190006, Jammu and Kashmir, India.

A. B. M. Khaldun
Bangladesh Agricultural Research Institute, Gazipur 1701, Bangladesh

Saima Khan
Plant Biotechnology Division, Indian Institute of Integrative Medicine, Jammu 180001, Jammu and Kashmir, India

Nisha Khatri
Department of Botany, University of Delhi, New Delhi, India

Divya Kilam
Amity Institute of Microbial Technology, Amity University, Noida 201313, Uttar Pradesh, India

Yong Chul Kim
Department of Plant Bioscience, Life and Industry Convergence Research Institute, Pusan National University, Miryang 627707, South Korea

Sun Tae Kim
Department of Plant Bioscience, Life and Industry Convergence Research Institute, Pusan National University, Miryang 627707, South Korea. E-mail: stkim71@pusan.ac.kr

Manoj Kumar
Amity Institute of Microbial Technology, Amity University, Noida 201313, Uttar Pradesh, India

Ram Kumar
National Research Centre on Plant Biotechnology, Indian Agriculture Research Institute, New Delhi 110012, India

Sajad Ahmad Lone
Plant Biotechnology Division, Indian Institute of Integrative Medicine, Jammu 180001, Jammu and Kashmir, India

Reetika Mahajan
School of Biotechnology, S K University of Agricultural Sciences and Technology of Jammu, Chatha, Jammu 180009, Jammu and Kashmir, India

Bisma Malik
Department of Bioresources, University of Kashmir, Hazratbal, Srinagar, Jammu and Kashmir, India

Malik Muzafar Manzoor
Plant Biotechnology Division, Indian Institute of Integrative Medicine, Jammu 180001, Jammu and Kashmir, India

Khalid Z. Masoodi
Centre for Plant Biotechnology, Division of Biotechnology, S K University of Agricultural Sciences and Technology of Kashmir, Shalimar, Srinagar 190025, Jammu and Kashmir, India

Chul Woo Min
Department of Plant Bioscience, Life and Industry Convergence Research Institute, Pusan National University, Miryang 627707, South Korea

Pragya Mishra
National Research Centre on Plant Biotechnology, Indian Agriculture Research Institute, New Delhi 110012, India

Vagish Mishra
National Research Centre on Plant Biotechnology, Indian Agriculture Research Institute, New Delhi 110012, India

Yashwanti Mudgil
Department of Botany, University of Delhi, New Delhi, India

Imtiyaz Murtaza
Division of Post-Harvest Technology, SKAUST-K, Shalimar, Srinagar, Jammu and Kashmir, India

Roohi Mushtaq
Department of Biotechnology, S P College Srinagar, Jammu and Kashmir, India

Karaba N. Nataraja
Department of Crop Physiology, University of Agricultural Sciences, GKVK, Bangalore, India. E-mail: nataraja_karaba@yahoo.com

Muslima Nazir
Department of Botany, Jamia Hamdard University, New Delhi, India

Pankaj Pandotra
School of Biotechnology, Sher-e-Kashmir University of Agricultural Science and Technology of Jammu, Jammu and Kashmir, India

Tanveer Bilal Pirzadah
Department of Bioresources, University of Kashmir, Hazratbal, Srinagar, Jammu and Kashmir, India

Irfan Qureshi
Department of Biotechnology, JMI, New Delhi

Rashmi Rai
School of Biotechnology, Faculty of Science, Banaras Hindu University, Varanasi 221005, India

Vandna Rai
National Research Centre on Plant Biotechnology, Indian Agriculture Research Institute, New Delhi 110012, India. E-mail: vandnarai2006@gmail.com

Meenakshi Raina
School of Biotechnology, Sher-e-Kashmir University of Agricultural Sciences & Technology of Jammu, Chatha, Jammu 180009, Jammu and Kashmir, India

Vaseem Raja
Plant Molecular Biology Laboratory, Department of Botany, University of Kashmir, Hazratbal, Srinagar 190006, Jammu and Kashmir, India

Randeep Rakwal
Research Laboratory for Biotechnology and Biochemistry (RLABB), GPO Box 13265, Kathmandu, Nepal; GRADE (Global Research Arch for Developing Education) Academy Private Limited, Adarsh Nagar 13, Main Road, Birgunj, Nepal; Faculty of Health and Sport Sciences and Tsukuba International Academy for Sport Studies (TIAS), University of Tsukuba, 1-1-1 Tennoudai, Tsukuba 3058574, Ibaraki, Japan

Reiaz Ul Rehman
Department of Bioresources, University of Kashmir, Hazratbal, Srinagar, Jammu and Kashmir, India

Parvathi M. S
Department of Crop Physiology, University of Agricultural Sciences, GKVK, Bangalore, India

R. K. Salgotra
School of Biotechnology, S K University of Agricultural Sciences and Technology of Jammu, Chatha, Jammu 180009, Jammu and Kashmir, India

Shruti Sharma
Department of Botany, University of Delhi, New Delhi 110007, India

B. D. Singh
School of Biotechnology, Faculty of Science, Banaras Hindu University, Varanasi 221005, India. E-mail: brahmadsingh@gmail.com

A.K. Singh
Collage of Agriculture and Research Station, 497335, Chhattisgarh, India

Nagendra Kumar Singh
National Research Centre on Plant Biotechnology, Indian Agriculture Research Institute, New Delhi 110012, India

Neer K. Singh
Department of Botany, University of Delhi, New Delhi 110007, India

Nisha Singh
National Research Centre on Plant Biotechnology, Indian Agriculture Research Institute, New Delhi 110012, India

Inayatullah Tahir
Department of Bioresources, University of Kashmir, Hazratbal, Srinagar, Jammu and Kashmir, India

Mohammed Shalim Uddin
Bangladesh Agricultural Research Institute, Gazipur 1701, Bangladesh

Mubashir Ahmad Wani
Plant Molecular Biology Laboratory, Department of Botany, University of Kashmir, Hazratbal, Srinagar 190006, Jammu and Kashmir, India

Umer Majeed Wani
Plant Molecular Biology Laboratory, Department of Botany, University of Kashmir, Hazratbal, Srinagar 190006, Jammu and Kashmir, India

Sajad Majeed Zargar
Centre for Plant Biotechnology, Division of Biotechnology SKUAST-K, Shalimar, Srinagar190025, Jammu and Kashmir, India. E-mail: smzargar@gmail.com

LIST OF ABBREVIATIONS

ACE	angiotensin converting enzyme
2D	two-dimensional
2DE	two-dimensional gel electrophoresis
ABA	abscisic acid
AB-QTL	advanced-backcross QTL
ABREs	ABA-responsive elements
ACE	angiotensin converting enzyme
ACMV	African cassava mosaic virus
ACN	acetonitrile
Adh	alcohol dehydrogenase
AFLP	amplified fragment length polymorphism
AGI	above ground index
ALSV	Apple latent spherical virus
amiRNA	artificial miRNA
AP	aecial pustules
APOX	antioxidant enzymes ascorbate peroxidase
APS	ATP sulfurylase
APXs	ascorbate peroxidases
ARF	auxin response factor
AS	alternative splicing
AsA	ascorbic acid
ASI	anthesis-silking interval
ATAF	*Arabidopsis* transcription activation factor
ATP	adenosine triphosphate
BAC	bacterial artificial chromosome
BaMMV	Barley mild mosaic virus
BaYMV	Barley yellow mosaic virus
BBS	bacterial brown spot
BGM	botrytis grey mold
bHLH	basic helix-loop helix
BiFC	bimolecular fluorescence complementation
BLSOM	batch learning self-organizing mapping
BMV	Brome mosaic virus
BnCBL1	*Brassica napus Calcineurin* B-Like 1

BnCIPK6	*Brassica napus* CBL-interacting protein kinase
bp	base-pair
BPH	brown planthopper
BPMV	Bean pod mottle virus
BSA	bulked segregant analysis
BSMV	Bean common mosaic virus
bZIP	basic region leucine zipper
C	carbon
CA	carbonic anhydrase
cAPX	cytosolic ascorbic peroxidase
CAs	carbonic anhydrases
CAT	cata-lase
CBB	common bacterial blight
CCD4	carotenoid cleavage dioxygenase 4
CCoAOMT	caffeoyl-CoA O-methyltransferase
cDNAs	complementary DNAs
CDPK	calcium-dependent protein kinase
CE	capillary electrophoresis
CGIAR	Consultative Group on International Agricultural Research
cgSSRs	candidate gene based simple sequence repeats markers
CH42	chlorinA42
CHI	chalcone synthase isomerase
CHL	chlorophyll level
CIAT	International Center for Tropical Agriculture
CID	collision-induced dissociation
CIM	composite interval mapping
CLCuBuV	*Cotton leaf curl Burewala virus*
CLCVX	*Cotton leaf curl virus x*
CLS	cercospora leaf spot
CMS	cytoplasmic male sterile
CMV	*Cucumber mosaic virus*
CNR	colorless nonripening
COD	centers of domestication
CRoPS	complexity reduction of polymorphic sequences
CTR4	constitutive triple response 4
CUC	cup-shaped cotyledon
CymMV	*Cymbidium mosaic virus*
DAF	days after fertilization
DArT	diversity arrays technology
DBP	diastolic blood pressure

DD	differential display
DDBJ	DNA Data Bank of Japan
ddNTPs	differentially dyed dideoxynucleotide triphosphates
DDS	dammarenediol synthase
DGE	digital gene expression
DHAR	dehydroascorbate reductase activity
DHPLC	denaturing high performance liquid chromatography
DNMTs	DNA Methyl-transferases
dNTP	dyed deoxynucleotide triphosphate
DPPH	1,1-diphenyl-2-picrylhydrazyl
DRE	dehydration response element
DRl	disease rating on leaves
DRM2	domain rearranged methyltransferase 2
DRst	disease rating on DRst
DS	disease severity
dsRNAs	double stranded RNAs
DWL	dried weight losses
ED	ear diameter
EH	ear height
EL	ear length
EMBL	European Molecular Biology Laboratory
EMS	ethyl-methane sulfonate
ENCODE	encyclopedia of DNA elements
eQTLs	quantitative trait loci
ERF	element binding factor
ESI	electrospray ionization
ESTs	expressed sequence tags
ETC	electron transport chain
FA	fatty acid
FBP	field-based phenotyping
FHB	*Fusarium* head blight
FN	fibrous root number
FT	flowering time
GA	gibberellic acid
GABA	gamma-aminobutyric acid
GBS	genotyping by sequencing
GC	guanine-cytosine
GEBVs	genomic estimated breeding values
GEI	genotypes-environment interaction
GEMs	gene expression

GEO	gene expression omnibus
GFLV	*Grapevine fanleaf virus*
GM	genetically modified
GMOs	genetically modified organisms
GO	gene ontology
GORK	gated outwardly-rectifying K$^+$
GPMDB	Global Proteome Machine
GS	genomic selection
GSH	glutathione
GUS	beta-glucuronidase
GVA	*Grapewine virus A*
GWAS	genome wide association studies
GWS	genome-wide selection
GY	grain yield
H$_2$O$_2$	hydrogen peroxide
ha	hectares
HAPs	highly abundant proteins
HB	*phabolusa*
HDGS	homology dependent gene silencing
HD-ZIP	homeodomain-leucine Zipper
HI	harvest index
HKT	histidine kinase transporter
hp	hunchback
HPLC	high-performance liquid chromatography
hpRNAs	hairpin RNAs
HSPs	heat stress proteins
ICAT	isotope-coded affinity tags
ICS1	isochorismate synthase 1
IDS1	indeterminate spikelet1
IF	infection frequency
IITA	International Institute of Tropical Agriculture
IM	interval mapping
IPCC	Intergovernmental Panel on Climate Change
IRGA	infrared gas analysis
IRGSP	International Rice Genome Sequencing Project
IRRI	International Rice Research Institute
IRT	infrared thermography
iTRAQ	isobaric tags for relative and absolute quantification
KEGG	Kyoto encyclopedia of genes and genomes
KN	kernel number per row

LA	leaf area
LAPs	low abundance proteins
LASP	leaf area covered by sporulating pustules
LC–MS/MS	liquid chromatography coupled with tandem mass spectrometry
LD	linkage disequilibrium
LEA	late embryogenesis abundant
LG	linkage group
LL	leaf length
LLS	late leaf spot
LOD	limit of detection
LP	latent period
LP	low-Phosphorous
LPO	lower lipid peroxidation
LRN	lateral root number
LRNP	lateral root number plasticity
lsiRNA	long siRNAs
LW	leaf width
M	mallikong
MABC	marker-assisted backcrossing
MALDI	matrix assisted laser desorption ionization
MALDI-TOF	matrix assisted laser desorption ionization-time of flight
MAPK	mitogen-activated protein kinase
MAS	marker-assisted selection
MBGP	Multinational Brassica Genome Project
MBP	mean blood pressure
MDHAR	monodehydroascorbate reductase activity
MFLP	microsatellite-anchored fragment length polymorphism
MIM	multiple-interval mapping
miRNAs	microRNAs
MM	mallikong mutant
mRNA	messenger ribonucleic acid
MS	mass spectrometry
MS medium	Murashige and Skoog medium
MSG	multiplexed shotgun genotyping
Mt	metric tons
MudPIT	multidimensional protein identification technology
MUFAs	monounsaturated fatty acids
MYC	myelocytomatosis oncogene
NAM	nested association mapping

NAM	no apical meristem
NASA	National Aeronautics and Space Administration
Nat-siRNA	natural antisense siRNA
NBS-LRR	nucleotide binding site-leucine-rich repeat
NCBI	National Center for Biotechnology Information
NCED	nine-cis-epoxycarotenoid dioxygenase
NERICA	new rice for Africa
nESI-LC– MS/MS	nano-electrospray ionization liquid chromatography-tandem mass spectrometry
NGS	next-generation sequencing
NILs	near isogenic lines
NMR	nuclear magnetic resonance
NP	normal-Phosphorous
npc-RNAs	non-protein-coding RNAs
NPT	new plant type
ORFs	open reading frames
OsIFR	rice isoflavone reductase
OsmiR 397	*O. sativa* miRNA 397
P	phosphorus
PAE	Pi acquisition efficiency
PAE	phosphorus absorption efficiency
PAZ	Piwi/Argonaute/Zwille
PCA	principal component analysis
PCR	polymerase chain reaction
PE	phosphorus efficiency
PEBV	Pea early browning virus
PGRFA	plant genetic resources for food and agriculture
PH	plant height
PhBSMT	*Petunia* hybrid BSMT gene
PHI	pod harvest index
PHV	*phavoluta*
Pi	phosphate
piRNAs	Piwi-interacting RNAs
PLS-DA	partial least squares-discriminant analysis
PMN	plant metabolic network
POD	peroxidase
Pol IV	polymerase-IV
PopMV	*Poplar mosaic virus*
PPBD	plant proteome database
PPIs	protein–protein interactions

PPO	polyphenol oxidase
PRIDE	proteomics identifications database
PSR	phosphate starvation responsive
PSTOL1	phosphorus starvation tolerance 1
PTGS	post transcriptional gene silencing
PTMs	posttranslational modifications
PUE	phosphorus utilization efficiency
PUFAs	polyunsaturated fatty acids
PVX	*Potato virus* X
PVY	*Potato virus* Y
QTL	quantitative trait loci
Q-TOF	quadrupole time-of-flight
RAA	renin-angiotensin-aldosterone
RAD	restriction-site associated DNA
RADseq	restriction-associated DNA sequencing
RAPD	random amplified polymorphic DNA
RAS	root architecture system
rasiRNA	repeat associated RNA
RBE	rice starch branching enzyme
RBW	relative above-ground biomass dry weight
RCHL	relative chlorophyll level
RdDM	RNA-directed DNA Methylation
rDNA	recombinant deoxyribonucleic acid
REs	restriction enzymes
REV	*revoluta*
RFLP	restriction fragment length polymorphism
RFN	relative fibrous root number
RGA	root growth angles
RIL	recombinant inbred line
RISC	RNA-induced silencing complex
RITS	RNA induced transcriptional silencing
RLA	relative leaf area
RLC	RISC loading complex
RLW	relative leaf width
RNAi	ribonulceic acid interference
ROS	reactive oxygen species
RPH	relative plant height
RPK1	receptor-like protein kinase 1
RRI	root rot index
RRLs	reduced-representation libraries

rRNA	ribosomal RNA
RRW	relative root dry weight
RS	resistant starch
RSR	root to shoot ratio
RTBV	*Rice tungro bacilliform virus*
RTSV	*Rice tungro spherical virus*
RTW	relative topsoil root dry weight
RV	root volume
RW	root weight
SA	salicylic acid
SAGE	serial analysis of gene expression
SBP	systolic blood pressure
SBP	squamosa promoter-binding proteins
SBP	sequence based polymorphic
SCAR	sequence characterized amplified region
SCL6	scarecrow-like transcription factor 6
SDS	sudden death syndrome
SFAs	saturated fatty acid
SGS	Second-generation sequencing
SHH1	*sawadee homeodomain homolog* 1
SHMV	*Sunnhemp mosaic virus*
SHRs	spontaneously hypertensive rats
SILAC,	stable isotope labeling by amino acids in cell culture
siRNA	short interfering RNA
SMA	single marker analysis
SNP	single-nucleotide polymorphism
SOD	superoxide dismutase
SOLiD	sequencing by oligonucleotide ligation and detection
SOS	salinity overly sensitive
SPL	squamosa promoter-binding protein-like
SPUE	shoot phosphorus utilization efficiency
SQS	squalene synthase
SRAP	sequence-related amplified polymorphism
SRL	seminal root length
SRN	seminal root number
SSH	suppression subtractive hybridization
SSPs	seed storage proteins
SSR	simple sequence repeat
ssRNA	single stranded RNA
SSRs	simple sequence repeats

STMS	sequences tagged microsatellite site
STMV	Satellite tobacco mosaic virus
STS	sequence tagged sites
SW	shoot weight
tasiRNA	trans-acting RNAs
ta-siRNAs	trans acting siRNAs
TBSV	Tomato bushy shunt virus
TCA	tricarboxylic acid
TE	transpiration efficiency
TEV	Tobacco etch virus
TF	transcription-factor
TFA	trifluoroacetic acid
TFs	transcription factors
TGMV	Tomato golden mosaic virus
TGS	transcriptional gene silencing
TGS	third generation sequencing
TILLING	targeting induced local lesions in genomes
TMTs	targeted mass tags
TMV	Tobacco mosaic virus
TOE3	target of eat 3
TRL	taproot length
TRV	Tobacco rattle virus
TYMV	Turnip yellow mosaic virus
Ub	ubiquitin
UDP	uridine diphosphate
UGPase	UDP-glucose phosphorylase
VIGS	virus induced gene silencing
VOCs	volatile organic compounds
WAF	weeks after flowering
WGS	whole genome shotgun sequencing
WM	white mold
WPUE	whole phosphorus utilization efficiency of plant
WSG	whole genome sequencing
WUE	water use efficiency

PREFACE

Due to the advent of the state-of-the-art technologies in the field of biotechnology, a lot of progress has been achieved during the last decade. Omics technologies are being extensively used to address various issues pertaining to agriculture. Recent advances in genomics, transcriptomics, proteomics, and metabolomics techniques have revolutionized the understanding of genetic response of plants to various biotic and abiotic stresses. Strategic application of this revolutionary technology will eventually lead toward attaining sustainability in agriculture. Updating the researchers/students with these new biotechnological developments will, therefore, be necessary.

In this book, we have attempted to cover various omics approaches (genomics, transcriptomics, and proteomics) that have been and can be used for mining potential candidate molecules to address the unsolved problems related to agriculture. Moreover, we have also included some chapters on breeding approaches that are being operated for improving crop plants.

The book is unique as it covers both basic research (plant omics) and applied research (crop breeding). In plant omics, we have gathered chapters covering most of the modern biotechnological tools that will serve as reference for students, researchers, and academicians. The crop breeding section will help agriculture scientists and researchers in designing the applied research in a better way. It will also assist the students in understanding concepts of crop breeding for the improvement of agricultural crops and means of using high-throughput techniques in the applied field.

All the chapters are unique. The recent advances made in plant genomics, transcriptomics, proteomics, and their utility in addressing the unresolved problems in agriculture, which include biotic as well as abiotic stresses, will attract a large number of readers.

FOREWORD

 It gives me immense pleasure to write the foreword of this book, *Plant OMICS and Crop Breeding*. In the field of plant biotechnology a lot of progress in terms of technological advances have been made since the last decade. Precise breeding by use of molecular markers and other genomics tools have helped breeders in introgression of desirable genomics regions in various crop plants. The availability of complete and high-quality map-based sequence of various crop plants will help in functional genomics and molecular breeding programs that aim at unraveling intricate cellular processes and improving crop productivity. Further, the advances in other Omics approaches will lead to mining molecular switches regulating tolerance mechanisms and other desirable traits in crop plants. This will eventually lead toward attaining sustainability in agriculture.

This book, with diverse coverage of various plant biotechnological aspects, covers both basic research (plant Omics) as well as applied research (crop breeding). As such it will serve scientists, researchers, and students working on basic as well as applied aspects of crop improvement. This book will also lead to better understanding of Omics tools for mining potential trait regulators and their role in attaining sustainability in agriculture.

I congratulate the editors for their efforts in compiling this book with valuable chapters covering various Omics techniques and molecular breeding approaches.

—Dr. Nazeer Ahmed
Vice-Chancellor
Sher-e-Kashmir University of Agricultural Sciences
& Technology of Kashmir

PHENOMICS SCIENCE: AN INTEGRATED INTER-DISCIPLINARY APPROACH FOR CROP IMPROVEMENT

BISMA MALIK[1], TANVEER BILAL PIRZADAH[1], IRFAN QURESHI[2], INAYATULLAH TAHIR[1], IMTIYAZ MURTAZA[3], and REIAZ UL REHMAN[1*]

[1]Department of Bioresources, University of Kashmir, Hazratbal, Srinagar, Jammu and Kashmir, India

[2]Department of Biotechnology, Jamia Milia Islamia, New Delhi, India

[3]Division of Post-Harvest Technology, SKAUST-K, Shalimar, Srinagar, Jammu and Kashmir, India

*Corresponding author. E-mail: rreiazbiores@gmail.com; reiazrehman@yahoo.co.in

CONTENTS

ABSTRACT

The plant phenotype (plant performance and appearance) is determined by its genetic composition (genotype) and the environment it grows in. The plants growing in different environments can have different phenotypes even though their genotype being the same. Normally, the study of the traits (morphological appearance) can take many days, weeks, or even months depending on the life cycle of the plant. However, with phenomics science (study of phenotyping) by automated technology and imaging systems in combination with high efficiency computing machines can be accomplished in seconds. This approach of rapid phenotyping would enable plant breeders to select crop varieties with high yield and field performance. Thus, the varieties with different qualitative and quantitative traits (such as abiotic better stress tolerant, biotic stress resistance, and photosynthetic efficient) could be available in shorter period of time, rather than in decades. This is of utmost importance in terms of food and fuel requirements for the ever-increasing population which is going to double in next 50 years. Moreover, there is a growing concern about the global climate changes and there is a need to select and identify plants which can adapt to varied environmental changes.

1.1   INTRODUCTION

The phenotypic character of a particular plant is influenced by numerous factors namely its own genetic make-up (genotype) and the environment it grows in. It is very interesting to understand how a particular genotype determines specific phenotype and permits the progress of organisms with commercially beneficial traits. Although, it is very difficult to predict phenotype from genotype because many genes or gene products are involved in determining phenotypic trait in conjugation with numerous environmental influences. Previously, it was not possible to understand the whole genome of an organism because of quite large size containing millions or billions of nucleotides but recently due to more advancement in molecular biology it broadens our vision to better understand the genotype of an individual. Nowadays, it is possible to determine the whole genome of an individual and entire genomes are now sequenced at a very fast pace and minimum cost. With the evolution of next generation sequencing technologies, it is now possible to determine whole genotype and epigenotype of not only a single member of a genus or species, but of numerous delineative of an evolutionary array or population. For instance, high-density single nucleotide polymorphism genotyping (first launched in the human HapMap Project) becomes amenable for every organism and nowadays employed to both plants and microbes for describing the natural genetic diversity and supports in attribute-driven attempts to clone and recognize particular type of genes. The phenotypic trait is quite difficult to study as compared to genotype because phenotypic traits are characterized at distinct levels such as from molecules to effective metabolic web to complicated processes (cellular developmental and physiological aspect) and the integrity of the whole system or the social behavior of complex populations.[1] Phenotypic complexity is also influenced by different ecosystem interactions (such as, interactions with symbionts, pathogens, or competing organisms). Besides, phenotypes are dynamically effective and the time-period in which they modify changes enormously. For example, abrupt changes of a bacterium to nutrient cycles[2] or the effective fluctuations in the photosynthetic process of a leaf as a single cloud passes over the sun,[3] which can be compared to the slow-going morphological events in old plants or even the long lasting modifications in the outside imprint of a human being. Previously, phenomics science was confined to data gathering and examination while phenome is the actual list of dimensions. Although, it allocates features with traditional mutant screening or measurable attribute examination; however, it is prominent from these classical techniques in scale and scope.[4,5] First, it employs extensive populations of genetically diverse with the aim of sampling alteration in several or whole genome. Second, individual genotype is evaluated for enough number of attributes, generally by utilizing perfectly examined and high-throughput functional approaches with systems in place to increase accuracy of sample tracing and data multiplication. Third, principle attributes of the growth states are well described and closely observed. Finally, the phenotypic data and metadata observations of the investigation circumstances are captured in formats that provide descriptive data examination. These examinations could not only sort out the solution to identify relationships between genotype and phenotype but affirm correlations between seemingly unrelated phenotypes[6] or genetic loci.[7] As environmental factors greatly influence the phenotypic traits; therefore, it is necessary to collect samples from multiple environments at different stages of development. It is not so easy to collect data of all alleles for each measurable attribute under every desirable growth conditions; however, the design of experiments and data analysis procedures must be correlated to the desired result.[8] For instance, the study of interactions between organisms and their respective environments under particular self-restraint abiotic components (controlled conditions) like photoperiod, temperature, availability of nutrients, and so forth can be frequently utilized or their response to well-known mutualistic, parasitic, or competitor organisms. Besides this it is very important to identify crop plant genotypes with highly desirable attributes such as productivity or nutritional quality which can be better studied by understanding the whole environmental conditions like climate, soil, and biotic stress factors comparable to where the ultimate variants will be produced commercially. Furthermore, microorganisms can be phenotypically evaluated under different types of conditions, depending upon the question pertinently being enquired which ranges from a simple dimension like growth rate in a monoculture under controlled conditions or to examine the features in the context of complicated communities of microbes or the association with plants or animals. In order to interpret high-dimensional phenomic data, especially when they spend many degrees of organization, it is essential to use phenomic chimerical support-structure (framework). However, this support-structure can be produced on well-organized intellectual folklore for examining the phenotypic data, which include quantitative genetics, evolutionary biology, epidemiology, and physiology. These domains offer tools to describe the different origins of diversity[9] and to unravel causes from correlations.[10] Phenomic research and development and molecular biotechnology holds a good promise for the improvement of crop productivity. For example, molecular breeding has huge capability for altering permanently the science and art of plant breeding.[11]

Molecular breeding is a novel tool that encompasses molecular outlines to choose breeding materials and the utilization of recombinant deoxyribonucleic acid (rDNA) procedure to add value to plant genetic resources for food and agriculture (PGRFA). Besides, there are other numerous emerging technologies that play a pivotal role in plant breeding programs. Crop varieties with superior agronomic attributes are the direct outcome of plant breeding as concluded by the Columbia Encyclopedia "the science of modifying the evolutionary models of plants to enhance their quality."[12] For sustainable development, it is the need of the hour to generate green and novel ways of food and energy production with minimal ecological footprints.[13] Although, phenotyping can be done under controlled or field conditions but under laboratory (controlled) conditions, environmental parameters may be controlled as per the desired level to investigate a particular factor that influences the phenotype trait. Besides, this field conditions are varying with space and time.[14,15] Currently, large data can be generated for statistical analysis by using novel techniques such as high-throughput phenotyping in association with climatic monitoring at high dimensional and ephemeral resolution. For instance, imaging spectroscopy is the latest tool to furnish elevated resolution images from base and airborne scaffold that includes large resolution spectral information of millions of pixels.[16] Single spectra can be ascribed to one plant or experimental plots that are associated to functional plant attributes.[17] Thus, phenomics holds a great promise in the state-of-art to best apprehend the genetic environmental interactions and uphold procreation for upgraded resource utilizing efficacy of significant crop plant.

In this chapter, we discuss the phenomics science in plant phenotyping and identify the scientific rationales in phenomics research and development. Then we describe the role of phenomics in crop improvement, phenomics technologies, and future challenges (Fig. 1.1).

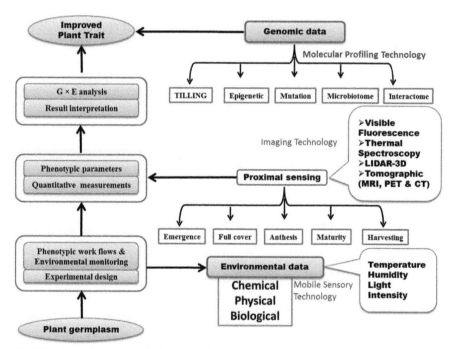

FIGURE 1.1 Overview of plant genetic and environmental interface analysis (G × E; genetic and environmental).

1.2 WHAT IS NEEDED TO ADVANCE PHENOMICS?

Phenotyping is considered as an essential multiple-scale characterization of an organism trait disclosed in space and through time. Using novel modern approaches to gather data information regarding phenotyping would help in mapping of genetic component to biological purposes at the desired level of detail.

1.2.1 THE ROLES OF "REFERENCE" AND "MODEL" SYSTEMS

Previously, model organisms were used to study diverse parameters since there is a match between the question being asked and the qualities of an organism. However, this archetype drifts, when researchers' approach toward reference organisms and model organisms with simple genetics during the final half of the 20th century (Fig. 1.2).

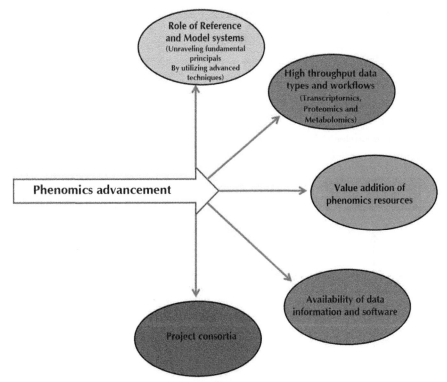

FIGURE 1.2 Pre-requisites for the advancement of phenomics research (the success of phenomics will be multiplied when different datasets can be combined and correlated across different studies, allowing increases in statistical power and the scope of analyses).

Nowadays, there are authenticated repositories of these model organisms which include diverse germplasm, developmental, physiological, as well as genomic databases. Hallmarks of these reference organisms now include the availability of broadly useful genetics resources such as diverse germplasm, detailed knowledge of developmental, biochemical, and physiological networks, advanced genomics tools, and large collaborative communities. Examples include microbes such as *Escherichia coli* and *Saccharomyces cerevisiae* (baker's yeast), the plants *Arabidopsis thaliana*, *Zea mays* (corn), and *Oryzasativa* (rice), and animals such as *Caenorhabditiselegans (*a soil roundworm), *Drosophila melanogaster* (fruit fly), and *Musmusculus* (mouse). These reference organisms have traditionally been a strong focus of research and funding because the available assets make them such efficient systems for primary investigations into a wide variety of biological phenomena. The large amount of information available about these systems—for example, high quality gene annotation, deep understanding of developmental, physiological, and metabolic networks—provides a valuable point of reference for researchers studying even distantly related organisms. Although, these reference models possess great significance in biological world but at the same time they only represent the observable phenotypic variation of the biological world. For instance, apprehending of *Arabidopsis* root progress serves as a delineative for numerous additional plant species,[18] but it cannot be used to study various ecological interactions such as symbiosis between mycorrhiza and plants—an important phenomenon in nutrient accumulation in diverse plant species.[19] Moreover, the traits which are externally distributed (like flowering time, rate of biomass assimilation) are even controlled by a network of genes.[20,21] Besides, other traits such as seed yield, growth, and so forth are guarded by multiple factors that they are much better studied in the target organism in their normal producing environments.

With the advent of novel sequencing and phenotyping technologies it is now possible to use non-reference organisms for the examination of some particular biological parameters. Transcriptomics and genomics technology provides an important platform to better characterize the organism at the gene, protein, as well as at messenger ribonucleic acid (mRNA) level. These omics technologies unravel the sequence that leads to the invention of the desirable genes,[22] and allows considerable proteomics (digital report of growth and development, physiological components, and protein-metabolite abundance) and genomic analysis.[23] Re-sequencing of genetic variants helps us to furnish huge figures of molecular markers that can be utilized as an important tool to determine the genes donating to target phenotypes by conventional genetic mapping or extra modern techniques (genome-wide correlated considerations).[24]

1.2.2 HIGH-THROUGHPUT DATA TYPES AND WORKFLOWS

Recently, main factors responsible for phenotype description (namely gene expression profiles) can be accomplished by relatively available, high-throughput technologies. Actually, large-scale transcript profiling studies in reference organisms have previously validated of high quality (EcoCyc for *E. coli*: http://ecocyc.org/; At Gen Express for *Arabidopsis*: http://www.weigel-world.org/resources/microarray/AtGenExpress/.[25] Besides this, technologies employed in characterizing downstream phenomics traits are growing rapidly but are presently far-off from daily utilization.

1.2.2.1 PROTEOME AND METABOLOME

An omics technology such as proteomics and metabolomics is nowadays a current science that is employed for the determination and identification of novel proteins and metabolites.[26] But at the same time there are few challenges (such as expensive equipments, diverse chemical properties of the molecules) that make these technologies inaccessible to the broad scientific community. For instance, it is much easier to analyze soluble, abundant, proteins and soluble metabolites than rare, labile, and insoluble molecules. To analyze protein and metabolites it needs tissue extraction, but there are well established techniques used for non-invasive dimensions like near infrared transmittance for seed metabolites[27] and more investigational procedures such as matrix assisted laser desorption/ionization time of flight (MALDI-TOF), mass spectrometry for spatial resolution of metabolites[28], and isobaric tags for relative and absolute quantification (iTRAQ technology) which was employed for the proteome studies (like redox systems, carbon metabolism, photosynthesis, signaling, and amino acid metabolism) which have been found to be associated with various stress responses.[29–32]

1.2.2.2 PHYSIOLOGICAL ATTRIBUTES

Physiological attributes (namely photosynthesis measurements, nutrient uptake, and transport) can be recreated and modified,[33] but attaining the fundamental throughput is challenging. Latest imaging technologies find great application in determining spatial variation of physiological parameters. This route opens a novel method of investigation, because in numerous illustrations, the diverseness of the reactions in space and time is a principle characteristic of the phenotype, which exhibits remarkable data about the fundamental biological characters.[34] For example, water availability is a major constraint of crop yield[35] and is the single most important factor limiting food production, with significant yield losses reported under water deficit.[36] It is essential that new techniques and approaches are developed for phenotyping[37] and to screen for limitations in water use efficiency (WUE). Traditional techniques used for the measurement of physiological attributes include infrared gas analysis (IRGA),[38,39] Carbon isotope discrimination[40] and Chlorophyll fluorescence that have long been used to examine various photosynthetic parameters in leaves.[41] Infrared thermography (IRT) provides a powerful imaging tool for rapidly, non-invasively, and remotely measuring leaf temperature.[42] Previous researchers have used combined chlorophyll fluorescence and thermal imaging approaches to evaluate photosynthetic performance in relation to stomatal behavior,[43] but the majority of these studies have been carried out at the leaf or tissue scale.[44] A. thaliana plants are demonstrated using the system, with spatial and temporal heterogeneity in different traits that have been observed in the images. It is important to take into account such heterogeneity as it is well established that photosynthesis was not uniform over a leaf surface[45] and that such heterogeneity is also dynamic[46] often being driven by variations in the microenvironment.[47]

1.2.2.3 PLANT GROWTH AND DEVELOPMENT

Organism's growth and development can be either measured by classical quantitative parameter such as biomass content or by using high-throughput technologies like high-resolution cameras, image software analysis, and novel computer algorithms which have ability of measuring the characteristic or process of concern. Currently, these high throughput technologies are readily available for some laboratories and greenhouse settings; however, their use is far from widespread. (http://www.lemnatec.com/; http://www.plantaccelerator.org.au/). Quantification of root structure and behavior is a serious bottleneck, but culture in or on gelled media makes image-based methods workable.[48,49] Tomographic systems can furnish deep understanding in structural dynamics of roots[50] but are presently not capable to deal with high throughput because of some technical errors and the reproducibility of enough data.

1.2.2.4 PHENOMICS IN SITU: MEASUREMENTS IN THE FIELD

Phenomics is not a new concept but many ecologists, systematists, and plant breeders have been practicing it for past many decades. In plant breeding programs high-throughput phenotyping for firmly aimed and associated attributes is a usual task, where thousands of matchless genotypes are calculated periodically. Nowadays, technique like remote sensing has gained much importance to study canopy spectral reflectance in plant breeding programs for measuring nitrogen or water utilizing efficacy.[51] But, there are numerous essential attributes that are challenging or valuable to calculate and however, phenomics technology would induce novel pathways that could not only increase the identification of better genotypes but efficiently train prediction miniature.

1.2.3 INTENSIFYING THE QUALITY OF PHENOMICS RESOURCES

Phenomics projects are resources intensive to create huge amount of data set, besides, these projects can be largely cost benefiting if the following data is of high value and is explored by more number of researchers. There are numerous elements that participate to the extending duration of achievement of phenomics projects: the origin of genetic variation applied and whether it is conserved for future use:

- Characteristics of the growth conditions;
- The phenotypic test investigated;
- Repository, storage, and elucidation of data.

1.2.4 GENETIC DIVERSITY

Germplasm collections are the primary repositories for numerous phenotypic investigations and their importance is going to enhance as novel phenotyping technologies emerges. Phenotyping tools play an essential role in phenomics to better understand the genetic variation and function when applied to study well-curated germplasm. These collections include lines or accessions generated either natural or induced mutations.[20] Creation of huge set of germplasm is time consuming and is not cost-effective. For instance, germplasm that can be used to study enough parameters or phenotypes may be expensive than custom designed for particular projects. It is an essential step to capture distinct phenotypic data for a germplasm collection in a central data repository.[52] This is because it allows members of the community to query multiple phenotypes and relate these traits to genotype. Finally, for a collection of germplasm to be of widest utility it should come without intellectual property restrictions or with a material transfer agreement that is simple and not onerous. This will help in exploring more and more phenotypic traits and depicts these traits in different genotypes without limitations.

1.3   PHENOMICS IN PLANT PHENOTYPING

One of the important strategy for apprehending the processes underlying in the phenotype and for the up-gradation of a phenotype is the identification of its genetic variation which is the key factor for its improvement (like plant disease resistance or identification of plants resistant to abiotic stress, modifying plant cell walls for biomass conversion technology). The approach of high-throughput phenomics facilitates the recognition of genetic differences in natural (varied germplasm collections, associated genetic panels) and derived (e.g., recombinant innate, mutant, wide introgression lines, etc.) populations. The successful applicability of a phenomic approach depends on the design and adaptation of the strong screens to permit high throughput, responsible and useful comparison which is the biggest challenge for its worldwide application. For effective phenotyping, capabilities of making quick, exact, and reproducible measurements are not only important but to know what phenotypes to measure. Unlocking of gene–phenotype relationship by a quality desegregation of high-throughput phenotyping technology and gene discovery is an important step forward for the better understanding of genetic make-up of such features. Besides, the production of a computable database on phenotype that permits the easy access of information about genes and its associated phenotypes would help in the easy process of data mining and management. The presentation of phenotypic information is a complex process; although, there are some data standards which are created for maintaining phenotypes and depositories within the species.[53] For high impact, several analysis tools of phenomic data are available publically like web interface tools including experts for specified investigations although phenomics data can also be analyzed with different data types. Schematic representation of phenomics approach using a common germplasm is given in Figure 1.3[54] which explains the data collection and interpretation by different analysis tools. Data integration holds a great promise in the process of phenotype–gene relationship as it correlates diverse phenotypic data with genetic data at different stages and this is applicable in reverse genetics process where noticeable external changes are rarely observed or slight modification in the phenotype by the mutations.[55]

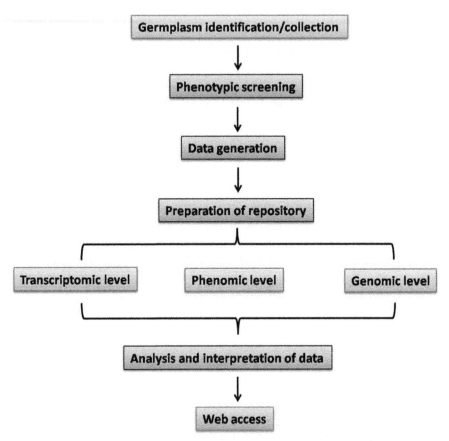

FIGURE 1.3 Schematic representation of phenomics using common germplasm.

The approach of targeting induced local lesions in genomes (TILLING) is a very promising tool in reverse genetics process which is utilized for the functional analysis of high-throughput phenotyping in corn, wheat, rice, soybean, tomato, and lettuce.[56] Basic TILLING method allows for high-throughput identification of single-base-pair (bp) allelic variations.[57] TILLING combines chemical mutagenesis[58] with a sensitive mutation detection instrument. In a pilot experiment, DNA from a collection of ethyl-methane sulfonate (EMS)-mutagenized *Arabidopsis* plants was pooled, subjected to polymerase chain reaction (PCR) amplification, and screened for mutations using denaturing high performance liquid chromatography (DHPLC). DHPLC detects mismatches in heteroduplexes created by melting and annealing of heteroallelic DNA. Among the lesions detected were base transitions causing missense and nonsense changes that can be used for phenotypic analyses. Many TILLING projects are ongoing in diverse plant species, such as *Arabidopsis* and *Brassica oleracea*,[59] lotus,[60] maize,[61] and wheat.[62] Like the other reverse genetics approach, this technique also needs considerable and accurate documentation of the collected material from which functional genetic data are generated[63] which in turn becomes the biggest challenge to such approach which permits the researchers to pheno-typic database management across populations and also the correlation of phenotypic data with genetic data. Therefore, the need of the hour is to develop such technologies which allow the scientists to encode and organize phenotypic data for high-throughput analysis.[64] In this consideration, one of the biggest problems is the consistently insufficient understanding of the underlying reaction mechanisms of plant adaptation to environmental stress. Thus, better apprehending of reaction and processes involved is required to enhance the accuracy of phenotyping, and there is also a need to understand the type of data collected and where and when to collect it and it is becoming unobservant at different stages of phenotyping, whether that is molecular, biochemical, or morphological which is the reason behind unsuccessful "physiological breeding." Regarding this issue, there are some lacunas that results in the inadequate information for the proper understanding of the mechanism involved in the phenotyping. In order to combat with this limita-tion, one of the major challenges that come in the path of the high-throughput plant phenotyping is the sturdy and instinctive analysis of plant data.[65] Plant phenotyping involves the measurement of remarkable plant characteristic which in turn defines the function of different genes affected by the envi-ronmental variation.[66] Besides, modern phenotyping procedures are widely utilized to investigate growth and development of different plant genotypes sets under varied stress conditions.[67,68] Phenotyping is a fast developing field of study in plant sciences that utilizes no or minimum invasive sensors to observe plant functions and unravels the genetic variants of increased agro-nomic attributes. It is a well-known fact that an intricate plant physiological characteristic, such as development, biomass accumulation, and produc-tivity, is the consequence of combined effect of the genetic and environ-mental variants throughout the plant developmental stage. In this context, it is very important to note that the environmental factors regulate the expression of genes in expanse and throughout plant growing season which directly affects the phenotype. As a consequence, significant physiological characteristics continuously change and follow a dose-response pattern. These reactions can be commonly applicable to the individual trait-environmental factor together by using a quantitative meta-analytical framework.[69] During the past 20 years, molecular profiling and classical sequencing technologies enabled significant advances toward the large-scale characterization of plant genomes.[70] Today, one of the promising technology that can modify the plant genomics further is the next generation sequencing technologies,[71] and molecular technologies have provided valuable tools for plant breeding such as marker-assisted selection and more recently genomic selection which inserts useful alleles at different loci that have less genetic impact when utilized specifically.[72] However, the plant phenotypic landscape needs to be populated at a faster pace to accelerate research in model organisms,[73] biotechnology industrial pipelines[74] and plant breeding programs for yield and resource efficiency gain in major crops.[75] In a nutshell, there are some hurdles that come in the way of plant phenotyping research which hampers the advancement both in the field of information and application.[73,76] Houle et al.[73] recently described a general study on the phenomic challenges and statistical expression of the gene-environment interplay. In this regard, 3D laser scanning (which include 3D-cameras, photogrammetric methods, or laser scanners) permits a non-destructive evaluation of different plant features under controlled conditions; these features include the plant structure, length, and magnitude of a particular plant organs/volume.[77–79] Overall assessment of these features or characteristics over time will results in the identification of correlation of variation in plant growth to stress resistance or to identify possible determinant of yield of various genotypes. Plant growth and development including the response of plants to various environmental stresses (abiotic and biotic) can be determined by a significant technique of structural geometrical analysis of plants[80] which is possible only by the highly resolved investigations and the production of 3D organ-based structural models (such as operational-structural plant models)[77,81] and the observation of precise alterations can be utilized to correlate geometrical alterations and changes to environmental effectiveness.[78] This technique has been utilized in several plant phenotyping methods for the analysis as it satisfied all the requirements for its implementation in plant phenotyping analysis approach.[78,82] One of the primary requirements in plant phenotyping is the identification of shapes from point clouds including geometrical standard shapes (cylinders, spheres, planes, or cones) as well as associations and modifications are well expressed in several research fields.[83,84] The general approach of 3D plant analysis utilizes the 3D mesh processing.[78] In this context, Paproki[85] employs a 3D point cloud produced from 64 images of a cotton plant to identify a single plant organ. This technique needs those primary plant organs which have inevitable smoothness coercions (e.g., leaf surface). Currently a technique has been developed which utilizes surface feature histograms for automated plant organ parameterization. This technique was proved to be independent of the point-to-point distance and can be applied to several plant species. Its capability, elasticity, and high order of automation made this technique much suited for the current demands of high-throughput phenotyping.[86]

The goal of modern plant phenotyping is to deliver quantitative data on the dynamic responses of plants to the environment. Much of the recent progress has been driven by increasingly available technologies to non-invasively measure plant growth and physiological status. Imaging phenotyping is becoming common. Achieving good accuracy with various types of sensors and tracking this accuracy over time depend on a correct understanding of the sensor physics and the sensitivity that can be obtained when image phenotyping is applied to quantitative screening protocols. Additionally, sensors (including camera systems) need regular calibration. Depending on the specific cameras, it can be important to perform both geometric and radiometric calibrations for each imaging setup.[87] Routine use of cameras for large-scale analyses in phenotyping platforms requires special care because many factors can influence the performance of the cameras and of the automated algorithms for image feature extraction. In this respect, it is good practice to include internal reference objects that mimic plants and can be used to track the performance of imaging setups. Experiment design depends on many considerations. Two important factors to consider are the experimental layout itself and the appropriate number of replications to address the desired questions.[88] It is also important to stress that recording and capturing environmental conditions should not be optional, but instead should be seen as a necessary step enabling meaningful data interpretation. Similar considerations apply to good-practice definitions for field experiments.[89] As global standardization is not feasible and may even be detrimental, this implies the need to capture environmental variables and protocol details in a systematic way as part of defining a minimum set of information for plant phenotyping experiments. This could be done on the basis of, for instance, the scheme of Poorter et al.[88] Quantitative measurements of plants and the environment during experiments are two sides of the same coin and should become easily available in plant phenotypic databases in the future.

1.4   PHENOMICS RESEARCH AND DEVELOPMENT IN CROP IMPROVEMENT

Food and Agriculture Organization of the United Nations[90] calculated that the human population growth rates increase world widely at a very faster pace which in turn forces the increase in food production so that 70% of additional food must be developed in the coming years in order to feed sufficiently the population which is over-running 9 billion by the year 2050. In order, to achieve such an increase, which needs considerably at least 37% increase in the expansion of existing annual food production[72] is tangibly lowered because of the effect of climate change and modifications in crop production system.[91] The remarkable progress in biotechnological fields proved to be advantageous and holds extensive potential in the process of crop improvement.[92] One of the key attribute or feature of phenotype of the crop plants is the yield which is actually the speculation of biomass or a measure of biomass that is transformed into crop yield; in case of grains it is called harvest index. Yield is actually the outcome of various physiological processes together in environment over time that might be changing or increasingly alterable. The mutable or alterable phenotypes results from the interrelationship between the genotype of the organisms and the environment. In order to utilize and reduce these variabilities, several designs have been formulated. One of the noticeable strategies is to reduce the environmental alterations by breeding the organism under highly controlled conditions. In addition to this, for the crop improvement attempts should be made to get the required amount of crop yield in the coming years[90] crop scientists and breeders should combine phenotype to genotype with huge efficacy.[93] This combination makes it easier through remarkable benefits in biotechnology along with marker-assisted selection, association mapping, and the elevating acquirablity of inexpensive DNA sequence information.[94] Nevertheless, the progress in biotechnological fields does not become analogue to supplementary pattern to efficiently and effectively phenotype at the crop level. Presently, field phenotyping of intricate attributes linked with the progress and production of biomass is a difficult process, frequently results in wrong or disastrous dimensions taken from a part of the whole experimental plot, which may not exactly elucidated the whole plot and is liable to human error. Although, efforts have been made to reduce the error and some ideal protocols have been formulated for phenotyping in crops such as a standard procedure in wheat are available[95] but the dimensions are costly as it requires a huge labor and for this reason it was rarely utilized by those commercial breeding companies who are well established. Besides, evaluating of field germplasm for intricate attributes is challenging, because of variability of environmental characteristics in duration of time and space. One of the recent examples was the regular occurrences of drought and floods in the Horn of Africa that constantly resulted in critical shortage of food[96] and is the signs suggesting the ultimate bad climatic conditions are detrimental for crop production and ultimately food security. Intergovernmental Panel on Climate Change (IPCC) described that an extra 40–170 million more people will suffer insufficient food supply in the coming years as a result of extreme climate change.[97] Positively, the staggering prognostication is that the bad climate events such as heat waves, heavy precipitation, and increasing of sea level will frequently occur in different parts of the world during the 21st century[98] resulting in frequent floods, drought, and salt stress as a crucial consequence. The proper layout for formulating the solutions to such problems actually varies[99] (Table 1.1) as the descriptions and measurements of these constraints are variable across geographical regions.

TABLE 1.1 Some Expected Negative Impacts of Climate Change on Crop Production by Regions Adapted from the Second Report on the State of the World's Plant Genetic Resources for Food and Agriculture (FAO, 2010).

Asia

- Crop yields could decrease by up to 30% in Central and South Asia
- More than 28 million hectares (ha) in arid and semi-arid regions of South and East Asia will require substantial (at least 10%) increases in irrigation for a 1 °C increase in temperature

Africa

- One of the most vulnerable continents to climate change and climate variability
- With many semi-arid regions and projected increase of 5–8% by the 2080s, likely reduction in the length of growing seasons will render further large regions of marginal agriculture out of production
- Projected reductions in crop yields of up to 50% by 2020
- Fall in crop net revenues by up to 90% by 2100
- Population of 75–250 million people at risk of increased water stress by the 2020s and 350–600 million people by the 2050s

Australia and New Zealand

- Agricultural production may decline by 2030 over much of Southern and Eastern Australia, and over parts of Eastern New Zealand, due to increased drought and fire
- Change land use in Southern Australia, with cropping becoming non-viable at the dry margins

TABLE 1.1   *(Continued)*

- Production of Australian temperate fruits and nuts will drop on account of reduced winter chill
- Geographical spread of a major horticultural pest, the Queensland fruit fly (Bactrocera tryoni), may spread to other areas including the currently quarantined fruit fly-free zone

Europe

- Crop productivity is likely to decrease along the Mediterranean and in South-eastern Europe
- Differences in water availability between regions are anticipated to increase
- Much of European flora is likely to become vulnerable, endangered, or committed to extinction by the end of this century

North America

- Increased climate sensitivity is anticipated in the Southeastern United States and in the United States corn belt making yield unpredictable
- Yields and/or quality of crops currently near climate thresholds (e.g., wine grapes in California) are likely to decrease
- Yields of cotton, soybeans, and barley are likely to change

Latin America

- Risk of extinctions of important species
- By the 2050s, 50% of agricultural lands in drier areas may be affected by desertification and salinization
- Generalized reductions in rice yields by the 2020s
- Reductions in land suitable for growing coffee in Brazil, and reductions in coffee production in Mexico
- The incidence of the coffee leaf miner (Perileucoptera coffeella) and the nematode Meloidogyne incognita are likely to increase in Brazil's coffee production area
- Risk of *Fusarium* head blight in wheat is very likely to increase in Southern Brazil and in Uruguay

Small island

- Subsistence and commercial agriculture on small islands will be adversely affected by climate change
- In mid-and high-latitude islands, higher temperatures and the retreat and loss of snow cover could enhance the spread of invasive species including alien microbes, fungi, plants, and animals

Indeed, the most liable part of the community is the poverty-stricken countries specifically the South Asia and Sub-Saharan Africa, as they have tolerated the extreme results of these climatic variations to their food security.[100] One of the recent studies concluded by Ejeta[101] evaluated a reduction in yield by 10–20% for the most significant crop of Africa in the coming years. A similar study carried out by Tester and Langridge[72] deduced that the demand for additional amount of yield increases as the world population increases particularly in the developing countries, in this regard, Foresight[102] declared that the utilization of conventional knowledge and technology can increase the yield by 2–3 folds in the developing countries of the world. Success in achieving the requirement of generating additional food under worst weather events and with limited natural resources forms the basis for elevating efficacy that achieves additional productivity per unit of input. In this context, growing of varied crop varieties called "smart crops" which have ability to produce more with less input is in correspondence with the "greener" outlook and is crucially important for attaining the absolute pre-requisite for increasing productivity. This will obviously demand the re-orientation of various fields of crop production pattern including plant breeding and culturing of high producing, fully accommodated, minimal effort and stress tolerant crop diversity that creates a key factor in these interventions. In this regard, Beddington et al.[91] appropriately supposed that the innovative interventions act as the main drivers for achieving the food security and environmental stability.

The crop productivity illustrated the overall consequences of the complex interactions between the two major crucial determining factors that have appropriately equal participatory efficiencies namely the genetic make-up of the crop and the breeding practices.[103] Positively, around the preceding seven decades in the United States, the contributory percentage of biotechnological profits to field productivity elevates in maize ranges from 33 to 94% with an average range from 50 to 60%.[104] Genetic engineering or biotechnological tools make it possible to harness the genetic gains by encoding the genetic make-up of the PGRFA, and make remarkable contribution in achieving the demand of 70% increase in food production. Illustration of the spectacular outcome of the genetic engineering on crop productivity involves the progress and enormous distribution of high productivity and stress tolerant cereal crop varieties around the globe starting from the green revolution in the late 1960s. The remarkable result and increase in yield productivity in several food lacking countries (especially in Asia) were acknowledged with the economical life and disburden from insufficient nutrition.[105] Currently, the application of biotechnological tool leads to the ingress of high producing rice varieties, such as the new rice for Africa (NERICA), Sub-Saharan Africa that has also been acknowledged with the considerable rise in the food productivity of these regions.[106] A recent study described by Foresight[13] suggesting the utilization of systematic and developmental tools to address considerable exception of generating abundantly more food with reduced environmental resources, particularly identified "plant breeding" utilizing ethnic and modern techniques to upgrade productivity; elevation of water, nutrients, and other input efficiencies, in order to attain this target. The World Economic Forum[107] also placed the procreation or breeding of current crop diversities as the first line up for its industry partners, that is, combination of worldwide companies to address the insecurity of food. In this regard, scientists are aware of the restrictions or limitations in the genetic resources and hence elevate the susceptibility of crops. To address this limitation, wild crop species, land races, and other non-adaptive genetic resources, which are commonly less productive and exhibiting objectionable attributes should be utilized regularly in genetic upgradation[72] The utilization of such non-adaptive resources in crop breeding would prove a good investment. This case involves the utilization of genes located on a translocated chromosome arm of rye in the genetic modification of wheat.[108] One of the study carried out by Gur and Zamir[109] also validate that the ingression of genes from the tomato (wild relative) into the drought-tolerant *Solanum pennelli* (green-fruited) enhanced the productivity up to 50%. It has been found that the two centers of the Consultative Group on International Agricultural Research (CGIAR) (namely the International Institute of Tropical Agriculture, Ibadan, Nigeria and the International Center for Tropical Agriculture, Cali, Colombia) have separately utilized the wild relatives of cassava to increase disease tolerance, nutritional trait and enhance life span of the fresh roots of the crop.[110] The incredible benefaction of the short height gene from the Japanese wheat variety (Norin 10), to the Green Revolution is globally chronicled and inevitable, other attempts also have produced considerable consequences as well. One of the important crop modification approaches is the recombinant DNA technology, which involves the usage of DNA sequence molecules obtained from different sources to produce novel genetic variant and the process is called as transformation or genetic modification and the new variety is known as genetically modified organisms (GMOs) or commonly as transgenic organism. Only four modified/transgenic crop varieties have been produced commercially (canola, maize, cotton, and soybean) exhibiting two transformation events; that is herbicide resistance and insect tolerance or their combinations since the first approvals in 1996, James[111] evaluated that there had been a 94-fold rise in hectarage from 1.7 million ha in 1996 to 160 million ha in 2011 (16 years) in the commercialization of transgenic crops. It has also been observed that the value of seed market of transgenic crops produced in 29 countries (19 developing and 10 industrial countries) at the United States was $13.2 billion in 2011 while the yield for transgenic soybean, maize, and cotton were valued high more than US$160 billion for the same year. In addition, due to less numbers of commercialized transgenic crops and the translation processes that bestowed the modified agronomic attributes, respectively, the progress and the classification of transgenic crops which is a remarkable trend in the modification of crops was no longer neglected. That is why the commercialization of transgenic crops was approved in 31 countries with their disposal in the environment.[111] Several of these techniques can be employed at high throughput and hence comes under the epithet of phenomics. Recently, Tester and Langridge[72] revealed that though there was the greater benefaction of non-transgenic crops to crop improvement events during this decade, but the development and the estimation of transgenic crops exits as an active theme of research with some ethical issues (public negative perceptions for the technology) that restricts the wideness of this technology by producers in many countries.

1.5   FUTURE CHALLENGES

The challenges for plant science to enable requisite increases in crop productivity while conserving the natural resource base for agriculture are daunting, but there are promising avenues to greatly enhance our capabilities to exploit genetic diversity through integration of phenomics with genomics. By providing powerful new capabilities for phenotyping large numbers of field plots, there is a clear research path that can transform field-based phenotyping (FBP). While not all of the components of an FBP system are in place, the research problems appear tractable, having numerous potential solutions. However, due to the need for highly integrative approaches, efforts to advance FBP cannot rely solely on individual researchers or small groups iteratively pursuing local solutions. Attaining the phenotyping capability that will allow agriculture to address climate change, food security, and bioenergy requires co-ordinated and sustained efforts with adequate resources to test and develop the necessary infrastructure and procedures. Recently established Australian and European phenomics centers (e.g., http://www.plantphenomics.org.au and http://www2.fz-juelich.de/icg/icg-3/jppc) are indicative of the growing interest in high-throughput technologies and of the potential for developing the relatively large, integrated research infrastructure required for FBP. As per the above information, we have avoided emphasizing specific traits. This reflects our certainty that searching for single indicator that correlate strongly with yield is unlikely to provide more information than simply analyzing yield differences. A given yield level is often attainable through multiple mechanisms, and the optimal combination of traits for one environment often differs from that required in another. The challenge of phenotyping is to provide data on those underlying mechanisms. In modern crop improvement, management of intellectual property is a continuous concern. There are surprisingly broad US patents relating to methods for inferring plant characteristics. US patent 5,764,819 awarded to Orr et al.[112] describes "techniques for catagorizing plants for determination and procreation programs by utilizing remote sensing and image analysis technology." While, patents can allow inventors to recover their research investments through royalties, Heller[113] emphasized that patenting component technologies can inhibit innovation in systems that require assembling multiple components, such as the instruments and software required for FBP. The basic principles and applications of proximal sensing appear to be well established and to lack the novelty required to justify patents. Innovation is needed; however, in design of specific instruments. If sufficiently novel, these might justify application for patents. Outputs from specific instruments often are in proprietary formats, which also complicate integration of components. Obviously, for public research, instruments are preferred whose outputs and controls are readily accessed without use of additional proprietary hardware or software, and an overall philosophy of open architectures and software using "off the shelf" electronic components will help stimulate collaborative development. The Cubesat program, which promotes development of low cost satellites that conform to an open design standard,[114] may provide a useful model. Technically, the limitation that comes in the way of development of genetically modified (GM) varieties globally involves the lack of effective genotype dependence free regeneration systems for majority of crops. Furthermore, the prolonged technical problems associated with the transformation incidents highly restrict the applicability of genetic transformation in breeding for polygenetic attributes such as tolerance to environmental stress (abiotic stress like salinity and drought) because of climate change and modifications. Nevertheless, the presence of the genes that possesses the capability to combat with the biotic and abiotic stresses (insect resistance and herbicide tolerance) shows the possibility to address this restriction. Furthermore, scientific research must aims at the increasing range of agronomical attributes which can be modified by this technique, although these two transformation incidents in commercial varieties are absolutely not enough to make GM technology a powerful crop improvement technique.

The emerging science of phenomics will be central to realizing this vision once several challenges are overcome. Improved methods are needed for high-throughput collection of diverse phenotypic measures, in both natural and laboratory environments. Phenomic datasets can be large and complex, likely dwarfing the size of genomic datasets. Tools are needed not only for generating these datasets, but also for storing, analyzing, and interrogating them, efficiently and affordably. The power of phenomics will be multiplied when datasets can be combined and correlated across different studies, allowing increases in statistical power and the scope of analyses. For this to bear fruit, it will be critical for the field to develop formalized methods for data quality control, and for describing phenotypic measures and the circumstances under which they were collected. Many of these challenges can be met through creative application of computational and engineering technologies, further advancing the mission of research at the intersection of the physical and life sciences.

KEYWORDS

- climate change
- crop improvement
- phenomics
- plant breeding

REFERENCES

1. Bilder, R. M.; Sabb, F. W.; Cannon, T. D.; London, E. D.; Jentsch, J. D.; Parker, D. S.; Poldrack, R. A.; Evans, C.; Freimer, N. B. Phenomics: The Systematic Study of Phenotypes on a Genome-Wide Scale. *Neuroscience.* **2009,** *164,* 30–42.
2. Segall, J. E.; Block, S. M.; Berg, H. C. Temporal Comparisons in Bacterial Chemotaxis. *Proc. Natl. Acad. Sci. USA.* **1986,** *83,* 8987–8991.
3. Murchie, E. H.; Niyogi, K. K. Manipulation of Photoprotection to Improve Plant Photosynthesis. *Plant Physiol.* **2011,** *155,* 86–92.
4. Lango Allen, H.; Estrada, K.; Lettre, G.; Berndt, S. I.; Weedon, M. N.; Rivadeneira, F., et al. Hundreds of Variants Clustered in Genomic Loci and Biological Pathways Affect Human Height. *Nature.* **2010,** *467,* 832–838.
5. Nichols, R. J.; Sen, S.; Choo, Y. J.; Beltrao, P.; Zietek, M.; Chaba, R.; Lee, S.; Kazmierczak, K. M.; Lee, K. J.; Wong, A.; Shales, M.; Lovett, S.; Winkler, M. E.;

Krogan, N. J.; Typas, A.; Gross, C. A. Phenotypic Landscape of a Bacterial Cell. *Cell.* **2011,** *144,* 143–156.

6. Lu, Y.; Savage, L. J.; Ajjawi, I.; Imre, K. M.; Yoder, D. W.; Benning, C.; Dellapenna, D.; Ohlrogge, J. B.; Osteryoung, K. W.; Weber, A. P.; Wilkerson, C. G.; Last, R. L. New Connections across Pathways and Cellular Processes: Industrialized Mutant Screening Reveals Novel Associations between Diverse Phenotypes in *Arabidopsis*. *Plant Physiol.* **2008,** *146,* 1482–1500.

7. Gerke, J.; Lorenz, K.; Cohen, B. Genetic Interactions between Transcription Factors Cause Natural Variation in Yeast. *Science.* **2009,** *323,* 498–501.

8. Shasha, D. E.; Kouranov, A. Y.; Lejay, L. V.; Chou, M. F.; Coruzzi, G. M. Using Combinatorial Design to Study Regulation by Multiple Input Signals. A Tool for Parsimony in the Post-Genomics Era. *Plant Physiol.* **2001,** *127,* 1590–1594.

9. Fisher, R. The Correlation between Relatives on the Supposition of Mendelian Inheritance. *Trans. R. Soc. Edinb. Earth Sci.* **1918,** *52,* 399–433.

10. Wright, S. Correlation and Causation. *J. Agric. Res.* **1921,** *20,* 557–555.

11. Stam, P. In *Marker-Assisted Introgression: Speed at any Cost?: Eucarpia Leafy Vegetables 2003,* Proceedings of the EUCARPIA Meeting on Leafy Vegetables Genetics and Breeding, Noordwijkerhout, The Netherlands, Mar 19–21, 2003; Van Hintum, T. L., Lebeda, A., Pink, D., Schut, J. W., Eds.; Centre for Genetic Resources (CGN): Wageningen, The Netherlands, 2003; pp 117–124.

12. *The Columbia Encyclopedia,* 6th ed.; Columbia University Press: New York, 2008. [http://www.encyclopedia.com/topic/plant_breeding.aspx].

13. Foresight; *The Future of Food and Farming*; Final Project Report, The Government Office for Science: London, 2011.

14. Rascher, U.; Nedbal, L. Dynamics of Photosynthesis in Fluctuating Light – Commentary. *Curr. Opi. Plant Biol.* **2006,** *9,* 671–678.

15. Mittler, R.; Blumwald, E. Geneti Engineering for Modern Agriculture: Challenges and Perspectives. *Annu. Rev. Plant Biol.* **2010,** *61,* 443–462.

16. Rascher, U.; Agati, G.; Alonso, L.; Cecchi, G.; Champagne, S.; Colombo, R.; Damm, A.; Daumard, F.; de- Miguel, E.; Fernandez, G.; Franch, B.; Franke, J.; Gerbig, C.; Gioli, B.; Gómez, J. A.; Goulas, Y.; Guanter, L.; Gutiérre-De-La-Cámara, Ó.; Hamdi, K.; Hostert, P.; Jiménez, M.; Kosvancova, M.; Lognoli, D.; Meroni, M.; Miglietta, F.; Moersch, A.; Moreno, J.; Moya, I.; Neininger, B.; Okujeni, A.; Ounis, A.; Palombi, L.; Raimondi, V.; Schickling, A.; Sobrino, J. A.; Stellmes, M.; Toci, G.; Toscano, P.; Udelhoven, T.; van der Linden, S.; Zaldei, A. CEFLES2: The Remote Sensing Component to Quantify Photosynthetic Efficiency from the Leaf to the Region by Measuring Sun-Induced Fluorescence in the Oxygen Absorption Bands. *Biogeosciences.* **2009,** *6,* 1181–1198.

17. Ustin, S.; Gamon, J. A. Remote Sensing of Plant Functional Types. *New Phytol.* **2010,** *186,* 795–816.

18. Beney, P. N.; Bennett, M.; Schiefelbein, J. Getting to the Root of Plant Biology: Impact of the *Arabidopsis* Genome Sequence on Root Research. *Plant J.* **2010,** *61,* 992–1000.

19. Smith, S. E.; Smith, F. A. Roles of Arbuscular Mycorrhizas in Plant Nutrition and Growth: New Paradigms from Cellular to Ecosystem Scales. *Annu. Rev. Plant Biol.* **2011,** *62,* 227–250.

20. Buckler, E. S.; Holland, J. B.; Bradbury, P. J.; Acharya, C. B.; Brown, P. J.; Browne, C.; Ersoz, E.; Flint Garcia, S.; Garcia, A.; Glaubitz, J. C.; Goodman, M. M.; Harjes, C.; Guill, K.; Koon, D. E.; Larsson, S.; Lepak, N. K.; Li, H.; Mitchell, S. E.; Pressoir,

G.; Peiffer, J. A.; Rosas, M. O.; Rocheford, T. R.; Romay, M. C.; Romero, S.; Salvo, S.; Sanchez Villeda, H.; da Silva, H. S.; Sun, Q.; Tian, F.; Upadyayula, N.; Ware, D.; Yates, H.; Yu, J.; Zhang, Z.; Kresovich, S.; McMullen, M. D. The Genetic Architecture of Maize Flowering Time. *Science.* **2009,** *325,* 714–718.

21. Salome, P. A.; Bomblies, K.; Laitinen, R. A.; Yant, L.; Mott, R.; Weigel, D. Genetic Architecture of Flowering-Time Variation in *Arabidopsis thaliana. Genetics.* **2011,** *188,* 421–433.

22. Cocuron, J. C.; Lerouxel, O.; Drakakaki, G.; Alonso, A. P.; Liepman, A. H.; Keegstra, K.; Raikhel, N.; Wilkerson, C. G. A Gene from the Cellulose Synthase-Like C Family Encodes A Beta-1,4 Glucan Synthase. *Proc. Natl. Acad. Sci. USA.* **2007,** *104,* 8550–8555.

23. Schilmiller, A. L.; Miner, D. P.; Larson, M.; McDowell, E.; Gang, D. R.; Wilkerson, C.; Last, R. L.. Studies of a Biochemical Factory: Tomato Trichome Deep Expressed Sequnce Tag Sequencing and Proteomics. *Plant Physiol.* **2010,** *153,* 1212–1223.

24. Rounsley, S. D.; Last, R. L. Shotguns and SNPs: How Fast and Cheap Sequencing is Revolutionizing Plant Biology. *Plant J.* **2010,** *61,* 922–927.

25. Kilian, J.; Whitehead, D.; Horak, J.; Wanke, D.; Weinl, S.; Batistic, O.; D'ngelo, C.; Bornberg-Bauer, E.; Kudla, J.; Harter, K. The AtGenExpress Global Stress Expression Data Set: Protocols, Evaluation and Model Data Analysis of UV-B Light, Drought and Cold Stress Responses. *Plant J.* **2007,** *50,* 347–363.

26. Last, R. L.; Jones, A. D.; Shachar-Hill, Y. Towards the Plant Metabolome and Beyond. *Nat. Rev. Mol. Cell. Biol.* **2007,** *8,* 167–174.

27. Velasco, L.; Pérez-Vich, B.; Fernández-Martínez, J. M. Estimation of Seed Weight, Oil Content and Fatty Acid Composition in Intact Single Seeds of Rapeseed (*Brassica napus* L.) by Near-Infrared Reflectance Spectroscopy. *Euphytica.* **1999,** *106,* 79–85.

28. Shroff, R.; Vergara, F.; Muck, A.; Svatos, A.; Gershenzon, J. Nonuniform Distribution of Glucosinolates in *Arabidopsis thaliana* Leaves has Important Consequences for Plant Defense. *Proc. Natl. Acad. Sci. USA.* **2008,** *105,* 6196–6201.

29. Zhen, Y.; Qi J. L.; Wang S. S.; Su J.; Xu G. H.; Zhang M. S., et al. Comparative Proteome Analysis of Differentially Expressed Proteins Induced by Al Toxicity in Soybean. *Physiol. Plant.* **2007,** *131,* 542–554.

30. Aghaei, K.; Ehsanpour, A.; Shah, A.; Komatsu, S. Proteome Analysis of Soybean Hypocotyl and Root under Salt Stress. *Amino Acids.* **2009,** *36,* 91–98.

31. Yamaguchi, M.; Valliyodan, B.; Zhang, J.; Lenoble, M. E.; Yu, O.; Rogers, E. E., et al. Regulation of Growth Response to Water Stress in the Soybean Primary Root. I. Proteomic Analysis Reveals Region–Specific Regulation of Phenylpropanoid Metabolism and Control of Free Iron in the Elongation Zone. *Plant Cell Environ.* **2010,** *33,* 223–243.

32. Qin, J.; Gu, F.; Liu, D.; Yin, C.; Zhao, S.; Chen, H., et al. Proteomic Analysis of Elite Soybean Jidou17 and its Parents Using Itraq-Based Quantitative Approaches. *Proteome Sci.* **2013,** *11,* 12.

33. Baxter, I. R.; Vitek, O.; Lahner, B.; Muthukumar, B.; Borghi, M.; Morrissey, J.; Guerinot, M. L.; Salt, D. E. The Leaf Ionome as a Multivariable System to Detect a Plant's Physiological Status. *Proc. Natl. Acad. Sci. USA.* **2008,** *105,* 12081–12086.

34. Walter, A.; Silk, W. K.; Schurr, U. Environmental Effects on Spatial and Temporal Patterns of Leaf and Root Growth. *Annu. Rev. Plant Biol.,* **60,** 279–304 2009.

35. Sinclair, T. R.; Rufty, T. W. Nitrogen and Water Resources Commonly Liit Crop Yield Increases, not Necessarily Plant Genetics. *Glob. Food Sec.* **2012,** *1,* 94–98.

36. Mueller, N. D.; Gerber, J. S.; Johnston, M.; Ray, D. K.; Ramankutty, N.; Foley, J. A. Closing Yield Gaps through Nutrient and Water Management. *Nature.* **2012,** *490,* 254–257.

37. Fiorani, F; Schurr, U. Future Scenarios for Plant Phenotyping. *Annu. Rev. Plant Biol.* **2013,** *64,* 267–294.

38. Meidner, H.; Mansfield, T. *Physiology of Stomata;* McGraw-Hill: New York, 1968.

39. Jones, H. G. Stomatal Control of Photosynthesis and Transpiration. *J. Exp. Bot.* **1998,** *49,* 387–398.

40. Farquhar, G. D.; O'Leary, M.; Berry, J. On the Relationship between Carbon Isotope Discrimination and the Intercellular Carbon Dioxide Concentration in Leaves. *Funct. Plant Biol.* **1982,** *9,* 121–137.

41. Baker, N. R. Chlorophyll Fluorescence: A Probe of Photosynthesis *In Vivo. Annu. Rev. Plant Biol.* **2008,** *59,* 89–113.

42. Jones, H. G. Use of Thermography for Quantitative Studies of Spatial and Temporal Variation of Stomatal Conductance over Leaf Surfaces. *Plant Cell. Environ.* **1999,** *22,* 1043–1055.

43. Glenn, D. M. Infrared and Chlorophyll Fluorescence Imaging Methods for Stress Evaluation. *Hort. Sci.* **2012,** *47,* 697–698.

44. Messinger, S. M.; Buckley, T. N.; Mott, K. A. Evidence for Involvement of Photosynthetic Processes in the Stomatal Response to CO2. *Plant Physiol.* **2006,** *140,* 771–778.

45. Kamakura, M.; Kosugi, Y.; Takanashi, S.; Tobita, H.; Uemura, A.; Utsugi, H. Observation of the Scale of Patchy Stomatal Behavior in Leaves of Quercus Crispula Using an Imaging-PAM Chlorophyll Fluorometer. *Tree Physiol.* **2012,** *32,* 839–846.

46. Lawson, T.; Weyers, J. Spatial and Temporal Variation in Gas Exchange Over the Lower Surface of *Phaseolus vulgaris* L. Primary Leaves. *J. Exp. Bot.* **1999,** *50,* 1381–1391.

47. Lawson, T.; Kramer, D. M.; Raines, C. A. Improving Yield by Exploiting Mechanisms Underlying Natural Variation of Photosynthesis. *Curr. Opin. Biotechnol.* **2012,** *23,* 215–220.

48. Brooks, T. L.; Miller, N. D.; Spalding, E. P. Plasticity of *Arabidopsis* Root Gravitropism Throughout a Multidimensional Condition Space Quantified by Automated Imag Analysis. *Plant Physiol.* **2010,** *152,* 206–216.

49. Clark, R. T.; MacCurdy, R. B.; Jung, J. K.; Shaff, J. E.; McCouch, S. R.; Aneshansley, D. J.; Kochian, L. V. Three-Dimensional Root Phenotyping with a Novel Imaging and Software Platform. *Plant Physiol.* **2011,** *156,* 455–465.

50. Jahnke, S.; Menzel, M. I.; van Dusschoten, D.; Roeb, G. W.; Bühler, J.; Minwuyelet, S., et al. Combined MRI-PET Dissects Dynamic Changes in Plant Structures and Functions. *Plant J.* **2009,** *59,* 634–644.

51. Gutierrez, M.; Reynolds, M. P.; Klatt, A. R. Association of Water Spectral Indices with Plant and Soil Water Relations in Contrasting Wheat Genotypes. *J. Exp. Bot.* **2010,** *61,* 3291–3303.

52. Lu, Y.; Savage, L. J.; Larson, M. D.; Wilkerson, C. G.; Last, R. L. Chloroplast 2010: A Database for Large-Scale Phenotypic Screening of *Arabidopsis* Mutants. *Plant Physiol.* **2011,** *155,* 1589–1600.

53. Lussier, Y.; Liu, Y. Computational Approaches to Phenotyping High-Throughput Phenomics. *Proc. Am. Thorac. Soc.* **2007,** *4,* 18–25.

54. Massonnet, C.; Vile, D.; Fabre, J.; Hannah, M. A.; Caldana, C.; Lisec, J.; Beemster, G. T.; Meyer, R. C.; Messerli, G.; Gronlund, J. T.; Perkovic, J.; Wigmore, E.; May, S.; Bevan, M. W.; Meyer, C.; Rubio-Diaz, S.; Weigel, D.; Micol, J. L.; Buchanan-Wollaston,

V.; Fiorani, F.; Walsh, S.; Rinn, B.; Gruissem, W.; Hilson, P.; Hennig, L.; Willmitzer, L.; Granier, C. Probing the Reproducibility of Leaf Growth and Molecular Phenotypes: A Comparison of Three *Arabidopsis* Accessions Cultivated in Ten Laboratories. *Plant Physiol.* **2010,** *152,* 2142–2157.

55. Bouch, N.; Bouchez. D. *Arabidopsis* Gene Knockout: Phenotypes Wanted. *Curr. Opin. Plant Biol.* **2007,** *4,* 222–227.

56. McCallum, C. M.; Comai, L.; Greene, E. A.; an Henikoff, S. Targeted Screening for Induced Mutations. *Nat. Biotechnol.* **2000,** *18,* 455–457.

57. Till, B. J.; Reynolds, S. H.; Greene, E. A.; Codomo, C. A.; Enns, L. C.; Johnson, J. E.,; Burtner, C.; Odden, A. R.; Young, K.; Taylor, N. E., et al. Large-Scale Discovery of Induced Point Mutations with High-Throughput TILLNG. *Gen. Res.* **2003,** *13,* 524–530.

58. Koornneef, M.; Dellaert, L. W.; van der Veen, J. H. *Mutat. Res.* **1982,** *93,* 109–123.

59. Gilchrist, E. J.; Haughn, G. W. TILLING without a Plough: A New Method with Applications for Reverse Genetics. *Curr. Opin. Plant Biol.* **2005,** *8,* 211–215.

60. Perry, J. A.; Wang, T. L.; Welham, T. J.; Gardner, S.; Pike, J. M.; Yoshida, S.; Parniske, M. A TILLING Reverse Genetics Tool and a Web Accessible Collection of Mutants of the *Legume lotus japonica. Plant Physiol.* **2003,** *131,* 866–871.

61. Till, B. J.; Reynolds, S. H.; Weil, C.; Springer, N.; Burtner, C.; Young, K.; Bowers, E.; Codomo, C. A.; Enns, L. C.; Odden, A. R., et al. Discovery of Induced Point Mutations in Maize Genes by TILLING. *BMC Plant Biol.* **2004,** *4,* 12.

62. Slade, A. J.; Fuerstenberg, S. I.; Loeffler, D.; Steine, M. N.; Facciotti, D. A Reverse Genetic, Nontransgenic Approach to Wheat Crop Improvement by TILLING. *Nat. Biotechnol.* **2005,** *23,* 75–81.

63. Donofrio, N.; Rajagopalon, R.; Brown, D.; Diener, S.; Windham, D.; Nolin, S.; Floyd, A.; Mitchell, T.; Galadima, N.; Tucker, S.; Orbach, M. J.; Patel, G.; Farman, M.; Pampanwar, V.; Soderlund C.; Lee, Y. H.; Dean, R. A. 'PACLIMS': A Component LIM System for High-Throughput Functionalgenomic Analysis. *BMC Bioinformatics.* **2005,** *6,* 94.

64. Exner, V.; Hirsch-Hoffmann, M.; Guissem, W.; Hennig, L. PlantDB – a Versatile Database for Managing Plant Research. *Plant Methods.* **2008,** *4,* 1.

65. Paproki, A.; Fripp, J.; Salvado, O.; Sirault, X.; Berry, S.; Furbank, R. In *Automated 3D Segmentation and Analysis of Cotton Plants,* International Conference on Digital Image Computing: Techniques and Applications (DICTA), Noosa QLD, Australia, Dec 6–8, 2011; IEEE: New York, 2011; pp 555–560.

66. Setter, T. Analysis of Constituents for Phenotyping Drought Tolerance in Crop Improvement. *Front. Plant Physiol.* **2012,** *3,* 1–12.

67. Furbank, R.; Tester, M. Phenomics–Technologies to Relieve the Phenotyping Bottleneck. *Trends Plant Sci.* **2011,** *16* (12), 635–644.

68. Mahlein, A. K.; Oerke, E. C.; Steiner, U.; Dehne, H. W. Recent Advances in Sensing Plant Diseases. *Eur. J. Plant Pathol.* **2012,** *133,* 197–209.

69. Poorter, H.; Niinemets, Ü.; Walter, A.; Fiorani, F.; Schurr, U. A Method to Construct dose–Response Curves for f Wide Range of Environmental Factors and Plant Traits by Means of a Meta-Analysis of Phenotypic Data. *J. Exp. Bot.* **2010,** *61,* 2043–2055.

70. Yano, M.; Tuberosa, R. Genome Studies Andmolecular Genetics—from Sequence to Crops: Genomics Comes of Age. *Curr. Opin. Plant Biol.* **2009,** *12,* 103–106.

71. Schneeberger, K.; Weigel, D. Fast-Forward Genetics Enabled by New Sequencing Technologies. *Trends Sci.* **2011,** *16,* 282–288.

72. Tester, M.; Langridge, P. Breeding Technologies to Increase Crop Production in a Changing World. *Science.* **2010,** *327,* 818–822.

73. Houle, D.; Govindaraju, D. R.; Omholt, S. Phenomics: The Next Challenge. *Nat. Rev. Genet.* **2010,** *11,* 855–866.

74. Deikman, J.; Petracek, M.; Heard, J. E. Drought Tolerance through Biotechnology: Improving Translation from the Laboratory to Farmers' Fields. *Curr. Opin. Biotechnol.* **2012,** *23,* 243–250.

75. Araus, J. L.; Serret, M. D.; Edmeades, G. O. Phenotyping Maize for Adaptation to Drought. *Front. Plant Physiol.* 2012, *3,* 305.

76. Finkel, E. With "Phenomics," Plant Scientists Hope to Shift Breeding into Overdrive. *Science.* **2009,** *325,* 380–381.

77. Frasson, R. P. D. M.; Krajewski, W. F. Three-Dimensional Digital Model of a Maize Plant. *Agric. For. Meteorol.* **2010,** *150,* 478–488.

78. Omasa, K.; Hosoi, F.; Konishi, A. 3D Lidar Imaging for Detecting and Understanding Plant Responses and Canopy Structure. *J. Exp. Bot.* **2007,** *58* (4), 881–898.

79. Hosoi, F.; Nakabayashi, K.; Omasa, K. 3-Dmodeling of Tomato Canopies Using a High-Resolution Portable Scanning Lidar for Extracting Structural Information. *Sensors.* **2011,** *11* (2), 2166–2174.

80. Schurr, U.; Heckenberger, U.; Herdel, K.; Walter, A.; Feil, R. Leaf Development in *Ricinus communis* during Drought Stress: Dynamics of Growth Processes, of Cellular Structure and of Sink Source Transition. *J. Exp. Bot.* **2000,** *51* (350), 1515–1529.

81. Dornbusch, T.; Wernecke, P.; Diepenbrock, W. A Method to Extract Morphological Traits of Plant Organs from 3D Point Clouds as a Database for an Architectural Plant Model. *Ecol. Model.* **2007,** *200* (1–2), 119–129.

82. Gartner, H.; Wagner, B.; Heinrich, I.; Denier, C. 3D-Laser Scanning: A New Method to Analyse Coarse Tree Root. *ISSR. Symp. Root Res. Appl.* **2009,** *106,* 95–106.

83. Beder, C.; Forstner, W. In *Direct Solutions for Computing Cylinders from Minimal Sets of 3D Points,* Proceedings of the 9th European Conference on Computer Vision - ECCV'06, Berlin, Heidelberg, Springer-Verlag: Berlin, 2006; Vol. I, pp 135–146.

84. Rabbani, T.; Van Den Heuvel, F. In *Efficient Hough Transform for Automatic Detection of Cylinders in Point Clouds,* Proceedings of ISPRS, Workshop, Laser Scanning 2005, Enschede, The Netherlands, Sept 12–14, 2005; ISPRS: The Netherlands. 2005; Vol. 36, pp 60–65.

85. Paproki, A.; Sirault, X.; Berry, S.; Furbank, R.; Fripp, J. A Novel Mesh Processing Based Technique for 3D Plant Analysis. *BMC Plant Biol.* **2012,** *12,* 63.

86. Paulus, S.; Dupuis, J.; Mahlein, A. K.; Kuhlmann, H. Surface Feature Based Classification of Plant Organs from 3D Laser Scanned Point Clouds for Plant Phenotyping. *BMC Bioinformatics.* **2013,** *14,* 238.

87. Matsuda, O.; Tanaka, A.; Fujita, T.; Iba, K. Hyperspectral Imaging Techniques for Rapid Identification of *Arabidopsis* Mutants with Altered Leaf Pigment Status. *Plant Cell. Physiol.* **2012,** *53,* 1154–1170.

88. Poorter, H.; Climent, J.; Van Dusschoten, D.; Bühler, J.; Postma, J. Pot Sizematters: A Meta-Analysis on the Effects of Rooting Volume on Plant Growth. *Funct. Plant Biol.* **2012,** *39,* 839–850.

89. Tuberosa, R. Phenotyping Drought-Stressed Crops: Key Concepts, Issues and Approaches. In *Drought Phenotyping in Crops: From Theory to Practice;* Monnevaux, P., Ribaut, J. M., Eds.; CGIAR, Generation Challenge Programme: Texcoco, Mexico,

pp 3–35, 2011. http://www.generationcp.org/component/content/article/86-communications/books/ 594-drought-phenotyping-in-crops-from-theory-to-practice.

90. Bruinsma, J. In *The Resource Outlook to 2050. By how much do Land, Water Use and Crop Yields Need to Increase by 2050?*, Proceedings of the FAO Expert Meeting on How to Feed the World in 2050, Jun 24–26, 2009; Rome, FAO: Rome, Italy, 2009.

91. Beddington, J.; Asaduzzaman, M.; Fernandez, A.; Clark, M.; Guillou, M.; Jahn, M.; Erda, L.; Mamo, T.; Van Bo, N.; Nobre, C. A.; Scholes, R.; Sharma, R.; Wakhungu, J. In *Achieving Food Security in the Face of Climate Change: Summary for Policy Makers from the Commission on Sustainable Agriculture and Climate Change*, CGIAR Research Program on Climate Change, Copenhagen, Denmark, 2011; Agriculture and Food Security (CCAFS): Copenhagen, 2011.

92. Moose, S. P.; Mumm, R. H. Molecular Plant Breeding as the Foundation for 21st Century Crop Improvement. *Plant Physiol.* **2008,** *147,* 969–977.

93. Hall, A.; Wilson, M. A. Object-Based Analysis of Grapevine Canopy Relationships with Wine Grape Composition and Yield in Two Contrasting Vineyards Using Multi Temporal High Spatial Resolution Optical Remote Sensing. *Int. J. Remote Sens.* **2013,** *34,* 1772–1797.

94. Ingvarsson, P. K.; Street, N. R. Association Genetics of Complex Traits in Plants. *New Phytol.* **2011,** *189,* 909–922.

95. Pask, A.; Pietragalla, J.; Mullan, D.; Reynolds, M. *Physiological Breeding II: A Field Guide to Wheat Phenotyping*; Technical Report; CIMMYT: Mexico, DF, 2012.

96. International Food Policy Research Institute (IFPRI); *Global Hunger Index: The Challenge of Hunger. In Taming Price Spikes and Excessive Food Price Volatility,* Welthungerhilfe: Washington, DC, 2011.

97. Evans, A. *The Feeding of the Nine Billion: Global Food Security for the 21st Century;* Chatham House: London, 2009.

98. Field, C. B.; Barros, V.; Stocker, T. F.; Qin, D.; Dokken, D. J.; Ebi, K. L.; Mastrandrea, M. D.; Mach, K. J.; Plattner, G. K.; Allen, S. K.; Tignor, M.; Midgley, P. M. Intergovernmental Panel on Climate Change (IPCC): Summary for Policymakers. In *Managing the Risks of Extreme Events and Disasters to Advance Climate Change Adaptation;* Cambridge University Press: Cambridge, 2012.

99. Chikelu, M. B. A.; Guimaraes, E. P.; Ghosh, K. Re-orienting Crop Improvement for the Changing Climatic Conditions of the 21st Century. *Agric. Food Sec.* **2012,** *1,* 7.

100. Hertel, T. W.; Burke, M B.; Lobell, D. B. The Poverty Implications of Climate-Induced Crop Yield Changes by 2030. *Global. Environ. Change.* **2010,** *20,* 577–585.

101. Ejeta, G. Revitalizing Agricultural Research for Global Food Security. *Food Sec.* **2009,** *1,* 391–401.

102. Foresight; *The Future of Food and Farming*; Final Project Report, The Government Office for Science: London, 2011.

103. Fernandez-Cornejo, J. *The Seed Industry in US Agriculture: An exploration of Data and Information on Crop Seed Markets, Regulation, Industry Structure, and Research and Development;* Agriculture Information Bulletin No. 786; Economic Research Service, US Department of Agriculture: Washington, DC, 2004; p 213.

104. Duvick, D. N. Genetic Progress in Yield of United States Maize (*Zea mays* L.). *Maydica.* **2005,** *50,* 193–202.

105. Ejeta, G. African Green Revolution Needn't Be a Mirage. *Science.* **2010,** *327,* 81–832.

106. Oikeh, S. O.; Nwilene, F.; Diatta, S.; Osiname, O.; Touré, A.; Okeleye, K. A. Responses of Upland NERICA Rice to Nitrogen and Phosphorus in Forest Agroecosystems. *Agron. J.* **2008,** *100,* 735–741.

107. World Economic Forum; *Realizing a New Vision for Agriculture: A Roadmap for Stakeholders*; World Economic Forum: Geneva, 2010.

108. Rabinovich, S. V. Importance of Wheat-Rye Translocations for Breeding Modern Cultivars of *Triticum aestivum* L. *Euphytica.* **1998,** *100,* 323–340.

109. Gur, A.; Zamir, D. Unused Natural Variation can Lift Yield Barriers in Plant Breeding. *PLoS. Biol.* **2004,** *2,* 1610–1615.

110. Chavez, A. L.; Sanchez, T.; Jaramillo, G.; Bedoya J. M.; Echeverry, J.; Bolanos, E. A.; Ceballos, H.; Iglesias, C. A. Variation of Quality Traits in Cassava Roots Evaluated in Landraces and Improved Clones. *Euphytica.* **2005,** *143,* 125–133.

111. Glover, D. *Monsanto and Smallholder Farmers: A Case-Study on Corporate Accountability*; IDS Working Paper No. 277. University of Sussex, Institute of Development Studies: Brighton, 2007; p 53.

112. Orr, P. M.; Warner, D. C.; O'Brien, J. V.; Johnson, G. R. Methods for Classifying Plants for Evaluation and Breeding Programs by Use of Remote Sensing and Image Processing. U.S. Patent US005, 764,819, p. 37 1998.

113. Heller, M. *The Gridlock Economy;* Basic Books: New York, 2008; p 259.

114. Woellert, K.; Ehrenfreund, P.; Ricco, A. J.; Hertzfeld, H. Cubesats: Cost-Effective Science and Technology Platforms for Emerging and Developing Nations. *Adv. Space Res.* **2010,** *47,* 663–684.

CHAPTER 2

GENOMIC AND TRANSCRIPTOMIC APPROACHES FOR QUALITY IMPROVEMENT IN OILSEED BRASSICAS

ABHA AGNIHOTRI*, MANOJ KUMAR, DIVYA KILAM, and
JASROOP K. ANEJA

*Amity Institute of Microbial Technology, Amity University,
Noida 201313, Uttar Pradesh, India*

Corresponding author. E-mail: agnihotri.abha@gmail.com

CONTENTS

ABSTRACT

Brassica oilseeds are source for 12% of world's edible vegetable oil production. Over the years, emphasis has been laid on achieving yield stability with
improved seed and oil quality. Fatty acid (FA) profile of "canola" is considered ideal for nutrition as phytosterols lower the LDL-cholesterol levels.
Tocopherols are lipid-soluble antioxidants essential to humans. High phosphate with low phytate levels, and good quality protein are desired in seed
meal. The ever increasing demand to improve present yield and quality standards poses a challenge for the breeders. The advent of molecular breeding
approaches has helped to explore the variation in gene pools and combine it
with conventional breeding methods. Study of genome wide single-nucleotide polymorphism (SNP) opens the field of comparative genetic mapping.
Detection of quantitative trait loci (QTL) in populations helps to locate the
linkage groups between genomes. Potential molecular markers can be used
for selection and cloning of important genes for desired trait(s). In Brassica,
expressed sequence tags (ESTs) are being used for deployment of microarray platforms for high-throughput transcriptome analysis. The ribonulceic
acid interference (RNAi) based post transcriptional gene silencing (PTGS)
helps toward inhibiting the genes ameliorating the FA pathways. The use of
seed specific antisense technology has brought the selective modulation of
key enzyme activities in the developing seed, while keeping the rest of the
genetic background of the plant intact. This chapter aims to outline the use
of genomic and transcriptomic information of Brassica species as tools for
trait specific breeding.

2.1   INTRODUCTION TO OILSEED BRASSICAS

Brassicaceae family comprising over 3700 species worldwide is one of the
major groups of the plant kingdom.[1] It comprises several oilseed (canola,
mustard) and vegetable crops (turnip, cabbage, broccoli) of agricultural
importance.[2] It is an inexpensive and healthy source of food, providing nutrients and phytochemicals such as phenolic compounds, vitamins, phytosterols, glucosinolates, soluble sugars, fiber, minerals, carotenoids, and so
forth.[3,4] The classical triangle of U (Fig. 2.1) depicts the three Brassica
genomes as Brassica interspecific hybrid plants.[5] It comprises three diploid
species: *Brassica rapa* (AA), $n = 10$; *Brassica nigra* (BB), $n = 8$; *Brassica oleracea* (CC), $n = 9$. Three digenomic allotetraploid species: *Brassica
juncea* (AABB), $n = 18$; *Brassica carinata* (BBCC), $n = 17$; *Brassica napus*

(AACC), $n = 19$ are a result of natural hybridization between diploid species. Inter-relationship of these six Brassica species has been established at the molecular level by studying genome evolution and comparative sequence analysis.[6]

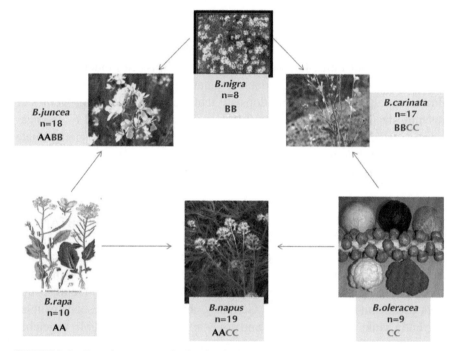

FIGURE 2.1   Brassica genomes in the classical "U" triangle.

Genetic resource of U's triangle can be used to create new hexaploid species with superior traits. Allohexaploid crops like wheat (*Triticum aestivum*) and oats (*Avena sativa*) have shown their importance in agriculture.[7,8] It is thus hypothesized that development of allohexaploid Brassica species will help to enhance yield and quality of Brassica crops. Till date no natural hexaploids of genus Brassica have been formed. Modern breeding and advanced molecular tools can be used to avoid genetic bottlenecks in the development of new allohexaploid Brassica species.[9] Artificially developed Brassica hexaploids (AABBCC) have shown promising agronomic traits.[10,11]

Brassica crop species have large and complex genomes as compared to their closest model plant *Arabidopsis thaliana*. Studies have been conducted to study genomic relations between these species to know the chromosomal basis of their evolution. Within Brassica species also diversity in gene pools can be seen, which is responsible for the variation in metabolite profiles. Modern varieties of rapeseed are being demonstrated as a potential outcome of the omics approach, that is, population genomics. Such available traits are a comparative small subset of the available genetic diversity, as breeding advancement suffers from low genetic pool caused by severe selection bottlenecks in recent decades. Rapeseed is thus likely to respond strongly to programs aimed at selectively enhancing genetic variation for key economic input and output traits. Growers can adapt the Brassica model plants with determinate growth habit of shoots terminating in pods, tagged as terminal flower mutant (TFL 1) in Arabidopsis,[12] that were identified in progenies of resynthesized *B. juncea*. A recent report analysis of F_1, F_2, and F_3 segregation revealed the monogenic recessive inheritance leading to determinate pod trait in both *B. juncea* and *B. carinata*. Further in *B. juncea*, the gene for determinacy (sdt1) was mapped to the linkage group B5.[13]

2.2   OIL YIELDING BRASSICA SPECIES AND THEIR IMPORTANCE

Brassica oilseeds, *B. napus* (canola); *B. rapa* (turnip rape or sarson); *B. juncea* (Indian mustard or brown mustard) comprise 12% of the total world's supply of edible vegetable oil[14] Major rapeseed producing countries are China, India, Canada, Germany, and France. Rapeseed (canola) currently ranks third after palm and soybean, as source of edible vegetable oil and third after soy and cotton, as source of oil meal. Its unsaturated fatty acid (FA) content makes it an excellent oil source. It is also a rich source of omega-3-unsaturated FAs.[15] Triacylglycerols are the main constituents (about 98%) of Brassica oils. The remaining 2% comprise sterols, phospholipids, and sphingolipids.[16] The value of Brassica crops is increased by their protein and free amino acid content that are used as animal feed.

Canola (*B. napus*) oil is known for its rich nutritional quality which makes it useful as a cooking oil and in margarine production. Oil extracted meal produce is used as animal feed due to its high protein content. The largest producers of canola oil are Canada and China.[17] *B. rapa* and *B. oleracea* are widely grown as cruciferous vegetable (turnips, Chinese cabbage, pak choi, cabbage, cauliflower, etc.).

B. juncea is an important oilseed crop as it can be grown in semi-arid regions. Brown mustard seeds obtained from *B. juncea* are used in medicines and culinary spices. They are also a rich source of minerals, vitamins, phytosterols, and antioxidants. Traditionally mustard seeds and its oil have been used to relieve rheumatic pain and also as a laxative to stimulate gastric mucosa.[18]

Agricultural importance of Brassicas has brought about considerable progress in Brassica biotechnology. Brassica crops and vegetables are known to provide health-promoting phytochemicals like phenolic compounds, glucosinolates, vitamins, minerals, carotenoids, and soluble sugars.[19, 20] Agronomically important genes have been identified and transferred in Brassica species. For introduction of desirable traits molecular breeding and transformation technique have also been used widely. Efficient transformation systems have been developed for all economically important species of Brassica leading to oil quality improvement and resistance against biotic and abiotic stresses. Tissue culture techniques have been widely exploited in Brassica for the development of efficient transformation protocols. Organogenesis has been accomplished in Brassica plants through cotyledons, hypocotyls, leaves, protoplasts, and roots. Brassica species are known to have varied chemical constituents, mainly in oil and glucosinolate content. Programs for Brassica improvement over the years have focused on reduction of undesirable erucic acid which is known to affect the oil quality. In 1970s development of canola quality lines was a major advancement; it contains less than 2% erucic acid. Conventional methods and biotechnological techniques like *in vitro* embryo rescue, double haploid, and genetic engineering have been used.[21]

There is a generalized consensus about the paleohexaploid ancestry of Brassica species. Study of the evolutionary events by comparative mapping between Arabidopsis and Brassica has revealed three major paleopolyploid (whole genome duplication) events.[22] The orthologous relation of the loci between two genomes is further complicated by chromosomal rearrangements, gene gain, and gene loss events. Advent of genome projects has helped in accumulating information about molecular determinants regulating various processes and traits.

2.2.1 OIL CONTENT AND FATTY ACID PROFILE

Oils are important part of our diet as they act as vehicle for transport of important metabolites. Brassica oilseeds have an oil content ranges from 35 to 44% and Indian mustard has oil content ranges from 39 to 42%.[23,24] They are composed of FA molecules that are covalently linked to a glycerol molecule. Based on their bond formations FAs are classified as saturated fatty acid (SFAs) and unsaturated fatty acids, monounsaturated fatty acids

(MUFAs) and polyunsaturated fatty acids (PUFAs). MUFAs (oleic acid) are thermostable and are known to reduce cholesterol levels in human beings.[25] PUFAs, linoleic acid, and linolenic acid are essential FAs required to be supplied in diet. Higher levels of PUFA make the oil amenable to oxidation thus reducing their shelf life. Rapeseed under the trade name "Canola" is considered to have ideal FA composition with SFAs (6%), oleic acid (65%), linoleic acid (20%), and linolenic acid (9%).[26] Studies have shown negative correlation of erucic acid with linoleic and oleic acid content in Brassicas.[27] Reduction in erucic acid has thus shown to increase oleic and linoleic acid levels, but it is known to be species specific. Downey[26] showed 60% increase in oleic acid content upon reduction of erucic acid in *B. napus*, whereas another study by Agnihotri and Kaushik[28] showed an increase of 45% in *B. juncea*. *B. carinata* with low erucic acid content showed an increase in linoleic and linolenic acids and reduction in oleic acid content.[29] Low erucic acid and high oleic acid have been transferred in *B. napus* and *B. juncea*.[30, 31] Oil quality modifications have also been achieved by RNAi based post transcriptional gene (PTGS) silencing or by adding new genes. These methods have enabled the development of traits like seed oil and protein content. In *B. napus* silencing of the Δ12 desaturase gene resulted in development of genotypes with 89% oleic acid.[32]

2.2.2 *GLUCOSINOLATES*

Glucosinolates are secondary metabolites found in Brassicaceae and related families. Based on structure of amino acid precursor glucosinolates are classified into three classes: aliphatic glucosinolates, indole glucosinolates, and aromatic glucosinolates. Environmental factors like climate, soil type, cultivation conditions, and plant position have shown to effect glucosinolate contents in Brassica crops.[33] Diversity based on glucosinolate composition is related to A, B, and C genomes in Brassica species. *B. nigra* contains glucosinolates with three carbon (C) side chains, *B. oleracea* contains either 3C side chains or 4C side chains and *B. rapa* contains glucosinolates with either 4C side chains or 5C side chains. *B. juncea, B. napus,* and *B. carinata* possess glucosinolate composition that consists of the profiles of two elementary species. Breeding objectives in Brassica vegetables have focused on enhancement of glucosinolates like glucoraphanin that have anti-carcinogenic properties and reduction of antinutritional compounds like progoitrin in *B. oleracea.*

2.2.3 PHENOLICS

Use of natural antioxidants has been lately envisaged for human consumption. Phenolic compounds are one of the important groups having antioxidant activity. Based on their carbon arrangement they are flavonoids (flavonols, anthocyanins, isoflavones, etc.) and non-flavonoids (hydroxycinnamates, phenolic acids). Brassica species have a diverse group of polyphenols mainly flavonols, anthocyanins, and hydroxycinnamic acid.[19] Major flavonols in Brassica crops are kaempferol, quercetin, and isorhamnetin. Among *B. oleracea* species, broccoli has been found to be a good source of flavonol and hydroxycinnamoyl, linked to high levels of phenolic compounds.[34] Cabbage and kale have also shown to have high phenolic content. Among Brassica oilseeds, canola seeds are the richest source of phenolic compounds. Major phenolics present in them are sinapic acid derivatives and minor phenolics include syringic, *p*-coumaric, ferulic, caffeic, *p*-hydroxybenzoic, and so forth.[35] Overexpression studies have been carried out in *B. napus* and leaf phenolics quantified in transgenic and non-transgenic plants. Dramatic increase in derivatives of quercetin, sinapic acid, cyanidin, and pelargonidin was reported in Arabidopsis by overexpression of PAP1gene, responsible for anthocyanin pigment production.[36]

2.2.4 STEROLS

They are the precursor of steroid hormones in human beings and phytohormones in plants.[37] Plants have about 40 well identified phytosterols which are majorly present in oilseed plants. Rapeseed is the richest source of natural phytosterols; 65% free phytosterol and 35% steryl ester fraction have been reported in rapeseed.[38] In another study phytosterol fraction of up to 10 g/kg oil was shown in rapeseed oil.[39]

2.2.5 VITAMINS

Vitamin E (α-tocopherol) is a lipid-soluble antioxidant. α- tocopherol has been reported to be predominant tocopherol in all Brassica vegetables.[40] Kale and broccoli are best source of lipid-soluble tocopherols. Moderate levels are found in brussels sprouts, whereas cauliflower and cabbage have shown to have relatively low amounts of α-tocopherol. A project under the name NAPUS 2000 was initiated to create novel tocopherol containing variants by breeding and genetic engineering techniques.[41] High carotenoid canola oil having 960 μg/g of beta-carotene has been developed by seed specific overexpression of phytoene synthase gene.[42] Integration of beta-carotene in Brassica oilseeds has been done by addition of lipophilic vitamin. Monsanto developed this high beta-carotene containing variety by overexpressing three genes; phytoene synthase (utilizes endogenous geranylgeranyl diphosphate to form phytoene) from daffodils, phytotene desaturase (phytoene to lycopene) from *Escherichia Coli,* and lycopene beta cyclase (lycopene to beta-carotene) from tomato.[43] Various biological activities in human body have been attributed to vitamin C (ascorbic acid). They act as enzyme cofactor, radical scavenger, electron donor/acceptor at the plasma membrane.[44] Vitamin C content in Brassica species varies significantly within subspecies.

2.2.6 PROTEIN

Brassica meal is an excellent source of protein, rich in essential amino acids, lysine, and methionine. Canola seeds are known to contain 17–26% protein, canola meal contains about 50% protein on dry weight basis.[45] Napin and cruciferin are major storage proteins present in canola meal and oleosin is a structural protein present in the oil fraction.[46] The protein composition of canola meal makes it favorable for human consumption. Its protein efficiency ratio was found to be 2.64, which is higher than that of soybean (2.19).[47] Canola protein extraction, development of canola protein, and limiting the presence of undesirable substances like phytates, phenols, and glucosinolates have been the focus of recent studies.

2.3 CONTRIBUTION OF GENOMICS AND TRANSCRIPTOMICS TOWARD QUALITY IMPROVEMENT

Evolution of novel technologies has enabled understanding of gene networks which are proving an asset to plant development and agronomy. Techniques involving study of traits having linkage with molecular markers, physical mapping, sequencing, and functional genomics provide major contribution.[48] Genomics can be a revolutionary tool for breeders as use of molecular breeding and direct mutagenesis enhances the breeding efficiency.[49] Genetic mapping can be done to map genes with unknown functions and then cloned. These genes can be good resources for transformation.[50] Next generation sequencing (NGS) and other genotyping platforms can also be used in various breeding programs. Genetic analysis of key agronomic traits in

Brassica species has seen in considerable progress over the years. Complex traits like yield and oil content are controlled by quantitative trait loci (QTL) that are present in different linkage groups. Quantitative inheritance of these complex traits is further complicated by environmental factors, plant development, and gene interactions. Transcript profiling on the other hand provides the candidate genes responsible for gene expression at different developmental stages and of agronomically important traits. The new era of systems biology, known as –omics, has brought about understanding of the whole organism and identification of genes involved in expression of a trait of interest to the breeders. An intact investigation on molecular evolution and transcriptional level has revealed the distinct role of proline dehydrogenase genes (BnaProDH1 and BnaProDH2). Such candidate genes have been over-expressed in pollen and roots at different phenological stages. The promoters of such gene sets were fused with uidA reporter gene (beta-glucuronidase, GUS) to characterize organ and tissue expression profiles in transformed *B. napus* plants. Data analysis indicates new elements to investigate the function of proline metabolism in plant development at evolutionary level.[51]

World's primary oilseed crop, *B. napus,* has been subjected to considerable improvement by breeders. Development of double-low (00) varieties (low erucic acid and low glucosinolate content) had considerable genetic bottlenecks. Thus the genetic diversity is to be increased to improve its yield and quality characteristics.

2.4   NGS TECHNIQUES IN BRASSICA

Classical molecular markers like restriction fragment length polymorphism (RFLP), random amplified polymorphic DNA (RAPD), amplified fragment length polymorphism (AFLP), and simple sequence repeats (SSRs) have been used by plant breeders for detection and utilization of genetic variation. Advances in NGS technology have facilitated simultaneous sequencing of selected genomic regions or targeted restriction fragments. Variant sites are identified to discover single nucleotide polymorphisms (SNPs) by aligning and comparing it with a reference genome sequence.[52] More than 20,000 SNPs were identified in *B. napus* using the restriction-associated DNA sequencing (RADseq).[53] First physical map for *B. rapa* A genome was constructed using 67,468 bacterial artificial chromosome (BAC) clones, with a physical length of 717 Mb. A3 chromosome of *B. rapa* was also obtained by 348 overlapping BAC clones.[54,55] Both were obtained using traditional sequencing method invented by Frederick Sanger (Sanger sequencing). With development of high-throughput and cost effective NGS technology, new products like 454 sequencing, sequencing by oligonucleotide ligation and detection (SOLiD), Solexa have come up. Roche 454 GS has proven efficiency in Brassica genome sequencing.[56] For sequencing of short Brassica target sequences, like transcriptome sequences and small RNAs, the illumina sequencing is highly suited and cost effective.[57] Since Brassica has a large, complicated genome third generation sequencing techniques like single-molecule real-time sequencing and nanopore sequencing can be promising.

2.5 APPLICATIONS OF NGS IN BRASSICA

2.5.1 DISCOVERY OF GENOME WIDE SNPS

Several methods have been developed for discovery of SNPs. Transcriptome analysis of *B. napus* using illumina platform was done to discover SNPs.[58] A total of 23,330–42,593 putative SNPs were detected. A total of 454 sequencing of two *B. napus* cultivars helped in discovery of 655 putative SNP markers. Genome wide SNP discovery by RAD sequencing technology lead to generation of more than 20,000 SNPs. Major QTL found on chromosome A9 through a combination of NGS and bulked segregant analysis (BSA) helped in discovering 70 SNPs associated with rapeseed pod shatter resistance. For large scale SNP analysis, SNP arrays are developed. They are effective for identification of unique genes. A high density illumina SNP array for Brassica was released in 2012.

QTL mapping of agronomic and nutritional traits in *B. napus* has been done previously.[59] Rapeseed oil quality is improved by low erucic acid content. Two genes involved in the synthesis of FAs, FAE1.1, and FAE1.2 in *B. napus* and *B. juncea* are closely linked to major QTL for seed erucic acid.[60] Other traits like glucosinolates, oil, and oleic acid content in seeds have been done in Brassica by QTL studies.[61] SNP marker based mapping has been used to reduce linkage drag when genes for useful traits are transferred from one population to another.[62]

2.5.2 LINKAGE MAPS AND COMPARATIVE GENOMICS

Linkage maps are essential for QTL analysis of agronomic traits, map based cloning, and comparative mapping. A number of genetic linkage maps have been developed in Brassica using different markers. In *B. napus* (AACC), more than 30 linkage maps have been developed using AFLP, RFLP, SSR, and sequence-related amplified polymorphism (SRAP) markers.[63–65] A high density SNP linkage map was developed to detect level of polymorphism across collections.[66] Linkage maps in Indian mustard, *B. juncea,* were developed with AFLP and RFLP markers. Recently high density linkage map has been developed with 1148 loci covering about 18 linkage groups.[67] Though linkage maps have been constructed for both *B. carinata* and *B. nigra,* limited genomic studies have been done on them. Linkage maps have also been developed for *B. oleracea,* representing C genome of Brassica. Genome sequencing has led to drafting of high density reference map including more than 600 SSR and SNP markers.[68] A linkage map for *B. rapa* was constructed with expressed sequence tag (EST)-based SNP markers and genes associated with leaf morphological traits and flowering time were identified.[69] Linkage maps enable identification of genes linked to a particular trait and provide marker-assisted crop breeding. In Brassicaceae, till date partial and complete sequencing of 10 species has been done. That includes the Brassica species, *B. rapa* and *B. oleracea,* and model plant *A. thaliana.* Brassica genome sequencing has revealed 85–90% identity of its exonic regions with *A. thaliana,* member of the Brassicacaeae family.[70] This has revealed a complex rearrangement of Brassica chromosomes in comparison to *A. thaliana* which might be presumed to be due to polyploidization.[71] *B. napus* is the most intensively studied Brassica amphidiploid. Sequencing of Brassica A genome (published in Nature Genetics by the *Brassica rapa* Genome Sequence Project Consortium 2011) was a major breakthrough for more targeted breeding in Brassica.

Brassica genome sequencing is essential to understand genome evolution and crop improvement. Sequencing of Brassica crops was initiated in 2002 by Multinational Brassica Genome Project (MBGP). First genome to be sequenced among Brassica species was *B. rapa* (Chinese cabbage) due to small size of its genome and low number of repetitive sequences. Recently, genome sequencing of *B. oleracea* was done using whole genome shotgun sequencing (WGS).[72]

2.5.3   LINKAGE DISEQUILIBRIUM MAPPING

Association mapping or linkage disequilibrium mapping is useful in identifying genes linked with natural variations in a population. Association analysis by QTL mapping has been used in canola for morphological, phenological, and quality traits.[73] Association mapping was also useful in detection of genetic loci for oil content in *B. napus.*[74] Two main methods for association mapping are candidate gene sequencing and whole-genome sequencing. When no reference genomic sequences are available, whole genome sequencing is usually carried out. Currently, construction of nested association mapping (NAM) populations for Brassica is underway. They provide advantage of both association mapping and linkage analysis as it can detect minor effect QTL. Association has been shown between tocopherol traits and allelic variations by candidate gene sequencing. These can be useful in marker-assisted breeding of varieties rich in tocopherol.[75]

2.5.4   TRANSCRIPTOME ANALYSIS

Advancement of NGS has helped in generation of a large amount of transcriptome data. A large amount of EST data has been generated from major Brassica crops. Transcriptome sequences have been generated in *B. napus* by NGS technique. Also, the genome has been dissected by transcriptome sequencing of parental and mapping population and an SNP linkage map has been constructed.[76] Genomic deletions in QTL regions of glucosinolate content have been explored in *B. napus* by associative transcriptomics.[77] Roche 454 was the first technology used in transcriptome sequencing. Later on, illumina and SOLiD technologies were developed that were high throughput. Bancroft et al.[76] used the illumina NGS to sequence leaf transcriptome of *B. napus, B. rapa,* and *B. oleracea.* Much important information with respect to gene discovery can be achieved by transcriptome sequencing. A NGS based method called digital gene expression is also used to quantify transcript levels during a biological process. Trick et al.[78] assembled 800,000 ESTs to analyze gene expression in *B. napus.* Assembly of RNA sequence reads can lead to gene discovery in species where the genome sequence is known. NGS also opens up gateway for discovery of novel microRNAs (miRNAs) and their targets. They are small non-coding RNAs that regulate gene expression and also have conserved functions. In *B. napus,* with varying oil content, 50 conserved miRNAs and 11 new miRNAs have been identified.[79] Fifty nine miRNAs were also detected at varying seed developmental stages.[80]

2.6   GENOME WIDE ASSOCIATIVE TRANSCRIPTOMICS TO DECODE THE GENETIC COMPLEXITY OF BRASSICA POLYPLOIDS

Genome wide association studies (GWAS) have been found to be an efficient definitive tool for the region specific Brassica genome, contributing to a range of phenotypes, generally using populations of highly diverse lines. Recently designed allopolyploid between two diploid species, namely *B. rapa* (A sub-genome) and *B. nigra* (B sub-genome), have illustrative genetic rib at large. At quantitative level SNP discovery has made a huge debut in decoding of Brassica genome where polyploids (complex) are found to have several blocks into the process of discovery. To circumvent this, a novel associative transcriptomics approach has been developed, the transcribed sequences comprise identifying SNPs which represent dissimilarity in gene sequences and gene expression (GEMs), and further correlate this variation with phenotypic traits.[76,77]

2.7 CONCLUSION

Advances in high-throughput sequencing and genotyping, together with – omics approaches can be used to identify trait specific genes or genomic regions. Genome sequencing of model plant *A. thaliana,* and later of *B. rapa* and *B. oleracea* has led to comparative mapping in important Brassicaceae crops. Since Brassica genomes are highly conserved, sequence information can be extended to other commercially important Brassica crops. Development of a common marker system will allow comparative QTL mapping, cloning of economically important genes, and marker-assisted selection to accelerate the breeding process. Thus, the integration of –omics approaches with conventional breeding is the way forward for developing high quality oilseed Brassicas.

KEYWORDS

- gene pool
- unsaturated fatty acid
- population genomics
- brassicas
- transcriptomics
- beta-carotene

REFERENCES

1. Warwick, S. I.; Francis, A.; Al-Shehbaz, I. A. Brassicaceae: Species Checklist and Database on CD-ROM. *Plant Syst. Evol.* **2006**, *259,* 249–258.
2. Gómez-Campo, C.; Prakash, S. Origin and Domestication. In *Biology of Brassica Coenospecies;* Gómez-Campo, C., Ed.; Elsevier: Amsterdam, 1993; pp 33–58.
3. Vallejo, F.; Tomas-Barberan, F. A.; Garcia-Viguera, C. Potential Bioactive Compounds in Health Promotion from Broccoli Cultivars Grown in Spain. *J. Sci. Food Agr.* **2002,** *82,* 1293–1297.
4. Heimler, D.; Vignolini, P.; Dini, M. G.; Vincieri, F. F.; Romani, A. Antiradical Activity and Polyphenol Composition of Local *Brassicaceae* Edible Varieties. *Food Chem.* **2005,** *99,* 464–469.
5. U, N. Genomic Analysis in *Brassica* with Special Reference to the Experimental Formation of *B. napus* and Peculiar Mode of Fertilization. *Jpn. J. Bot.* **1935,** *7,* 389–452.
6. Schmidt, R.; Bancroft, I. Genetics and Genomics of the Brassicaceae. In *Plant Genetics and Genomics: Crops and Models;* Springer: New York, 2011; p 9.
7. Murphy, D. J. *People Plants and Genes: the Story of Crops and Humanity;* Oxford University Press: UK, 2007.
8. Leitch, A. R.; Leitch, I. J. Genomic Plasticity and the Diversity of Polyploid Plants. *Science.* **2008,** *320,* 481–483.
9. Reif, J. C.; Zhang, P.; Dreisigacker, S.; Warburton, M. L.; Van Ginkel, M.; Hoisington, D.; Bohn, M.; Melchinger, A. E. Wheat Genetic Diversity Trends During Domestication and Breeding. *Theor. Appl. Genet.* **2005,** *110,* 859–864.
10. Pradhan, A.; Plummer, J. A.; Nelson, M. N.; Cowling, W. A.; Yan, G. Successful Induction of Trigenomic Hexaploid Brassica from a Triploid Hybrid of *B. napus* L. and *B. nigra* (L.) Koch. *Euphytica.* **2010,** *176*(1), 87–98.
11. Tian, E.; Jiang, Y.; Chen, L.; Zou, J.; Liu, F.; Meng, J. Synthesis of a Brassica Trigenomic Allohexaploid (*B. carinata* × *B. rapa*) De Novo and Its Stability in Subsequent Generations. *Theor. Appl. Genet.* **2010,** *121*(8), 1431–1440.
12. Alvarez, J.; Guli, C. L.; Yu, X.; Smyth, D. R. Terminal Flower: a Gene Affecting Inflorescence Development in *Arabidopsis thaliana. Plant J.* **1992,** *2*(1), 103–116.
13. Kaur, H.; Banga, S. S. Discovery and Mapping of *Brassica juncea* Sdt1 Gene Associated with Determinate Plant Growth Habit. *Theor. Appl. Genet.* **2015,** *128*(2), 235–245.
14. http://faostat3.fao.org/download/Q/QC/E (accessed August 28, 2015).
15. Vermunt, S. H. F.; Beaufrere, B.; Riemersma, R. A.; Sebedio, J. L.; Chardigny, J.; Mensink, R. P. Dietary *Trans*-α-Linolenic Acid from Deodorised Rapeseed Oil and Plasma Lipids and Lipoproteins in Healthy Men: the *Trans* LinE Study. *Brit. J. Nutr.* **2001,** *85,* 387–392.
16. Jahangir, M.; Kim, H. K.; Choi, Y. H.; Verpoorte, R. Health-Affecting Compounds in Brassicaceae. *Compr. Rev. Food Sci. Food Saf.* **2009,** *8*(2), 31–43.
17. UNFAO, http://www.fao.org/index_en.htm, Food and Agriculture Organization of the United Nations.
18. Sarwar, M. F.; Sarwar, M. H.; Sarwar, M.; Qadri, N. A.; Moghal, S. The Role of Oilseeds Nutrition in Human Health: A Critical Review. *J. Cereals Oilseeds.* **2013,** *4*(8), 97–100.
19. Cartea, M. E.; Francisco, M.; Soengas, P.; Velasco, P. Phenolic Compounds in Brassica Vegetables. *Molecules.* **2010,** *16*(1), 251–280.
20. Thiyam, U.; Kuhlmann, A.; Stöckmann, H.; Schwarz, K. Prospects of Rapeseed Oil By-products with Respect to Antioxidative Potential. *C. R. Chim.* **2004,** *7,* 611–616.

21. Agnihotri, A. Synergy of Biotechnological Approaches with Conventional Breeding to Improve Quality of Rapeseed-Mustard Oil and Meal. *Anim. Nutr. Feed Techn.* **2010,** *10 S,* 169–177.

22. Franzke, A.; Lysak, M. A.; Al-Shehbaz, I. A.; Koch, M. A.; Mummenhoff, K. Cabbage Family Affairs: the Evolutionary History of Brassicaceae. *Trends Plant Sci.* **2011,** *16,* 108–116.

23. Downey, R. K.; Rimmer, S. R. Agronomic Improvement in Oilseed. In *Advances in Agronomy;* Acadmic Press: California, 1993; Vol. 50, p 1.

24. Banga, S. K. Breeding for Oil and Meal Quality. In *Oilseed and Vegetable Brassicas: Indian Perspective;* Chopra, V. L., Prakash, S., Eds.; Oxford and IBH Publishing Company: New Delhi, 1996; pp 234–249.

25. Kris-Etherton, P. M.; Pearson, T. A.; Wan, Y.; Hargrove, R. L.; Moriarty, K.; Fishell, V.; Etherton, T. D. High–Monounsaturated Fatty Acid Diets Lower Both Plasma Cholesterol and Triacylglycerol Concentrations. *Am. J. Clin. Nutr.* **1999,** *70*(6), 1009–1015.

26. Downey, R. K. Brassica Oilseed Breeding-Achievements and Opportunities. *Plant Breed. Abstracts.* **1990,** *60,* 1165–1170.

27. Downey, R. K.; Craig, B. M. Genetic Control of Fatty Acid Biosynthesis in Rapeseed (*Brassica napus* L.). *J. Am. Oil Chem. Soc.* **1964,** *41,* 475–478.

28. Agnihotri, A.; Kaushik, N. Incorporation of Superior Nutritional Quality Traits in Indian *B. juncea. I.J.P.G.R.* **2000,** *12*(3), 352–358.

29. Rakow, G. In *Developments in the Breeding of Oil in Other Brassica Species,* Proceedings of the Ninth International Rapeseed Congress, Cambridge: UK, 1995; Vol. 2, pp 401–406.

30. Agnihotri, A.; Prem, D.; Gupta, K. The Chronicles of Oil and Meal Quality Improvement in Oilseed Rape. In *Advances in Botanical Research: Oilseed Rape Breeding;* Gupta, S. K., Ed.; Elsevier Publishers, Academic Press: New York, 2007; Vol. 45, pp 49–97.

31. Agnihotri, A.; Kumar, A.; Singh, N. B. In *The Oil and Meal Quality Improvements in Rapeseed-Mustard: Status in India,* Proceedings 13th International Rapeseed Congress, Brassica 2011, Prague, Czech Republic, June 5–9, 2011.

32. Smith, N. A.; Singh, S. P.; Wang, M. B.; Stoutjesdijk, P. A.; Green, A. G.; Waterhouse, P. M. Total Silencing by Intron-Spliced Hairpin RNAs. *Nature.* **2000,** *407,* 319–320.

33. Verkerk, R.; Schreiner, M.; Krumbein, A.; Ciska, E.; Holst, B.; Rowland, I.; Schrijver, R. D.; Hansen, M.; Gerhäuser, C.; Mithen, R. Glucosinolates in Brassica Vegetables: the Influence of the Food Supply Chain on Intake, Bioavailability and Human Health. *Mol. Nutr. Food Res.* **2009,** *53,* S219–S265.

34. Moreno, D. A.; Carvajal, M.; Lopez-Berenguer, C.; Garcia-Viguera, C. Chemical and Biological Characterisation of Nutraceutical Compounds of Broccoli. *J. Pharm. Biomed. Anal.* **2006,** *41,* 1508–1522.

35. Szydlowska-Czerniak, A.; Trokowski, K.; Karlovits, G.; Szlyk, E. Determination of Antioxidant Capacity, Phenolic Acids, and Fatty Acid Composition of Rapeseed Varieties. *J. Agric. Food Chem.* **2010,** *58,* 7502–7509.

36. Li, X.; Gao, M. J.; Pan, H. Y.; Cui, D. J.; Gruber, M. Y. Purple Canola: Arabidopsis PAP1 Increases Antioxidants and Phenolics in *Brassica napus* Leaves. *J. Agric. Food Chem.* **2010,** *58,* 1639–1645.

37. Lindsey, K.; Pullen, M. L.; Topping, J. F. Importance of Plant Sterols in Pattern Formation and Hormone Signalling. *Trends Plant Sci.* **2003,** *8,* 1360–1385.

38. Gordon, M. H.; Miller, L. A. D. Development of Steryl Ester Analysis for the Detection of Admixtures of Vegetable Oils. *J. Am. Oil Chem. Soc.* **1997,** *74,* 505–510.

39. Piironen, V.; Lindsay, D. G.; Miettinen, T. A.; Toivo, J.; Lampi, A. M. Review Plant Sterols: Biosynthesis, Biological Function and Their Importance to Human Nutrition. *J. Sci. Food Agr.* **2000,** *80,* 939–966.

40. Piironen, V.; Syvaoja, E. L.; Varo, P.; Salminen, K.; Koivistoinen, P. Tocopherols and Tocotrienols in Finnish Foods: Vegetables, Fruits, and Berries. *J. Agric. Food Chem.* **1986,** *34*(4), 742–746.

41. Leckband, G.; Rades, H.; Frauen, M.; Friedt, W. In *NAPUS 2000—a Research Programme for the Improvement of the Whole Rapeseed,* Proceedings of the 11th International Rapeseed Congress, The Royal Veterinary and Agricultural University, Copenhagen, Denmark, July 6–10, 2003, pp 209–211.

42. Shewmaker, C. K.; Sheehy, J. A.; Daley, M.; Colburn, S.; Ke, D. Y. Seed-Specific Overexpression of Phytoene Synthase: Increase in Carotenoids and Other Metabolic Effects. *Plant J.* **1999,** *20,* 401–412.

43. Agnihotri, A.; Prem, D. Oil Quality Improvement in Rapeseed-Mustard. In *Changing Global Vegetable Oils Scenario-Issues and Challenges before India;* Indian Society of Oil Seeds Research: Hyderabad. 2007b; pp 315–326.

44. Davey, M. W.; Montagu, M. V.; Inzé, D.; Sanmartin, M.; Kanellis, A.; Smirnoff, N.; Iris J.; Benzie, J.; John, J. S.; Favell, D.; Fletcher, J. Plant L-ascorbic Acid: Chemistry, Function, Metabolism, Bioavailability and Effects of Processing. *J. Sci. Food Agr.* **2000,** *80* (7), 825–860.

45. Bajjalieh, N. In *Proteins from Oilseeds,* Proceedings Protein Sources for the Animal Feed Industry-Expert Consultation and Workshop, Bangkok, April 29– May 3, 2002.

46. Uppstrom, B. Seed Chemistry. In *Brassica Oilseeds: Production and Utilization;* Kimber, D., McGregor, D. I., Eds.; CAB International: UK, 1995.

47. Aider, M.; Barbana, C. Canola Proteins: Composition, Extraction, Functional Properties, Bioactivity, Applications as a Food Ingredient and Allergenicity–a Practical and Critical Review. *Trends Food Sci. Tech.* **2011,** *22*(1), 21–39.

48. Varshney, R. K.; Tuberosa, R. Genomics-Assisted Crop Improvement. In *Genomics Approaches and Platforms;* Springer: the Netherlands, 2007; Vol. 1, p 386.

49. Varshney, R. K.; Graner, A.; Sorrells, M. E. Genic Microsatellite Markers in Plants: Features and Applications. *Trends. Biotechnol.* **2005,** *23*(1), 48–55.

50. Salvi, S.; Tuberosa R. To Clone or not to Clone Plant QTLs: Present and Future Challenges. *Trends Plant Sci.* **2005,** *10* (6), 297–304.

51. Faës, P.; Deleu, C.; Aïnouche, A.; Le Cahérec, F.; Montes, E.; Clouet, V.; Gouraud, A. M.; Albert, B.; Orsel, M.; Lassalle, G.; Leport, L.; Bouchereau, A.; Niogret, M. F. Molecular Evolution and Transcriptional Regulation of the Oilseed Rape Proline Dehydrogenase Genes Suggest Distinct Roles of Proline Catabolism During Development. *Planta.* 2015, *241*(2), 403–19.

52. Sharma, A.; Xiaonan, L.; Yong Pyo, L. Comparative Genomics of Brassicaceae Crops. *Breed. Sci.* **2014,** *64,* 3.

53. Bus, A.; Hecht, J.; Huettel, B.; Reinhardt, R.; Stich, B. High Throughput Polymorphism Detection and Genotyping in *Brassica napus* Using Next-Generation RAD Sequencing. *BMC Genom.* **2012,** *13,* 281.

54. Mun, J. H.; Kwon, S. J.; Yang, T. J.; Kim, H. S.; Choi, B. S.; Baek, S.; Jung S. K. The First Generation of a BAC-Based Physical Map of *Brassica rapa*. *BMC genom.* **2008,** *9*(1), 280.

55. Mun, J. H.; Kwon, S. J.; Seol, Y. J.; Kim, J. A.; Jin, M.; Kim, J. S.; Lim, M. H. Sequence and Structure of *Brassica rapa* Chromosome A3. *Genome biol.* **2010,** *11*(9), 94.

56. Wang, J.; Lydiate, D. J.; Parkin, I. A. P.; Falentin, C.; Delourme, R.; Carionl, P. W. C.; King, G. J. Integration of Linkage Maps for the Amphidiploid *Brassica napus* and Comparative Mapping with Arabidopsis and *Brassica rapa*. *BMC Genom.* **2011,** *12,* 101.

57. Wei, L.; Xiao, M.; Hayward, A.; Fu, D. Applications and Challenges of Next-Generation Sequencing in Brassica Species. *Planta.* **2013,** *238*(6), 1005–1024.

58. Trick, M.; Long, Y.; Meng, J.; Bancroft, I. Single Nucleotide Polymorphism (SNP) Discovery in the Polyploid *Brassica napus* Using Solexa Transcriptome Sequencing. *Plant Biotechnol. J.* **2009,** *7,* 334–346.

59. Parkin, I. A.; Clarke, W. E.; Sidebottom, C.; Zhang, W.; Robinson, S. J.; Links, M. G.; Karcz, S.; Higgins, E. E.; Fobert, P.; Sharpe, A. G. Towards Unambiguous Transcript Mapping in the Allotetraploid *Brassica napus*. *Genome.* **2010,** *53,* 929–938.

60. Qiu, D. C.; Morgan, J. S.; Long, Y.; Liu, J.; Li, R.; Zhuang, X. A Comparative Linkage Map of Oilseed Rape and Its Use for QTL Analysis of Seed Oil and Erucic Acid Content. *Theor. Appl. Genet.* **2006,** *114* (1), 67–80.

61. Snowdon, R. J.; Friedt, W. Molecular Markers in Brassica Oilseed Breeding: Current Status and Future Possibilities. *Plant Breed.* **2004,** *123,* 1–8.

62. Pradhan, A. K. In *Breeding Oilseed Mustard Brassica juncea Using Conventional and Molecular Approaches*, 14th international rapeseed congress, 2015, 8.

63. Radoev, M.; Becker, H.; Ecke, W. Genetic Analysis of Heterosis for Yield and Yield Components in Rapeseed (*Brassica napus* L.) by Quantitative Trait Locus Mapping. *Genetics.* **2008,** *179,* 1547–1558.

64. Parkin, I. A. P.; Sharpe, A. G.; Keith, D. J.; Lydiate, D. J. Identification of the A and C Genomes of Amphidiploid *Brassica napus* (Oilseed Rape). *Genome.* **1995,** *38,* 1122–1131.

65. Sun, Z.; Wang, Z.; Tu, J.; Zhang, J.; Yu, F.; McVetty, P. B. E.; Li, G. An Ultradense Genetic Recombination Map for *Brassica napus,* Consisting of 13551 SRAP Markers. *Theor. Appl. Genet.* **2007,** *114,* 1305–1317.

66. Delourme, R.; Falentin, C.; Fomeju, B.; Boillot, M.; Lassalle, G.; André, I.; Duarte, J.; Gauthier, V.; Lucante, N.; Marty, A. Highdensity SNP-Based Genetic Map Development and Linkage Disequilibrium Assessment in *Brassica napus*. *BMC Genom.* **2013,** *14,* 120.

67. Ramchiary, N.; Padmaja, K. L.; Sharma, S.; Gupta, V.; Sodhi, Y. S.; Mukhopadhyay, A.; Arumugam, N.; Pental, D.; Pradhan, A. K. Mapping of yield influencing QTL in *Brassica juncea*: implications for breeding of a major oilseed crop of dryland areas. *Theor. Appl. Genet.* **2007,** *115*(6), 807–817.

68. Wang, W.; Huang, S.; Liu, Y.; Fang, Z.; Yang, L.; Hua, W.; Yuan, S.; Liu, S.; Sun, J.; Zhuang, M. Construction and Analysis of a High-Density Genetic Linkage Map in Cabbage (*Brassica oleracea* L. var. capitata). *BMC Genom.* **2012,** *13,* 523.

69. Li, F.; Kitashiba, H.; Inaba, K.; Nishio, T. A. *Brassica rapa* Linkage Map of EST-Based SNP Markers for Identification of Candidate Genes Controlling Flowering Time and Leaf Morphological Traits. *DNA Res.* **2009,** *16*(6), 311–323.

70. Schmidt, R. Plant Genome Evolution: Lessons from Comparative Genomics at the DNA Level. *Plant Mol. Biol.* **2002,** *48,* 21–37.

71. Snowdon, R. J. Cytogenetics and Genome Analysis in Brassica Crops. *Chromosome Res.* **2007,** *15,* 85–95.

72. Yu, J.; Zhao, M.; Wang, X.; Tong, C.; Huang, S.; Tehrim, S.; Liu, Y.; Hua, W.; Liu, S. Bolbase: a Comprehensive Genomics Database for *Brassica oleracea*. *BMC Genom.* **2013,** *14,* 664.

73. Honsdorf, N.; Becker, H. C.; Ecke, W. Association Mapping for Phenological, Morpho-logical, and Quality Traits in Canola Quality Winter Rapeseed (*Brassica napus* L.). *Genome.* **2010,** *53,* 899–907.

74. Zou, J.; Jiang, C.; Cao, Z.; Li, R.; Long, Y.; Chen, S.; Meng, J. Association Mapping of Seed Oil Content in *Brassica napus* and Comparison with Quantitative Trait Loci Identified from Linkage Mapping. *Genome.* **2010,** *53,* 908–916.

75. Fritsche, S.; Xingxing, W.; Jinquan, L.; Benjamin, S.; Friedrich J. K.; Jessica,E.; Gunhild, L. A Candidate Gene-Based Association Study of Tocopherol Content and Composition in Rapeseed (*Brassica napus*). *Front. Plant Sci.* **2012,** *3,* 1–24

76. Bancroft, I.; Morgan, C.; Fraser, F.; Higgins, J.; Wells, R.; Clissold, L.; Baker, D.; Long, Y.; Meng, J.; Wang, X. Dissecting the Genome of the Polyploid Crop Oilseed Rape by Transcriptome Sequencing. *Nat. Biotechnol.* **2011,** *29,* 762–766.

77. Harper, A. L.; Trick, M.; Higgins, J.; Fraser, F.; Clissold, L.; Wells, R.; Hattori, C.; Werner, P.; Bancroft, I. Associative Transcriptomics of Traits in the Polyploid Crop Species *Brassica napus*. *Nat. Biotechnol.* **2012,** *30,* 798–802.

78. Trick, M.; Cheung, F.; Drou, N.; Fraser, F.; Lobenhofer, E. K.; Hurban, P.; Magusin, A.; Town, C. D.; Bancroft, I. A Newly-Developed Community Microarray Resource for Transcriptome Profiling in Brassica Species Enables the Confirmation of Brassica-Specific Expressed Sequences. *BMC Plant Biol.* **2009,** *9,* 50.

79. Zhao, Y. T.; Wang, M.; Fu, S. X.; Yang, W. C.; Qi, C. K.; Wang, X. J. Small RNA Profiling in Two *Brassica napus* Cultivars Identifies Micro-RNAs with Oil Production- and Development-Correlated Expression and New Small RNA Classes. *Plant Physiol.* **2012,** *158,* 813–823.

80. Korbes, A. P.; Machado, R. D.; Guzman, F.; Almerao, M. P.; de Oliveira, L. F.; Loss-Morais, G.; Turchetto-Zolet, A. C.; Cagliari, A.; Dos Santos-Maraschin, F.; Margis-Pinheiro, M.; Margis, R. Identifying Conserved and Novel MicroRNAs in Developing Seeds of *Brassica napus* Using Deep Sequencing. *PLoS ONE.* **2012,** *7,* 11.

CHAPTER 3

A RECAP ON QUANTITATIVE TRAIT LOCI ASSOCIATED WITH DISEASE RESISTANCE IN FOOD LEGUMES

RASHMI RAI[1], A. K. SINGH[2], SAJAD MAJEED ZARGAR[3], and B. D. SINGH[1*]

[1]*School of Biotechnology, Faculty of Science, Banaras Hindu University, Varanasi 221005, Uttar Pradesh, India*

[2]*College of Agriculture and Research Station, 497335, Chhattisgarh, India*

[3]*Center for Plant Biotechnology, Division of Biotechnology, S K University of Agricultural Sciences and Technology of Kashmir, Shalimar, Srinagar 190025, Jammu and Kashmir, India*

*Corresponding author. E-mail: brahmadsingh@gmail.com

CONTENTS

ABSTRACT

Legumes come across a number of diseases that are controlled by quantitative trait loci (QTLs) including ascochyta blight (chickpea, faba bean, lentil, and pea), molds (common bean), powdery mildew (pea), rusts (groundnut, pea, lentil, and faba bean), wilt (common bean, pigeonpea, and pea), root, or stem rots (pea, soybean). For most of the important legumes different linkage maps have been developed using different types of molecular markers to study the linkages between different genes with different traits of interest among legumes. The availability of molecular markers on linkage maps are of great value for QTL mapping studies and marker assisted programs in disease resistance legume breeding. QTLs mapping essentially requires knowledge of inheritance pattern of the diseases and segregation nature of molecular markers, along with use of suitable statistical tools. QTL detection is highly influenced by several factors such as type and size of mapping population, effect of QTLs involved (major/minor), heredity of QTLs, density of linkage map, and the technique used for QTL detection. The mapping is more accurate and decisive when diseases are scored using different parameters (components, pathogen strains, different organs, or stages of plants life cycle) in multiple environments. The consistent expression of QTLs by different parameters in multiple environments proves their usefulness for marker assisted breeding. In this chapter, we have reviewed the QTLs detected for different diseases, their causing organism, type and size of mapping population, type of molecular marker, and its further use in marker-assisted selection (MAS) of disease resistance QTLs in legumes.

3.1 INTRODUCTION

Legumes are members of family Leguminosae or Fabaceae, the third largest family of higher plants after Poaceae and Cucurbitaceae. This family consists of about 750 genera and 20,000 species, and is divided into three sub-families, Mimosoideae, Caesalpinioideae, and Papilionoideae.[1] Major food legume crops include, common bean (*Phaseolus vulgaris* L.), chickpea (*Cicer arietinum* L.), pea (*Pisum sativum* L.), cowpea (*Vigna unguiculata* L.), lentil (*Lens culinaris* Medik.), groundnut (*Arachis hypogaea*), soybean (*Glycine max*), pigeonpea (*Cajanus cajan* L. Millsp.), faba beans (*Vicia faba* L.), mungbean (*Vigna radiata*), mothbean (*Vigna acutifolia*), and urdbean (*Vigna mungo*). Primary regions of legume production include India, Canada, China, Myanmar, Brazil, and France (each producing >3.5 million metric tons (Mt)) followed by Mexico, Russian federation, Niger, Australia, and Ethiopia (each producing >1.0 million Mt). The total annual production of 61.5 million Mt is obtained from 70.6 million ha area worldwide.[2] Legumes have a unique ability to fix atmospheric nitrogen, and thereby have a major impact on agriculture, the environment, animal/human nutrition, and health.[3] Food legumes are rich source of protein and minerals and complement cereals in terms of amino acid balance of the diet. Legumes are also of high agronomic value as they supply the latter some amount of nitrogen and reduce soil pathogens responsible for diseases when they are used in crop rotation with cereals.[4] However, the adaptability and productivity of legumes are limited by major biotic and abiotic stresses, including fungal and viral diseases, insect pests, drought, heat, frost, chilling, water logging, salinity, and mineral toxicities although the type and the severity of the stresses depend on the specific crop location.[5]

Among the biotic stresses affecting legumes, fungal diseases are most important that includes, ascochyta blight (*Ascochyta*: chickpea, faba bean, lentil, pea), mold (*Sclerotinia*: common bean), rust (*Uromyces*: pea, lentil, faba bean; *Puccinia*: groundnut), wilt (*Fusarium*: common bean, chickpea), leaf spot (*Phaeoisariopsis*: groundnut; *Cercospora*: cowpea), and stem/root/ hypocotyl rot (*Rhizoctonia*: soybean; *Phytophthora*: soybean, pigeonpea), but viruses (*Beet curly top virus*: common bean; *Tobacco Ring Spot Viruses*: soybean) and bacteria (*Xanthomonas*: common bean) can also considerably reduce legume production. These diseases affect quality and quantity of legume produce in almost all legume-growing areas, and are responsible for substantial yield losses.[6–10]

Resistances to most of the above diseases are known, but they are generally quantitative in nature as they are governed by multiple genes representing QTLs. Quantitative disease resistance shows typical continuous variation due to polygenic segregation, influence of environmental factors, and genotype × environment interaction.[11–13] The QTL × environment interaction (Q × E) greatly affects the magnitude and sometimes even the direction of QTL effects.[14–15] In legumes, genotype × environment interaction is high because of which major QTLs are detected in all the environments and categorized as constitutive whereas minor QTLs are environment-specific and categorized as adaptive.[16–17] QTL × QTL interaction is also important in expression of complex traits and sometimes accounts for a large proportion of phenotypic variance.[8,18] Therefore, selection of resistance is problematic due to poor relationship between phenotypic expression and genotype, as such it would be greatly facilitated by marker assisted selection (MAS). There are other factors affecting QTL mapping such as, type and size of mapping population, heritability and effect of QTL, and the technique used for QTL mapping.[19,20] It has been suggested that availability of molecular marker linkage maps in most of the legumes would facilitate detection and molecular mapping of disease resistance QTLs.[21] In view of this, many studies have focused on QTL mapping for disease resistance in legume crops, which are discussed in following sections, single or multiple QTLs associated with disease resistance in legumes have been summarized (Table 3.1), and some of them are discussed in details according to the respective legume crop and the concerned disease.

3.1.1   COMMON BEAN (PHASEOLUS VULGARIS)

QTL mapping has been carried out in common bean for resistance to common bacterial blight (CBB) (*Xanthomonas axonopodis* pv. *phaseoli*), bacterial brown spot (BBS) (*Pseudomonas syringae* pv. *syringae*), leafhoppers (*Empoasca fabae* and *Empoasca kraemeri*), wilt (*Fusarium oxysporium* Schlectend. Fr. f. sp. *phaseoli* J.B. Kendrick and W.C. Snyder), white mold (WM) disease (*Sclerotia sclerotiorum* (Lib.) de-Bary), and thrips (*Thrips palmi* Karny). CBB is a serious disease of common bean, which substantially reduces yields. Nodari et al. detected four QTLs in a recombinant inbred line (RIL) population using restriction fragment length polymorphism(RFLP) and random amplification of polymorphic DNA (RAPD) markers, which together explained 75% of phenotypic variation.[22] Subsequently several QTLs for CCB resistance were tagged by RAPD markers and some of them were converted into sequence characterized amplified region (SCAR) markers, such as $UBC420_{900}$ and $SU91$.[23,24] $UBC420_{900}$ was found to be efficient for MAS across different genetic backgrounds but it was tightly linked to the V gene, which causes undesirable black mottled seed coat color.[24,25] Yu et al. used a RIL population of 112 F_5 lines developed from cross HR67 × OAC95-4, and detected a single major QTL at the end of linkage group (LG) H7.[26] Simple sequence repeat (SSR) marker *PVtttc001* and RAPD marker *BC420* were situated at 6.2 and 7.1 cM away from CBB locus, respectively.[26] In order to identify more tightly linked markers, 81 F_5 and F_8 RILs derived from the same cross were screened in greenhouse and growth chamber, respectively.[27] The QTL was mapped through eight amplified fragment length polymorphism (AFLP) markers spanning a 13 cM region of chromosome 1. AFLP markers *E35M50.123, E38M41.100, E33M49.183,* and *E33M49.333* were linked to the major QTL that accounted for 30% of the phenotypic variance in both the environments.

TABLE 3.1 Molecular Mapping of Disease Resistance QTLs in Legumes.

Disease/pathogen	Parents and mapping population	QTL number/ name	LG/ chromosome	Associated markers/ marker interval/closest marker	Phenotypic variance (%)	Reference
Common bean						
Common bacterial blight (*Xanthomonas axonopodis* pv. *Phaseoli*)	HR-67 × OAC95-4 112 F_5 RILs	1	LG1	*PV-ttic001* $BAC420_{900}$	70	[26]
Common bacterial blight (*Xanthomonas axonopodis* pv. *Phaseoli*)	HR-67 × OAC95-4 81 F_8 RILs	1	LG1	*UBC420.900, PV-ttic001 E35M50.123, E38M41.100, E33M49.183 (STS183), E33M49.333 (STS333a), E35M50.330 (STS330) (repulsion)*	25–52	ᵞ[27
Bacterial brown spot (*Pseudomonas syringae* pv. *Syringae*)	Eagle × Puebla 152 RILs	1	B1	*P1.1500, H8.600*	13	[28]
		2	B3	*AN6.1600, AK6.1500*	19	
		3	B6	*O10.350, V10.450, F10.950*		
		4	B11	*H20.950, AL15.1300*		
Wilt (*Fusarium oxysporium* f. sp. *Phaseoli*)	Belneb RR-1 (race Durango) × A55 (race Mesoamerica)	1	LG10	*U20.750*	63.3	[30]
		2	LG 11	*K10.700*	10.8	
		3	LG3	*AD4.450*	10.1	
White mold (*Sclerotinia sclerotiorum* Lib. de-Bary)	76 F$_{6:10}$ RILs PC-50 × XAN-159	1	B8	*H19.1250*	12	[31]
		2	B7	*J09.950*	5-9	
		3	B5	*D05.1100*	11	
		4	B4a	*AI13.700*	5	
		5	B6	*Y07.1200*	2	
		6	B7	*O13.1350*	4	

TABLE 3.1 (*Continued*)

Disease/pathogen	Parents and mapping population	QTL number/ name	LG/ chromosome	Associated markers/ marker interval/closest marker	Phenotypic variance (%)	Reference
		7	B8	C	24	
		8	B3	$G03.1150$	7	
		9	B2	$W02.1000$	7	
		10	B4a	$U10.900$	5	
		11	B8	$Bc439.1000$	3	
		12	B8	$AO17.1050$	9	
		13	B11	$AE07.800$	3	
		14	B4a	$R20.1250$	5	
	Bunsi × Newport 98 F$_3$	1	B2	$BC20_{1600} - O15_{1800}$	13	[34]
		2	B7	$aggctt85\text{-}aacctt130$	15.6	
	Benton × NY6020-4 77 F$_{5:7}$ RILs	1	B2	$AU5._{1350}$ $R4._{950/1350}$ $AX6._{1000}$ $A8._{510}$	12–13	[37]
		2	B8	$B10._{1100}$ $AW9._{1200}$ $C5._{350/400}$ $V14._{775}$	26–38	
	Bunsi × Raven 98 F$_{4:7}$ RILs	1	B2	$O12.1600$	10.1	[35]
		2	B2	$O15.1800$	8.7	
		3	B5	$E_{ACT}M_{CAT}85$	10.7	
		4	B7	$E_{AAC}M_{CTT}223$	14.7	
		5	B7	$P9.1750$	14.2	
		6	B8	$E_{AGA}M_{CTG}190$	9.2	

TABLE 3.1 (Continued)

Disease/pathogen	Parents and mapping population	QTL number/ name	LG/ chromosome	Associated markers/ marker interval/closest marker	Phenotypic variance (%)	Reference
	G122 × CO72548 94 F$_{5:6}$ RILs	1	B1	*PatMaca300*	20.2	[6]
		2	B2b	*EacaMaga220*	14.8	
		3	B8	*EacaMaga228*	7.0	
		4	B8	*PatMaac500*	10.9	
		5	B9	*BM154*	12.7	
		6	B8	*PagMact254*	12.4	
	Benton × VA19 (BV) 79 F6 RILs	WM2.2	B2	*Me1Em5.50, F16R8.230*	13–36	[36]
		WM4.2	B4	*BMd15*	14	
	Raven × 19365-31(R31) 105 F5 RILs	WM5.3	B5	*BM138*	14–21	
		WM5.4	B5	*H19.725*	5–8	
		WM6.1	B6	*Pv163*	12	
		WM7.3	B7	*SF18R7.410/415*	22–51	
		WM8.3	B8	*F12R9.260/280*	11	
		WM8.4	B8	*Ae02.950*	8	
	Tacana × PI 318695 (TW) 89BC$_2$F$_{3:4}$ inbred backcross line	WM2.3[BR.GC.TW]	B2	*Mc2Em4.350*	14.3	[39]
		WM3.3[TW]	B3	*BM189*	13.2	
		WM9.2[TW]	B9	*F6R8.600*	15.7	
	Tacana × PI 313850 (TW) 75BC$_1$F$_{4:5}$ inbred backcross line	WM4.2[R31.TL]	B4	*F1R2.550*	22.0	
		WM7.5[TL]	B7	*F11R3.200*	32.8	
		WM11.1[TL]	B11	*BMd-33*	15.4	

TABLE 3.1 *(Continued)*

Disease/pathogen	Parents and mapping population	QTL number/name	LG/chromosome	Associated markers/ marker interval/closest marker	Phenotypic variance (%)	Reference
	Xana × Cornell 49242	$WM1.1^{XC}$	B1	*Fin, BM170*^{130}	12–23	[40]
	104 F$_7$ RILs	$WM3.2^{XC}$	B3	*OA18*^{1500}	8–12	
		$WM6.1^{XC}$	B6	*BM170*^{160}	12–23	
		$WM7.1^{XC}$	B7	*MCTGEAG*^{54}*, Pha*	10–15	
		$WM7.4^{XC}$	B7	*SpB*	12–14	
Chickpea						
Ascochyta blight (*A. rabiei*)	*C. arietinum* FLIP84-92C × *C. reticulatum* Lad. PI 59907 142 F5:6	1	6	*UBC733b - UBC181a*	50.3	[41]
		2	1	*UBC836b- Dia4*	45.0	
		3	4	*UBC681a-UBC858b*	5.0	
	FLIP 84-92C × PL 599072 142 RILs	1	4	*GAA47*	–	[42]
		2	4	*TA72, GA2*	–	
	ILC 1272 × ILC 3279	Ar1	2	*GA16*	–	[43]
		Ar2a	2	*GA16*	–	
		Ar2b	4	*TA130, TA72, TS72*	–	
	PI 359075(1) × FLIP84-92(2) 250F$_7$ RILs	ar1a	2B–6B	*GA20, GA16*	10.4–19.3	[44]
		ar1b	2B	*TA37, TA200*		
		ar2a	4A	*GA24, GAA47*		
	C. arietinum (ILC 72) × *C. reticulatum* (Cr5-10) 97 F$_{6:7}$ RILs	1	2	*OPAI09*$_{746}$*-UBC881*$_{621}$	28	[45]

TABLE 3.1 (Continued)

Disease/pathogen	Parents and mapping population	QTL number/ name	LG/ chromosome	Associated markers/ marker interval/closest marker	Phenotypic variance (%)	Reference
	kabuli cvILC3279 × desi cv. WR315, 106 F$_{6:7}$ RILs	Ar1	4A	SC/OPK13$_{603}$	34	[46]
		Ar2	4B	TA72, TA146	21	
	kabuli cv ILC3279 × desi cv. WR315, 111 F$_{6:7}$ RILs	Ar3	LG2	TA58-TS82	11.3–22.6	
	C. aretinum "ICCV96029" × "CDC Frontier", 186 F$_2$	1	3	TA64, TS19	13.4	[47]
		2	4	TA2, TA146	29.4	
		3	6	TA80, TA22	11.6	
	ICC 4991 (Pb 7) × ICCV 04516, 179 F$_2$	QTL 1	3	TR58	18.6	[48]
		QTL 2	4	TA146, TS54	7.7	
		QTL 3	4	TA2, TA172	9.6	
	Chickpea ICC 3996 × C. reticulatum ILWC, 184 F$_2$	QTL 3[9]	3	TA34, TA142	21.1	
		QTL 4[1]	4	STMS11, TAA174	3.6	
		QTL 4[7]	4	H3D09, H1A12	16.5	
Botrytis grey mold (Botrytis cinerea pers. Ex. Fr.)	kabuli cv. ICCV 2 × desi cv. JG 62	QTL1	LG6A LG3	SA14-TS71rts36r	12.8	[49]
		QTL2	LG3	TA25-TA144	9	
		QTL3		TA159-TA118	48	
Wilt (F. oxysporium f. sp. Cicer) Race 0	CA2156 × JG62 F$_{6:7}$ RIL	1		TA59	34-37.8	[52]

TABLE 3.1 (Continued)

Disease/pathogen	Parents and mapping population	QTL number/ name	LG/ chromosome	Associated markers/ marker interval/closest marker	Phenotypic variance (%)	Reference
Race 02	C. arietinum (ICCL81001) × C. reticulatum (Cr5-9)	QTL_{Foc02}	2	TA59	48	[7]
Race 5	88 $F_{6:7}$ RIL	QTL_{Foc5}	2	TA59	46	
Cowpea						
Root rot (Macrophomina phaseolina)	IT93K-503-1 × CB46 RILs	Mac-1	2	1_0709, 1_0551	14.4–20.9	[10]
		Mac-2	3	1_0853	8.0–40.0	
		Mac-3	3	1_0604	9.7–10.6	
		Mac-4	3	1_0464, 1_0201	6.1–13.3	
		Mac-5	11	ACA-CAT13, 1_0079, 1_0496	6.7–18.1	
		Mac-6	5	1_0699, 1_0804, ACT-CAT8	7.7–18.2	
		Mac-7	5	1_0678, 1_0153	13.3–19.4	
		Mac-8	6	AAG-CTC9, 1_0030	8.6–18.0	
		Mac-9	6	1_0032, 1_1533,	18.3–12.1	
Faba bean						
Ascochyta blight (A. fabae)	Vf6 × Vf36 196 F_2	Af1	3	$OPA11_{104}$ - $OPAB07_{102}$	25.5	[81]
	29H × VF136 159 $F_{2:3}$	Af2	2	$OPE171_{272}$ - $OPJ18_{626}$	21.0	
		Af3	2	$OPD16_{1732}$, $OPG04_{1131}$, $OPB19_{644}$ $OPD20_{1007}$	6.3–22.4	#,ψ,μ [82]

TABLE 3.1 (Continued)

Disease/pathogen	Parents and mapping population	QTL number/name	LG/chromosome	Associated markers/ marker interval/closest marker	Phenotypic variance (%)	Reference
		Af4	6	$OPJ18_{655}$, $OPG11_{1118}$, $OPB13_{1133}$, $OPL12_{492}$, $OPH17_{884}$	4.0–35.2	
		Af5	15	$OPJ1_{1592}$, $OPO12_{822}$	30.1–36.1	
		Af6	12	$OPD12_{368}$, $OPM09_{503}$	33.6	
		Af7	3	$OPQ19_{1605}$, $OPD12_{1797}$, $OPG04_{810}$, $OPL09_{1161}$, $OPM09_{752}$, $OPO12_{563}$	6.2–44.4	
		Af8	14	$OPG11_{1350}$, $OPG11_{1066}$, $OPK16_{389}$	44.7	
Groundnut						
Late leaf spot (*Phaeoisariopsis personata*)	TAG 24 × GPBD 4 268 $F_{6:7}$ RILs	$QTL_{LLS}01$	1	PM436-Lec-1	3.70–6.50	#,7[9]
		$QTL_{LLS}02$	11	TC9F10-GM660	3.10–4.40	
		$QTL_{LLS}03$	9	TC2G05-TC9H09	3.80–4.80	
		$QTL_{LLS}04$	10	TC1A01-pPGSseq18G1	1.80	
		$QTL_{LLS}05$	1	gi-1107-pPGSseq7G2	1.70–2.90	
		$QTL_{LLS}06$	2	IPAHMS24-TC4D09	2.90–4.40	
		$QTL_{LLS}07$	5	PM179-GM633	3.00–4.80	
		$QTL_{LLS}08$	8	pPGSseq13E6-PM	4.90	
		$QTL_{LLS}09$	13	TC5A07-IPAHM395	4.30	
		$QTL_{LLS}10$	6	IPAHM103-pPGSseq19D6	2.60	
		$QTL_{LLS}11$	12	TC7H11-IPAHM176	2.00	

TABLE 3.1 *(Continued)*

Disease/pathogen	Parents and mapping population	QTL number/ name	LG/ chromosome	Associated markers/ marker interval/closest marker	Phenotypic variance (%)	Reference
Rust (*Puccinia arachidis*) Speg.	TAG 24 × GPBD 4	QTLrust01	6	IPAHM103-pPGSseq19D6	6.90–55.20	
		QTLrust02	1	PM436-Lec-1	2.20–4.50	
		QTLrust03	2	TC11A04-IPAHM524	2.10	
		QTLrust04	3	TC1B02-TC9F04	1.70–5.20	
		QTLrust05	7	C4E09-IPAHM121	2.60	
		QTLrust06	8	pPGSseq13E6-PM3	4.90	
		QTLrust07	8	pPGSseq19G7-TC2C07	2.00	
		QTLrust08	9	TC2G05-TC9H09	2.30	
		QTLrust09	9	GM624-TC4G10	2.80–7.00	
		QTLrust10	8	PM434-TC4F02	6.80	
		QTLrust11	9	TC9H09-GM624	6.00	
		QTLrust12	10	PM377-TC1A01	3.90	
Lentil						
Ascochyta blight (*A. lentis*)	ILL5588 × ILL7537 F$_2$	QTL1	2	$W03_{1050}$-$S01_{750}$	11.4	[83]
		QTL2	2	$G04_{530}$-$AC02_{480}$ $P08_{1200}$-$G04_{530}$	6.8–9.3	
		QTL3	4	$T16_{500}$-$C04_{580}$	7.4	
		QTL4	5	$5U14_{560}$-$B08_{520}$	51.8–69.3	
		QTL5	1	$B18_{1100}$-$W08_{800}$	55.4	
	ILL7537 × ILL6002 F$_2$	QTL6	1	C-CTA/M-ACC_{190}-C-TTA/M-AC_{285}	7.8–16.4	

TABLE 3.1 (Continued)

Disease/pathogen	Parents and mapping population	QTL number/ name	LG/ chromosome	Associated markers/ marker interval/closest marker	Phenotypic variance (%)	Reference
		QTL7	II	*TTA/M-AC$_{285}$- TTA/M-AC$_{165}$*	26.8–30.7	
		QTL8	III	*M20$_{700}$-C-GTA/M-GC$_{191}$*	6.2–10.3	
Mung bean						
Cercospora leaf spot (*Cercospora canescens* Illis & Martin)	KPS1 × V4718 155 F$_2$	*qCLS-1*	3	*CEDG117–VR393*	65.15–78.22	[76]
	(KPS1 × V4718) × KPS1 76 BC$_1$F$_1$				80.53	
		qCLS-2	3	*CEDG117–VR393*		
Powdery mildew (*Erysiphe polygoni* D.C.)	VC1210A × TC1966	*PMR1*	–	*Mac71a, Mac114*	63.16	[78]
	96 F$_2$	*PMR2*	–	*Bng065*	10.00	
	Kamphaeng Saen 1 × VC6468-11-1A	*PMR1*	I	*CEDG282, CEDG191*	14.64–20.1	[79]
	190 F$_7$ RILs	*PMR2*	II	*MB-SSR238, CEDG166*	38.53– 57.81	
	Kamphaeng Saen 1 × VC6468-11-1A	*qPMR-1*	I	*CEDG282-CEDG191*	20.10	[80]
	190 F$_7$ RILs	*qPMR-2*	II	*MB-SSR238-CEDG166*	57.81	
Pea						
Aphanomyces root rot (*A. euteiches* Drechs.)	Puget × 90-2079 F$_{10}$ 127 RILs	*Aph1*	IVb	*N14.950, E7M4.251, E2M4.292* / *E3M3.167, E7M2.254*	10–47	[111]
		Aph2	V	*U370.900*	8–32	
		Aph3	Ia	*E2M4.249*	14	

TABLE 3.1 *(Continued)*

Disease/pathogen	Parents and mapping population	QTL number/name	LG/chromosome	Associated markers/ marker interval/closest marker	Phenotypic variance (%)	Reference
		Aph4	Ib	E2M4.249	16	
		Aph5	Ib	E3M3.167	13	
		Aph6	VII	Pgdp	6	
		Aph7	B	E1M4.174	7	
	Puget × 90-2079 127 RILs	Aph1	IVb	E7M4.108, U326.190	11.0	[60]
		Aph2	V	E3M3.167	11.0–14.0	
		Aph3	Ia	af, U370.900, E1M3.154	12.0–16.0	
		Aph8	III	E7M4.183S	7.0	
		Aph9	IVb	E6M4.108	8.0	
		Aph10	IVb	U530.700	9.0	
		Aph11	IVb	U226.150	11.0	
		Aph12	VI	E1M2.145	10.0	
		Aph13	VII	E8M2.268	12.0	
		Aph14	A	E3M2.182	7.0	
	Baccara 9 × 552 and Baccara 9 × PI 180693 RIL	Ae-Ps1.1	I	N14_600	6.1	[17]
		Ae-Ps1.2	I	AF0164458, Af, D21, AB56, AA179	3.7–13.8	
		Ae-Ps2.1	II	AA473, G10_700	8.7–15.4	
		Ae-Ps2.	II	AA372.1, AB112, PSLEGKP, A19_800, AB128b, A	5.6–26.9	

TABLE 3.1 *(Continued)*

Disease/pathogen	Parents and mapping population	QTL number/name	LG/chromosome	Associated markers/ marker interval/closest marker	Phenotypic variance (%)	Reference
		Ae-Ps2.3	II	*E12_150*	4.5–8.4	
		Ae-Ps3.1	III	*AB70, A08_2000, G10_1200, X03_1000, AD57, X03_1000, AD180c*	5.6–26.7	
		Ae-Ps3.2	III	*AA5, L13_1050, K04_350, AB54*	5.3–8.6	
		Ae-Ps4.1	IV	*E12_1100, AC39c, AA174b, AD186, AA219, AA386a,*	5–20.9	
		Ae-Ps4.2	IV	*AB36c*	17.0	
		Ae-Ps4.3	IV	*AA378, AB145*	5.6–6.3	
		Ae-Ps4.4	IV	*AA122, PSAJ3318*	5.3–6.9	
		Ae-Ps4.5	IV	*AD171*	8.8	
		Ae-Ps5.1	V	*AB23, AA81, S01_1300*	6.6–15.7	
		Ae-Ps5.2	V	*J14_1500*	11.7	
		Ae-Ps5.3	V	*AA255a*	5.9	
		Ae-Ps6.1	VI	*E09_1400*	11.2–49.4	
		Ae-Ps6.2	VI	*AA200, X17_2100*	7.2–8.7	
		Ae-Ps7.1	VII	*E11_900*	5.2	
		Ae-Ps7.2	VII	*AD70, AD90*	4.9–6.4	
		Ae-Ps7.3	VII	*S01_2500, AA416*	5.1–6.5	
		Ae-Ps7.4	VII	*N14_1500*	8.1–12.2	
		Ae-Ps7.5	VII	*AA446, AB114, AA317*	6.3–9.9	

TABLE 3.1 (Continued)

Disease/pathogen	Parents and mapping population	QTL number/name	LG/chromosome	Associated markers/ marker interval/closest marker	Phenotypic variance (%)	Reference
		Ae-Ps7.6	VII	AA07_1700, AB122b, AB136, AD53b, AA387, B12_850, AA07_1700	5.9–19.9	
		Ae-Ps7.6a	VII	AB27, AB122b, AB136, V07_900, G04_500, AA07_1700	6.0–13.7	
		Ae-Ps7.6b	VII	AA176, AB101, AA07_1700	4.9–42.2	
Ascochyta blight (M. pinoides and P. medicaginis var. pinodella)	3148-A88 × Rover 225 F_2	Asc1.1	I	Q407	35	[53]
		Asc2.1	II	CD40	16	
		Asc3.1	III	—	16	
		Asc4.1	IV	P628	8<16	
		Asc4.2	IV	P357-P9	8<16	
		Asc4.3	IVb	P346	8<16	
		Asc5.1	V	sAFP2P2c	8<16	
		Asc7.1	VII	P202	8<16	
Mycosphaerella blight (M. pinoides)	Carneval × MP1401 88 RILs	1	II	ccta2	5	[54]
		2	IV	cccc1	19.1	
		3	VI	acct1	16.8	
	DP × JI 296 131 $F_{2:7}$ RILs	mpII-1	II	AD12-800	6	#,χ,ψ[55]

TABLE 3.1 (Continued)

Disease/pathogen	Parents and mapping population	QTL number/name	LG/chromosome	Associated markers/marker interval/closest marker	Phenotypic variance (%)	Reference
		mpII-2	II	A	9	
		mpIII-1	III	E08-980, V03-1200	6.0–42.0	
		mpIII-2	III	PSP40SG	9.0	
		mpIII-3	III	V03-1000, PSMPSAA175	6.0	
		mpIII-4	III	F09-1900	29.0	
		mpIII-5	III	PSMPSAA374a	11.0	
		mpVa-1	Va	PSMPSAA163.2	7.0–10.0	
		mpVa-2	Va	TA14-2200	16.0	
		mpVII-1	VII	PSMPSAA399	5.0–9.0	
		mpVII-2	VII	Z17-550	8.0	
	P665 × Messire 111 F$_{6:7}$ RILs	mpII.1	II	OPE5_1345/OPK3_1143	13	Y,W[53]
		mpIII.1	III	OPW5_387, OPM6_598	9–29	
		mpIII.2	III	OPM15_431P, OPBI1_1477	9–14	
		mpIII.3	III	OPAI14_1353/OPW2_1157P, OPAI14_1353P OPAI14_1273/ OPAI14_1353P	31–52	
		mpIV.1	IV	OPRS4_782	14	
		mpV.1	V	OPK6_818	29	
	Pisum sativum subsp. syriacum accession P665 × cv. Messire 111 F$_7$ RILs	MpII.1_DRseedI	I	OPE5_1345/OPK3_1143	12	[58]
		MpIII.1_DRseedI	III	OPW5_387	18	
		MpIII.1_DRseedI	III	OPM15_431	8	

TABLE 3.1 (Continued)

Disease/pathogen	Parents and mapping population	QTL number/ name	LG/ chromosome	Associated markers/ marker interval/closest marker	Phenotypic variance (%)	Reference
		MpV.1_DRseed1	V	OPK6_818	28	
		MpIII.2_DRl_05	III	OPR3_1068/OPB11_1477	9	
		MpIII.3_DRl_05	III	OPAI14_1353/AA175	37	
		MpIII.1_DRst_05	III	A6	9	
		MpIII.3_DRst_05	III	OPAI14_1353/AA175	37	
		MpIII.3_DS_05	III	AA175	36	
		MpVI.1_DS_05	VI	OPAB5_498	8	
		MpIII.3_DRl_06	III	OPAI14_1353/AA175	47	
		MpIII.3_DRst_06	III	OPAI14_1273/OPAI14_1353	39	
		MpIII.3_DS_06	III	OPAI14_1353/AA175	39	
		MpIV.1_DS_06	IVb	AA315	9	
Rust (Uromyces pisi Pers. Wint.)	IFPI3260 × IFPI325 94 F$_3$	Up1	03	OPYI11316 and OPV171078	63	#[62]
Rust (U. fabae Pers. de-Bary)	FC 1 × HUVP-1 136 F$_{6:7}$ RILs	Qruf	VII	AA505-AA446	22.2–46.1	#,γ[64]
		Qruf1	VII	AD146 - AA416	11.2–12.4	
Fusarium root rot (F. solani F. sp. Pisi)	Carman × Reward 71 F$_8$ RILs	1	VII	AA160 – AD53	32.4–39.0	γ[65]
Fusarium with (Fusarium oxyspo- rium F. sp. pisi) Race 2	Shawnee × Bohatyr 187 F$_7$ RILs	Fnw3.1	III	AD81_420-AB70_302, AB70_302-AD180_60	2.8–5.4	#,γ[112]
		Fnw3.2	III	Le-UBC88	3.4	
		Fnw4.1	IV	AC22_185-AD171_197	68–80	

TABLE 3.1 *(Continued)*

Disease/pathogen	Parents and mapping population	QTL number/ name	LG/ chromosome	Associated markers/ marker interval/closest marker	Phenotypic variance (%)	Reference
Soybean						
Phytophthora root and stem rot (*P. sojae*)	Conard × Sloan,	2		*Satt252/Satt374*	32.4, 10.6	[67]
	Conrad × Harosoy, and Conrad × Williams			*Satt252/Satt423*	35.0, 15.9	
				Satt252/Satt149	21.4, 20.7	
	Conrad × OX760-6-1 62 F_6 RILs	Qsatt414-569	LG J	*Satt414, Satt569*	13.7–21.5	[68]
	Conrad × OX760-6-1 112 $F_{2:7}$ RILs	QGP1	F	*Satt509–Satt030*	6.7–13.2	[69]
		QGP2	F	*Satt343–OPG16600*	2.4–8.2	
		QGP3	D1b + W	*OPL18800–Satt274*	9.6	
		QFP1	D1b + W	*OPL18800–Satt274*	16.71–21.55	
Phytophthora root rot	Conard × Hefeng 25	QPRR-1	F	*Satt325–Satt343*	9.23–10.24	[68]
		QPRR-2	D1b + w	*Satt005–Satt600*	11.28–21.98	
		QPRR-3	D1b + w	*Satt579–Sat_089*	5.47–27.98	
		QPRR-4	A2	*Satt233–Satt437*	4.98–17.02	
		QPRR-5	B1	*Satt484–Satt453*	4.98–14.80	
		QPRR-6	C2	*Satt489–Satt100*	5.35–21.78	
		QPRR-7	C2	*Satt277–Satt365*	9.34–21.76	
		QPRR-8	C2	*Satt460–Satt307*	4.24–7.76	

TABLE 3.1 (Continued)

Disease/pathogen	Parents and mapping population	QTL number/ name	LG/ chromosome	Associated markers/ marker interval/closest marker	Phenotypic variance (%)	Reference
Sudden death syndrome (SDS) (*Fusarium solani* f. sp. *Glycines*)	Essex × Forrest 100 RILs	1	G	*BARC-Satt214*	24.1	[75]
		2	G	*BARC-Satt309*	16.3	
		3	G	*BARC-570*	19.2	
		4	G	*OEO2$_{1000}$*	12.6	
		5	C2	*BARC-Satt317*	12	
		6	I	*BARC-Satt354*	11.5	
Brown stem rot *Phialophora gregata* (Allington and Chamberlain) W. Gams	BSR101 × PI 437.645 320 RILs	1	J	*AAGATG152E - ACAAGT260*	–	#,γ,ψ[66]
Rhizoctonia root (*Rhizoctonia solani*) and hypocotyl rot (*Thanatephorus cucumeris*)	PI 442031 × Sterling F$_2$ and F$_{4:5}$	1	C2	*Satt281*	11–39	γ[72]
		2	A2	*Satt177*	7–23	
		3	M	*Satt245*	6.8–14	
Sclerotinia stem rot (*S. sclerotiorum*)	S19-90 × Williams 82 152 F$_3$	1	C2	*IaSU-A226H-1*	6.4–11.9	#,γ,ψ
		2	K	*BARC-Satt46, OW13$_{900}$*	7.5–12.5	
		3	M	*OAA09$_{600}$ OAA15$_{750}$ OF20$_{1100}$ OIO4$_{500}$ OQ18$_{550}$ OYI1$_{370}$ OX03$_{500}$*	5.3–11.0	
	Williams 82 × Corsoy 79	1	A1	*Satt545*	4.0–10.0	γ[73]
	Williams 82 × Dassel,	2	A2	*Satt 424*		

TABLE 3.1 *(Continued)*

Disease/pathogen	Parents and mapping population	QTL number/ name	LG/ chromosome	Associated markers/ marker interval/closest marker	Phenotypic variance (%)	Reference
	Williams 82 × Vinton 81,	3	A2	*Satt 233*		
	Williams 82 × DSR 173,	4	B1	*Satt 197*		
	Williams 82 × NK	5	B2	*Satt 438*		
	"S19-90"	6	D1a	*Satt 147, Satt 129*		
		7	D1b	*Satt 172*		
		8	D1b	*Satt 459*		
		9	D2	*Satt 458*		
		10	D2	*Satt 154*		
		11	D2	*Satt 543, Satt301*		
		12	D2	*Satt 256*		
		13	E	*OP_m12*		
		14	F	*Satt 114*		
		15	F	*Satt 510, Satt 335*		
		16	G	*Satt 394*		
		17	G	*Satt 472, Satt 191*		
		18	I	*Satt 451*		
		19	K	*Satt 273*		
		20	K	*Satt 260*		
		21	K	*Satt 588*		
		22	L	*Satt 143, Satt_134*		

TABLE 3.1 *(Continued)*

Disease/pathogen	Parents and mapping population	QTL number/ name	LG/ chromosome	Associated markers/ marker interval/closest marker	Phenotypic variance (%)	Reference
		23	L	*Satt 481*		
		24	N	*Satt 009*		
		25	N	*Satt 387*		
		26	O	*Satt 478*		
		27	O	*Satt 477, Satt 123,*		
		28	O	*Satt 243, Satt _108, Satt _109*		
	PI 194639 × Merit	1	A2	*Sat_138*	12.1	#[74]
	155 F$_{4.5}$ RILs	2	B2	*Satt126*	11.2	
		3	K	*Satt273*	5.5	
		4	L	*Satt182*	5.5	

#QTLs were detected using several resistance scoring criteria/components.
ʸQTLs were detected in different environments (year/location).
ʷQTLs were detected by scoring disease/losses on different parts of plants (stem, leaf, root, etc.).
ᴾQTLs were detected using different strains of pathogen.

Pseudomonas syringae pv. *syringae*, an epiphytic and ice nucleation active bacterium, causes BBS in common bean. It produces blemishes on pods, making them unsuitable for canning and other market uses. Navarro et al. developed a RIL population, from Eagle (susceptible) × Puebla 152 (resistant), and a BC_1F_1 population ((Eagle × Puebla 152) × Eagle), and tested them for ice nucleation temperature in test tube and for number of BBS lesions was assessed in the field.[28] In the RIL population, composite interval mapping (CIM) detected four QTLs, one each on LGs B1, B3, B6, and B11. The QTLs that mapped near RAPD markers P1.1500 and H8.600 on LG B1 and AN6.1600 and AK6.1500 on LG B6 explained 13 and 19%, phenotypic variation, respectively, for BBS. Three RAPD markers namely O10.350, V10.450, and F10.950 were found to be linked with the QTL for resistance to BBS on LG B3. The QTL on B11 was flanked by markers H20.950 and Al15.1300. However, the QTLs on LG3 and LG141 were not confirmed in the backcross population involving the same resistant and susceptible parent.

Wilt disease in common bean is caused by vascular pathogen *Fusariumoxysporum* Schlectend. Fr. f. sp. *phaseoli* J.B. Kendrick and W.C. Snyder. In previous reports, fusarium wilt was considered as a qualitative trait.[29] Fall et al. analyzed 76 $F_{6:10}$ RILs and identified a major locus for wilt resistance, while single factor analysis detected three loci, linked to markers $U20_{750}$ on LG10, $AD4_{450}$ on LG 3, and $K10_{700}$ on LG11, associated with disease severity (DS) index.[30] Using CIM only one major QTL was detected on LG10 that was tightly linked to RAPD marker $U20_{750}$ (LOD score 23.9) and accounted for 63.3% of phenotypic variance, this marker would be a good candidate for SCAR development if it is confirmed and validated.

WM disease, caused by necrotrophic fungus *S. sclerotiorum* (Lib.) de-Bary reduces yield by reducing number, weight, and quality of seeds. QTLs reported for resistance to WM ranged from single major QTL to many weak QTLs. Park et al. analyzed a RIL population (70 lines) from a cross between PC-50 and XAN-159. QTLs for WM resistance mapped on LGs B2, B3, B4, B5, B7, B8, and B11 on an integrated RFLP linkage map of common bean.[31] Among them three QTLs were specific for isolate 152 and seven QTLs for isolate 279, while one QTL (linked to marker J09.950) was specific to both the isolates. Seven of these QTLs were specific to partial field resistance; among them six were at same location as QTLs for partial physiological resistance indicated by porosity over the furrow, plant height, and so forth. Miklas et al. found that a single QTL on LG B7 explained 38% of phenotypic variance for WM resistance.[32] Kolkman and Kelly mapped two major QTL near markers $BC20_{1600}$ and $O15_{1800}$ on LG B2 (11.6% of PV), and aggctt85 and aacctt130 on LG7 (16.8% of phenotypic variance), conferring resistance to WM in a mapping population of 98 F_3 plants derived from the cross Bunsi (resistant) × Newport.[33] The QTLon B2 was confirmed in another population of 28 RILs derived from the cross Huron × Newport; this QTL is also involved in conferring avoidance to WM.[33] Miklas et al. detected two QTLs on LG B6 and B8 in a mapping population of 77 $F_{5:7}$ RILs derived from the cross Benton × NY6020-4.[34] Ender and Kelly evaluated 98 $F_{4:7}$ RILs derived from the cross Bunsi × Raven and identified QTLs for WM resistance on LG B2, B5, B7, and B8 of the integrated bean map, which accounted for 9.2–14.7% of the phenotypic variance.[35] The QTL on B2 in this study was detected in the same genomic region that was identified in two other crosses involving the resistant parent Bunsi.[33] This genomic region also contains QTLs associated with resistance to CBB and root rot as well as several genes for host defense which indicates its importance in disease resistance in common bean.[35] In addition, the QTL detected in NY6020-4 on B8 (26% of phenotypic variation in field) was located adjacent to a QTL from PC 50 specifying partial resistance to WM, but the QTL detected by Ender and Kellyon LG 8 is located in a different region at 30 cM.[35] The QTL detected on B7 appears to be located in the same genomic region as the QTL detected by Kolkman and Kelly;[33] this QTL is closely linked to marker $E_{AAC}M_{CTT}130$.

Maxwell et al. studied 94 $F_{5:6}$ RILs derived from a cross between resistant "G122" and susceptible "Colorado pinto" breeding line CO72548.[6] Five QTLs on LG B1, B2b, B8, and B9 were detected through CIM for WM resistance, based on greenhouse straw test, which explained 48% of phenotypic variability. A major QTL, explaining 20% of phenotypic variance, on B1 between AFLP markers *Patmaac239* and *Patmaca300* and a moderate QTL accounting for 14.8% of phenotypic variance near marker *Eacamaat200* on B2b. Two minor QTLs were detected on LG B8 (near markers *EacaMaga228* and *PatMaac00*) that had favorable alleles from the susceptible parent. QTL detected on LG B9 ($R^2 = 12\%$ for average severity index) mapped nearest to SSR marker *BM154*. Only one QTL (*PagMact254*) on LG8 was detected for field DS. In this study the QTL on LG B7 (38% of phenotypic variance) by Miklas et al. did not validate the same level of effect and explained only 8% of the variance for WM resistance.[32] Soule et al. evaluated two RIL populations, namely "Benton"/VA19 (BV) and "Raven"/I9365-31 (R31) consisting of 79 F_6 and 105 F_5 RILs, respectively, for WM reaction in multiple greenhouse and field tests.[36] Two QTLs were found in BV, WM2.2 ($R2 = 33\%$ greenhouse, 13% field), and WM8.3 (11% field only), and in R31 three expressed in greenhouse tests and four in the field, ranging in phenotypic variance from

5 to 52% (WM2.2, WM4.2, WM5.3, WM5.4, WM6.1, WM7.3, and WM8.4). They compared these QTLs with 26 previously identified QTLs (in different mapping populations derived from same or different parents by different group of researchers), resulting in a comparative linkage map of 35 QTL, which coalesced into 21 distinct regions across nine LGs. Among comparatively mapped QTLs six were expressed only in field, seven were only in greenhouse straw test, and eight in both. Three QTLs (WM1.1, WM3.1, and WM5.3) were associated only with disease avoidance traits (canopy height, lodging, and harvest maturity). Further, previously identified QTLs WM2.2 were detected (on LG2;[33,34,37]), WM2.3,[6,35] WM6.1,[34] WM7.1,[31,32] WM8.1[6,38] and WM8.3 (LG8;[6,34]), WM8.4.[6,34] These eight resistance QTLs (WM2.2, WM2.3, WM6.1, WM7.1, WM7.2, WM8.1, WM8.3, and WM8.4) validated in multiple populations would be excellent for marker-assisted breeding. Mkwaila et al. used two inbred backcross line populations and detected four new QTLs namely WM3.3[TW], WM7.5[TL], WM9.2[TW], and WM11.1[TL] and validated two QTLs namely WM2.3[BR, GC, TW] and WM4.2[R, TL],[31,39] previously detected by various groups.[6,35,36] In a mapping population of 104 F_7 RILs derived from the cross Xana × Cornell 49,242, using greenhouse straw test, Pérez-Vega et al. detected QTLs (WM1.1[XC], WM3.2[XC], WM6.1[XC], WM7.1[XC], and WM7.4[XC]) against five local isolates of *Sclerotinia* named A, B, C, D, and E.[40] The specificity of plant–pathogen interaction is not clear, except one QTL (WM3.2[XC] associated with only one isolate C) others lack isolate specificity. They found that morphological traits (plant height, first internode length, and width) also influence the disease score and therefore percentage of phenotypic variance explained by the QTLs.

3.1.2 CHICKPEA (CICER ARIETINUM L.)

Ascochyta blight caused by necrotrophic pathogen, *Ascochyta rabiei* (Pass.) in chickpea is reported to be controlled by a number of QTLs. Santra et al. tagged one QTL each on LG VI (*QTL 1*) and LG I (*QTL 2*) using RAPD and ISSR markers in a RIL population derived from the interspecific cross *C. arietinum* L. × *Cicer reticulatum* Lad.[41] The two QTLs together accounted for 45 and 50% of total phenotypic variance in the two successive years. Tekeoglu et al. used the same population and integrated six co-dominant sequences tagged microsatellite site (STMS) markers to the two already known QTLs conferring resistance.[42] *QTL 1* was linked to *GAA*$_7$ (SSR) and *ubc733* (RAPD), whereas five STMS markers (*TA72, TA2, TS54, TA46,* and *GA2*) were linked to *QTL 2*. Udupa and Baum detected and mapped one major isolate specific QTL on LG2 named ar1 (linked to *GA16*) for resistance to pathotype I, and two minor QTLs on LG2 and 4 for resistance to pathotype II, namely ar2a (linked to *GA16*) and ar2b (linked to *TA130, TA72,* and *TS72*), respectively.[43] The QTLs on LG2 were mapped in same region but segregation data showed them two loci. Later, Cho et al. elucidated pathotype-specific blight resistance using two isolates of pathotype I and ten isolates of pathotype II in a mapping population of 250 F_7 RILs derived from the intraspecific cross PI 359075 (1) × FLIP84-92 (2).[44] A single QTL for resistance to pathotype II (ar1b on LG2, flanked by markers *TA37, TA200*) and two QTLs for resistance to pathotype I were detected on LG4A (ar2a, flanked by markers *GA24, GAA47*) and LG2 + 6 (ar1a, flanked by markers *TA37, TA200*), respectively. A single recessive resistance gene, *Ar19* or *Ar21d*, on LG2 + 6 appeared to explain the majority of resistance provided by the two QTLs for pathotype I, and generated partial resistance to pathotype II. Most of the above QTL studies were based either on partial map with a low coverage of molecular markers or on comparatively small population size or both. Cobos et al. detected a new putative QTL in a mapping population of 96 $F_{6:7}$ recombinant RILs derived from an interspecific cross *C. arietinum* × (ILC 72) and *C. reticulatum* (Cr5-10).[45] The QTL was on LG2 and flanked by markers $UBC881_{621}$ and $OPA109_{746}$ (14.1 cM apart). Two QTLs, QTL_{AR1} and QTL_{AR2} detected on LG4a and 4b and one QTL, QTL_{AR3} on LG2 by Iruela et al.,[46] in a mapping population of $F_{6:7}$ RILs derived from the intra specific cross between kabuli cvILC3279 × desi cv. WR315. QTL_{AR1}, QTL_{AR2}, and QTL_{AR3} were flanked by markers $SC/OPK13_{603}$, *TA72, TA146,* and *TA58-TS82*, respectively. The QTL detected on LG2 is possibly the same as reported by Cobos et al.,[45] *TA103* the indicative marker, and reported by Udupa and Baum[43] and Cho et al.,[44] indicative marker *GA16*. Tar'an et al. analyzed 186 F_2 plants from a cross between *C. aretinum* "ICCV96029" × "CDC Frontier" with 144 SSR markers and one morphological marker (flower color, *fc*).[47] The plants were inoculated by single-spore-derived culture of *A. rabiei* isolate *ar68-2001*. CIM identified one QTL each on LG3, LG4, and LG6, accounting for 13, 29, and 12% phenotypic variance, respectively. The QTL on LG4, located between markersTA2 and TA146, was co-localized with the QTL detected by Tekeoglu et al.[42] and Cho et al.[44] The QTL on LG6, flanked by *TA80* and *TA22*, was located close to the region where a gene for resistance was mapped by Cho et al.[44] SSR markers *TA64* and *TS19* flanked the QTL on LG3, which was unique to this population. Kottapalli et al. identified three QTLs for resistance to an Indian isolate of ascochyta blight in a mapping population of 179 F_2 from the cross ICC 4991 (Pb 7) × ICCV 04516.[48] *QTL 1* is only for adult plant resistance on LG3 linked with SSR marker TR58 (9.2 cM), whereas *QTL 2* and *QTL* both are for seedling resistance specific located on LG4 flanked by markers TA146 and TS54, and TA2 and TAA170, respectively. The marker TA146 linked to *QTL 2* was validated in half-sib population from ICCV 10 × ICCV 04516. Three QTLs (*QTL3* [9] on LG6A, *QTL4*[1], and *QTL*[3] on LG3) explained 9–48% of phenotypic variance, identified in a mapping population derived from the cross of kabuli cv. ICCV 2 × desi cv. JG 62. The *QTL3*[9] flanked by marker TA34 and TA142 looks to be adjacent to the QTL on LG3 found by Tar'an et al.[47] and Kottapalli et al.[48] There have been many reports in development of genome map and molecular mapping of QTLs for ascochyta blight resistance in chickpea, some of the QTLs are expressed in many mapping populations at same genomic regions others are specifically expressed only in single or few mapping populations. These variations may be because of the different resistant or susceptible or both parents, strength of infection in different environments, different pathogen isolates or different resistance criteria used, and so forth.

Botrytis grey mold (BGM), an important foliar disease of chickpea caused by *Botrytis cinerea* Pers. ex. Fr., may cause complete yield loss during years of extensive rains and high humidity. Anuradha et al.[49] detected three QTLs responsible for resistance to BMG in a mapping population of 126 F_{10} RILs from a cross between kabuli chickpea cultivar ICCV 2 (moderately resistant) and desi chickpea cultivar JG 62 (highly susceptible). These three QTLs, located on LG6A (*QTL1*), LG3 (*QTL2* and *QTL3*), explained 12.8, 9, and 48% phenotypic variance, respectively. QTL 1 was found tightly linked with markers SA14 (<1 cM) and TS71rts36r (<1 cM), QTL 2 flanked by markers TA25 (1 cM) and TA144 (6 cM). QTL3 flanked by TA159 (12 cM) and TA118 (4 cM). These QTLs need to be validated before they could be used for MAS.

In earlier reports, fusarium wilt race 0 (*FOC-0*) resistance was reported to be due to one or two independent genes.[50,51] Cobos et al.[52] developed a linkage map of *C. arietinum* based on two $F_{6:7}$ RIL populations derived from two intraspecific crosses, namely CA2156 × JG62 and CA2139 × JG62. In this study, resistance was considered to be a quantitative trait and interval mapping (IM) detected one QTL on LG3 in mapping population CA2139 × JG62 for resistance to *FOC-0*; this QTL explained 34.8–37.8% of phenotypic variance and was flanked by markers $OPJ2_{600}$ and *TA59*. Subsequently, Cobos et al.[7] mapped one QTL for resistance to FOC race 02 (QTL_{FOC02}) and a tightly linked QTL for resistance to FOC race 5 (QTL_{FOC5}) on LG2; these QTLs were closely linked (<1 cM) to *TA59* STMS marker. QTL_{FOC02} and QTL_{FOC5} explained 48 and 46% phenotypic variance, respectively, and were tightly linked, but they were considered since distinct loci as all the lines resistant to race 0 were not necessarily resistant to race 5.

3.1.3   PEA (PISUM SATIVUM L.)

Ascochyta blight is a serious foliar disease of pea in almost all pea-growing regions; it is caused by a complex of three closely related pathogens, *Ascochyta pisi, Phoma medicaginis* var. *pinodella,* and *Mycosphaerella pinodes* (Berk. and Blox). Timmerman-Vaughan et al.[53] mapped 13 QTLs for ascochyta resistance on seven LGs (LG I, II, III, IV, V, VII and Group A) which explained phenotypic variance ranging from 8 to 35% against 24 isolates of ascochyta blight pathogens (11 isolates of *M. pinoides*, five isolates of *Ascochyta pinodes* and 12 isolates of *P. medicaginis* var. *pinodella*) in a population of 225 F_2 plants from the cross 3148-A88 (resistant) × Rover (susceptible). Three of these QTLs, namely, *Asc1.1, Asc2.1,* and *Asc3.1* had high (35, 16, and 16%, respectively) contribution to phenotypic variance. Eight QTLs, namely *Asc1.1* (LG I), *Asc2.1* (LG II), *Asc3.1* (LG III), *Asc4.1* (LG IV), *Asc4.2* (LG IV), *Asc4.3* (LG IVb), *Asc5.1* (LG V), and *Asc7.1* (LG VII), were detected by more than one trait measurement of disease score (stem1, stem2, pod1, and pod2) and in multiple environments. Some of the QTLs were linked to candidate genes including disease response genes and resistance gene analogs (RAG1.1, RAG2.97, and RAG3A). Tar'an et al.[54] assessed 88 RILs from cross Carneval × MP1401, across 11 different environments for resistance to ascochyta under natural infection. Three QTLs, namely *ccta2* (LG II), *ccccl* (LG IV), and *acctl* (LG VI), were detected, which together explained 36% of the total phenotypic variance. Prioul et al.[55] mapped and characterized QTLs for partial resistance to *M. pinodes* at seedling and adult plant stages using 135 RILs from the cross DP (partially resistant) × JI296 (susceptible) evaluated under growth chamber (seedling stage resistance) and field (adult plant resistance) conditions. They detected 11 QTLs on five LGs against four single-spore isolates (Mp94.14.6, Mp94.76.3, Mp94.56.1, and Mp96.35.1) of *M. pinodes*. Six QTLs (*mpIII-1, mpIII-2, mpIII-3* (LGIII), *mpVa-1* (LGV), *mpVI-1* (LGVI), and *mpVII-1* (LGVII)) were responsible for resistance at seedling stage, their contribution to phenotypic variance ranged from 5 to 20%, while together they explained 73 and 74% of phenotypic variance on stipules and stem, respectively. All the QTLs, except *mpIII-2* had resistance alleles in the resistant parent. Of the 10 QTLs detected for adult stage resistance, three were specific to stipules (*mpII-2, mpIII-1,* and *mpIII-3*), four were specific to stem (*mpIII-4, mpIII-5, mpVa-1,* and *mpVII-1*), and three (*mpII-1, mpVII-1,* and *mpVII-2*) were common to both. Four stable QTLs (*mpIII-1, mpIII-3, mpVa-1,* and *mpVII-1*) specified resistance at seedling as well as adult stages, the closest markers to them were *E08-980* (RAPD), *V03-1000* (RAPD), *PSMPSAA163.2* (SSR), and *PSMPSAA399* (SSR), respectively. Locus *Asc2.1*[53] and *mpII-2*[55] were linked to marker CD40 and locus *a,*[56] respectively. Since CD40 and *a* are closely linked, it was suggested that *Asc2.1* and *mpII-2* could represent the same QTL. Similarly, *Asc.3.1* and *mpIII-1* could represent a common QTL because they were linked to marker M27 and L109, respectively, which are closely linked to each other.[57] In a mapping population of 111 F_7 RILs derived from the cross between wild *P. sativum* subsp. *syriacum* accession P665 × cv. Messire. Fondevilla et al.[58] detected QTL *MpII.1* on LG II, three QTLs (*MpIII.1, MpIII.2,* and *MpIII.3*) on LGIII, one QTL *MpIV.1* on LG IV, and *MpV.1* on LG V which, collectively explained 30–75% of phenotypic variance. Among them *MpIII.1* and *MpIII.2* were detected for both disease rating of seedling scored under growth chamber (DRseedl), and DS, disease rating on leaves (DRl) stems (DRst) scored under field conditions. QTLs *MpII.1* and *MpV.1* were specific for DRseedl and *MpIII.3* and *MpIV.1* were specific for field conditions. The QTLs *MpII.1, MpIII.1, MpIII.2,* and *MpIII.3* coincide with resistance QTLs earlier described in *P. sativum* by Timmerman-Vaughanet al.[53] and Prioulet al.[55] Fondevilla et al.[58] perform inclusion of 38 additional SSR markers to the P665 × cv. Messiremap and favor comparative mapping and QTL validation. They detected 14 QTLs and named them according to the controlled and field condition resistance criteria used and year of the scoring. They confirmed the position of most of the QTLs detected by Fondevilla et al.[59] but three QTLs detected earlier *MpIII.3* (for DS in year 05 and for DRst year 06) and *MpIII.2* (for DRst 05) were not detected in the new map. Three additional QTLs associated with resistance under field conditions *MpIII.1_DRst_05, MpVI.1_DS_05,* and *MpVI.I_DS_05* were detected after inclusion of the additional SSR markers. These differences may be because of the re-arrangements of LGs after inclusion of SSR markers that result in an increase or decrease of the corresponding *limit of detection* (LOD) values required to reach the critical threshold declaring a QTL significant. The mapping of common markers AB64, AA175, and AD174 confirmed that three QTLs, *MpIII.1, MpIII.3,* and *MpIII.2* identified in this study, corresponded respectively with the QTLs *mpIII-1, mpIII-3,* and *mpIII-5* in the *P. sativum* ssp. *Sativum* population, DP × JI296 reported by Priouland et al.[55]

The other devastating disease of pea is the common root rot is a caused by the soil-borne fungus *Aphanomyces euteiches* Drechs. F. sp. *Pisi* W.F.

Pfender and D.J. Hagedorn. A population of 127 RILs from the crosses of "Puget"(susceptible) × partially "90-2079" (resistant) was analyzed at two locations in two years for resistance to root rot.[60] A total of seven QTLs (*Aph1–Aph7*) associated with partial resistance were identified; three of these major QTLs, namely *Aph1, Aph2,* and *Aph3* explaining 7, 32, and 11%, phenotypic variance, respectively, were located on LG IVb, LG V near gene "*r*" (seed shape) and LG Ia near gene "*f*" (leaf morphology), respectively. The other four minor QTLs were located on three LGs (LG Ib (*Aph4, Aph5*), VII (*Aph6*), and B (*Aph7*)) specific to particular environment and resistance criteria. The three resistance criteria used, namely above ground index (AGI), root rot index (RRI), and percentage of dried weight losses (DWL). The QTLs *Aph1, Aph2,* and *Aph3* were not used for a specific resistance criteria whereas *Aph4* was specific for RRI, and *Aph5* and *Aph6* were specific for AGI. Pilet-Nayel et al.[60] later evaluated the same population in greenhouse against two pure isolates, namely, SP7 (United States) and Ae106 (France). Major QTLs, *Aph1* and *Aph3* were detected with both the isolates, but *Aph2* was detected with only the French isolate. $U326_{190}$ and $E7M4_{251}$ were the closest markers for *Aph1*. Seven additional QTLs (*Aph8–Aph14*) showed isolate specificity, that is, they were detected with only one of the two isolates, and were not responsible for partial resistance in the field. Hamon et al.[17] identified a total of 135 QTLs in 23 additive effect genomic regions in two mapping populations of 178 RILs each derived from the crosses between 552 or IP 180693 (partially resistant) and Baccara (susceptible). Among 23 genomic regions, five were consistently detected and showed high stability for all of the strains of pathogen, environments, resistance criteria, and RIL populations used for study.

Pea rust caused by *Uromyces fabae* (Pers.) de-Bary is a major problem in warm humid regions, especially in tropical regions, and is responsible for huge economic losses.[61] A mapping population of 136 $F_{6:7}$ RILs derived from the cross between, HUVP 1 (susceptible) and FC 1 (resistant) was evaluated for two years in polyhouse as well as under field conditions by Rai et al.[62] CIM identified one major (*Qruf*) and one minor (*Qruf1*) QTL for rust resistance on LGVII. *Qruf* was flanked by SSR markers, *AA505* and *AA446* (10.8 cM) and explained 22.2–42.4% and 23.5–58.8% of the total phenotypic variation for infection frequency (IF) and AUDPC, respectively. The minor QTL *Qruf1* was environment-specific, as it was detected only in the polyhouse, it was flanked by SSR markers, *AD146* and *AA416* (7.3 cM), and explained 11.2–12.4% of the total phenotypic variation. Barilli et al.[63] reported a single QTL (*Up1*, 63% phenotypic variance) on LG03 responsible for rust resistance caused by *Uromyces pisi* (Pers.) Wint. in wild pea (*Pisum fulvum* L.). Rai et al.,[64] by using the same mapping population as reported in Rai et al.,[62] evaluated disease severity percent (DS%) and three components of slow rusting, number of aecial pustules per leaf (AP), leaf area covered by sporulating pustules (LASP), and number of aecial cups per leaf (TNAC) in two crop seasons under polyhouse and field conditions. They found that, slow rusting components were governed by four QTLs, two major (*Qruf* on LGVII, *Qruf2* on LGI), and two minor QTLs (*Qruf1* on LG VII and *Qruf3* on LGVI). This confirmed the positions of one each of the major (*Qruf*) and minor (*Qruf1*) QTLs and also detected two new QTLs *Qruf2* and *Qruf3*. The new major QTL *Qruf2* (phenotypic variance 21.3–29.6%) appeared to be the most important component-specific QTL and played key role in deciding disease resistance. The minor QTL *Qruf3* appeared environment-specific and contributed by the susceptible parent.

Fusarium solani (Mart.) Sacc.f. sp. *pisi* (F. R. Jones) W. C. Synder and H. N. Hans is responsible for root rot disease of pea in Canada. Feng et al.[65] identified a major QTL on LG VII flanked by SSR markers AA416 and AB60 that explained 39% of phenotypic variance. Vascular wilt disease of pea is caused by *F. oxysporum* f. sp. *pisi* and leads to significant yield losses. One major QTL on LGIV (*Fwn4.1*: 68–80% phenotypic variance) and two minor QTLs on LGIII (*Fwn3.1*: 2.8–5.4, *Fwn3.2*: 3.4% phenotypic variance) are involved in wilt resistance.[57]

3.1.4   SOYBEAN (GLYCINE MAX L.)

Brown stem rot caused by soil-borne fungus *Phialophora gregata* (Allington and Chamberlain) W. Gams damages the crop by necrosis, inter-venial chlorosis, premature leaf abscission, and internal stem browning. A population of 320 RILs from the cross BSR101 × PI 437-645 was evaluated for brown stem rot resistance in a growth chamber. QTL analysis using CIM detected one major and one minor QTL was on LG J, for both foliar DS and stem browning. The major QTL was located between AFLP markers *AAGATG152E* and *ACAAGT260*, and the minor QTL was situated between the resistance gene analogs RAG31-3 and RAG31-2.[66]

Phytophthora root and stem rot in soybeans is caused by *Phytophthora sojae* Kauffmann and Gerdemann. Burnham et al.[67] identified two putative QTLs on linkage groups F and D1b + W for partial resistance to this disease in three mapping populations derived from the crosses Conard (tolerant) × Sloan (susceptible), Conrad × Harosoy, and Conrad × Williams, which were segregating for root lesion and growth rate after inoculation.

The QTL on LG F was located in an interval of 15.9 cM (SSR markers *Satt252/Satt374*), 11.7 (*Satt252/Satt423*), and 11.0 cM (*Satt252/Satt149*), respectively, and explained 32.4, 35.0, and 21.4% of phenotypic variation for the crosses Conard × Sloan, Conrad × Harosoy, and Conrad × Williams, respectively. The second QTL located on D1b + W identified in the cross Conard × Sloan and Conrad × William population between markers *Satt579* and *Satt600I*, the interval was small only 3 and 6 cM, and explained 10.6 and 20.7% of phenotypic variance, in the crosses, respectively. In the cross Conrad × Harosoy, this QTL fall between marker interval *Satt579* and *Satt600*, explained, 15.9% of phenotypic variance. Another mapping population of 62 F_6 RILs derived from cross between Conrad × OX760-6-1 evaluated at two field sites Woodslee and Weaver, Weng et al.[68] identified single QTL (Qsatt414-569; 13.5–21.5% of the phenotypic variation) on LG J flanked by markers *Satt414-Satt569* in field analysis. Han et al.[69] evaluated resistance to four isolates of *P. sojae* from Northeastern China (JiXi, JianSanJiang, and ShuangYaShan) and Canada (Woodslee) in 112 F_7RILs derived from the cross Conrad (tolerant) × OX760-6-1 (susceptible), and detected three QTLs namely QGP1, QGP2, and QGP3 for phytophthora root rot (PRR) tolerance in the greenhouse. The QTL QGP1 (*Satt509*) and QGP2 (*Satt334*) were located on LG F and explained 13.2, 5.9, and 6.7% of the phenotypic variance for tolerance to the JiXi, JianSanJiang, and Shuang-YaShan isolates, and 5.1 and 2.4% of the phenotypic variance for JiXi and ShuangYaShan isolates, respectively. QGP3 was located on LG D1b + W ($OPL18_{800}$-$SCL18_{659}$) explained 10.2 of the phenotypic variance for Wood-slee isolate. Only one QTL (QPF1 on LG D1b + W) was detected for tolerance to PRR in the field (Woodslee and Weaver), flanked by the markers $OPL18_{800}$-*Satt274*. They validated the QTLs, identified on LGF (QGP2) and LGD1b + W (QGP3).[67] Similarly, Li et al.[70] detected eight QTLs underlying tolerance to PRR, located on five linkage groups (F, D1b + w, A2, B1, and C2) and explained phenotypic variation ranged from 4.24 to 27.98%. Both the parents (Conrad × Hefeng 25) were tolerant to the PRR and the accumulation of tolerant loci is positively correlated with decreases in disease loss percentage. Three QTLs (*QPRR-1, QPRR-2,* and *QPRR-3*) were in genomic regions comparable to the loci identified by Burnham et al.[67] and Han et al.[69] Both used the same Conrad cultivar as the PRR tolerant parent. In a mapping population of 298 RILs derived from the cross V71-370 × PI407162, one major (on LGJ) and two minor QTLs (on LGI and LGG) identified in greenhouse analysis using lesion length as resistance criteria.[71] The major QTL accounted for 22–42% of phenotypic variance and does corresponds to the QTL identified on LGJ.[68]

Rhizoctonia rot and hypocotyl rot is a highly destructive and difficult to control soil-borne disease caused by *Rhizoctonia solani* Kühn (telemorph *Thanatephorus cucumeris* (Frank) Donk). Zhao et al.[72] reported that root and hypocotyl rot resistance against AG-4 isolate is quantitatively inherited and controlled by additive effects of major and minor genes. Mapping populations were developed by crossing moderately resistant parent (PI 442031) to four moderately susceptible cultivars RCAT Staples, Sterling, Evans, and PS 83. QTLs for resistance were found to be associated with SSR markers *Satt281* (LG C2), *Satt177* (LG A2), and *Satt245* (LG M) in F_2 and $F_{4:5}$ populations from cross PI 442031 × Sterling. Linkage between *Satt281* and resistance was successfully verified in another ($F_{4:5}$) population derived from the cross RCAT Staples × PI 442031.

Sclerotinia stem rot in soybean is caused by *S. sclerotiorum* (Lib.) de-Bary. Arahana et al.[73] developed five different RIL populations (100 RILs in each population) from crosses between Williams 85 (susceptible) and five partially resistant cultivars Corsoy 79, Dassel, Vinton 81, DSR 173, and NK S19-90. The RILs were evaluated for stem rot using the detached leaf inoculation technique. Twenty-eight putative QTLs, each explaining 4–10% of the phenotypic variance, were detected on 15 LGs. Later, Han et al.[69] mapped seven QTLs on LGs A2, B2, D2, and L within the same intervals identified by Arahana et al.[73] in different mapping populations using stem inoculation technique. Vuong et al.[74] identified two major and two minor QTLs controlling stem rot resistance in the soybean plant introduction (PI) 194,639. Major QTLs controlling leaf lesion were mapped on linkage groups A2 (12.1%) and B2 (11.2%), and these were linked to SSR marker *Sat_138* and proximal to marker *Satt123*, respectively. The two minor QTLs mapped on LG K and LG L each of them explained 5.5% of the total phenotypic variance.

Resistance to sudden death syndrome (SDS) caused by *F. solani* f. sp. *glycines* is quantitative in nature and is governed by multiple QTLs.[75] A mapping population comprising 100 RILs from a cross between Essex (susceptible) and Forrest (resistance) was evaluated in six environments for SDS. Six QTLs were detected on three different linkage groups. Four QTLs located on LG G together explained 50% of the variance and were detected by SSR markers *BARC-Satt214* ($R^2 = 24.1\%$), *BARC-Satt309* ($R^2 = 16.3\%$), and *BARC-570* ($R^2 = 19.2\%$), and a RAPD marker $OEO2_{1000}$. Two other QTLs were reported on LG C2 ($R^2 = 12\%$) and LG I ($R^2 = 11.5\%$) linked to SSR markers *BARC-Satt317* and *BARC-Satt354*, respectively.

3.1.5 MUNGBEAN (VIGNA RADIATE L. WILCZEK)

Cercospora canescens Illis and Martin is the casual organism of Cercospora leaf spot (CLS) in mung bean and is responsible for up to 50% loss in seed yield. Chankaew et al.[76] evaluated F_2 and BC_1F_1 ((KPS1 × V4718) 9 KPS1) populations developed from cross V4718 (resistant) × Kamphaeng Saen 1 (KPS1) (susceptible) for CLS resistance. CIM identified one major QTL, *qCLS* on LG3 in both the mapping populations. The *qCLS* was flanked by SSR markers CEDG117 and VR393 (24.91–34.67 cM, apart), and explained 65.5–80.53% of phenotypic variance.

Powdery mildew is an important disease of mung bean caused by the obligate fungus *Erysiphe polygoni* D.C. Young et al.[77] reported powdery mildew resistance to be controlled by three loci which together explained 58% of the phenotypic variance in a mapping population of 58 F_2 plants derived from the cross VC3980A (resistant) × TC1966 (susceptible). Two different group of researchers, Chaitieng et al.[78] and Humphury et al.,[79] separately reported only one major locus (*qPMR-2*) resistance to this disease that explained 65 and 80% of phenotypic variance, respectively, and one minor QTL *qPMR-1* explained phenotypic variance upto 20%. Kasettrananet al.[80] analyzed 190 F_7 RILs from the cross Kamphaeng Saen 1 (susceptible) × VC6468-11-1A (resistant), and using CIM detected two QTLs, one major QTL (*qPMR-2i*) on LGII and one minor (*qPMR-1*) on LGI, which explained 57.81 and 20.10%, respectively of the total phenotypic variation. The QTLs*qPMR-1* and *qPMR-2* were flanked by SSR markers CEDG282 and CEDG191 (5.2 cM apart) and MB-SSR238 and CEDG166 (0.6 cM apart), respectively.

3.1.6 FABA BEAN (VICIA FABA L.)

Ascochyta fabae Speg causes ascochyta blight in *V. faba*, and may lead to up to 40% yield reduction. Roman et al.[81] mapped two QTLs, *Af1* and *Af2*, on chromosome 3 and 2, respectively, conferring partial resistance to one isolate of *A. fabae*, in a mapping population of 196 F_2 plants from the cross Vf6 (resistant) × Vf136 (susceptible). QTLs *Af1* and *Af2* acted in an additive manner, did not exhibit isolate and organ specificity, and explained 25.5 and 21.0% phenotypic variability, respectively, RAPD markers $OPA11_{1045}$ and $OPAB07_{102}$, and $OPE171_{272}$ and $OPJ18_{626}$ flanked *Af1* and *Af2*, respectively. A new mapping population of 159 $F_{2:3}$ plants derived from 29H (resistant) × Vf136 (susceptible) was used to develop a linkage map with 103 markers located on 18 linkage groups.[82] Two different isolates (CO99-01 and LO98-01) of *A. fabae* were used to study disease reaction on stems and leaves. Six isolate and organ specific QTLs were reported for resistance to *A. fabae*, and named *Af3–Af8*. QTLs *Af3* and *Af4* were detected in both stem and leaf against both the strains of the pathogen, *Af6* was effective only in leaves and *Af8* only in stem. *Af5* was effective only against isolate CO99-01, while *Af6*, *Af7*, and *Af8* were effective only against isolate LO98-01.

3.1.7 GROUNDNUT (ARACHIS HYPOGAEA L.)

Late leaf spot (LLS) (caused by *Phaeoisariopsis personata* (Berk. and Curt.)) and rust (caused by *Puccinia arachidis* Speg.) are two major diseases that occur together and are responsible for 50–70% yield losses in groundnut. In a mapping population (TAG 24 × GPBD 4) of 286 RILs, Khedikar et al.[9] detected 11QTLs (explaining 1.70– 6.50% phenotypic variance) for LLS and 12 QTLs (accounting for 1.70–55.20% of phenotypic variance) for rust. A major QTL (*QTLrust01*, 6.90–55.20% phenotypic variance) was identified for rust resistance; this QTL was expressed in all the environments, developmental stages, and resistance components. A molecular marker *IPAHM 103* acts as candidate marker for MAS as it is found nearest to QTL *QTLrust01* (position, 0–12 cM from LOD peak on LG VI). This marker was validated in a set of 46 resistant and susceptible germplasm lines with different genetic backgrounds and in another mapping population of 146 RILs (TAG 26 × GPBD 4) where it contributes 24.10–48.90% phenotypic variance by CIM and 27.98–51.96 by single marker analysis (SMA).

3.1.8 COWPEA (VIGNA UNGUICULATA L.)

Macrophomina phaseolina (Tassi) Goid is responsible for charcoal rot, seedling damping- off, and ashy stem blight diseases of many crops. For cowpea under high temperature and drought it is a major yield-suppressing factor. Muchero et al.[10] mapped nine QTLs (*Mac-1–Mac-9* explaining 6.1–40% of phenotypic variance) distributed over four LGs of consensus cowpea linkage map, associated with resistance to Macrophomina infection in a RIL population, IT93K-503-1 × CB46. Based on synteny with *G. max* and *Medicago truncatula* and genic single nucleotide polymorphism (SNP), candidate resistance genes were detected in mapped QTL intervals; major QTL *Mac-2* collocated with SNP marker derived from pectin esterase inhibitor encoding gene. The two QTLs *Mac-2* and *Mac-2* collocated with maturity QTLs *Mat-2* and *Mat-1*, and three QTLs, *Mac-4*, *Mac-5*, and *Mac-9* collocated with seedling drought response QTL *Dro-5*, *Dro-10*, and *Dro-7*, respectively.

3.1.9   LENTIL (LENS CULINARIS MEDIKUS.)

Ascochyta blight of lentil is an important disease caused by *Ascochyta lentis* Vassilievsky. QTLs for ascochyta blight resistance were analyzed by Rubeena et al.[83] using two F_2 populations from crosses ILL5588 × ILL7537 and ILL7537 × ILL6002. In the F_2 from cross ILL5588 × ILL7537, five QTLs (*QTL1–QTL5*) were detected by CIM on four different linkage groups, namely LG1, LG2, LG4, and LG5). In the F_2 populations from cross ILL7537 × ILL6002, three QTLs (*QTL6* and *QTL7* on LGI and *QTL8* on LGII) were detected by CIM. AFLP marker $C\text{-}TTA/M\text{-}AC_{285}$ was located at 3.4 cM from *QTL6* and at 12 cM from *QTL7*; these two QTLs together explained 47% of the phenotypic variance. *QTL8* explaining 10% of phenotypic variance was detected on LGII and linked to $M20_{700}$ RAPD marker. *QTL7* and *QTL8* identified by multiple-interval mapping (MIM) were found to be dominant and partially dominant, respectively, and together explained 50% of the phenotypic variance.

3.2   MARKER-ASSISTED SELECTION (MAS) IN LEGUMES

DNA markers that are tightly linked to important genes may be used as molecular tools for MAS in plant breeding.[84] In MAS, presence or absence of a marker is used as an indication for the presence of the desired gene; MAS can also supplement phenotypic selection to make it more efficient, effective, dependable, and cost-effective.[85] MAS will be more useful, in such cases where the trait of interest is expressed late in plant development, for example, fruit/flower/seed characteristics or adult features particularly in species that have a juvenile phase, the target allele is recessive, the trait expression has a threshold requirement, such as, presence of the concerned pathogen/insect for disease/pest resistance,[86] the trait is difficult measure, and/or is affected by the environment. MAS also helps accelerate backcross breeding, and enables pyramiding of several QTLs or many major genes for single complex traits, including disease/insect resistance.[5,87]

There are several examples of use of molecular markers for selection and evaluation of breeding materials, in backcross programs and gene pyramiding in legumes.[88–90] MAS was successfully used for the breeding of soybean resistant to cyst nematode,[91] common bean resistant to CBB pathogen,[92] and narrow-leafed lupin (*Lupinus angustifolius* L.) resistant to phomopsis stem blight pathogen[93] and anthracnose pathogen.[94] There are numerous examples of introgression and pyramiding of favorable alleles and QTLs in legumes facilitated by MAS. Most relevant work has been carried out in common bean breeding for rust and anthracnose resistance pathogen.[95] RAPD markers linked to the 11 genes (*Ur-1–Ur-11*) conferring rust resistance are being used to incorporate and pyramid these genes into common bean cultivars, and/or to combine rust resistance with resistance to other diseases, such as bean common mosaic virus (BSMV), CBB (pathogen; and anthracnose).[96] Similarly, molecular markers linked to the majority of genes conferring anthracnose resistance (*Co-1–Co-10*) provide an opportunity to pyramid them in a resistant cultivar through MAS.[97] Gene pyramiding combining QTLs for resistance with monogenic traits have been carried out by Walker et al. in soybean.[98,99]

Two major QTLs associated with resistance to white mold (pathogen) in common bean have been introgressed using MAS.[100] One RAPD marker, $BC20_{1800}$ (on LG B2), and two AFLP markers *EAACMCTT130* and *EAGGMCTT85* (on LG B7), were used for MAS and, potential of resistant parent Bunsi as a genetic donor of QTLs for white mold resistance was confirmed in two bean populations. QTLs controlling tolerance to PRR have been pyramided using MAS in soybean.[70] Seven QTLs, namely *QPRR-1–QPRR-7*, were identified from RILS evaluated under multiple environments. The RILs having the maximum number of PRR tolerance loci from both the parents showed much lower percentage loss (7.78–23.33%) due to the disease than the parents (parental range) of the RILs.

Usually, only major QTLs are considered by plant breeders for MAS because of inconsistency of estimated QTL effects.[9,17,101] Sometimes linkage drag due to tight linkage of desirable QTL alleles with undesirable genes governing other traits or insufficient linkage between marker and the desired QTL also affect efficiency of MAS.[102] The major reasons behind inconsistency of QTLs are expression of different QTLs in different environments, interaction between QTLs and effect of the genetic background.[15,103] Over the last two decades, a huge amount of investment has been directed toward research on MAS for plant breeding worldwide. However, examples of large-scale marker implementation in plant breeding are few, and MAS leading to release of new cultivars is limited.[104,105] Key challenge in MAS for molecular plant breeding is to develop markers linked to genes of interest, which are applicable to multiple breeding populations, and environments.

In most of the grain legumes, resistance against majority of the diseases is of quantitative nature and governed by multiple QTLs. Disease reaction is markedly influenced by environmental factors like temperatures, humidity, pathogens, races, other pathogens, and so forth. These factors may have strong effect on the detection of QTLs responsible for resistance and dependable quantification of their effects. Environmental influences are attempted to be resolved into individual Q × E using replicated phenotyping in different environments. Involvement of one or few major QTLs with multiple supporting minor QTLs is a common feature for disease resistance in legumes.[7,9,17,64,67] The major QTLs usually consistently detected in multiple environments but the magnitude of their effect varies considerably across the environments. However, minor QTLs may or may not be detected in different environments for different isolates or may have opposite effect in different environments possibly due to their small contribution to phenotypic variance effect and large Q × E and/or QTL × QTL interactions of additive, epistatic, and/or of negative epistatic type.[8,17] Therefore, for better understanding of disease resistance different disease phenotyping criteria/ components are scored on different parts of infected plants in controlled and field conditions during different growth stages such as seedling or adult stage.[55,64] Some of the disease resistance criteria/components generally used to score diseases are DS% (the proportion of plant part affected by disease showing symptoms), IF (number of pustules per cm^2), latent period (LP), harvest index (HI), yield losses reduction in test weight, and so forth. Whenever multiple components are used to score disease some of the QTLs are expressed by all of them but several are specific to particular components or environment. This may be because of the pleiotropic effect of same gene or at least few of the genes responsible for these components are common. Sometimes these components are so much effective that they individually identify the resistant/susceptible nature of plants. For those disease having only partial resistance such components are very important and the molecular mapping of QTLs responsible for them play important role in determining the nature of plant.

The architecture of plant can influence the microclimate and hence the spread of disease; therefore, in a number of above studies along with QTL detection for disease resistance, QTLs for morphological traits were also detected and as anticipated they found co-localized. These co-localizations may be due to the direct effect of plant architecture, canopy structure or earliness on the epidemiology of the pathogen, genetic linkages between genes controlling resistance to disease and plant architecture (such as height or flowering or maturity time etc.), and/or pleiotropic effects of genes.

Similarly, the QTLs specificity either to the particular organ or to the growth stage may be because of the expression of genetic factors specific to a given stage or organ, interactions with environmental effects, such as climate effects, differences in scoring methodologies such that under controlled conditions the focus is on resistance to infection only and under field conditions an overall response, including both resistance to infection and resistance to fungus progress upwards on the plant, is taken into account.

The study of specificity of QTLs for different isolates or races of the pathogen is also important, because resistance genes present in resistant lines might be only effective against some of them. Therefore, some QTLs might be effective against wider isolate spectrum where as others are effective against single or few isolates determined by host–pathogen interaction. In most of the studies known strain/isolate or strains of the pathogens are used to inoculate host plant but in some unknown or locally available inoculums with multiple strains/isolates has been used to study QTL analysis. In studies were multiple isolates of pathogens are used isolate-non-specific and isolate-specific QTLs are often expressed as major and minor effects, respectively. Many of the isolate-specific resistance QTL identified to date in this study have minor effects. In addition, unusually the QTL that is contributed by susceptible parents have also been found to respond in an isolate-specific fashion, and their expression was influenced by environmental conditions. Sometimes the distribution pattern of disease response against different isolates varies suggesting different genetic mechanism controlling pathogenicity. In multiple pathogen disease complex the study of specificity of QTLs for different pathogens is a necessary pre-requisite. As ascochyta blight disease of pea is caused by three related pathogens M. pinodes, A. pisi, and P. medicaginis, different groups of researchers performed QTL analysis using different isolates of the three pathogens.[53,55] The increasing use of large set of transferable markers and consensus genetic linkage maps in legumes provide a faster and more detailed investigations of disease resistance QTLs across different sets of mapping populations and among species, validation of expression-QTL across variable genetic backgrounds and positioning of a growing number of candidate genes to be tested in association mapping experiments.[10,106]

Some of the detected QTLs are validated or confirmed their presence in another mapping population/populations or set of genotypes although their contribution to phenotypic variance always varies. Usually the non-consistent, minor QTLs are specific to particular population but sometimes the non-consistent QTL in one population acting as major and consistent QTL in other population. In legumes genome, the disease resistance genes are commonly organized in complex clusters or loci. Therefore, the regions on linkage maps are often rich in co-localized QTLs conferring resistance to different pathogens and/or to different races of the same pathogen. Based on few common molecular markers studies the QTLs responsible for one disease is either found in the same region or close to the QTLs responsible for other disease.

3.3   FUTURE PROSPECTS

Conventional practices for yield improvement in legumes are not so effective owing to the complex physiology and susceptibility to several diseases caused by biotic stresses.[107–109] Disease resistance in legumes is usually quantitative in nature, governed by single or multiple QTLs.[6,7,9,10] Now, functionally characterized genes, expressed sequence tags (ESTs), and coding genome sequences are becoming available for many legumes including medicago, chickpea, common bean, cowpea, pigeon pea, and soybean.[44,109] Therefore, perfect or functional markers such as SSR, SNP, and conserved orthologous set markers should be employed for QTL mapping of disease resistance and subsequent use of the identified markers in MAS. The availability of total or euchromatic genome sequences of *M. truncatula, G. Max,* and *Lotus japonicas* will greatly aid in this area of researches these are taxonomically closer to a number of legume crops like pea, lentil, faba bean, and chickpea.[110] Moreover, the recently published genome of *P. vulgaris* L. will further in understanding genomic similarities for disease resistance among different legume crops.

KEYWORDS

- legumes
- QTLs
- molecular markers
- linkage maps
- marker-assisted selection

REFERENCES

1. Lewis, G. P.; Schrire, B. D.; Mackinder, B. A.; Lock, M. *Legumes of the World;* Royal Botanic Garden: Kew, UK, 2005.
2. Food Agriculture Organization Database; http://www.faostat.fao.org. *FAO Database;* (Accessed on September 2014).
3. Graham, P. H.; Vance, C. P. Legumes: Importance and Constraints to Greater Use. *Plant Physiol.* **2003,** *131,* 872–877.
4. Chakraborty, U.; Sarkar, B.; Chakraborty, B. N. Protection of Soybean Rot by *Brady-rhizobium japonicum* and *Trichoderma harzianum* Associated with Changes in Enzyme Activities and Phytoalexin Production. *J. Mycol. Plant Patho.* **2003,** *33,* 21–25.
5. Dita, M. A.; Rispail, N.; Prats, E.; Rubiales, D.; Singh, K. B. Biotechnology Approaches to Overcome Biotic and Abiotic Stress Constraintsin Legumes. *Euphytica.* **2006,** *147,* 1–24.
6. Maxwell, J. J.; Brick, K. A.; Byrne, P. F.; Schwartz, H. F.; Shan, J.; Ogg, J. B.; Hensen, R. A. Quantitative Trait Loci Linked to White Mold Resistance in Common Bean. *Crop Sci.* **2007,** *47,* 2285–2294.
7. Cobos, M. J.; Winter, P.; Kharrat, M.; Cubero, J. I.; Gil, J.; Millan, T.; Rubio, J. Genetic Analysis of Agronomic Traits in a Wide Cross of Chickpea. *Field Crops Res.* **2009,** *111,* 130–136.
8. Wu, X.; Blake, S.; Sleper, D. A.; Shannon, J. G.; Cregan, P.; Nguyen, H. T. QTL, Additive and Epistatic Effects for SCN Resistance in PI 437654. *Theor. Appl. Genet.* **2009,** *118,* 1093–1105.
9. Khedikar, Y. P.; Gowda, M. V. C.; Sarvamangala, C.; Patgar, K. V.; Upadhyaya, H. D.; Varshney, R. K. A QTL Study on Late Leaf Spot and Rust Revealed One Major QTL for Molecular Breeding for Rust Resistance in Groundnut (*Arachis hypogaea* L.). *Theor. Appl. Genet.* **2010,** *121,* 971–984.
10. Muchero, W.; Ehlers, J. D.; Close, T. J.; Roberts, P. A. Genic SNP Markers and Legume Synteny Reveal Candidate Genes Underlying QTL for *Macrophomina phaseolina*Resistance and Maturity in Cowpea [*Vigna unguiculata* (L) Walp.]. *BMC Genomics.* **2011,** *12,* 8.
11. Pereira, M. G.; de Olivera, L. O.; Lee, M. QTL Mapping and Disease Resistance in Cereals. *J. New Seeds.* **2000,** *2,* 1–21.
12. Xu, J. Global View of QTL. In *Quantitative Genetics Genomics and Plant Breeding;* Kang, M. S., Ed.; CABI Publications: Boston, MA, 2002; pp 109–134.
13. Ding, J. Q.; Wang, X. M.; Chander, S.; Yan, J. B.; Li, J. S. QTL Mapping of Resistance to Fusarium Ear Rot Using a RIL Population in Maize. *Mol. Breed.* **2008,** *22,* 395–403.
14. Li, Z. K.; Yu, S. B.; Lafitte, H. R.; Huang, N.; Courtois, B.; Hittalmani, S.; Vijayakumar, C. H. M.; Liu, G. F.; Wang, G. C.; Shashidhar, H. E.; Zhuang, J. Y.; Zheng, K. L.; Singh, V. P.; Sidhu, J. S.; Srivantaneeyakul, S.; Khush, G. S. QTL × Environment Interactions in Rice. I. Heading Date and Plant Height. *Theor. Appl. Genet.* **2003,** *108,* 141–153.
15. Collard, B. C. Y.; Mackill, D. J. Marker-Assisted Selection: An Approach for Precision Plant Breeding in the Twenty-First Century. *Phil. Trans. R. Soc. Lond. B. Biol. Sci.* **2008,** *363,* 557–572.
16. Collins, N. C.; Tardieu, F.; Tuberosa, R. Quantitative Trait Loci and Crop Performance Under Abiotic Stresses: Where do we Stand? *Plant Physiol.* **2008,** *147,* 469–486.
17. Hamon, C.; Baranger, A.; Coyne, C. J.; McGee, R. J.; Goff, I. L.; L'Anthoëne, V.; Esnault, R.; Riviére, J. P.; Klein, A.; Mangin, P.; McPhee, K. E.; Roux-Duparque,

M.; Porter, L.; Miteul, H.; Lesne,' A.; Morin, G.; Onfroy, C.; Moussart, A.; Tivoli, B.; Delourme, R.; Pilet-Nayel, M. L. New Consistent QTL in Pea Associated with Partial Resistance to *Aphanomyces euteiches* in Multiple French and American environments. *Theor. Appl. Genet.* **2011,** *123,* 261–281.

18. Carlborg, O.; Haley, C. S. Epistasis: Too often Neglected in Complex Trait Studies? *Nat. Rev. Genet.* **2004,** *5,* 618–624.

19. Asins, M. Present and Future of Quantitative Trait Locus Analysis in Plant Breeding. *Plant Breed.* **2002,** *121,* 281–291.

20. Collard, B. C. Y.; Jahufer, M. Z. Z.; Brouwer, J. B.; Pang, E. C. K. An Introduction to Markers, Quantitative Trait Loci (QTL) Mappingand Marker-Assisted Selection for Crop Improvement: The Basis Concept. *Euphytica.* **2005,** *142,* 169–196.

21. Vounge, T. D.; Wu, X.; Pathan, M. S., Valliyodan, B.; Nguyen, H. T. Genomics Approaches to Soybean Improvement. In*Genomics Assisted Crop Improvement;* Varshney, R. K., Tuberosa, R., Eds.; Springer: Berlin, Germany, 2007; pp 243–249.

22. Nodari, R. O.; Tsai, S. M.; Guzman, P.; Gilbertson, R. L.; Gepts, P. Towards an Integrated Linkage Map of Common Bean. III. Mapping Genetic Factors Controlling Host-Bacteria Interactions. *Genetics.***1993,** *134,* 341–350.

23. Bai, Y.; Michaels, T. E.; Pauls, K. P. Identification of RAPD Markers Linked to Common Bacterial Blight Resistance Genes in *Phaseolus vulgaris* L. *Genome.* **1997,** *40,* 544–551.

24. Yu, K.; Park, S. J.; Poysa, V. Marker-Assisted Selection of Common Beans for Resistance to Common Bacterial Blight: Efficacy and Economics. *Plant Breed.* **2000,** *119,* 411–415.

25. Liu, S.; Yu, K.; Park, S. J.; Conner, R. L.; Balasubramanian P.; Mundel H. H.; Kiehn, F. A.; Marker-Assisted Selection of Common Beans for Multiple Disease Resistance. *Annu. Rep. Bean Improv. Coop.* **2005,** *48,* 82–83.

26. Yu, K.; Park, S. J.; Zhang, B.; Haffner, M.; Poysa, V. An SSR Marker in the Nitrate Reductase Gene of Common Bean is Tightly Linked to a Major Gene Conferring Resistance to Common Bacterial Blight. *Euphytica.* **2004,** *138,* 89–95.

27. Liu, S.; Yu, K.; Park, J. Development of STS Markers and QTL Validation for Common Bacterial Blight Resistance in Common Bean. *Plant Breed.* **2008,** *127,* 62–68.

28. Navarro, F.; Sass, M. E.; Nienhuis, J. Identification and Confirmation of Quantitative Trait Locus for Root Rot Resistance in Snap Bean (*Phaseolus vulgaris* L.). *Crop Sci.* **2008,** *48,* 962–972.

29. Salgado, M. O.; Schwartz, H. F.; Brick, M. A. Inheritance of Resistance to a Colorado Race of *Fusarium oxysporum* f. sp. *phaseoli* in Common Beans. *Plant Dis.* **1995,** *79,* 279–281.

30. Fall, A. L.; Byrne, P. F.; Jung, G.; Coyne, D. P.; Brick, M. A.; Schwartz, H. F. Detection and Mapping of a Major Locus for Fusarium Wilt Resistance in Common Bean. *Crop Sci.* **2001,** *41,* 1494–1498.

31. Park, S. O.; Coyne, D. P.; Steadman, J. R.; Skroch, P. W. Mapping of QTL Resistance to White Mold in Common Bean. *Crop Sci.* **2001,** *41,* 1253–1262.

32. Miklas, P. N.; Johnson, W. C.; Delorme, R.; Gepts, P. QTL Conditioning Physiological Resistance to White Mold in Dry Bean. *Crop Sci.* **2001,** *41,* 309–315.

33. Kolkman, J. M.; Kelly J. D. QTL Conferring Resistance and Avoidance to White Mold in Common Bean. *Crop Sci.* **2003,** *43,* 539–548.

34. Miklas, P. N.; Kelly, J. D.; Singh, S. P. Registration of Anthracnose-Resistant Pinto Bean Germplasm Line USPT-ANT-1. *Crop Sci.* **2003,** *43,* 1889–1890.

35. Ender, M.; Kelly, J. D. Identification of QTL Associated with White Mold Resistance in Common Bean. *Crop Sci.* **2005,** *45,* 2482–2490.
36. Soule, M.; Porter, L.; Medina, J.; Santana, G. P.; Blair, M. W.; Miklas, P. N. Comparative QTL Map for White Mold Resistance in Common Bean, and Characterization of Partial Resistance in Dry Bean Lines VA19 and I9365–31. *Crop Sci.* **2011,** *51,* 123–139.
37. Miklas, P. N.; Larsen, K. M.; Terpstra, K.; Hauf, D. C.; Grafton, K. F.; Kelly, J. D. QTL Analysis of ICA Bunsi Derived Resistance to White Mold in a Pinto X Navy Bean Cross. *Crop Sci.* **2007,** *47,* 174–179.
38. Park, S. O.; Coyne, D. P.; Steadman, J. R.; Skroch, P. W. Mapping of the Ur-7 Gene for Specific Resistance to Rust in Common Bean. *Crop Sci.* **2003,** *43,* 1470–1476.
39. Mkwaila, W.; Terpstra, K. A.; Ender, M.; Kelly, J. D. Identification of QTL for Resistance to White Mold in Wild and Landrace Germplasm of Common Bean. *Plant Breed.* **2011,** *130,* 665–672.
40. Pérez-Vega, E.; Pascual, A.; Campa, A.; Giraldez, R.; Miklas, P. N.; Ferreira, J. J. Mapping QTL Conferring Partial Physiological Resistance to White Mold in the Common Bean RIL Population Xana/Cornell 49242. *Mol. Breed.* **2012,** *29,* 31–41.
41. Santra, D. K.; Tekeoglu, M.; Ratnaparkhe, M.; Kaiser, W. J.; Muehlbauer, F. J. Identification and Mapping of QTLs Conferring Resistance to Ascochyta Blight in Chickpea. *Crop Sci.* **2000,** *40,* 1606–1612.
42. Tekeoglu, M.; Rajesh, P. N.; Muehlbauer, F. J. Integration of Sequence Tagged Microsatellite Sites to the Chickpea Genetic Map. *Theor. Appl. Genet.* **2002,** *105,* 847–854.
43. Udupa, S. M.; Baum, M. Genetic Dissection of Pathotype Specific Resistance to Ascochyta Blight Resistance in Chickpea (*Cicer arietinum* L.) Using Microsatellite Markers. *Theor. Appl. Genet.* **2003,** *106,* 196–1202.
44. Cho, S.; Chen, W.; Muehlbauer, F. J. Pathotype-Specific Genetic Factors in Chickpea (*Cicer arietinum* L.) for Quantitative Resistance to Ascochyta Blight. *Theor. Appl. Genet.* **2004,** *109,* 733–739.
45. Cobos, M. J.; Rubio, J.; Strange, R. N.; Moreno, M. T.; Gil, J.; Millan, T. A New QTL for Ascochyta Blight Resistance in an RIL Population Derived from an Interspecific Cross in Chickpea. *Euphytica.* **2006,** *149,* 105–111.
46. Iruela, M.; Rubio, J.; Barro, F.; Cubero, J. I.; Millan, T.; Gil, J. Detection of Two Quantitative Trait Loci for Resistance to Ascochyta Blight in an Intra-Specific Cross of Chickpea (*Cicer arietinum* L.): Development of SCAR Markers Associated with Resistance. *Theor. Appl. Genet.* **2006,** *112,* 278–287.
47. Tar'an, B.; Warkentin, T. D.; Tullu, A.; Vandenberg, A. Genetic Mapping of Ascochyta Blight Resistance in Chickpea (*Cicer arietinum* L.) Using a Simple Sequence Repeat Linkage Map. *Genome.* **2007,** *50,* 26–34.
48. Kottapalli, P.; Gaur, P. M.; Katiyar, S. K.; Crouch, J. H.; Buhariwalla, H. K.; Pande, S.; Gali, K. K. *Mapping and Validation Of Qtls for Resistance to an Indian Isolate of Ascochyta Blight Pathogen in Chickpea. Euphytica.* **2009,** *165,* 79–88.
49. Anuradha, C.; Gaur, P. M.; Pande, S.; Gali, K. K.; Ganesh, M.; Kumar, J.; Varshney, R. K. Mapping QTL for Resistance to Botrytis Grey Mould in Chickpea. *Euphytica.* **2011,** *182,* 1–9.
50. Tekeoglu, M.; Santra, D. K.; Kaiser, W. J.; Muehlbauer, F. J. Ascochyta Blight Resistance in Three Chickpea Recombinant Inbred Line Populations. *Crop Sci.* **2000,** *40,* 1251–1256.

51. Rubio, J.; Hajj-Moussa, E.; Kharrat, M.; Moreno, M. T.; Millán, T.; Gil, J. Two Genes and Linked RAPD Markers Involved in Resistance to *Fusarium oxysporum* f. sp. *ciceris*Race 0 in Chickpea. *Plant Breed.* **2003,** *122,* 188–191.

52. Cobos, M. J.; Fernandez, M. J.; Rubio, J.; Kharrat, M.; Moreno, M. T.; Gil, J. A Linkage Map of Chickpea (*Cicer arietinum* L.) Based on Populations from Kabuli × Desi Crosses: Location of Genes for Resistance to Fusarium Wilt Race 0. *Theor. Appl. Genet.* **2005,** *110,* 1347–1353.

53. Timmerman-Vaughan, G. M.; Frew, T. J.; Russell, A. C.; Khan, T.; Butler, R.; Gilpin, P.; Murray, S.; Falloon, K. QTL Mapping of Partial Resistance to Field Epidemics of Ascochyta Blight of Pea. *Crop Sci.* **2002,** *42,* 2100–2111.

54. Tar'an, B.; Warkentin, T.; Somers, D. J.; Miranda, D.; Vandenbreg, A.; Blade, S.; Woods, S.; Bing, D.; Xue, A.; DeKoeyer, D.; Penner, G. Quantitative Trait Loci for Lodging Resistance, Plant Height and Partial Resistance to Mycosphaerella Blight in Field Pea (*Pisum sativum* L.). *Theor. Appl. Genet.* **2003,** *107,* 1482–1491.

55. Prioul, S.; Frankewitz, A.; Deniot, G.; Morin, G.; Baranger, A. Mapping of Quantitative Trait Loci for Partial Resistance to *Mycosphaerella pinodes* in Pea *(Pisum sativum* L.), at the Seedling and Adult Plant Stages. *Theor. Appl. Genet.* **2004,** *108,* 1322–1334.

56. Weeden, N. F.; Ellis, T. H. N.; Timmerman-Vaughan, G. M.; Swiecicki, W. K.; Rozov, S. M.; Berdnikov, V. A. A Consensus Linkage Map for *Pisum sativum. Pisum Genet.* **1998,** *30,* 1–4.

57. McPhee, K. E. Genome Mapping and Molecular Breeding in Plants. In *Pulses, Sugar and Tuber Crops;* Kole, C., Ed.; Springer-Verleg: Berlin, Heidelberg, Pullman, 2007; pp 33–48.

58. Fondevilla, S.; Küster, H.; Krajinski, F., et al. Identification of Genes Differentially Expressed in a Resistant Reaction to *Mycosphaerellapinodes* in Pea Using Microarray Technology. *BMC Genom.* **2011,** *12,* 28.

59. Fondevilla, S.; Satovic, Z.; Rubiales, D.; Moreno, M. T.; Torres, A. M. Mapping of Quantitative Trait Loci for Resistance to *Mycosphaerella pinodes* in *Pisum sativum* subsp. syriacum. *Mol. Breed.* **2008,** *21,* 439–454.

60. Pilet-Nayel, M. L.; Muehlbauer, F. J.; McGee, R. J.; Kraft, J. M.; Baranger, A.; Coyne, C. J. Consistent Quantitative Trait Loci in Pea for Partial Resistance to *Aphanomyces euteiches* Isolates from the United States and France. *Phytopathology.* **2005,** *95,* 1287–1293.

61. Kushwaha, C.; Srivastava, C. P.; Chand, R.; Singh, B. D. Identification and Evaluation of a Critical Time for Assessment of Slow Rusting in Pea against *Uromyces fabae. Field Crops Res.* **2007,** *103,* 1–4.

62. Barilli, E.; Satovic, Z.; Rubiales, D.; Torres, A. M. Mapping of Quantitative Trait Loci Controlling Partial Resistance against Rust Incited by *Uromyces pisi* (Pers.) Wint. in a *Pisum fulvum* L. Intraspecific Cross. *Euphytica.* **2010,** *175,* 151–159.

63. Rai, R.; Singh, A. K.; Chand R.; Srivastava C. P.; Joshi A. K.; Singh B. D. Genomic Regions Controlling Components of Resistance for Pea Rust Caused by *Uromyces fabae* (Pers.) De-Bary. *J. Plant Biochem. Biotechnol.* **2015,** *25* (2), 1–9. DOI 10.1007/s13562-015-0318-6.

64. Rai, R.; Singh, A. K.; Singh, B. D.; Joshi, A. K.; Chand, R.; Srivastava, C. P. Molecular Mapping for Resistance to Pea Rust Caused by *Uromyces fabae* (Pers.) De-Bary. *Theor. Appl. Genet.* **2011,** *123,* 803–813.

65. Feng, J.; Hwang, R.; Chang, K. F.; Conner, R. L.; Hwang, S. F.; Strelkov, S. E.; Gossen, B. D.; McLaren, D. L.; Xue, A. G. Identification of Microsatellite Markers Linked to

Quantitative Trait Loci Controlling Resistance to Fusarium Root Rot in Field Pea. *Can. J. Plant Sci.* **2011**, *91*, 199–204.

66. Lewers, K. S.; Crane, E. H.; Bronson, C. R.; Schupp, J. M.; Keim, P.; Shoemaker, R. C. Detection of Linked QTL for Soybean Stem Rot Resistance in "BSR 101" as Expressed in a Growth Chamber Environment. *Mol. Breed.* **1999**, *5*, 33–42.

67. Burnham, K. D.; Dorrance, A. E.; VanToai, T. T.; Martin, S. K. Quantitative Trait Loci for Partial Resistance to *Phytophthora sojae* in Soybean. *Crop Sci.* **2003**, *43*, 1610–1617.

68. Weng, C.; Yu, K.; Anderson, T. R.; Poysa, V. A Quantitative Trait Locus Influencing Tolerance to Phytophthora Root Rot in the Soybean Cultivar Conrad. *Euphytica.* **2007**, *158*, 81–86.

69. Han, Y. P.; Teng, W.; Yu, K.; Poysa, V.; Terry, A.; Qiu, L.; Lightfoot, D. A.; Li, W. Mapping QTL Tolerance to *Phytophthora* Root Rot in Soybean Using Microsatellite and RAPD/SCAR Derived Markers. *Euphytica.* **2008**, *162*, 231–239.

70. Li, X.; Han, Y.; Teng, W.; Zhang, S.; Yu, K.; Poysa, V.; Anderson, T.; Ding, J.; Li, W. Pyramided QTL Underlying Tolerance to Phytophthora Root Rot in Mega-Environments from Soybean Cultivars Conrad and Hefeng 25. *Theor. Appl. Genet.* **2010**, *121*, 651–658.

71. Tucker, D. M.; Saghai Maroof, M. A.; Mideros, S.; Skoneczka, J. A.; Nabati, D. A.; Buss, G. R.; Hoeschele, I.; Tyler, B. M.; St Martin, S. K.; Dorrance, A. E. Mapping Quantitative Trait Loci for Partial Resistance to *Phytophthora sojae* in a Soybean Interspecific Cross. *Crop Sci.* **2010**, *50*, 628–635.

72. Zhao, G.; Ablett, G. R.; Anderson, T. R.; Rajcan, I.; Schaafsma, A. W. Inheritance and Genetic Mapping of Resistance to Rhizoctonia Root and Hypocotyl Rot in Soybean. *Crop Sci.* **2005**, *45*, 1441–1447.

73. Arahana, V. S.; Graef, G. L.; Specht, J. E.; Steadman, J. R.; Eskridge, K. M. Identification of QTLs for Resistance to *Sclerotinia sclerotiorum* in Soybean. *Crop Sci.* **2001**, *4*, 180–188.

74. Vuong, T. D.; Diers, B. W.; Hartman, G. L. Identification of QTL for Resistance to Sclerotinia Stem Rot in Soybean Plant Introduction 194639. *Crop Sci.* **2008**, *48*, 2209–2214.

75. Iqbal, M. J.; Meksem, V. N.; Njiti, V. N.; Kassem, M. A.; Lightfoot, D. A. Microsatellite Markers Identify Three Additional Quantitative Trait Loci for Resistance to Soybean Sudden-Death Syndrome (SDS) in Essex × Forrest RILs. *Theor. Appl. Genet.* **2001**, *102*, 187–192.

76. Chankaew, S.; Somta, P.; Sorajjapinun, W.; Srinives, P. Quantitative Trait Loci Mapping of Cercospora Leaf Spot Resistance In Mungbean, *Vigna radiata* (L.) Wilczek. *Mol. Breed.* 2010, *28*, 255–264.

77. Young, N. D.; Danesh, D.; Menancio-Hautea, D.; Kumar, L. Mapping Oligogenic Resistance to Powdery Mildewin Mungbean with RFLPs. *Theor. Appl. Genet.* **1993**, *87*, 243–249.

78. Chaitieng, B.; Kaga, M.; Han, O. K.; Wang, X. W.; Wongkaew, S.; Laosuwan, P.; Tomooka, N.; Vaughan, D. A. Mapping a New Source of Resistance to Powdery Mildew in Mungbean. *Plant Breed.* **2002**, *121*, 521–525.

79. Humphry, M. E.; Magner, T.; McIntyre, C. L.; Aitken, E. A. B.; Liu, C. J. Identification of a Major Locus Conferring Resistance to Powdery Mildew (*Erysiphe polygoni* DC) in Mungbean (*Vigna radiata* (L.) Wilczek) by QTL Analysis. *Genome.* **2003**, *46*, 738–744.

80. Kasettranan, W.; Somta, P.; Srinives, P. Mapping of Quantitative Trait Loci Controlling Powdery Mildew Resistance in Mungbean (*Vigna radiata* (L.) Wilczek). *J. Crop Sci. Biotech.* **2010,** *13,* 155–161.

81. Roman, B.; Satovic, Z.; Avila, C. M.; Rubiales, D.; Moreno, M. T.; Torres, A. M. Locating Genes Associated with *Ascochyta fabae*Resistance in *Vicia faba* L. *Aust. J. Agric. Res.* **2003,** *54,* 85–90.

82. Avila, C. M.; Satovic, Z.; Sillero, J. C.; Rubiales, D.; Moreno, M. T.; Torres, A. M. Isolate and Organ-Specific QTLs for Ascochyta Blight Resistance in Faba Bean (*Vicia faba* L). *Theor. Appl. Genet.* **2004,** *108,* 1071–1078.

83. Rubeena, P.; Taylor, W. J.; Ades, P. K.; Ford, R. QTL Mapping of Resistance in Lentil (*Lens culinaris* s sp. *culinaris*) to Ascochyta Blight (*Ascochyta lentis*). *Plant Breed.* **2006,** *125,* 506–512.

84. Ribaut, J. M.; Hoisington, D. Marker-Assisted Selection: New Tools and Strategies. *Trends Plant Sci.* **1998,** *3,* 236–239.

85. Rafalski, J.; Tingey, S. Genetic Diagnostics in Plant Breed: RAPDs, Microsatellites and Machines. *Trends Genet.* **1993,** *9,* 275–280.

86. Arus, P.; Moreno-Gonzalez, J. Marker-Assisted Selelction. In *Plant Breeding: Principles and Prospects;* Hayward, M. D., Bosemark, M. D., Romagosa, I., Eds.; Champman & Hall: London, 1993; pp 314–331.

87. Xu, Y.; Crouch, J. H. Marker-Assisted Selection in Plant Breeding: From Publications to Practice. *Crop Sci.* **2008,** *48,* 391–407.

88. Faleiro, F. G.; Vinhadelli, W. S.; Ragagnin, V. A.; Correa, R. X.; Moreina, M. A.; Barros, E. G. RAPD Markers Linked to a Block of Genes Conferring Rust Resistance to the Common Bean. *Mol. Bio.* **2000,** *123,* 399–402.

89. Miklas, P. N. Marker Assisted Selection for Disease Resistance in Common Beans. *Annu. Rev. Rep. Bean Improv. Coop.* **2002,** *45,* 1–3.

90. de Oliveira, E. J.; Alzate-Marin, A. L.; Borem, A.; de Azeredo, F. S.; de Barros, E. G.; Moreira, M. A. Molecular Marker Assisted Selection for Development of Common Bean Lines Resistant to Angular Leaf Spot. *Plant Breed.* **2005,** *124,* 572–575.

91. Diers, B. W. In *Soybean Genetic Improvement through Conventional and Molecular Based Strategies,* Proceedings of Second International Conference Legume Genome and Genetics, 2004; Dijon, France, 2004.

92. Mutlu, N.; Miklas, P. N.; Steadman, J. R.; Vidaver, A. K.; Lindgren, D. T.; Reiser, J.; Coyne, D. P.; Pastor-Corrales, M. A. Registration of Common Bacterial Blight Resistant Pinto Bean Germplasm Line ABCP-8. *Crop Sci.* **2005,** *45,* 806–807.

93. Yang, H.; Shankar, M.; Buirchell, B. J.; Sweetingham, M. W.; Caminero, C.; Smith, P. M. C. Development of Molecular Markers Using MFLP Linked to a Gene Conferring Resistance to *Diaporthe toxica* in Narrow-Leafed Lupin (*Lupinus angustifolius* L.). *Theor. Appl. Genet.* **2002,** *105,* 265–270.

94. Yang, H.; Boersma, J. G.; You, M. P.; Buirchell, B. J.; Sweetingham, M. W. Development and Implementation of a Sequence Specific PCR Marker Linked to a Gene Conferring Resistance to Anthracnose Disease in Narrow-Leafed Lupin (*Lupinus angustifolius* L.). *Mol. Breed.* **2004,** *14,* 145–151.

95. Faleiro, F. G.; Ragagnin, V. A.; Moreira, M. A.; de Barros, E. G. Use of Molecular Markers to Accelerate the Breeding of Common Bean Lines Resistant to Rust and Anthracnose. *Euphytica.* **2004,** *138,* 213–218.

96. Stavely, J. R. Pyramiding Rust and Viral Resistance Genes Using Traditional and Marker Techniques in Common Bean. *Ann. Rep. Bean Improv. Coop.* **2000,** *43,* 1–4.

97. Kelly, J. D.; Vallejo, V. A. A Comprehensive Review of the Major Genes Conditioning Resistance to Anthracnose in Common Bean. *Hort. Sci.* **2004,** *39,*1196–1207.

98. Walker, D.; Boerma, H. R.; All, J.; Parrott, W. Combining Cry1Ac with QTL Alleles from PI 229358 to Improve Soybean Resistance to Lepidopteran Pests. *Mol. Breed.* **2002,** *9,* 43–51.

99. Walker, D. R.; Narvel, J. M.; Boerma, H. R.; All, J. N.; Parrott, W. A. A QTL that Enhances and Broadens BtInsect Resistance in Soybean. *Theor. Appl. Genet.* **2004,** *109,* 1051–1957.

100. Ender, M.; Terpstra, K.; Kelly, J. D. Marker-Assisted Selection for White Mold Resistance in Common Bean. *Mol Breed.* **2008,** *21,* 149–157.

101. Singh, A. K.; Rai, R.; Singh, B. D.; Srivastava, C. P. Validation of SSR Markers Associated with Rust (*Uromyces fabae*) Resistance in Pea (*Pisum sativum* L.). *Physiol. Mol. Biol. Plants.* **2015,** *21,* 243–247.

102. Milkas, P. N. Marker-Assisted Backcrossing QTL for Partial Resistance to *Sclerotinia* White Mold in Dry Bean. *Crop Sci.* **2007,** *47,* 935–942.

103. Bernardo, R. Molecular Markers and Selection for Complex Traits in Plants: Learning from the Last 20 Years. *Crop Sci.* **2008,** *48,* 1649–1664.

104. Young, N. D. A Cautiously Optimistic Vision for Marker Assisted Breeding. *Mol. Breed.* **1999,** *5,* 505–510.

105. Koebner, R.; Summers, R. The Impact of Molecular Markers on the Wheat Breeding Paradigm. *Cell. Mol. Biol. Lett.* **2002,** *7,* 695–702.

106. Millan, T.; Winter, P.; Jungling, R.; Gil, J.; Rubio, J.; Cho, S.; Cobos, M. J.; Iruela, M.; Rajesh, P. N.; Tekeoglu, M.; Kahl, G.; Muehlbauer, F. J. A Consensus Genetic Map of Chickpea (*Cicer arietinum* L.) Based on 10 Mapping Populations. *Euphytica.* **2010,** *175,* 175–189.

107. Miklas, P. N.; Kelly, J. D.; Beebe, S. E.; Blair, M. W. Common Bean Breeding for Resistance against Biotic and Abiotic Stresses: From Classical to MAS Breeding. *Euphytica.* **2006,** *147,* 105–131.

108. Muehlbauer, F. J.; Cho, S.; Sarker, A.; McPhee, K. E.; Coyne, C. J.; Rajesh, P. N.; Ford, R. Application of Biotechnology in Breeding Lentil for Resistance to Biotic and Abiotic Stress. *Euphytica.* **2006,** *147,* 149–165.

109. Torres, A. M.; Roman, B.; Avila, C. M.; Satovic, Z.; Rubiales, D.; Sillero, J. C.; Cubero, J. I.; Moreno, M. T. Faba Bean Breeding for Resistance Against Biotic Stresses: Towards Application of Marker Technology. *Euphytica.* **2006,** *147,* 67–80.

110. Young, N. D.; Udvardi, M. Translating *Medicago truncatula*Genomics to Crop Legumes. *Curr. Opin. Plant Biol.* **2009,** *12,* 193–201.

111. Pilet-Nayel, M. L.; Muehlbauer, F. J.; McGee, R. J.; Kraft, J. M.; Baranger, A.; Coyne, C. J. Quantitative Trait Loci for Partial Resistance to Aphanomyces Root Rot in Pea. *Theor. Appl. Genet.* **2002,** *106,* 28–39.

112. McPhee, K. E.; Inglis, D. A.; Gundersen, B.; Coyne, C. J. Mapping QTL for Fusarium Wilt Race 2 Partial Resistance in Pea (*Pisum sativum*). *Plant Breed.* **2012,** *131,* 300–306.

CHAPTER 4

GENOMIC AND PROTEOMIC TOOLS FOR UNDERSTANDING MYSTERIOUS PROTEIN DIOSCORIN FROM *DIOSCOREA* TUBER

SHRUTI SHARMA and RENU DESWAL*

Department of Botany, University of Delhi, New Delhi 110007, India

*Corresponding author. E-mail: rdeswal@botany.du.ac.in; renudeswal@yahoo.co.in

CONTENTS

ABSTRACT

Yam (*Dioscorea* spp) is a tuberous crop extensively found in Eastern Africa and Asia constituting 6% of the world's tuber production. It serves as a nutritional supplement with a high amount of protein (2.8%). A lot of medicinal relevance is also associated with the tubers. It is used as herbal medicine for the cure of diseases such as diabetes, aging, neurodegenerative diseases, and cancers. Dioscorin, the major protein of *Dioscorea* is a multi-functional enzyme possessing carbonic anhydrase (CA), dehydroascorbate reductase activity (DHAR), and monodehydroascorbate reductase activity (MDHAR). Despite of its wide utility the tuber remains an orphan crop as its genome and proteome have not been analyzed in detail. Efforts are under way to obtain the complete genome sequences to improvise yam molecular breeding. The availability of new genomic markers would help to identify duplicates accessions, to conduct diversity analysis and association mapping. The combination of proteomics and genomics, proteo-genomics would be useful for refining the genome annotation using the proteomics data. This chapter will review the advances made in understanding the tuber proteomics and outline the direction for future research.

4.1   INTRODUCTION

Plants contribute to worldwide economic sustainability and security as the sources of food, feed and fuel. However, the increasing population has created a huge gap between the demand and supply of these resources. The conventional breeding strategies are handicapped to satisfy the huge gap. Therefore, the acceleration of plant productivity and nutritional security are of paramount importance. The plant biologists are paying close attention to the unexplored flora to look for solutions that can cater the nutritional requirements. The food material obtained from plants can be categorized either as the staple food or as fruits and vegetables. Root and tuber crops are subsidiary to the staple crop like cereals and legumes accounting for 20% of the world's production.[1] The tubers have unique ability to produce large amount of dietary energy. These are stable under conditions where other plants may not survive. Many of the world's poorest producers are dependent on the root and tubers as a contributing source of food and nutrition. These are looked upon as an alternative to meet the food demand as their nutritional coefficient is equivalent to that of staple crops. Five tuberous crops are considered as staple and account for 90% of the total tuber production.[2] These include potato, cassava, yam, taro, and sweet potato. The differences between potato, yam, taro, and sweet potato in terms of protein content and biological activities are shown in Table 4.1.

Yams are the third most cultivated tuber crop after potato and cassava.[3] However, they still remain the least explored tuberous crop in comparison with potato and cassava. More than 600 species of *Dioscorea* are known all over the world out of which only six are edible. The genus is widely distributed into the rhizomatous and tuberous crops.[4] The main taxa of the tuberous group are *Dioscorea rotundata*, *Dioscorea alata*, and *Dioscorea cyanensis*. Out of these *D. rotundata*, *D. caynensis* (guinea yam) that are the most abundant and economically important species while *D. alata* (water yam) in the most vastly distributed and agronomically stable species.[5] In the rhizomatous group the most studied taxon is the *Dioscorea toroko* owing to its pharmaceutical relevance. It is a staple food in many parts of the world and regarded as a famine crop. Nigeria accounts for 71% of the world's total yam production (Fig. 4.1).

The past decade marked a significant increase in yam production from 19 to 39 million tons.[6] Before summarizing the potential genomics and proteomics approaches for yam production, it is important to understand the tuber biochemistry and its wide application in pharmaceutical industry, to emphasize yam's significance.

TABLE 4.1   Comparison of the Four Most Cultivated Tuber Crops of the World.

	Potato (Solanum tuberosum)	Sweet potato (Ipomea batata)	Taro (Colocassia esculenta)	Yam (Dioscorea spp.)
Storage protein	Patatin (40 kDa)	Sporamin (25 kDa)	Tarin (GNA related lectin1, 40%) Trypsin inhibitor (GNA related lectin2, 40%)	Dioscorin (31 kDa)
Origin	Swollen stem	Swollen root	Swollen stem	Swollen hypocotyl
Protein content	2–3%	1–3%	1–4%	1–3%
PTMs	N-Glycosylation	O-Glycosylation	No report	Glycosylation Nitrosylation
Location	Vacuoles/ leaves	Vacuoles	Vacuoles	Vacuoles
Types	Class I & Class II. Class I is more abundant in tubers.	Sporamin A &B present in 2:1 ratio.	G1a, G1c, G1b & G1d G2a (24 kDa) G2b (22 kDa)	Dioscorin A & B
Isoforms	40–41 kDa	3	10	5–6
Associated enzyme activity	Esterase activity Acyl hydrolases β 1-3 glucanase (defense-related role) Polyphenol oxidase activity, GR, SOD, MDHAR, DHAR, APx, catalase	Kunitz type inhibitor DHAR MDHAR 1,1 diphenyl-2 picryhydrazl radicals Hydroxyl radical, protease activity	Agglutination of erthrocytes Trypsin inhibitor activity Lectin presents show defense-related role	CA, DHAR, MDHAR, TIA, amylase, 1,1-diphenyl-2-picryl-hydrazyl (DPPH), hydroxyl radical scavenging, peroxynitrite radical scavenging

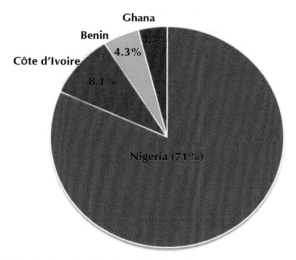

FIGURE 4.1   The pie chart showing the cultivation of *Dioscorea* in Africa (leading producer of yams).

4.1.1   BIOCHEMICAL PROPERTIES DIOSCORIN

Dioscorin is a storaenzyme is divided into two major groups: α-carbonic anhydrases (CAs) found in animals and β-CAs found in plants. The clustral analysis of humanα-CA with dioscorin showed the presence of identical amino acids' sequences indicating that CA in *Dioscorea* is more closely related with human than plants. Some other biological activities associated with dioscorin apart from CA are: dehydroascorbate reductase (DHAR), monodehydroascorbate reductase (MDHAR), trypsin inhibitor (TIA), and α-amylase.

4.1.1.1   CARBONIC ANHYDRASE

CA is ubiquitous in nature with extremely high turnover number[7] catalyzing the simple interconversion of carbon-di-oxide to bicarbonate. In mammals it plays an important role in respiration.[8] The actual role of the CA activity in plants is not known; however, it is anticipated that it is important in plant defense. The α-CA contain Zn ion co-ordinated by the imidazole rings of three histidine residues whereas β-CA has zinc at its active site. Five different families of carbonic anhydrases have been reported (Table 4.2) with no significant amino acid similarity and thus, presents an example of convergent evolution. The dioscorins from the crude extract of *D. alata,* *Dioscorea batatas,* and *Dioscorea pseudojaponicum* were shown to possess carbonic anhydrase activity[9] with no appreciable amount of zinc.[9]

TABLE 4.2   Distribution of Different Types of CA in the Living Organisms.

α-Carbonic anhydrase	Animals
β-Carbonic anhydrase	Plant chloroplast & prokaryotes
γ-Carbonic anhydrase	Methane producing bacteria
δ-Carbonic anhydrase	Diatoms

4.1.1.2   ANTIOXIDANTS AND MEDICINAL RELEVANCE OF DIOSCORIN

Dioscorin has been used from ancient times for making Chinese herbal medicine as it has antioxidant ability. Antioxidants are molecules that prevent the oxidation of other molecules. The oxidative damages to physiological substances are associated with diseases, such as aging, neurodegenerative diseases, and cancers.[10] Dioscorin exhibited the scavenging activity against both 1,1-diphenyl- 2-picrylhydrazyl (DPPH) and hydroxyl radical in a dose-dependent manner as reported by Hou.[11] It was observed that some cys-containing synthetic peptides derived from dioscorin hydrolysis show DPPH and hydroxyl radical scavenging activities.

Hypertension is a high prevalence chronic disease in the world. Angiotensin converting enzyme (ACE) inhibitors are being used in hypertensive patients and patients with other complications, like diabetes and chronic kidney disease. ACE is a part of the renin-angiotensin-aldosterone (RAA) system. It converts angiotensin I to angiotensin II (potent vasoconstrictor). Hsu et al.[12] reported for the first time that native dioscorin purified from *D. alata* exhibited ACE inhibitor activity in a dose-dependent manner. The IC50 of dioscorin on ACE was 6.404 μM compared to 0.00781 μM of captopril, the commercial ACE inhibitor. Hypertensive rats (SHRs) were fed with TN1-dioscorin and its peptic hydrolysates which showed significant lowering on mean blood pressure (MBP), systolic blood pressure (SBP), and diastolic blood pressure (DBP). These pressure-lowering effects were equal to that shown by the captopril group. These facts establish the medicinal relevance of dioscorin and its application in pharmaceutical industry.

4.1.1.3   DEFENSE-RELATED ROLE

It is a well-established fact that tuber storage proteins exhibit biological activities contributing to pest resistance, pathogens, or abiotic stresses tolerance.[13] For example, acyl hydrolase and esterase activity exhibited by patatin, antioxidant properties of dioscorin contribute to the pathogen resistance.[11] Chitinase and lectins are some other defense-related proteins present in tubers. In *Dioscorea japonica* fungicide and insecticidal property of chitinase have been established. Lectins are carbohydrate-binding proteins that are present in almost all living organisms. Most of the known plant lectins can be characterized into seven families of structurally and evolutionary related proteins. These seven families are legume lectins, amaranthine group of lectins, chitin-binding lectins, type 2 ribose-inactivating proteins, monocot mannose-binding lectins, cucurbitaceae phloem-binding lectins, and jacalin related lectins. From *D. batatas* four major proteins namely: DB1, DB2, DB3, and DB4 were isolated. DB1 is a mannose-binding lectin protein which accounts for 20% of the total protein. DB3 is a maltose-binding lectin protein. The biological role of the DB3 lectin is however not known. DB4 had chitinase activity incomplete.

4.1.2   NUTRITIONAL RELEVANCE

Apart from the medicinal relevance, the tuber also has many nutritional benefits associated with it. It has high protein content (2.8%) and low fat content (0.25%). Thus, it serves as a famine crop with high protein content and low fat (Fig. 4.2, Table 4.1).

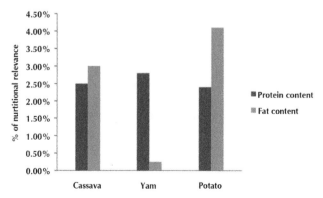

FIGURE 4.2   The protein and fat content of *Dioscorea* tuber as per the information provided by FAO.[14]

In spite of its multiple uses, its average yield is far below the potential yield. For instance, in Nigeria (the leading yam producer) the average yield is about 14% of its potential yield. The production of the crop is constrained by various biotic and abiotic factors. Thus, genetic improvement could make considerable contribution to minimize the yield gap in its production.

4.1.3 BREEDING YAM TUBERS

The conventional tools and methodologies have proved beneficial in developing improved varieties. The availability of genomic tools and resources has led to a new revolution in plant breeding as they facilitate the study of genotypes and its relationship with the phenotypes. Breeding of the tuber to obtain higher yield has gained importance as the food security has become the top most priority of researchers. Figure 4.3 provides the flow chart of potential yam breeding strategies.

Yam breeding strategy depends on control of seed production

↓

Seeds sown in the nursery

↓

First tubercles are produces

↓

1st Clonal selection done on the basis of traits with high heritability

↓

Selection on the basis of tuber yield is done at a later stage

↓

Elite clones reused as parents provided they produce flowers

FIGURE 4.3 A flow diagram to highlight the yam breeding strategy (adapted from Mignouna).[15]

4.1.3.1   TRAITS THAT CALL FOR ATTENTION

The breeding of yam has several constraints that include:

The long growth cycle (about 8–9 months). Dioecious plants that flower poorly. For improving the cultivation, the most important constraint is the flowering of yam tubers. Thus, the genetic diversity needs to be exploited in the non-flowering polyploidy genome, vegetative propagation, poor seed set and not much knowledge about the genetics of the tuber, tolerance to diseases, pests, and other abiotic factors.

Progress has been made in breeding yams using tissue culture techniques. Regeneration of whole plant using Murashige and Skoog medium (MS medium) modified with phyto hormones has now been established.[16] Somatic embryogenesis in the callus has been achieved from leaf, nodes, and isolated protoplast in *D. alata*.[17] There are also reports of sub cellular manipulations in *Dioscorea bulbifera* tuber. It has been successfully transformed using *Agrobacterium tumefaciens*by pre-incubating with wound substances from *Solanum* leading to crown gall tumor formation.[18] DNA fingerprinting was done to determine genetic variability within yam and its major fungal pathogens.

4.1.4   CONTRIBUTION OF GENOMICS IN YAM BREEDING

The genome sequences are important for understanding the functions of individual genes and their network. The combination of conventional breeding techniques with genomics tools has led to new genomics-based plant breeding. The knowledge of yam genomics is scarce in spite of its nutritional and medicinal relevance. This could be due to the absence of any convenient model system that can form the basis of yam genomics. The *Dioscorea* c-DNA values database contains c-value for 11 yam species. The Mbp value ranges from 466 in *Dioscorea togoensis* to 2352 in *Dioscorea villosa*. The huge difference is attributed to the varying ploidy level in *Dioscorea*species.[19] The basic chromosome number in yam species is x = 10 or x = 9. Diploids, hexaploids, and octaploids species are found but tetraploids are the most common. This is confirmed without doubt that ploidy is most common amongst *Dioscorea* species which contributes to the large difference in c-DNA values. Thus, making the genomic study complex. In an era where more than 50 genomes have been sequenced (*Arabidopsis* and rice the first model and crop plant, respectively, to be sequenced) while few others like lotus, potato have just been sequenced completely. *Dioscorea* remains one of the most under studied crop. Most of the information is available from the wild *Dioscorea* species for which the plastid genome has been sequenced.[20] This could be the reason that most of the accessions represent housekeeping genes or photosynthesis related genes that are non-nuclear in origin. Very few nucleic acid (1015) and protein sequences (1555) of *Dioscorea* species are available in public databases.[21] Major sequences are partial sequences of housekeeping genes obtained from organelle genome (chloroplast & mito-chondria). The most cultivated species of *Dioscorea* like *D. cyanensis, D. rotundata,* and *D. alata*have significantly low number of entries.

Molecular markers such as restriction fragment length polymorphism (RFLP), amplified fragment length polymorphism (AFLP), random ampli-fication of polymorphic DNA (RAPD), and simple sequence repeats (SSR) are used to examine the genetic diversity of the cultivated and wild yam species. Some progress has been made to develop molecular markers and to assess their potential for germplasm characterization and phylogenetic studies. SSR markers have been developed and used for the characterization of genetic diversity of *D. alata* germplasm.[22] These markers have been used to select genetically distant parents so as to maximize heterozygosity and heterosis in the progenies. Progress has been made in developing molecular tools and in generating linkage maps for mapping population for two most important species *D. alata* and *D. rotundata*.[15,23,24] Initially efforts were to develop polymorphic DNA markers. However, these are not cost efficient thus RAPD & AFLP were used as the molecular markers.[25,24] RAPD markers could be used for studying the genetic diversity but the polymorphism detected was low. Thus, AFLP were preferred for yam genome analysis and mapping. In *D. alata* 338 AFLP markers on 20 linkage groups and 107 AFLP markers on 12 linkage groups have been mapped. The development of genomic resources and technology is one of the major focuses in yam breeding. Efforts are being directed to develop additional genomic infor-mation as these gaps have hampered the gene characterization and genetic linkage mapping studies. The germplasm characterization would accel-erate the yam breeding programs. Thus, efforts are directed for expressed sequence tag (EST) development from the cDNA libraries. In an initial attempt EST from floral tissues have been developed.[26] However, most of these were either ribosomal or housekeeping genes. A collaborative project of International Institute of Tropical Agriculture (IITA) and University of Virginia aims to generate cDNA libraries from leaf tissue infected with *Colletotrichum gloeosporiodes.* It is a fungal pathogen responsible for yam anthracnose disease. The sequence of the cDNA clone was done to subsequently identify the ESTs with varied gene expression so as to aid in marker development. This resulted in the identification of more than 800,000 EST sequences of which around 1100 EST-SSR have been generated in *D. alata* for crop improvement. The data obtained from the ESTs would be helpful in comparative mapping effort, candidate gene discovery, and gene expression analysis. To date, several high yielding varieties with pest and disease resistance, wide adaptability, and good organoleptic attributes have been developed.

4.1.5 FUTURE PROSPECTS

4.1.5.1 COMPARATIVE GENOMICS—A WAY TO UNDERSTAND YAMS

The deficiency in the gene level knowledge could be overcome by adopting approaches such as resistance gene analogs (RGAs) that can be exploited to identify the plant defense-related genes.[27] Comparative genomics may assist in transferring the knowledge from highly studied model plants like *Arabidopsis* to the less studied plants like yams. One example of comparative genomics is the utilization of the genes involved in flowering in model plants to identify the homologous genes in garlic[28] and cauliflower.[29] The same approach could also be utilized for understanding the flowering-signaling pathway in yam. The phenomenon of dormancy is a persistent challenge in stored yams and could also be tackled with the help of comparative genomics intervention. Once the candidate genes are recognized, it becomes utmost important to develop microarray system for the functional gene analysis of yam. The generation of sufficient nucleotide sequence would pave the way for global gene expression via microarray. The current EST projects anticipate generating ample ESTs to build microarray chips for transcript analysis. Northern analysis and differential polymerase chain reactions (PCRs) have been used but with limited success.[15,23] Thus, it calls to search for reliable approaches to decipher the yam genomic information. In addition to the genomic information efforts are also being made for the transciptome profiling of yams based on differential gene expression. This would not only enrich the genomic resources but also contribute to the improvement in yam breeding. Therefore, efforts should be made to generate DNA, mRNA, and protein data for accelerated development of genomics/proteomics tools for analyzing the yam species.

4.1.5.2   REVERSE GENETICS—TILLING AND NEXT GENERATION SEQUENCING

The advances made in the study of functional genomics of model plants have capitalized the application of targeting induced local lesions ingenomes (TILLING) in yam research. Based on the knowledge in highly studied plants light could be shed to understand the genetic mechanism and pathways of key physiological traits in yams. Some traits that can be targeted by reverse genetics are resistance to diseases, like yam mosaic virus (YMC), flowering, and dormancy.

One of the main objectives of genomic breeding is the development of high throughput DNA sequencing technologies, collectively known as next generation sequencing (NGS) methods. These new technologies provide genome wide molecular tools for breeders that can be incorporated into existing breeding methods for accelerating the breeding process.

4.1.6   GRAY AREAS FOR YAM GENOMICS

Understanding the complexity of yam genome for improved yield and quality remains a big challenge. Some of the issues that need to be addressed include:

1. Construction of genetic maps of important cultivated species for breeding and germplasm enhancement.
2. Candidate gene identification must be geared up in order to use microarray and other approaches so as to identify genes involved in important agronomic traits.

Research to better understand the biology and agronomy of yam will facilitate efforts aimed at unlocking the hidden potential of yam germplasm. However, only genomics could not completely unravel the mystery. The rapidly growing field of plant proteomics could also prove to be an indispensible tool in understanding the complexity of this non-model plant—*Dioscorea.*

4.1.7 *PROTEOMIC ANALYSIS—A TOOL TO UNFOLD THE MYSTERY*

In the past, researchers have focused mainly on the functional genomics studies of the model plant species. Numerous non-model plants that are essential as food, feed, or energy source have been neglected. These may possess certain unique features that could not be analyzed using a model system, for example, woody plants are perennials and possess a long life cycle. They can survive in harsh conditions and provide a very good system to study stress tolerance genes.[30,31] The ability of all proteomics or transcriptomics approach for identification of candidate genes is lost due to lack of genomic information for such plants. Nevertheless, a proteomics approach has a great potential to study non-model plants.

This chapter basically aims to analyze the power of proteomics and genomics tools to improvise the breeding and yield of one of the non-model plant *Dioscorea*. It is important both as a food source and can be promoted as a commercial crop, but is still poorly characterized.

An alternative to the genomics approach is to study the end products of genes, the proteins. The sequences obtained from proteins are more conserved and by comparison with the well-known orthologous proteins could be of great help in understanding the non-model system. There are mainly two complementary approaches that can be used to explore the yam proteome—the gel-based approach and the gel free approach. The gel-based approach is a cornerstone that can be used to analyze the intact proteins and provide an overview of different isoforms and post translational modifications. The complete proteome study would provide a list of targets that would help in understanding the process of yam tuberization. This would help to exploit these targets for crop improvement. Till date, most of the proteomic-based studies have only focused on the purification and characterization of dioscorin.[32,33] Few attempts have also been made to study the various isoforms of dioscorin.[34,35] A comprehensive understanding of the process of tuberization, the biochemical changes associated with it, and also the changes at the proteome level is still incomplete. This knowledge would help in a better understanding of the system and would provide a direction for crop improvement and breeding.

On the other hand the gel free approach is totally automated and is coupled with the use of liquid chromatography (LC/MS/MS). It is a bottom up strategy where the proteins are first digested and then separated using a reverse phase column. But no qualitative and quantitative information of the protein isoforms can be obtained using the gel free technique. This approach has been applied to study the potato proteome but not the yams. However, there are some gel free quantitative proteomics techniques that have great potential to unravel low abundance proteins.

The major obstacle in the study of yam proteome is the non-availability of its complete genome information resulting in poor identification of the targets obtained using both gel-based or gel free approach.[36,37] The cross species identification is the sole option for protein identification. Another problem in studying the yam proteome is the high abundant storage protein that constitutes 85% of the major protein.[38] The fractionation of dioscorin and enrichment of the other proteins would greatly increase the proteome coverage and help in identifying more targets and characterizing them. The agronomic related targets can thus be identified and analyzed to improve the yam breeding.

4.1.8 GRAY AREAS FOR YAM PROTEOMICS

1. To devise an optimum extraction procedure to overcome the interfering components of the tuber like carbohydrates, phenolics, and so forth.
2. The fractionation of highly abundant proteins (HAPs), dioscorin, to improve the proteome coverage.
3. The analysis of the complete proteome would help in better understanding of the tuber physiology and the properties associated with it.

4.1.9 POTENTIAL APPROACH FOR YAM IMPROVEMENT

Yams remain an orphan crop till date with little genome and proteome information. To establish it as a main tuber crop, it is very important to understand the process of tuber growth and development. Proper staging must be done to assign morphological, biochemical, and molecular markers to each growth stage (Fig. 4.4). The assignment of markers would provide a schematic basis for detailed analysis of the process of tuberization. It would provide a new insight into dynamics of genome and proteome. It may provide certain potential target proteins/genes that can be exploited to improve the crop.

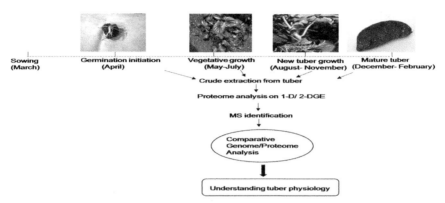

| Sowing
(March) | Germination initiation
(April) | Vegetative growth
(May–July) | New tuber growth
(August- November) | Mature tuber
(December- February) |

Crude extraction from tuber

↓

Proteome analysis on 1-D/ 2-DGE

↓

MS identification

↓

Comparative
Genome/Proteome
Analysis

↓

Understanding tuber physiology

FIGURE 4.4   An illustration to show the potential approach that can be used for future work.

4.2 CONCLUSION

To meet the increasing food demand, the viable approach is to adopt the innovative plant breeding strategies for yam that integrate the latest innovations in molecular technologies with conventional breeding practices. Given the present diversity, there is a need to use biotechnology and other botanical innovations for resolving poor flowering in elite clones of yams and to search for disease and pest resistance from wild species of yam. Yams are polyploidy species where separate male and female parents are present. Due to the slow multiplication time it becomes difficult to synchronize the male and female flowering time. Thus, molecular markers may help in selection of potential plants at an early stage of plant growth. These can also be used to elucidate the genetic control of yield potential, tuber quality, pest, and disease resistance genes. Efforts are under way to obtain the complete genome sequences and the development of additional genomic resources for improvising yam molecular breeding. The availability of new genomic markers would make it possible to fingerprint yam germplasm to identify duplicates accessions, to conduct diversity analysis and association mapping. The combination of proteomics and genomics, proteo-genomics would be useful for refining the genome annotation using the proteomics data. A cumulative effort would definitely help to improve the yam production.

KEYWORDS

- yam
- **breeding germplasm**
- **tuber crops**
- **antioxidants**

REFERENCES

1. Orkwor, G. C.; Asiedu, R.; Ekanayake, I. J. *Food Yams. Advances in Research*; IITA and NRCRI: Nigeria, 1998; p 249.
2. Shewry, P.; Napier, J. A.; Tatham, A. S. Seed Storage Proteins: Structures and Biosynthesis. *Plant Cell*. **1995**, *7*, 945–956.
3. FAO. *Production Year Book*; Food and Agriculture Organization of the United Nations: Rome, 1999; Vol. 53.
4. Barrau, J. Histoire Etprehis to Irehorticole de l'Oceanietropicale. *J. Soc. Oceaniste*. **1965**, *2*, 2155–2178.
5. Eastwood, R. B; Steele, W. M. The Conservation of Yam Germplasm in West Africa. *Plant Foods Man*. **1978**, *2*, 153.
6. FAO. FAOSTAT Agriculture Data. Food and Agriculture Organization of the United Nations. 2002. http://apps.fao.org/collections.
7. Tashian, R. E. The Carbonic Anhydrases; Widening Perspectives on Their evolution, Expression and Function. *Bioassays*. **1989**, *10*, 186–192.
8. Hewett-Emmett, D.; Tashian, R. E. Functional Diversity, Conservation and Convergence in the Evolution of the α, β and γ Carbonic Anhydrase Gene Families. *Mol. Phylogenet. Evol.* **1996**, *5*, 50–77.
9. Hou, W.; Lin, H. Dioscorins, the Major Tuber Storage Proteins of the Yams (*Dioscoreabatatas*), with Dehydoascorbate Reductase and Mono Dehydroascorbate Reductase Activities. *Plant Sci.* **1999**, *149*, 151–156.
10. Moon, J. K.; Shibamoto, T. Antioxidant Assays for Plant and Food Components. *J. Agric. Food Chem.* **2009**, *57*, 1655–1666.
11. Hou, W. C.; Lee, M. H.; Lin. Antioxidant Activities of Dioscorin, the Tuber Storage Protein of Yam (*DioscoreabatatasDecne*). *J. Agric. Food Chem.* **2001**, *49*, 4956–4960.
12. Hsu, F. L.; Lin, Y. H.; Lee, M. H.; Lin, C. L.; Hou, W. C. Both Dioscorin, the Tuber Storage Protein of Yam (*Dioscoreaalata cv. Tainong No. 1*), and Its Peptic Hydrolysates Exhibited Angiotensin Converting Enzyme Inhibitory Activities. *J. Agric. Food Chem.* **2000**, *50*, 6109–6113.
13. Shewry, P. R. Tuber Storage Protein. *Ann. Bot.* **2003**, *917*, 61–67.
14. FAO. *How to Feed the World in 2050, High-Level Expert Forum;* Food and Agriculture Organization of the United Nations: Rome, 2009.
15. Mignouna, H. D.; Mank, R. A.; Ellis, T. H. N.; van den Bosch, N.; Asiedu, R.; Abang, M. M.; Peleman, J. A Genetic Linkage Map of Water Yam (*Dioscoreaalata*L.) Based

on AFLP Markers and QTL Analysis for Anthracnose Resistance. *Theor. Appl. Genet.* **2002d**, *105*, 726–735.

16. Batygina, T. B.; Vinogradova, G. Phenomenon of Polyembryony. Genetic Heterogeneity of Seeds. *Russ. J. Dev. Biol.* **2007**, *38*, 126–151.

17. Kahl, G.; Ramser, J.; Kaemmer, D.; Kost, S.; Knobloch, I.; Rompf, R.; Huttl, B.; Geistlinger, J; Weigland, F.; Weisieng, K. In *Gene Technology and Yam Improvement*, Proceedings of the Biotechnology Workshop, IITA, Ibadan, Nigeria, 1990.

18. Dansi, A.; Mignouna, H. D.; Zoundjihékpon, J.; Sangaré, A.; Asiedu, R.; Ahoussou, N. Using Isozyme Polymorphism to Assess Genetic Variation within Cultivated Yam (*Dioscoreacayenensis/Dioscorearotundata* Complex) of the Benin Republic. *Genet. Resour. Crop Evol.* **2000a**, *47*, 371–383.

19. Bennett, M. D.; Leitch, I. J. Plant Genome Size Research: A Field in Focus. *Ann. Bot.* **2005**, *95*, 1–6.

20. Hansen, D. R.; Dastidar, S. G.; Cai, Z.; Penaflor, C.; Kuehl, J. V.; Boore, J. L.; Jansen, R. K. Phylogenetic and Evolutionary Implications of Complete Chloroplast Genome Sequences of Four Early-Diverging Angiosperms: *Buxus* (Buxaceae), *Chloranthus* (Chloranthaceae), *Dioscorea* (Dioscoreaceae), and *Illicium* (Schisandraceae). *Mol. Phylogenet. Evol.* **2007**, *4*(2), 47–63.

21. Mayer, E. S.; Michael, T. B.; Rotem, N.; Faigenboim, A. D.; Kosmala, A.; Perlikowski, A.; Amir Sherman, A; Kamenetsky, R. Garlic Development. *Front. Plant Sci.* **2015**, *6*, 271.

22. Arnau, G.; Nemorin, A.; Maldeon, E.; Abraham, K. Revision of Ploidy Status of *Dioscoreaalata L* (Dioscoreaceae) by Cytogenetics and Microsatellite Segregation Analysis. *Theor. Appl. Genet.* **2009**, *118*, 1239–1249.

23. Mignouna, H. D.; Mank, R. A.; Ellis, T. H. N.; van den Bosch, N.; Asiedu, R.; Abang, M. M.; Peleman, J. A Genetic Linkage Map of Guinea Yam (*Dioscorearotundata*L.) Based on AFLP Markers. *Theor. Appl. Genet.* **2000e**, *105*, 716–725.

24. Mignouna, H. D.; Ellis, N. T. H.; Asiedu, R.; Ng, Q. N. Analysis of Genetic Diversity in Guinea Yams (*Dioscorea*spp) Using AFLP Fingerprinting. *Trop. Agric.* **1998**, *75*, 224–229.

25. Dansi, A.; Mignouna, H. D.; Zoundjihékpon, J.; Sangaré, A.; Asiedu, R.;Ahoussou, N. Identification of Some Benin Republic's Guinea Yam (*Dioscoreacayenensis/Dioscorearotundata*) Cultivars Using Randomly Amplified Polymorphic DNA. *Genet. Resour. Crop Evol.* **2000b**, *47*, 619–625.

26. Debra, R.; Sayantani, G.; Zhengqiu, C; Cynthia, P.; Jennifer, V.; Boore, J.; Jansen, R. K. Phylogenetic and Evolutionary Implications of Complete Chloroplast Genome Sequences of Four Early-Diverging Angiosperms: *Buxus* (Buxaceae), *Chloranthus* (Chloranthaceae), *Dioscorea* (Dioscoreaceae), and *Illicium* (Schisandraceae). *Mol. Phylogenet. Evol.* **2000**, *45*, 547–563.

27. Moroldo, M.; Paillard, S.; Marconi, R.; Fabrice, L.; Canaguier, A.; Cruaud, C.; Berardinis, V. D.; Guichard, C.; Brunaud, V.; Clainche, I. L.; Scalabrin, S.; Testolin, R.; Gaspero, G. D.; Morgante, M.; Blondon, A. F. A. A Physical Map of the Heterozygous Grapevine 'Cabernet Sauvignon' Allows Mapping Candidate Genes for Disease Resistance. *BMC Plant Biol.* **2008**, *8*, 66.

28. Rotem, N.; Shemesh, E.; Peretz, Y.; Akad, F.; Edelbaum, O.; Rabinowitch, H. D.; Sela, I.; Kamenetsky, R. Reproductive Development and Phenotypic Differences in Garlic are Associated with Expression and Splicing of LEAFY Homologue galFY. *J. Exp. Bot.* **2007**, *58*, 1133–1141.

29. Lopez, A. B.; Van Eck, J.; Conlin, B.; Paolillo, D. J.; O'Neill, J.; Li, L. Effect of the Cauliflower or Transgene on Carotenoid Accumulation and Chromoplast Formation in Transgenic Potato Tubers. *J. Exp. Bot.* **2008,** *59,* 213–223.

30. Gion, J. M.; Lalanne, C.; Le Provost, G.; Ferry-Dumazet, H.; Paiva, J.; Chaumeil, P.; Frigerio, J. M.; Brach, J.; Barre, A.; De Daruvar, A.; Claverol, S.; Bonneu, M.; Sommerer, N.; Negroni, L.; Plomion, C. The Proteome of Maritime pine Wood Forming Tissue. *Proteomics.* **2005,** *5,* 3731–3751.

31. Carpentier, S. S.; Panis, B.; Vertommen, A.; Swennen, R.; Sergeant, K.; Renaut, J.; Laukens, K.; Witters, E.; Samyn, B.; Devreese, B. Proteome Analysis of Non-Model Plants: A Challenging but Powerful Approach. *Mass Spectom. Rev.* **2008,** *27,* 354–377.

32. Oluoha, U. Purification and Kinetic Characterisation of Lactate Dehydrogenase from *Dioscoreacyanensis. Biol. Plantarum.* **2001,** *44*(4), 535–539.

33. Hou, W.; Lin, H. X. Dioscorins, the Major Tuber Storage Proteins of the Yams (*Dioscoreabatatas*), with Dehydoascorbate Reductase and Mono Dehydroascorbate Reductase Activities. *Plant Sci.* **1999,** *149,* 151–156.

34. Tsai, P. J.; Lin, Y. H. 6th Federation of Asian and Oceanian Biochemisry Congress, Shanghai, 1992; p. 223.

35. Gaidamashvili, M.; Ohizumi, Y.; Iijima, S.; Takayama, T.; Ogawa, T.; Muramoto, K. Characterization of the Yam Tuber Storage Proteins from *Dioscoreabatatas* Exhibiting Unique Lectin Activities. *J. Biol. Chem.* **2004,** *279,* 26028–26035.

36. Hurkman, W. J.; Tanaka, C. K. Improved Methods for Separation of Wheat Endosperm Proteins and Analysis by Two-Dimensional Gel Electrophoresis. *J. Cereal Sci.* **2004,** *40,* 295–299.

37. Lester, P. J.; Hubbar, S. J. Comparative Bioinformatic Analysis of Complete Proteomes and Protein Parameters for Cross-Species Identification in Proteomics. *Proteomics.* **2002,** *2,* 1392–1405.

38. Harvey, P.; Boulter, D. Isolation and Characterization of the Storage Protein of Yam Tubers (*Dioscorearotundata*). *Phytochemistry.* **1983,** *22,* 1687–1693.

CHAPTER 5

SMALL RNAs-I: ROLE AS DEVELOPMENTAL AND ADAPTIVE REGULATORS IN PLANTS

SAURABH ANAND, NEER K. SINGH, and SANDIP DAS*

Department of Botany, University of Delhi, New Delhi110007, India

Corresponding author. E-mail: sdas@botany.du.ac.in; sandipdas04@ gmail.com

CONTENTS

ABSTRACT

RNA interference has emerged as a powerful approach to gain knowledge about the role and function of various genes in plants and animals. MicroRNAs (miRNAs) and short interfering RNAs (siRNAs) are two such small RNAs reported to be involved in silencing of genes involving gene/transcript–target complementarity. Small RNAs were discovered initially based on reverse and forward genetics approached as in *Petunia* and *C. elegans*; a number of subsequent discoveries were made based on direct cloning [selective ligation, purification, and sequencing of small RNA generated by dicer (characterized by 5'-mono-Phosphate and 3'-OH)], and now has gathered momentum with the advent of next generation small RNA sequencing technologies coupled with computational/bioinformatics prediction tools (e.g., miRU, miRNATools, findmiRNA), and comparative genomics strategies (e.g., miROrtho). The biogenesis of miRNAs involves single stranded RNA that forms typical hairpin precursors and are derived endogenously from host genome itself. SiRNAs originate from long dsRNAs involving precursors that are derived either from transposons, repeated elements or fromexogenously introduced transgenes or viruses. HEN1, HYL1, dicer/drosha, DCL1, SE, EXP, RdRP (RNA-dependent RNA polymerase), and AGO1, etc. are some of the critical players in biogenesis of miRNA and siRNA and their mutants (*dcl1, hen1-1,* and *hyl1*) that affect small RNA biogenesis and accumulation exhibit developmental defects. Small RNAs are reported to be involved in developmental regulation in plants via mRNA degradation, translational repression, and chromatin modification through auto-silencing or hetero-silencing. In plants, some developmental traits regulated by miRNA include seed germination (miR159), leaf development (miR164, miR165), root development (miR160, miR166, miR399), phase change/floral transition (miR156), flower development and timing (miR172), and tuberization (miR172). Small RNAs are also found to be involved in important adaptive roles such as plant responses to different types of biotic and abiotic stress (miR156, miR159, miR167, miR390, miR393, miR398, miR399, and several others). In addition to summarizing the discovery, biogenesis and role of small RNAs in development and adaptation, the chapter also highlights cross-talk between the various pathways involving miRNAs, and siRNA and miRNA.

5.1   INTRODUCTION

RNA silencing is a sequence-specific RNA regulatory mechanism that was initially discovered and termed as "post-transcriptional gene silencing" or "co-suppression" in plants, RNA interference (RNAi) in animals, and quelling in fungi[1]; subsequently, it has also been proven to mediate chromatin remodeling via DNA/cytosine methylation and histone modification, thus expanding its scope of regulation of RNA to also include DNA. Several functional genomics and computational based approaches have been employed to discover, investigate, and understand the diverse mechanisms involved in processing, biogenesis, mode of action, and different roles of various classes of small RNAs.[3] Among the various classes, microRNA (miRNA) and short interfering RNA (siRNA) have been widely studied to understand pathways of biogenesis and mechanism of action.[4] During the last decade, RNAi has also proven to be an extremely useful tool for elucidating gene function.[2]

Interestingly, the role of small RNA as a mobile signal that has a direct effect on the developmental patterning in plants has been unraveled using grafting experiments.[5–7] Such mobile silencing signal composed of small RNAs can have effects in long distances, away from source cell, exerting non-cell autonomous RNA silencing.[5,7,8] As revealed through high throughput sequencing, recipient cells of these trigger responses show epigenetic changes as an effect of the movement of these endogenous small RNAs.[8,9]

The two major classes of small RNAs are siRNA and miRNA, which differ primarily with respect to their origin and biogenesis.[5] siRNAs are generally[21–22] nucleotides that originate from perfect dsRNA, and are involved in silencing by guiding the cleavage of target genes with perfect complementary sequences.[10–12] It serves as an immune system to protect the genome against the effects of invading nucleic acids (i.e., viruses, transposons, or transgenes) or to stabilize the different states of chromatin.[13,14] Because the source and the target of the siRNA are often the same, mechanism of silencing is termed as auto-silencing.

miRNAs, on the other hand, can range between 19 and 25 nucleotides and are derived from single-stranded RNAs that form hairpin with imperfect pairing and are involved in post-transcriptional gene silencing (PTGS) by guiding the cleavage of target mRNAs;[4] because the miRNA and the target mRNA are produced from distinct and discrete genomic loci, the silencing is often termed as hetero-silencing. The pairing between miRNA and mRNA can allow two or three mismatches during complementary base pairing.[15] miRNAs regulate their targets either by cleaving the mRNA (transcript cleavage), through translation repression, or through chromatin remodeling.[15] Plant miRNAs show a higher degree of sequence specificity and allow less mismatches to their targets as compared to miRNAs that have been characterized from animal systems[15,16]; this is also correlated to their mechanism of action as transcript cleavage is the favored mode of action in plants compared to translational repression in animals.[15,17] Another major difference between mode of action of plant and animal miRNA is that plant miRNAs can target the duplicated members of same gene family while animal miRNAs can have multiple targets.[17] Various studies that have relied upon *in-silico* analysis and experimental data suggest conservation of miRNAs among different kingdoms. In plants, miRNAs that are conserved and present in all clades ranging from bryophytes to angiosperms have been identified.[18] Several approaches have been applied to identify plant miRNAs; these include analysis of mutants using forward and reverse genetic screens, sequencing, and computational prediction of small RNAs enriched libraries; other approaches include cloning and generation of small RNA libraries, use of computational algorithms to find homologous miRNA sequences from various taxa.[16,18] miRNA sequences which are conserved between different species are likely to be involved in the basic developmental processes while non-conserved miRNAs are found to be taxon specific has led to the hypothesis that these might have evolved much recently and are performing lineage or taxon-specific roles and conferring adaptive advantages.[4,5,16]

This chapter summarizes the knowledge on discovery, comparative account of biogenesis of small RNAs, their roles in various plant developmental pathways, and responses to different biotic and abiotic stress responses.

5.2   DISCOVERY

In 1990, the phenomenon of silencing was first reported in Petunia by Napoli and Jorgensen.[19] In an attempt to generate petunias with violet petals by over-expressing chalcone synthase (CHS), a key enzyme in flavonoid biosynthesis and a rate-limiting enzyme in anthocyanin biosynthesis, transgenic flowers having various degrees of silencing, were obtained. Silencing of the CHS was manifested in form of colorless or white petunias, and the "extent" of colorless zones was positively correlated to the copy number of the introduced copies of CHS. This observation led them to hypothesize and coin the phenomena as "co-suppression" of endogenous *CHS* gene as an effect of insertion of transgene; subsequently,[2] Matzke coined the term "homology dependent gene silencing (HDGS)".[20]

In 1992, Romano and Macino reported a similar phenomenon in *Neurospora crassa* and called it "quelling" of the endogenous gene.[21] An attempt was being made to produce carotenoid pigment in *N. crassa* mycelium and conidia through insertion of *albino-1 (al-1)*, *albino -2 (al-2)*, and *albino -3 (al-3)* genes. al-1, al-2, and al-3 encode phytoene dehydrogenase, phytoene synthase, and geranylgeranyl pyrophosphate synthase, respectively. As these genes are involved in carotenoid production that imparts orange color, any silencing of these could be immediately visualized. When *N. crassa* was transformed with any of the three genes, differential levels of silencing was observed.[21] The generalized nature of silencing was demonstrated when other genes such as qa-2 (responsible for quinic acid dehydrogenase),[22] *white collar* (*wc-1* and *wc-2*; transcription factors[23,24]), and *ad-9* (*adenine biosynthesis*[25] were also silenced upon introduction of foreign copies. This silencing phenomenon was termed quelling. Cogoni and Macino also identified "*quelling deficient/quelling defective*" mutant or qde through mutagenesis of an *albino* strain to identify reversible of silencing and genetic mapping.[22] Three classes of qde mutants were isolated and further analysis of the qde loci revealed that the gene product of qde-1 showed homology to RNA dependent RNA polymerase (RdRP);[26] qde-2 product was much difficult to identify when first discovered as it showed little similarity to then known proteins and was considered with a novel function;[27] however, qde-2 homologs in *Arabidopsis thaliana* encodes AGO1.[28,29] qde-3 loci was shown to encode RecQ-like DNA helicase.[30]

In animals, RNAi was first observed in *Caenorhabditis elegans* by Guo and Kempues. They observed degradation of *par-1* mRNA when sense and antisense RNAs to *par-1* were introduced. They suggested that the antisense RNA was causing inhibition and degradation of *par-1* by pairing with endogenous copies of *par-1* transcript but failed to provide a reason for silencing when sense strand of *par-1* was introduced.[31]

The proof of involvement of small RNA in silencing came in 1998, when Fire and Mello were able to explain these phenomena by identifying small temporal RNAs (stRNAs) encoded by the *lin-4* and *let-7* in *C. elegans* mutants exhibiting timing defects during larval development.[16]

Another important group of small RNA, the siRNAs, was first discovered in plants exhibiting transgene- or virus-mediated RNA silencing.[32]

5.3   BIOGENESIS

Biogenesis of small RNA requires either perfectly paired dsRNA in case of siRNA, or ssRNA that can fold with imperfect pairing to generate a hairpin like structure with stem and loop for miRNA.[2] In general, dicer, an enzyme with RNase III domains, acts on larger double-stranded RNA (dsRNA) precursors to generate small RNAs that are associated with gene silencing.[2,33] The sRNA duplexes, thereby formed after action of dicer, are reported to have 2-nucleotide 3'-overhangs and 5'-phosphates.[2] siRNAs are generated from a long dsRNA precursor having long duplex with perfect pairing.[18] In case of siRNA, several sources may contribute the initial ssRNA such as insertion of genome of ssRNA virus or availability of a "normal" transcript.[33] Both these can be converted into dsRNA by RdRP in a primer dependent or primer independent manner.[15,33] The conversion of ssRNA into dsRNA generates a perfectly paired dsRNA template for dicer activity which produces 20–21 nt short RNA.[5] dsRNA template can also be generated through aberrant transcription of inverted repeats (IRs) or if a coding region is flanked by two promoters on opposite strand such that a dsRNA template is produced.[34] The 20–21 nt dicer products unwind and can act as primers to drive subsequent rounds of dsRNA production by RdRP in a primer dependent manner, thus leading to an amplification cycle.[33,34] In animals, data from biochemical analysis revealed that multiple and closely present siRNAs (21 nt) are produced from exogenous, long, and perfectly complementary dsRNA molecules.[3] Function of these siRNAs is to behave as guides in RNA-induced silencing complex, a multi-protein complex called as RNA-induced silencing complex (RISC); and identify, pair complementarily and induce cleavage of target RNAs.[3,13,35] However, in plant data of biochemical data demonstrated that exogenous dsRNAs are transformed into two different classes of RNAs, 21 and 24 nt RNAs ; and 21 nt RNAs are associated with mRNA cleavage and despite being more abundant 24 nt RNAs have no role in mRNA degradation.[3,10,12] Additionally, 21, 22, and 24 nt RNAs procured from dsRNAs having promoter elements have critical roles in DNA methylation and transcriptional gene silencing.[36,37] siRNAs can be originated by transcription of a recombinant IR even in the absence of an RdRP.[37]

miRNAs, on the other hand, are transcribed from discrete genomic loci with their independent cis-regulatory units such as promoters and terminator sequences.[15] In animal systems, miRNAs are also known to be encoded by introns of protein-coding genes which imply sharing of cis-elements.[15,17] Several reports suggest that miRNA genes are transcribed by RNA polymerase II (pol II)[33]although a solitary report on transcription via RNA pol II also exists.[39] *miRNA* gene structures are like any other protein-coding genes and contain exon and intron, and are subjected to splicing;[33] in a few cases, alternative splicing of miRNA transcripts has also been observed.[40] RdRps are reported to show role in biogenesis of miRNAs in plants, fungi, and worms, but not in insects and mammals.[39]

Plant miRNAs are produced by the miRNA genes in the genome, they are processed in the same way as mRNA with cap and poly A tail; animal miRNAs are generally intron-derived or from untranslated regions or from primary transcripts.[37] Primary miRNA contains a stem loop precursor region having miRNA/miRNA* sequence and an unpaired loop.[2] In plants, studies using mutants have revealed that dicer-like1 (DCL1)(type III RNAse) along with HYL1 (dsRNA binding protein) and SERRATE (SE; C2H2 zinc finger protein) cleaves pre-miRNA. DCL1 and HYL1 co-localize in the sub-nuclear region termed as D-bodies, which might be involved in miRNA processing.[33] DAWDLE (DDL; an RNA binding protein) interacts with DCL1 to stabilize primary transcript, which is then cleaved by DCL1. HASTY (HST), the ortholog of exportin, exports the miRNA/miRNA* duplex to the cytoplasm.[40,41]

Argonaute (AGO) proteins, central component of RISC, are a part of a large superfamily and have been further sub-divided into three distinct clades—piRNAs binding PIWI clade, miRNAs and siRNA binding AGO clade, and a third clade that has only been associated with nematodes.[5,15] AGO effector protein in the RISC forms stable association with one of the two strands of dsRNA duplex.[29] This strand, named to be guide strand, is involved in target recognition by complementary base pairing, while the fate of the other strand, passenger strand, is also functional, in cases where both miRNA and miRNA* are known to target mRNAs.[42]

5.3.1 MIRNA BIOGENESIS

The biogenesis of miRNAs involves a series of tightly regulated events, each with unique biochemical signature.[15,33] Such events include transcription, dicing, nuclear export, and loading of the miRNA into an AGO protein to form the miRNA-induced silencing complex (miRISC),and these are mostly conserved between plants and animals.[17,33] The widespread and developmental roles of miRNAs are also indicative in the pleiotropic developmental defects displayed by the mutants defective in miRNA biogenesis and AGO[5,43–47] (additional evidences regarding the role of miRNA have been confirmed by numerous studies elucidating the functional aspects of miRNAs throughout plant development including embryo development,[48,49] organogenesis,[5] organ patterning,[48,50,51] phase transitions and control,[52,53] and reproductive development[54,55]).

miRNAs can be encoded from their gene sequences present either in the intergenic (between protein-coding genes) region or intragenic (with in protein-coding genes, from an exonic or intronic sequences) region.[43] However, most *miRNA* genes in plants are located in intergenic region as single units including independent cis-regulatory regions and transcribed by RNApol II;[56] in a number of instances, *miRNA* genes are also organized as poly-cistronic units and a single promoter can co-transcribe them.[57] Most of the *miRNA* genes in plants and animals have TATA box motif upstream of their transcription start sites (TSSs).[33] These *miRNA* genes are transcribed by RNA pol II, and undergo post-transcriptional processing such as splicing, 3'-poly-adenylation, and 5' capping as observed in protein-coding genes.[5,15,33]

miRNAs are synthesized from non-protein-coding *miRNA* genes as primary (pri)-miRNA transcript (~1 kb) that are characterized by multiple hairpin like structures or with complex intra-molecular pairing giving rise to multiple arms and loops;[58,59] the pri-miRNA is then processed by involvement of dicer enzyme, which generates a single hairpin like folded precursor (pre)-miRNA transcript with 5'cap and 3' poly A tail.[17,59,60] Animal pri-miRNA may contain multiple pre-miRNA while in plants one pri-miRNA forms single pre-miRNA.[17] miRNAs in animals are present in separate coding units which may often be clustered to form multiple miRNAs; however, some miRNAs are also encoded by intronic sequences and are termed as mirTRons.[15,61,62,63]

The processing of pre-miRNA to generate the mature miRNA:miRNA* can occur via two distinct mechanism (a) stem-to-loop processing and (b) loop-to-stem processing.[2,61] In animals, double-stranded stem loop of pri-miRNA which is then processed by drosha and pasha to form pre-miRNA.[59] In plants, pri-miRNA generated by RNA polII is stabilized by DDL , the RNA binding protein. Since the plants lack drosha, so the processing of pri-miRNA to precursor and from precursor to mature miRNA is performed by DCL1 (a plant homolog of dicer).[59] Due to sequential dicing events plant miRNAs exhibit greater diversity in their size and structure of the precursors which in turn determines the processing of mature miRNA.[61,62] Studies have shown the presence of 15 nt terminal stem below miRNA/miRNA* followed by an internal loop.[61] In stem- to-loop processing the first cut by DCL1 occurs at this terminal stem region and the second cut is produced at the 21 nt nucleotide from the first cut, thus releasing both terminal stem and loop region leaving mature miRNA; whereas, processing of long stem loop precursors like that of miR159 and miR 319 occur in loop-to-stem manner.[2] These long precursors consist of the terminal loop followed by the upper stem region that lies above the miRNA/miRNA*sequence and the lower stem with few internal loops.[2,61] In this, the first cleavage occurs in the loop region and the subsequent cuts take place at 20 to 22 nt intervals to release mature miRNA. The processing by DCL1 follows a loop-to-base or "loop first" strategy where DCL1 first cleaves closer to the loop region followed by a second cleavage closer to the base to release the mature miRNA duplex.[64] In contrast, stem-first processing involves the first cleavage occurring at the base of the stem followed by the second cleavage closer to the loop of the hairpin structure.[64] The entire process of generating miRNA duplex occurs inside the nucleus in plants in sub-nuclear regions termed as D-bodies.[65,66] SE and HYL1 proteins interact with DCL1 in their motif regions called dicing bodies (D-bodies) or SmD3/SmB-bodies in which stabilized pri-miRNAs are organized and thereby acts as nuclear processing centers.[65–69] Once the mature miRNA duplex is generated, HUA enhancer1 (HEN1), a 3′-methyltransferase, catalyzes 2'-O-methylation at the 3' nucleotides of each strand.[70,71] Methylation of miRNA/miRNA* duplex at the 3'-end with the formation of 2 nt overhang serves as the biochemical requirement and signal for dicer cleavage activity.[70] In addition, methylation is a critical step in miRNA biogenesis as it prevents miRNA uridylation and degradation from exonuclease activity of the small RNA degrading nuclease (SDN) family.[72,73] After methylation the miRNA duplex is exported out into the cytoplasm with the help of nuclear export protein HST, the ortholog of exportin5.[66,74] miRNA strand known as guide strand, one of the two strands of miRNA/miRNA* duplex, is inducted into the RISC, while the passenger strand or miRNA* atrophies.[37,38,58,74,75] Incorporation of a specific strand of a miRNA/miRNA* duplex into AGO-containing RISC is determined by following factors.[75–77]

i)   differential thermal stability of the miRNA/miRNA* duplex
ii) preferential binding of AGO1 to miRNA containing 5′terminal uridine residues
iii) activity of HYL1/DRB1, accessory proteins in D-bodies

miRNA* is short lived because of its quick degradation as compare to miRNA.[76] Once in the cytoplasm, the miRNAs are loaded onto AGO1 where it binds to complementary mRNA sequence for PTGS.[29,76] Unlike animals, miRNA/miRNA* duplex shows high degree of sequence-specific binding in plants.[75] miRISC contains AGO family protein, reported to have two conserved RNA binding domains; an N-terminal PAZ domain that can bind to the 3-end of single-stranded RNAs, and a PIWI domain (structurally similar to RNase H) in C-terminus, showing the endonuclease activity of RISC.[29,78,79] AGO family has 10 AGO members, among which AGO1 has been found to be one of the most relevant in regulating biogenesis and processing miRNA in plants.[29,80,81]

The biogenesis of miRNA in animal systems differs from that of plants in several respects.[17] First, nearly 30% miRNAs in animal systems are intron encoded and are termed as mirTRons[82] in contrast to a few examples that have been discovered so far in Arabidopsis and rice.[83,84] Presence of mirTRons may allow the miRNAs to utilize the cis-regulatory elements of the host protein-coding genes, thus ensuring co-expression of mRNA and miRNA.[33] So far mirTRons have been reported from mammals (humans, mice), nematodes (*C. elegans*),and flies (*Drosophila melanogaster*), but it appears to be universal in occurrence in animal kingdom but not evolutionary conserved.[33] Second, apart from being transcribed by RNA pol II, some of the animal miRNAs are transcribed by RNA pol III.[39] A survey of the miRBase and literature also reveals that the pre-miRNA in plants (*A. thaliana*) ranges from 60 up to 700 nt; pre-miRNA in animals are generally between 60 and 80 nt, thus displaying conservation in terms of length variation.[58,60] One of the features of animal miRNA biogenesis is partitioning of processing of pri-miRNA to pre-miRNA, and pre-miRNA to mature miRNA duplex between nucleus and cytoplasm.[18] The conversion of pri-miRNA to pre-miRNA is catalyzed by drosha coupled with DGCR8 (pasha) which are nuclear RNase III enzymes.[61,78] The pre-miRNA is then exported into the cytoplasm with the help of exportin5, where the cytoplasmic enzyme dicer catalyzes the processing of pre-miRNA to liberate the mature miRNA/miRNA* duplex. The miRNA is then loaded onto AGO2 which has slicer activity.[15,78]

mirTRons arise out of introns following splicing of introns using lariat de-branching enzyme (Ldbr) from pre-mRNA and thus, by-passes the canonical drosha mediated pathway.[63] Introns destined to give rise to mirTRons with 5' tail and 3'tail after splicing are acted upon by 5'–3' nuclease (yet unknown) or by 3'–5' exosome for trimming to generate the pre-mirTRons.[85,86] The presence and processing of mirTRons may thus imply their evolution much before the drosha dependent processing has been evolved.[84]

Each of the RISC is characterized by a specific type of AGO protein.[81] AGO proteins are part of a large evolutionarily conserved family and are found in plants, animals, fungi, protists, and some archaea.[87] For example, humans have eight AGO proteins whereas Arabidopsis has 10, four in *Drosophila*, 27 in *C. elegans* and only one in *Schizosaccharomyces pombe*.[87,88] The members of AGO family can be divided into three groups—AGO (class-1[89]), PIWI-AGO (P-element induced wimpy testis-AGO; class-2),[90] and WAGO (worm AGOs; class-3).[91] Four distinct domains are part of AGO proteins—N-terminal domain, PAZ domain, middle domain, and PIWI domain; of these four, the PAZ domain is responsible for binding to nucleic acid-RNA via the 3' end; the PIWI domain has an RNase H fold.[80,81,88] Depending on what kind of small RNA is bound, a RISC can be termed as miRISC or siRISC in *Drosophila*.[92]

The role of various AGO proteins in different small RNA biogenesis and gene silencing is still emerging, and seems to be related to the various dicer/DCL proteins.[43] Indeed the precise association between small RNAs, AGO and DICER/DCL are key to their biological role. For example, in *A. thaliana*, the 10 different AGO proteins seem to be associated with different classes of small RNAs.[76] DCL1 is involved in production of miRNAs and loading onto AGO1.[69,71] DCL2–4 is involved in biogenesis of a longer variant of small RNA which are loaded onto other AGO proteins.[93] AGO2 is bound to trans-acting siRNA (ta-siRNA) and repeat associated siRNA (ra-siRNA); AGO4 is involved in small RNA directed DNA methylation (RdDM) by associating with heterochromatic siRNA; AGO5 proteins are bound to siRNAs that have originated from various intergenic regions; and AGO7 is known to be involved in biogenesis of ta-siRNA.[94–101]

The process of transcription and processing of *miRNA* genes to generate the hairpin structure, and subsequent processing by dicer and other associated enzyme complexes are well understood; Table 5.1 lists some of the important the enzymes and their putative roles in the miRNA biogenesis process. Dicer cleaves the double-stranded pri-miRNA in polysomes within the cytoplasm into the single-stranded miRNA component of the RISC ready for final transition into the functional mature miRNA, heterochromatin maintenance and centromeric gene silencing have been associated with dicer function.[102,103] The subsequent regulation of target gene expression is accomplished by one of three methods[1]: direct cleavage of the target mRNA after binding to the complementary sequence in the 3' untranslated region (3'UTR),[2] cleavage-independent mRNA degradation, and[3] direct or indirect blockade of protein synthesis.[101] This section has been elaborated in the chapter elsewhere in this book (Neer et al.) and is thus not explained further.

TABLE 5.1 Enzymes Involved in MiRNA Biogenesis.

Protein	Structure	Function	Loss of function mutations	References
Dicer	One N- terminal helicase and two C- terminal RNaseIII domain	Nuclease carries out miRNA processing; processing of mirTRons	Completely defective in RNA interference (RNAi) and generation of microRNAs (miRNAs)	[103]
DCL (4 dicer homolog in plants i.e., DCL1, DCL2, DCL3, and DCL4)	containing PAZ (PIWI/AGO1/ ZWILLE) domain,	Involved in pri- and pri-miRNA production. Also process other small RNAs (rasi-, tasi-)	Double mutants *dcl1* and *dcl3* exhibit delay in flowering; dcl2 and dcl3 mutants developed normally; dcl4 and dcl3 dcl4 mutants had weak and severe rdr6 phenotypes, respectively	[33, 104]
HEN 1 (HUA ENHANCER 1)	Has two dsRNA-binding domains and a nuclear localization signal	Involved in methylation of the terminal sugar residues of miRNA	*hen1* mutants exhibit pleiotropic phenotypes such as reduced organ size, altered rosette leaf shape and increased number of co-florescences,	[105]
HASTY (Exportin 5 homolog in plants)	Ortholog of exportin 5/MSN5, interacts with RAN1; having N-acetylmethionine residue	Involved in transporting miRNA/miRNA* duplex from the nucleus to the cytoplasm	*hst* muants show many pleiotropic phenotypes; disrupting leaf shape, flower morphology, reducing fertility, accelerating vegetative phase change and disrupted phyllotaxy	[15, 33]
HYL1 (plant homolog of R2D2)	Two double-stranded RNA binding domains in its N-terminal half, closest homologs of RDE-4, a dicer associated protein	Regulate the first step of pri-miRNA processing	*hyl1* mutants are characterized by shorter stature, delayed flowering, leaf hyponasty, reduced fertility, decreased rate of root growth, and an altered root gravitropic response; sensitivity to auxin and cytokinin and hypersensitivity to abscisic acid (ABA).	[106]

TABLE 5.1 *(Continued)*

Protein	Structure	Function	Loss of function mutations	References
Argonaute	Have a PAZ ;PIWIdomain. PAZ domain may allow interaction with dicer and other proteins; help aligning miRNA with its target. Structure of PIWI domain is identical to RNAse H; involved in target cleavage	Involved in miRNA accumulation and PTGS.	*ago1* mutants exhibit numerous phenotypic abnormalities such as radialized leaves, abnormal infertile flowers with filamentous structures resembling small squid [thus the name Argonaute], and in some cases shoot apical meristem defects	[107]

5.4 CLASSIFICATION OF SMALL RNAS SIMILARITIES AND DIFFERENCES (DETAILED IN TABLES 5.2 AND 5.3)

5.4.1 MIRNAS AND SIRNAS

The following features are shared by both miRNAs and siRNAs.[79,94,107]

i) Single-stranded forms of both RNAs are found to be associated with effector complexes called RISCs.

ii) Dicer enzymes required by small RNAs to be excised from their precursors.

iii) AGO proteins required in association to form silencing function.

iv) Involved primarily in post-transcriptional regulation.

v) Target silencing based on Watson–Crick base pairing/complementary base pairing.

vi) Mechanisms of the silencing machinery are adaptive under change of circumstances, that is new miRNAs/siRNAs are synthesized and reprogrammed for action and older ones are discarded, establishing their role in defense responses

vii) Under new threats from novel invaders/foreign genes, these small RNAs can exploit these sequences themselves by co-opting them into the siRNA mechanism, thereby suppressing the invasive genes.

TABLE 5.2   Differences between miRNAs and siRNAs.

Types	miRNAs	siRNAs
Origin	Endogenous	Exogenous
Precursor	Stem-loop precursors	Long, fully complementary dsRNA precursors
Argounaute proteins involved	AGO1	AGO2
Site of biogenesis	Nucleus and cytoplasm	Cytoplasm
Functions	mRNA degradation, target cleavage Translational repression	mRNA destabilization and target cleavage

TABLE 5.3 Summary of Small RNA Comparison.[108,109]

Primary classifications	Secondary classifications	Tertiary classifications	Biogenesis
hpRNAs (includes small RNAs involving hair-pin shaped single-stranded precursors)	Other hpRNAs (other than miRNAs arising imprecisely processed precursors)		Two step process involving Drosha and dicer for regulation of hairpin precursors
	miRNAs (hairpin precursors to be precisely processed, giving rise to one or few miRNAs)	Lineage specific miRNA (restricted to one species or within few closely related species)	
		Long miRNAs (similar to heterochromatic siRNAs)	
siRNAs (includes small RNAs having precursor as dsRNA)	Heterochromatic siRNAs (originating from intergenic or repetitive regions associated with chromatin marks)		Cleavage of long endogenous dsRNAs by dicer
	Secondary siRNAs (small RNA trigger present upstream and RdRP affects precursor dsRNA synthesis)	Phased siRNAs (dsRNA precursor has a unique terminus)	
		trans-Acting siRNAs (having on or more targets away from the site of origin)	
	NAT-siRNAs (dsRNA precursor is synthesized due to the hybridization of independently transcribed RNAs complementary to each other)	cis-NAT-siRNAs (precursors are transcribed from overlapping genes in opposite direction)	
		trans-NAT-siRNAs (precursors are transcribed from non-overlapping genes whose mRNAs are complementary to each other)	

5.5 FUNCTIONS

5.5.1 *MIRNA-REGULATED PLANT DEVELOPMENT*

Several developmental pathways in plants involve small RNAs, especially miRNAs, such as in root formation and development,[97,110] vascular development,[111,112] leaf morphogenesis and polarity,[97,113] floral differentiation and development,[114] and developmental patterning during phase transition, that is vegetative growth to reproductive growth.[115,116] The functional importance of these miRNAs have been studied in detail by creating gain-in-function and loss-of-function of *miRNA* genes, mutating its complementary sites on target genes or genes involved in their biogenesis.[117] *dcl1* and *hasty* are important for plant miRNA biogenesis, so their loss-of-function causes abnormalities in leaf and flower development, such as altered leaf morphology, delayed floral transition, female sterility, and early stage embryo arrest.[79,118,119] Similarly, loss-of-function in the *hasty* causes of reduced levels of most miRNAs and results in several developmental defects, such as leaf shape alteration, deformed reproductive development, and defective vegetative phase transition.[120,121] Similarly, studies on miR164-resistant *cup-shaped cotyledon1* (*CUC1*) and *CUC2* resulting in abnormal embryonic, vegetative, and floral development led to the understanding of its role in controlling and regulating the boundary size in meristems and the formation and separation of different events during embryonic, vegetative, and floral organ development.[3,122] The abnormal embryonic, vegetative, floral development and a boundary enlargement phenotype caused by miR164-resistant versions of *CUC1* and *CUC2* support that miR164 has critical role in the control of boundary size in meristems and the formation and separation of embryonic, vegetative, and floral organs.[122]

Some miRNA families, strongly conserved across plant kingdom, regulate essential plant processes like modulation of leaf polarity (miR166) and floral organ identity (miR172).[79] These miRNAs regulate homologous mRNAs in basal plants having distinct vegetative and reproductive morphology, supporting the idea of conserved regulatory modules.[119]

5.5.1.1 *PHASE TRANSITION AND FLOWERING TIME CONTROL*

In plants, transition from vegetative phase to reproductive phase is a complex process involving integrated network of genes and small RNAs. Phase change and flowering initiation both are under control of various native transcription factors, growth regulators, environmental cues, and hormone signaling.[117,123] A key morphological feature of this phase transition is alteration in apical dominance, constancy in the number of rosette leaves, that is new leaves are not formed during phase transition, and conversion of vegetative meristem into reproductive/floral meristem.[123] Among some of the genetic factors, miRNAs have been shown to modulate these transition events as both positive and negative regulators.[118]

miR156 targets a group of genes encoding squamosa promoter-binding protein-like (SPL) transcription factors, such as SPL3, SPL4, and SPL5.[124] Analysis of overexpression mutants of miR156 in Arabidopsis,[125,126] rice,[127]and poplar[117,128] revealed late flowering and delayed vegetative phase change phenotype; while overexpression of miR156-resistant *spl3/4/5* genes causes early flowering in transgenics than wild type.[129] Expression analysis revealed that the endogenous level of miR156 is temporally regulated, which is very high in the early juvenile vegetative phase but decreases rapidly before the onset of the adult juvenile phase.[130] *SPL9*and *SPL15*genes, direct targets of miR156, are responsible for shoot maturation from shoot meristem.[124] Phase transition from vegetative to reproductive phase was found to be repressed by accumulation of SPL13, when miR156-resistant *spl13* (r*spl13*) was used to generate transgenic *A. thaliana*.[130] As the plant receives various cues (environmental, hormonal, nutritional) and progresses toward transition from vegetative to reproductive phase, miR156 level declines with a concomitant increase in SPL level that promotes flowering through activation of FT (flowering locus T), *MADS-box* genes (MCM1, AGAMOUS, DEFICIENS, SRF), and LFY (leafy).[129,131–133]

SPL is also known to induce the expression of another miRNA, miR172, which targets APETALA2 (AP2).[133–137] Since the phase transition is estimated by age and number of rosette leaves, in maize, *ap2* (target of miR172) mutant, *glossy15,* show mature leaves in early phases of development, confirming that the phase transition is strongly regulated by miR172.[138] The endogenous level of miR172 was detectable only when leaves started to enter in their mature phase and increased in number.[138,139] Activity of feed-forward loop of these genes and miRNAs ensures phase transition in irreversible direction.[117] In a recent study, miR172 has been shown to regulate phase transition and floral patterning by repressing the expression of *target of eat3* (*TOE3*) gene that encodes AP2 and AP2-like transcription factors.[140] Overexpression of the miR172-resistant *TOE3* gene results in delayed flowering.[140]

Based on the various experimental data, role of miRNA in guiding phase transition in plants has been shown to be conserved across a range of species including Acacia, Eucalyptus, Quercus, and Populus[128] and a general model proposes inverse relation between miR156 and miR172.[135] Levels of miR156 start to reduce as the plants move from juvenile to adult phase, with a concomitant increase in levels of miR172 which reaches its peak during flowering/flower development.[131,135,136] During phase transition, overexpression of miR172 causes upregulation of *FT* and floral meristem identity genes *LFY* and *AP.*[136]

In Arabidopsis, three members of miR159 family, miR159A, miR159B, and miR159C, regulate expression of three GA-*MYB* transcription factors*MYB33, MYB65,*and *MYB101*;[137,141–143] and this module of miR159-*MYB* has been shown to be involved in promoting flower induction.[137,144,145] Constitutive overexpression of miR159 leads to late flowering in transgenic plants as compared to wild type.[146]

Like miR159, miR319 family also has three members, mIR319a–c, which regulate the expression of *TCP* family members *TCP2, TCP3, TCP4, TCP10,* and *TCP24.*[137,147,148] Overexpression mutants of miR319 exhibit delayed flowering whereas loss-of-function mutants of TCP4 also generated late flowering phenotypes; showing that miR319 has a role in flowering time control.[137,147,149]

5.5.1.2   SEED DEVELOPMENT

Seed development in Arabidopsis involves distinct phases including seed and silique/fruit development and maturation, seed dispersal. Several studies have confirmed the role of miRNAs in embryogenesis and seed development by analysis of loss-of-function and gain-of-function mutants of the genes involved in biogenesis of miRNAs.[117,143,150–152, 153] For example, loss-of-function mutant of*DCL1–15* shows early maturation during seed development, due to increase in expression of two seed maturation genes *LEC2* and *FUS3.*[117,142] In 2010, Nodine and Bartel studied miRNA mediated regulation of embryo development in *A. thaliana* and showed that the *dcl1* mutant has higher levels of ~50 miRNA targets.[51] Nine of these genes encode transcription factors upregulated at the 8-cell-embryo stage and are predicted to be the targets of miR156, miR159, miR160, miR166, miR319, and miR824; additional six targets of miR168, miR393, miR400, or miR778 were upregulated at the globular stage of embryo development and patterning.[151]

Members of no apical meristem (*NAM*),*ATAF1/2*, and *CUC2* (*NAC*)-domain transcription factor family are targets of miR164, important for spatio-temporal regulation of embryo patterning in early embryogenesis.[3]

MiR164-resistant *CUC1* exhibits alterations in embryonic development.[3] Seeds of miR159abdouble mutant show reduced seed size and irregular shape.[141] miR159 regulates seed germination and development by controlling levels of its downstream transcription factor GA-*MYB33* and GA-*MYB101*.[142] Another miRNA, miR166, that targets homeodomain-leucine Zipper (HD-ZIP) transcription factors, shows seedling arrest at early maturation phases when overexpressed in Arabidopsis.[151]

In an initial study, 350 small RNAs from tomato fruit were characterized and most of them were found to be novel and unique.[152] miR1917, one of the tissue-specific miRNAs found in tomato fruit targets constitutive triple response 4 (CTR4), a member of the CTR family, reported to be negative regulator of ethylene responses required during tomato fruit maturation.[154]

miR156/157 is known to target colorless non-ripening (SBP homolog), whereas miR172 downregulates SlAP2 in tomato.[155] Overexpression mutants of miR156 exhibit vegetative and reproductive trait defects in tomato having smaller fruit and abnormal inflorescence structure; it further leads to nearly 70% lower fruit yield.[156] In grapevine, several tissue-specific (such as flower and fruit) and novel miRNAs have been described using transcriptome data and miR088 and miR096 have been reported to express in berries,[157] however, functional characterization of these miRNAs has to be experimentally validated.

In a recent study in barley, variation in interaction between alleles of AP2 and miR172 was shown to regulate the density of grains, thereby affecting the grain yield, suggesting the role of miR172 and AP2 in determining size and shape of spike.[158]

Evidences from transcriptomics and deep sequencing data have been helpful in identifying different miRNAs in seed development. For example, in a study on *Zea mays* using deep sequencing data, 125 miRNAs from 24 miRNA families have been identified and characterized during seed development; out of which 54 have been found to be novel, and miR319, miR166, andmiR167 show high expression in developing seeds.[159] In soybean (*Glycine max*),deep sequencing data reveal26 new miRNAs to be expressed in developing seeds, among which gma-miR398 and gma-miR1530 were highly expressed.[160] Deep sequencing data in *Brasssica napus* seeds unveiled 219 miRNAs (24 novel) from 45 plant miRNA families to be involved in seed development; and miR156, miR167, miR169, miR171, miR319, and miR396 were found to regulate lipid metabolism during embryonic development in Arabidopsis.[161] In a recent study in *B. napus,* overexpression mutants of miRNA394 show altered silique and seed development, seeds with higher contents of protein and glucosinolates and lower level of oil accumulation as compare to wild type.[162] However, studies to exactly pinpoint role of each miRNA in embryo and seed developments are lacking.

5.5.1.3   ROOT INITIATION AND DEVELOPMENT

Root system in plants is critical for anchorage, stability, and helps in uptake of water and nutrition. Some of the miRNAs that are involved in developmental aspects of root with hormonal cues such as auxin signaling[163] are miR164 targeting *NAC1*[3,164] for lateral root initiation, miR160 targeting *ARF10/ARF16/ARF17* in regulation of development,[165] miR390 targeting *TAS3-ARF*,[110,166] miR167 targeting *ARF8*, and miR393 targeting *AFB3* in root elongation.[167]

miR164 regulates lateral root emergence by targeting *NAC1* and auxin signaling,*NAC1* transcript levels were found to be high in loss-of-function miR164 mutants, which resulted in more lateral roots; while miR164 overexpression mutants had low *NAC1* transcript level, with concomitant fewer lateral roots.[164] miR160 targets three critical genes of auxin signaling pathway, *ARF10, ARF16,* and *ARF17*.[168] The expression levels of these genes were found to be increased in loss-of-function miR160Arabidopsis-mutants and such mutants were reported to have several developmental defects including root formation and growth.[168] Plants expressing miR160-resistant targets were found to have defects in root tips.[169] Two miRNAs, miR160 that targets *ARF10, ARF16,* and *ARF17* and miR167 that targets *ARF6* and *ARF8,* have been reported to have antagonistic roles in auxin regulation.[170] miR160 is induced by *ARF6* and represses *ARF17*. While, *ARF17* may activate miR167 to coordinate *ARF6* and *ARF8* expression and supports root growth and development.[167] These feedback circuits form an integrated network and unraveling of such inter-connected regulatory roles of different miRNAs are critical to understanding normal root development and architecture.[167] In another example, *miRNA165a* and *miR166b* genes upon activation of scarecrow-like (SCR) proteins repress the expression of plant transcription factors HD-ZIPIII for proper root development.[171,172]

Lateral root development is mediated by miR390, known to express at lateral root initiation sites.[110] miR390 targets three auxin response factors—ARF2, ARF3, and ARF4—identified as repressors of lateral root growth development; miR 390 negatively regulates the activity of these repressors and promotes lateral root growth.[110,160] miRNAs involved in nutrient signaling such as miR156, niR169, miR395, miR398, miR399, miR778, miR827, and miR2111 in phosphate signaling, miR164 and miR167 in nitrogen signaling, miR395 in sulfate metabolism, and miR398 in Cu/Zn homeostasis have been described to be affecting root growth in response to specific nutrient availability.[172]

5.5.1.4   LEAF DEVELOPMENT

Leaf development and morphogenesis have been studied in detail in context of regulation by small RNAs. A number of small RNAs have been shown to be involved in leaf morphogenesis, development, and senescence.[117] One such example is of miR396 in *A. thaliana,* which is a transcriptional regulator of plant-specific transcription factor—growth-regulating factor/GRF.[173] Overexpression mutants of miR396 in Arabidopsis show smaller leaf size while the miR396-resistant form resulted in bigger leaf size as compared to the wild type.[174,175] Overexpression of miR396 leads to lower expression of GRF2in the distal part of the leaf (the expression domain of GRF), causing attenuation in cell proliferation in developing leaves.[175]

Another miRNA, miR319, also called miRJAW, was shown to regulate a subset of *TCP* genes, a plant-specific transcription factor, and control leaf morphogenesis development.[53,176,177] Overexpression of miR319 led to leaves with crinkly margin and altered surface curvature. In several parallel studies, homologs of *TCP* (CIN) and miR319 have been shown to influence leaf shape and architecture in Antirrhinum[178] and tomato.[179] The overexpression of *MIR-JAW* in *jaw-D* mutants resulted in cotyledon epinasty, serration of leaf margins, flowering delay, and fruit formation defects.[53,123] Expression of miR319-resistant *TCP4* results in aberrant seedlings with fused cotyledons and without forming apical meristems.[53]

Another aspect of leaf morphogenesis is the correct adaxial–abaxial identity of leaf surface which is governed by activity of miR165/166 through direct regulation of *HD-ZIPIII* class of transcription factors, *REVOLUTA (REV), PHABOLUSA (HB), PHAVOLUTA (PHV), ATHB8,* and *ATHB15.*[119]

A legume-specific miRNA, miR3508, is expressed in the leaves and regulates *polyphenol oxidase (PPO)* genes post-transcriptionally. *PPO* is found to be on the thylakoids of chloroplasts and in vesicles or other bodies in nongreen plastids, and functions to catalyze the oxygen-dependent oxidation of phenols to quinones.[180] Loss-of-function mutants of miR3508 showed altered leaf morphology and defects in leaf patterning.[180]

In Arabidopsis, overexpression of miR156 leads to change in shape and leaf number.[145] Increase in leaf number was observed to be nearly 10 times higher in miR156 overexpression plants than that of wild-type plants, thereby increasing total plant biomass. As a consequence, delay in flowering time, decreased apical dominance, and more number of branches were observed in transgenic plants.[145] miR156 targets squamosa promoter-binding proteins (SBP) or SPL.[124,129,132,145,181,182] The role of miR156 in determining leaf characters was also conserved in *Panicum virgatum,* where transgenic plants show increase in leaf number and leaf biomass; such traits are considered as important for future prospects in biofuel production.[183] In addition, overexpression of miR156 exhibited decreased apical dominance and increase in branch number and leaf biomass in tomato and rice.[184]

5.5.1.5 VASCULAR DEVELOPMENT

Vascular development in plants is a coordinated event as it involves different structural modifications during the formation of provascular cells to procambium, from which, xylem and phloem are differentiated.[112,185] miR165/miR166 also are known to regulate vascular development in plants by targeting a subset of HD-ZIP III transcription factors in Arabidopsis— *ATHB8,ATHB15,PHABULOSA* (PHB), *PHV,* and*REV.*[112] Overexpression of miR166a led to the lower *ATHB15*mRNA levels and this resulted in early and rapid vascular cell differentiation from cambial/procambial cells, defects in xylem tissues and inter-fascicular cambium.[185] This miR166-mediated regulation of *HD-ZIP III* gene and vascular development has been found to be conserved across plants having differentiated vasculature.[108,119]

5.5.1.6 FLOWER DEVELOPMENT

miRNAs are involved in flower development by regulating both developmental timing and/or phase transition, and by controlling the genes involved in developmental pathway. Loss-of-function and overexpression mutants of the genes involved in these events exhibit altered phenotypes or defects in reproductive organ formation.[117] For example, overexpression mutants of miR172 (targeting *AP-2* like target genes) show early flowering and defects in floral organ identity.[136] Several *AP2*-like genes such as *TOE1, TOE2,* and *TOE3* genes in Arabidopsis, and *indeterminate spikelet1* (*IDS1*) and *GLOSSY15* in *Z. mays* are also regulated by miR172.[138] The loss-of-function *TOE1-1*mutants show early flowering and the *TOE1-1 TOE2-1* double mutant flowers mature much earlier than that of the wild type.[138] In contrast, overexpression of *TOE1* (*TOE1-1D*) results in delayed flowering, while overexpression of miR172 from the constitutive promoter induces early flowering, and overcomes the late flowering phenotype of *TOE1-1D*.[115,116]

Overexpression of miR319, which downregulates *TCP* mRNAs expression, forms plants with uneven leaf shape and delayed flowering time.[53] Overexpression of miR159a specifically reduces *MYB* mRNA accumulation and results in male sterility.[142,145]

Auxin signaling is reported to be closely associated with floral organogenesis. miR167 targets ARF6 and ARF8, which play a role in fertility in plants.[115,186] Overexpression mutants of miR167 exhibit defects in floral organ development such as aborted ovules and indehiscent anthers in Arabidopsis;[57] and shortened petals, stamens, and styles, aborted pollens and ovules in tomato.[186] In opium poppy, miR169b and miR169c target CCAAT-binding transcription factor (Cbf-b/Nf-ya) regulating flowering time, and miR171 regulates a plant-specific scarecrow-like transcription factor 6 (SCL6) involved in floral development.[187]

miR164 promotes boundary formation in a variety of organs through targeting plant-specific transcription factors *NAC*-related genes that regulate initial events of morphogenesis.[188] Out of these transcription factors, *CUC1* and *CUC2* regulate the organization of shoot apical meristem and axillary meristem and modulate formation of floral organs.[189] Experimental studies show that *CUC1* and *CUC2* transcripts are directly targeted by miR164.[3,122,190–196] miR164 also regulate the carpel fusion and number of petals during differentiation of floral organs.[190] In recent studies, two members of miR164 family, miR164a and mi4164b, have been found to be functionally redundant in flower development.[3,190,197,198]

miR165/166 regulates floral organ development and polarity as is evident from overexpression phenotype of miR165/166 that results in defective floral development.[151,199]

Another miRNA, miR444 in wheat, is reported to regulate flower patterning through regulation of transcription factor MADS box.[117,200]

5.5.2 MIRNAS INVOLVED IN STRESS TOLERANCE

Plants are affected by a range of abiotic stresses such as dehydration or drought, salinity, alkalinity, heat and cold, and biotic stresses, that is, pathogenic infections causing mild to severe lethal diseases. In response, plants show different adaptive mechanisms operating at transcriptional, post-transcriptional, translational, and post-translational levels to cope up with these conditions.[210,202] At post-transcriptional level, several miRNAs have been reported to be involved in regulating these adaptive responses.[117,202,203] Expression of several of these miRNAs under given stress conditions has been studied in detail and interestingly, their expression has been detected both as upregulated and downregulated. Plants, when subjected to extreme adverse conditions, exhibit yield and productivity loss. Studies on these miRNAs have helped in understanding their functioning and thereby can be effective in generating transgenics, advantageous for specific stress-tolerant traits. Table 5.4 summarizes role of different miRNAs involved in these adaptive responses.

5.5.2.1   DROUGHT AND SALINITY STRESS

Drought and high salinity stress are two major concerns that are an impediment toward sustaining and increasing agricultural productivity around the world. Numerous drought and high salinity stress responsive miRNAs have been identified in different models such as Arabidopsis[111,199] and rice[204] and crops such as soybean,[205] cotton,[206,207] barley,[201,208] Brassica,[209] wheat,[210, 211, 212, 213, 214] maize,[215] peach,[202] and switchgrass.[216] miR156, miR159, miR165, miR168, miR169, miR319, miR393, miR395, miR396, miR398, miR399, and miR402 were found to be induced under both drought and/or salinity in different plant species,[196] implying a conserved common pathway involved in management of two stress responses.

To date, several genes related to water-deficit stress conditions and several miRNAs have also been identified regulating some of these genes have been discovered.[217] To cite an example, *NFYA5* transcription factor is down regulated under water deficiency, promotes drought resistance in Arabidopsis, and is regulated post-transcriptionally by miR169.[218,219] Overexpression of *NFYA5* transgene in Arabidopsis induces stomatal closure under drought stress, significantly improving drought tolerance.[220] In rice, miR169 targets eight NF transcription factor Y subunit mRNAs, and is implicated in playing a role in managing drought tolerance.[220] Liu et al. reported miR167, miR168, miR171, and miR396 also to be involved in drought stress responsive miRNAs in Arabidopsis.[221] *miR167*, an abscisic acid (ABA) responsive gene, is found to be upregulated under drought stress.[221] In another example in Arabidopsis, under dehydration stress, ABA gets accumulated and promotes the activity of miR159.[222] In a recent study, miR169, miR319, miR397, miR118, miR393, and miR167 also have been described to be both ABA and drought responsive.[208,223] On the other hand, miR166 has been found to be downregulated in response to drought.[203,224]

TABLE 5.4 A Representative List of miRNA and Their Role in Plant Development.

miRNA	Target gene	Plant part involved	Plant	Role in plant development	References
miR390	TAS3, ARF	Root	Arabidopsis	Root development	[97, 110, 166]
miR164	NAC1,CUC2	Root, leaf, flower, seed	Arabidopsis	Lateral root emergence, leaf morphology, petal development regulation, floral meristem activity, seed development	[3, 122,164, 185, 191]
miR160	ARF10, ARF16	Root, seed	Arabidopsis, rice, barley	Lateral root initiation, seed development	[51, 165, 168, 201]
miR171	SCL6	Root, flower	Arabidopsis, rice, barley, opium poppy	Root and flower development	[165, 187, 201]
miR165/166	HD-ZIP III GENE FAMILY, PHB, PHV	Leaf, vascular system, flower, seed	Arabidopsis	Leaf development and morphology, meristem, seed development	[3, 108, 119, 151]
miR159	CIN, TCP, GAMYB	Leaf, seed, flower	Antirrhinum majus, Arabidopsis	Leaf morphogenesis, seed development and germination, flowering time regulation, reproductive development	[53, 76, 142, 185]
miR156 and miR157	SBP BOX CNR, SlAP2	Leaf, flower, seed	Arabidopsis, rice, poplar, tomato	Leaf morphogenesis, phase transition, flowering time regulation, seed development	[51, 124, 125, 129, 155]
miR396	GRF	Leaf	Arabidopsis	Leaf development	[174, 175]
miR3508	PPO	Root and leaf	Peanut	Root and leaf development	[180]
miR172	TOE1, TOE2, TOE3, IDS1, GLOSSY15	Flower	Arabidopsis, rice, maize	Flowering time, vegetative phase change	[114, 115, 138, 139]
miR169	CBF-B/NF-YA	Flower	Opium poppy	Flowering time	[187]

TABLE 5.4 *(Continued)*

miRNA	Target gene	Plant part involved	Plant	Role in plant development	References
miR167	ARF6, ARF8	Flower, root, seed	Arabidopsis, rice	Root elongation, gynoecium and stamen development, seed development	[165, 167, 170, 185]
miR444	MADS-BOX	Flower	Arabidopsis, rice	Flower development	[117, 200]
miR397 and miR528	AO	Seed	Rice	Seed development	[165]
miR319	TCP4, LANCEOLATE	Seed, leaf	Arabidopsis, tomato	Seed development, leaf development and morphology	[51, 53, 79, 177]
miR1530	TKT	Seed	Soybean	Seed development	[160]
miR1917	CTR4	Seed, fruit	Tomato	Fruit maturation	[154]
miR398 and miR1350	Not known	Seed	Soybean	Seed development	[160]
miR394	LCR	Seed	*Brassica napus*	Seed development and maturation	[162]
miR165a and miR165b	SCR	Root	Arabidopsis	Root initiation and development	[171, 172]
miR156	SBP	Leaf	*Panicum virgatum*, rice, tomato	Leaf development and leaf biomass	[183, 184]
miR444	MADS box	flower	Wheat	Flower patterning	[200]

Some of the crops have been extensively studied to understand the role of small RNAs involved in stress responses. In *Medicago truncatula*, miR398 and miR408 levels increase upon water deficiency which then causes down-regulation of *cox5b* and plantacyanin, respectively, thus playing regulatory role in drought stress.[219] In barley, miR156, miR166, miR171, and miR408 were observed to be differentially expressed upon dehydration.[202] In *Phaseolus vulgaris,* under drought stress and ABA treatment, miR159.2, miR1514, miR2118, and miR2119 levels were differentially expressed.[225] In poplar, miR102, miR156, miR162, miR167, and miR473 were upregulated under water scarcity conditions.[226] Overexpression of miR319 results in drought and salinity stress tolerance in transgenic plants generated in *Agrostis stolonifera* background.[133,196] Constitutive expression of miR394 in Arabidopsis leads to increase in drought tolerance due to decline in leaf water loss[227] and sensitivity to salinity and this process is mediated in an ABA-dependent manner.[228] According to miRBase data released on June 2013, 69 miRNAs from barley and 43 miRNAs from wheat have been described to be drought responsive.

Plants growing under salt stress conditions affect growth and leads to loss in productivity. miR396, miR168, miR167, miR165, miR319, miR159, miR394, miR156, miR393, miR171, miR158, and miR169 have been found to be expressed in response to salt treatment.[221] Overexpression mutants of miR396 in rice and Arabidopsis show reduced salt and alkali stress tolerance that to wild-type plants.[180] Levels of miR398, targeting two Cu/Zn superoxide dismutases (CSD1 and CSD2) show differential expression in response to salt stress.[229] In addition, miR395 targeting *APS1* gene shows regulatory role in sulfur accumulation and allocation.[230] In poplar, miR530, miR1445, miR1446, miR1447, and miR171 were found to be downregulated, while miR482.2 and miR1450 were upregulated under salinity stress.[231] In radish (*Raphanus sativus*), 49 known and 22 novel miRNAs have been found to be salt stress responsive; among which 41 (31 known and 10 novel) miRNAs are upregulated and 30 (18 known and 12 novel) miRNAs to be downregulated.[232]

5.5.2.2   NUTRIENT DEFICIENCY

Nutrient availability for the plant is a critical limiting factor for optimal growth and development. Plants have devised several metabolic and molecular mechanisms to cope up with any nutrient deficiency. Under given circumstances of nutrient-limitation in plants, miRNA-mediated regulation of gene expression of phosphate deficiency,[233–235] sulfate deficiency,[234,235] and copper deficiency[234,236,237]have been studied in detail.

miR156, miR169, miR395, miR398, miR778, miR827 miR828, and miR2111 have been reported to express differentially upon phosphate deprivation.[204]In addition, miR395, complementary to mRNA of ATP sulfurylase (APS) proteins, plays critical role in sulfur metabolism, as expression of miR395 is induced by low external sulfate concentrations.[234] Similarly, in response to low Cu^{2+} availability, miR398 regulates levels of *CSD* transcripts in order to restore Cu^{2+} ions for plastocyanin;[238] other miRNAs such as miR397, miR408, and miR857 have also been detected to be copper responsive.[236] miR393 targeting AFB3 and miR167 targeting ARF8 are responsible for controlling nitrogen metabolism and regulating root system architecture in plants.[167] In *M. truncatula*, miR169a targets *hap2-1* encoding a subunit of hetero-trimeric CCAATBOX-BINDING transcription factor expressing in nodule meristematic zone; overexpression of miR169a leads to suppression of *hap2-1*, resulting in ceasation of nodule development, thereby affecting nitrogen assimilation.[239]

5.5.2.3   COLD STRESS

Cold stress and frost damage severely affect plant productivity and growth. Knowledge on the regulation of adaptive mechanism in plants can be helpful in devising strategies to cope with cold stress. Several reports have appeared that have found miRNAs to be involved in regulating plant growth under low temperature stress. A total of 15 miRNA families have been described that are differentially regulated in response to cold stress.[240] In this study, expression of members of miR171 family was analyzed in response to cold stress. It was found that the various members of miR171 family are differently regulated; for instance transcript levels of pri-miR171a/e/f/i were downregulated, whereas pri-miR171c/d/h levels were upregulated.[240]

In separate studies, expression levels of miR165/166, miR169, miR172, miR393, miR396, miR397, miR402, and miR408 were detected to be increased, and miR398 to be decreased after exposing plants of *A. thaliana* to low temperature.[203,241]

In *Brachypodium*, 25 miRNAs, including a few novel miRNAs, displayed differential expression upon cold treatment.[242] Roots and shoots of sugarcane show decreased levels of miR319, however, strong induction was observed in plantlets.[196]In beans, one novel miRNA, miR2118, was recognized to be expressed under cold stress.[225]

5.5.2.4   OXIDATIVE STRESS

Under biotic and abiotic stress conditions, elevated reactive oxygen species (ROS) including hydrogen peroxide (H_2O_2) and free radicals are found in plant cells. In response to oxidative stress, expression level of miR398 was found to be decreased, as it targets copper/zinc superoxide dismutases (Cu/Zn-SODs), cytosolic *CSD1,* and chloroplast localized *CSD2,* and results in increase in the accumulation of CSD1 and CSD2 in Arabidopsis[243] and rice;[220] miR169, miR397, miR528, miR1425, miR827, miR319a.2, and miR408-5p were reported to be H_2O_2 responsive and expressed differentially in rice seedlings.[220] Evidences from high throughput sequencing confirmed that the expression levels of miR169, miR397, miR827, and miR1425 were detected to be increased whereas that of miR528 were decreased upon external treatment of H_2O_2.[244]

5.5.2.5   RESPONSE TO BIOTIC STRESSES

Plants respond to different pathogenic attack of bacteria, viruses, fungi, and insects by activating their defense mechanisms which otherwise severely affect crop yield and plant productivity. Role of miRNAs in resistance responses against any pathogenic infections has been unveiled in different plant systems.[197,245–250]

miR393 responds to bacterial inoculation in plants; as bacterial elicitor flg22 causes induction of miR393 expression and consequently suppression of auxin signaling.[246] Expression level of miR393 has been estimated to increase nearly 10-fold in Arabidopsis, when leaves were inoculated with non-pathogenic *Pseudomonas syringae* pv. *tomato* (*Pst*) DC3000 hrcC mutant, defective in type III secretion system (T3SS).[246,249] Expression levels of miR160 and miR167 were also detected to be elevated upon exposure to same non-pathogenic inoculation.[246]

Transcripts of miR156, miR158, miR164, and miR1885 have also been shown to be induced by bacterial infection in Arabidopsis.[251]

High throughput small RNA deep sequencing experiment using three different inoculations including *Pst* DC3000 hrcC (non-pathogenic), *Pst* DC3000 EV (virulent), and *Pst* DC3000 avrRpt2 (avirulent) was performed and 15, 27, and 20 differentially expressed miRNA families, respectively, were identified in Arabidopsis.[248] Levels of miR398 were found to be reduced upon treatment with bacterial pathogen causing oxidative stress, when Arabidopsis plants were inoculated with *Pst*a virulent strains (*Pst*

DC3000 carrying avrRpm1 and avrRpt2), but not with the virulent strain.[229] miR393* shows antibacterial response in plants by targeting and thereby causing suppression of expression of MEMB12, a golgi-localized SNARE protein. Suppression of expression of MEMB12 leads to increased exocytosis of pathogenesis-related protein PR1, and consequently, it inhibits pathogen growth.[252] In response to fungal pathogen inoculation, expression level of miRNAs was estimated in wheat infected with powdery mildew, and a total of 125 putative stress responsive long non-protein-coding RNAs (npc-RNAs) have been identified.[253] Out of these, miR675, miR167, and miR2004 have been identified to function in response to fungal infection.[117] miR482 targets nucleotide binding site-leucine-rich repeat (NBS-LRR) defense genes for regulating plant defense system against diseases.[254]

Table 5.5 is a comprehensive list of miRNAs that have been shown to be involved in stress responses.

TABLE 5.5   A Representative List of miRNA Involved in Stress Response in Plants.

miRNA	Target gene	Stress	Plant	References
miR156	SBP family TFs	Drought, salt, cold, fungal pathogen	Arabidopsis, barley, rice, Populus	[201, 221, 226, 231, 255]
miR159	*MYB33* and *MYB101, MYB33* and *MYB65* TFs	Drought, ABA, salt, bacterial pathogen	Arabidopsis, bean	[222]
miR162	DCL1	Drought, salt	Populus, maize, rice	[226, 255]
miR164	*NAC TFs*	Mechanical stress	Populus	[231]
miR165/ miR166	HD-ZIP III TFs, PHABULOSA, homeobox genes	Salt, cold, drought heat, fungal and bacterial pathogen	Arabidopsis, barley, soybean, wheat, maize	[201, 221, 226, 231, 240, 241, 248, 255]
miR167	ARF6 and ARF8 TFs	Drought, salt, cold, nutrient deprivation, bacterial pathogen	Arabidopsis, maize, rice	[202, 221, 226, 248]
miR168	AGO1	Drought, salt, cold, heat	Arabidopsis, rice, Populus, wheat	[221, 231, 255]
miR169	NF subunit Y, CCAAT-BOX Binding factors	Drought, salt, cold	Arabidopsis, rice	[204, 226, 242]
miR171	SCL TFs	Drought, mechanical, salt, cold	Arabidopsis, Populus, barley, rice	[201, 221, 222, 240]

TABLE 5.5 *(Continued)*

miRNA	Target gene	Stress	Plant	References
miR172	AP2 TF	Cold, drought	Arabidopsis, Populus, barley, rice	[216, 238, 249]
miR319	TCP family TFs	Salt, drought, cold, bacterial pathogen	Arabidopsis, rice	[221, 240, 248]
miR390, miR476	GLUTAMATE RECEPTOR proteins	Cold, bacterial pathogen, drought	Arabidopsis, Populus	[221, 238, 240, 248]
miR393	F-BOX protein, TRANSPORT INHIBITOR RESPONSE1 protein, tir1, afb2, and afb3 genes	Salt, heat, cold, ABA, drought, nutrient deprivation, bacterial and fungal pathogen	Arabidopsis, rice, wheat	[202, 221, 222, 226, 245]
miR394	LCR	Drought	Arabidopsis	[227]
miR395	APS	Nutrient deprivation, drought	Arabidopsis, rice	[15, 173, 255]
miR396	GRF family TFs	Drought, salt, cold	Arabidopsis, barley, rice	[180, 181, 201,221]
miR397b	LACCASEs	Cold, ABA, salt	Arabidopsis, rice, soybean, Brachypodium	[203, 204, 221, 242]
miR398	CSD1, CSD2, CCS1, COX5B.1	Nutrient deprivation, oxidative stress, bacterial pathogen	Arabidopsis, *Medicago*	[202, 219, 220, 237, 243, 248]
miR402	DML3, Argonaute 2	Cold, ABA, salt	Arabidopsis	[203, 241]
miR408	PLANTACYANIN	Mechanical stress, drought	*Populus*, barley, *Medicago*	[201, 219, 231]
miR474	PPRs	Cold	*Populus*	[238]
miR166	Homeodomain-leucine Zipper (HD-ZIP) TFs	Cadmium stress	Rice	[255]
miR528	F-BOX/ LRR-MAX2	Aluminum stress	Rice	[172]
miR825	REMORINA, ZF, homeobox	Bacterial	Arabidopsis	[256]
miR2118	Unknown gene	Drought, ABA	Arabidopsis	[225]

5.6   CONCLUDING REMARKS

The discovery of role of small RNAs such as miRNAs and siRNAs in mediating development and adaptation has completely altered our understanding of regulatory processes and networks. These small RNA molecules are the common element that unites gene regulation operating at the level of DNA (chromatin remodeling) and RNA (PTGS and translational repression). They are responsible for conferring protection not only against foreign invaders such as viruses, but also from internal genetic elements such as transposons or retrotransposons. A comparison between plant and animal miRNAs reveals that the basic biochemical and enzymatic machinery involved in biogenesis and function is conserved; however, differences do occur. For example, mirTRons are commonly found in mammals but rarely in plants; similarly, the biogenesis is differently partitioned in plants and animals. The role of miRNAs and siRNAs as "silencers" has been exploited by researchers to develop transgenic plants with both loss-of-function and gain-of-function phenotype for basic and applied studies. Indeed a number of such transgenic plants exhibit desirable traits such as stress tolerance or increased biomass or male sterility, and thus has opened new avenues to develop transgenic crop plants. Such plants are also being subjected to high throughput next-generation sequencing (NGS) based deep sequencing and detailed analysis using the data to unravel the regulatory network involved in various biosynthetic pathways and their underlying mechanisms. In spite of the tremendous advances made in gaining insights, a lot of efforts need to be made to fully understand the role played by small RNAs in order to harvest the complete benefits of gene silencing technologies.

KEYWORDS

- miRNA
- siRNA
- gene silencing
- PTGS
- development
- adaptation

REFERENCES

1. Agrawal, N.; Dasaradhi, P. V. N.; Mohmmed, A.; Malhotra, P.; Bhatnagar, R. K.; Mukherjee, S. K. RNA Interference: Biology, Mechanism, and Applications. *Microbiol. Mol. Biol. Rev.* **2003**, *67* (4), 657.

2. Bologna, N. G.; Schaphire, A. L.; Palatnik, J. F. Processing of Plant Microrna Precursors. *Brief. Funct. Genomics.* **2012**, *12*, 1–9.

3. Mallory, A. C.; Vaucheret, H. Micrornas: Something Important between the Genes. *Curr. Opinion Plant Biol.* **2004**, *7*, 120–125.

4. Xie, Z.; Johansen, L. K.; Gustafson, A. M.; Kasschau, K. D.; Lellis, A. D.; Zilberman, D.; Jacobsen, S. E.; Carrington, J. C. Genetic and Functional Diversification of Small RNA Pathways in Plants. *PLoS Biol.* **2004**, *2* (5), e104.

5. Chen, X. Small RNAs in Development – Insights from Plants. *Curr. Opin. Genet. Dev.* **2012**, *22* (4), 361–367.

6. Palauqui, J. C.; Elmayan, T.; Pollien, J. M.; Vaucheret, H. Systemic Acquired Silencing. Transgene-Specific Post-Transcriptional Silencing is Transmitted by Grafting from Silenced Stocks to Non-Silenced Scions. *EMBO J.* **1997**, *16*, 4738–4745.

7. Voinnet, O.; Vain, P.; Angell, S.; Baulcombe, D. C. Systemic Spread of Sequence-Specific Transgene RNA Degradation in Plants is Initiated by Localized Introduction of Ectopic Promoter Less DNA. *Cell.* **1998**, *95*, 177–187.

8. Dunoyer, P.; Himber, C.; Voinnet, O. DICER-LIKE 4 is Required for RNA Interference and Produces the 21-Nucleotide Small Interfering RNA Component of the Plant Cell-to-Cell Silencing Signal. *Nat. Genet.* **2005**, *37*, 1356–1360.

9. Molnar, A.; Melnyk, C. W.; Bassett, A.; Hardcastle, T. J.; Dunn, R.; Baulcombe, D. C. Small Silencing RNAs in Plants are Mobile and Direct Epigenetic Modification in Recipient Cells. *Science.* **2010**, *328*, 872–875.

10. Hamilton, A. J.; Voinnet, O.; Chappell, L.; Baulcombe, D. Two Classes of Short Interfering RNA in RNA Silencing. *EMBO J.* **2002**, *21*, 4671–4679.

11. Llave, C.; Kasschau, K. D.; Rector, M. A.; Carrington J. C. *Endogenous andsilencing-*Associated Small RNAs in Plants. *Plant Cell.* **2002**, *14*, 1605–1619.

12. Tang, G.; Reinhart, B. J.; Bartel D. P.; Zamore P. D. A Biochemical Framework for RNA Silencing in Plants. *Genes Dev.* **2003**, *17*, 49–63.

13. Hannon, G. J. RNA Interference. *Nature.* **2002**, *418*, 244–251.

14. Finnegan, E. J.; Matzke, M. A. The Small RNA World. *J. Cell Sci.* **2003**, *116*, 4689–469.

15. Bartel, D. P. MicroRNAs: Genomics, Biogenesis, Mechanism, and Function. *Cell.* **2004**, *116*, 281–297.

16. Reinhart, B. J.; Weinstein, E. G.; Rhoades, M. W.; Bartel, B.; Bartel, D. P. MicroRNAs in Plants. *Genes Dev.* **2002**, *16*, 1616–1626.

17. Millar, A. A.; Waterhouse, P. M. Plant and Animal MicroRNAs: Similarities and Differences. *Funct. Integr. Genomics.* **2005**, *5*, 129–135.

18. Axtell, M. J.; Snyder, J. A.; Bartel, D. P. Common Functions for Diverse Small RNAs of Land Plants. *Plant Cell.* **2007**, *19*,1750–1769.

19. Napoli, C.; Lemieux, C.; Jorgensen, R. Introduction of a Chimeric Chalcone Synthase Gene into Petunia Results in Reversible Co-Suppression of Homologous Genes in Trans. *Plant Cell.* **1990**, *2*, 279–289.

20. Matzke, A. J. M.; Neuhuber, F.; Park, Y. D.; Ambros, P. F.; Matzke, M. A. Homology-Dependent Gene Silencing in Transgenic Plants: Epistatic Silencing Loci Contain Multiple Copies of Methylated Transgenes. *Mol. Gen. Genet.* **1994**, *244*, 219–229.

21. Romano, N.; Macino, G. Quelling: Transient Inactivation of Gene Expression in *Neurospora crassa* by Transformation with Homologous Sequences. *Mol. Microbiol.* **1992,** *6,* 3343–3353.

22. Cogoni, C.; Macino, G. Isolation of Quelling-Defective (Qde) Mutants Impaired in Posttranscriptional Transgene-Induced Gene Silencing in *Neurospora crassa. Proc. Natl. Acad. Sci. USA.* **1997a,** *94,* 10233–10238.

23. Ballario, P.; Vittorioso, P.; Magrelli, A.; Talora, C.; Cabibbo, A.; Macino, G. White Colar 1, a Central Regulator of Blue-Light Responses in *Neurospora crassa. EMBO J.* **1996,** *15,* 1650–1657.

24. Linden, H.; Macino, G. White Collar 2, a Partner in Blue-Light Signal Transduction, Controlling Expression of Light-Regulated Genes in *Neurospora crassa. EMBO J.* **1997,** *16,* 98–109.

25. Cogoni, C.; Macino, G. Quelling: Transgene-Induced Silencing in *Neurospora crassa.* In *"Cellular Integration of Signalling Pathways in Plant Development"*; NATO ASI Series, Springer-Verlag: Berlin, Heidelberg, 1998; Vol. 104, pp 103–112.

26. Cogoni, C.; Macino, G. Gene Silencing in *Neurospora crassa* Requiresa Protein Homologous to RNA-Dependent RNA Polymerase. *Nature.***1999a,** *399,* 166–169.

27. Catalanotto, C.; Azzalin, G.; Macino, G. Cogoni, C. Gene Silencing in Worms and Fungi. *Nature.* **2000,** *404,* 24.

28. Fagard, M.; Boutet, S.; Morel, J. B.; Bellini, C.; Vaucheret, H. AGO1, QDE-2, and RDE-1 are Related Proteins Required for Post-Transcriptional Gene Silencing in Plants, Quelling in Fungi, and RNA Interference in Animals. *Proc. Natl. Acad. Sci. USA.***2000,** *97,* 11650–11654.

29. Bohmert, K.; Camus, I.; Bellini, C.; Bouchez, D.; Caboche, M.; Benning, C. AGO1 Defines a Novel Locus of *Arabidopsis* Controlling Leaf Development. *EMBO J.* **1998,** *17,* 170–180.

30. Cogoni, C.; Macino, G. Homology-Dependent Gene Silencing in Plants and Fungi: A Number of Variations on the Same Theme. *Curr. Opin. Microbiol.* **1999b,** *2,* 657–662.

31. Guo, S.; Kemphues, K. J. Par-1, a Gene Required for Establishing Polarity in C. Elegans Embryos, Encodes a Putative Ser/Thr Kinase that is Asymmetrically Distributed. *Cell.* **1995,** *81,* 611–620.

32. Eamens, A.; Wang, M. B.; Smith, N. A.; Waterhouse, P. M. RNA Silencing in Plants: Yesterday, Today, and Tomorrow. *Plant Physiol.* **2008,** *147,* 456–468.

33. Mishra, N. S.; Mukherjee, S. K. A Peep into the Plant MiRNA World. *Open Plant Sci. J.* **2007,** *1,* 1–9.

34. Foulkes, W. D.; Priest, J. R.; Duchaine, T. F. Dicer1, Mutations, MicroRNAs and Mechanisms. *Nat. Rev.* **2014,** 1–11.

35. Zamore, P. D. Ancient Pathways Programmed by Small RNAs. *Science.* **2002,** *296,* 1265–1269.

36. Mette, M. F.; Aufsatz, W.; van der Winden, J.; Matzke, M. A.; Matzke, A. J. Transcriptional Silencing and Promoter Methylation Triggered by Double-Stranded RNA. *EMBO J.* **2000,** *19,* 5194–5201.

37. Alan, H. J. Pathways Through the Small RNA World of Plants. *FEBS. Lett.* **2005,** *579,* 5879–5888.

38. Ha, M.; Kim, V. N. Regulation of MicroRNA Biogenesis. *Nat. Rev. Mol. Cell Biol.* **2014,** *15,* 509–524.

39. Borchert, G. M.; Lanier, W.; Davidson, B. L. RNA Polymerase III Transcribes Human Micrornas. *Nat. Struct Mol. Biol.* **2006,** *13,* 1097–1101.

40. Melamed, Z. A.; Levy, R.; Ashwal-Fluss, G.; Lev-Maor, K.; Mekahel, N.; Atias, S. G.; Sharan, R.; Levy, C.; Kadener, S.; Ast, G. Alternative Splicing Regulates Biogenesis of MiRNAs Located Across Exon-Intron Junctions. *Mol. Cell.* **2013**, *50*, 869–881.

41. Mette, M. F.; Aufsatz. W.; van der Winden, J.; Matzke, M. A.; Matzke, A. J. Transcriptional Silencing and Promoter Methylation Triggered by Double-Stranded RNA. *EMBO J.* **2000**, *19*, 5194–5201.

42. Guo, L.; Lu, Z. The Fate of MiRNA* Strand through Evolutionary Analysis: Implication for Degradation as Merely Carrier Strand or Potential Regulatory Molecule? *PLoS ONE.* **2010**, *5* (6), e11387.

43. Ha, M.; Kim, V. N. Regulation of MicroRNA Biogenesis. *Nat. Rev. Mol. Cell Biol.* **2014**, *15*, 509–524.

44. Liu, Q.; Yan, Q.; Liu, Y.; Hong, F.; Sun, Z.; Shi, L.; Huang, Y.; Fang, Y. Complementation of *Hyponastic* Leaves by Double Strand RNA-Binding Domains of Dicer-Like1 in Nuclear Dicing Bodies. *Plant Physiol.* **2013**, *163*, 108–117.

45. Chen, X.; Liu, J.; Cheng, Y.; Jia, D. HEN1 Functions Pleiotropicallyin *Arabidopsis* Development and Acts in C Function in the Flower. *Development.* **2002**, *129*, 1085–1094.

46. Clarke, J. H.; Tack, D.; Findlay, K.; Van Montagu, M.; Van Lijsebettens, M. The SERRATE Locus Controls the Formation of the Early Juvenile Leaves and Phase Length in *Arabidopsis. Plant J.* **1999**, *20*, 493–501.

47. Lu, C.; Fedoroff, N. A mutation in the *Arabidopsis* HYL1 Gene Encoding a dsRNA Binding Protein Affects Responses to Abscisic Acid, Auxin, and Cytokinin. *Plant Cell.* **2000**, *12*, 2351–2366.

48. Ray, S.; Golden, T.; Ray, A. Maternal Effects of the Short Integument Mutation on Embryo Development in *Arabidopsis. Dev. Biol.* **1996**, *180*, 365–369.

49. Bohmert, K.; Camus, I.; Bellini, C.; Bouchez, D.; Caboche, M.; Benning, C. AGO1 Defines a Novel Locus of *Arabidopsis* Controlling Leaf Development. *EMBO J.* **1998**, *17*, 170–180.

50. Grigg, S. P.; Canales, C.; Hay, A.; Tsiantis, M. SERRATE Coordinates Shoot Meristem Function and Leaf Axial Patterning in *Arabidopsis. Nature.* **2005**, *437*, 1022–1026.

51. Nodine, M. D.; Bartel, D. P. MicroRNAs Prevent Precocious Gene Expression and Enable Pattern Formation during Plant Embryogenesis. *Genes Dev.* **2010**, *24*, 2678–2692.

52. Chen, X. A MicroRNA as a Translational Repressor of APETALA2 in *Arabidopsis* Flower Development. *Science.* **2004**, *303*, 2022–2025.

53. Palatnik, J. F.; Allen, E.; Wu, X.; Schommer, C.; Schwab, R.; Carrington, J. C.; Weigel, D. Control of Leaf Morphogenesis by MicroRNAs. *Nature.* **2003**, *425*, 257–263.

54. Wang, J. W.; Czech, B.; Weigel, D. Mir156-Regulated SPL Transcription Factors Define an Endogenous Flowering Pathway in *Arabidopsis thaliana. Cell.* **2009**, *138*, 738–749.

55. Wu, G.; Park, M. Y.; Conway, S. R.; Wang, J. W.; Weigel, D.; Poethig, R. S. The Sequential Action of Mir156 and Mir172 Regulates Developmental Timing in *Arabidopsis. Cell.* **2009**, *138*, 750–759.

56. Millar, A. A.; Gubler F. The *Arabidopsis* GAMYB-Like Genes, MYB33 and MYB65, are MicroRNA-Regulated Genes that Redundantly Facilitate Anther Development. *Plant Cell.* **2005**, *17*, 705–721.

57. Wu, M. F.; Tian, Q.; Reed J. W. *Arabidopsis* Micro RNA167 Controls Patterns of ARF6 and ARF8 Expression, and Regulates Both Female and Male Reproduction. *Development.* **2006**, *133*, 4211–4218.

58. Nozawa, M.; Miura, S.; Nei, M. Origin and Evolution of MicroRNA Genes in Plants. *Genome Biol. Evol.* **2012,** *4*(3), 230–239.

59. Merchan, F.; Boualem, A.; Crespi, M.; Frugier, F. Plant Polycistronic Precursors Containing Non-Homologous Micrornas Target Transcripts Encoding Functionally Related Proteins. *Genome Biol.* **2009,** *10* (12), 1–14.

60. Carthrew, R. W.; Sontheimer, E. J. Origin and Mechanisms of MiRNAs and SiRNAs. *Cell.* **2009,** *136,* 642–655.

61. Czech, B.; Hannon, G. J. Small RNA Sorting: Matchmaking for Argonautes. *Nat. Genet.* **2011,** *12,* 19–31.

62. Melo, C. A.; Melo, S. A. *Biogenesis and Physiology of MicroRNAs, Non-Coding RNAs and Cancer;* Springer-Verlag: Berlin, Heidelberg, 2014; Ch.2, pp 5–24.

63. Westholm, J. O.; Lai, E . C. MirTRon*s*: MicroRNA Biogenesis via Splicing. *Biochimie.* **2011,** *93,* 1897–1904.

64. Yang, J. S.; Lai, E. C. *Alternative* MiRNA Biogenesis Pathways and the Interpretation of CoreMiRNA Pathway Mutants. *Mol. Cell.* **2011,** *43,* 892–903.

65. Kurihara, Y.; Watanabe, Y. *Arabidopsis* Micro-RNA Biogenesis through Dicer-like 1 Protein Functions. *Proc. Natl. Acad. Sci. USA.* **2004,** *101,* 12753–12758.

66. Bologna, N. G.; Mateos, J. L.; Bresso, E. G.; Palatnik, J. F. A Loop-to-Base Processing Mechanism Underlies the Biogenesis of Plant MicroRNAs MiR319 and MiR159. *EMBO J.* **2009,** *28,* 3646–3656.

67. Fang, Y.; Spector, D. L. Identification of Nuclear Dicing Bodies Containing Proteins for MicroRNA biogenesis in Living *Arabidopsis* Plants. *Curr. Biol.* **2007,** *17,* 818–823.

68. Rogers, K.; Chen, X. *Biogenesis,* Turnover, and Mode of Action of Plant MicroRNAs. *Plant Cell.* **2013,** *25,* 2383–2399.

69. Vaucheret, H. Plant ARGONAUTES. *Trends Plant Sci.* **2008,** *13* (7), 350–358.

70. Fujioka, Y., et al. Location of a Possible MiRNA Processing Site in Smd3/Smb Nuclear Bodies in *Arabidopsis. Plant Cell Physiol.* **2007,** *48,* 1243–1253.

71. Song, L. *Arabidopsis* Primary MicroRNA Processing Proteins HYL1 and DCL1 Define a Nuclear Body Distinct from the Cajal Body. *Proc. Natl. Acad. Sci. USA.* **2007,** *104,* 5437–5442.

72. Yang, Z.; Ebright, Y. W.; Yu, B.; Chen, X. HEN1 Recognizes 21-24nt Small RNA Duplexes and Deposits a Methyl Group onto the 2'OH of the 3' Terminal Nucleotide. *Nucleic Acids Res.* **2006,** *34*(2), 667–675.

73. Yu, L.; Yu, X.; Shen, R.; He, Y. HYL1 Gene Maintains Venation and Polarity of Leaves. *Planta.* **2005,** *221,* 231–242.

74. Ji, L.; Chen, X. Regulation of Small RNA Stability: Methylation and beyond. *Cell Res.* **2012,** *22,* 624–636.

75. Zhao, Y.; Mo, B.; Chen, X. Mechanisms that Impact MicroRNA Stability in Plants. *RNA Biol.* **2012,** *9*(10), 1218–1223.

76. Chen, X. MicroRNA Biogenesis and Function in Plants. *FEBS Lett.* **2005,** *579,* 5923–5931.

77. Baril, P.; Ezzine, S.; Pichon, C. Monitoring the Spatiotemporal Activities of MiRNAs in Small Animal Models Using Molecular Imaging Modalities. *Int. J. Mol. Sci.* **2015,** *16,* 4947–4972.

78. Xie, Z.; Khanna, K.; Ruan, S. Expression of MicroRNAs and it Regulation in Plants. *Semin. Cell Dev. Biol.* **2010,** *21* (8), 790–797.

79. Jones-Rhoades, M. W.; Bartel, D. P.; Bartel, B. MicroRNAs and their Regulatory Roles in Plants. *Annu. Rev. Plant Biol.* **2006,** *57,* 19–53.

80. Hock, J.; Meister, G. The *Argonaute* Protein Family. *Genome Biol.* **2008**, *9*(2), 210.1–210.8.

81. Hutvagner, G.; Simard, M. J. *Argonaute* Proteins: Key Players in RNA Silencing. *Nat. Rev. Mol. Biol.* **2008**, *9*, 22–32.

82. Rodriguez, A.; Griffiths-Jones, S.; Ashurst, J. L.; Bradley, A. Identification of Mammalian MicroRNA Host Genes and Transcription Units. *Genome Res.* **2004**, *14*, 1902–1910.

83. Meng, Y.; Shao, C. Large-Scale Identification of MirTRons in *Arabidopsis* and Rice. *PLoS ONE.* **2012**, *7*, e31163.

84. Joshi, P. K.; Gupta, D.; Nandal, U. K.; Khan, Y.; Mukherjee, S. K.. Sanan-Mishra, N. Identification of MirTRons in Rice Using MirTRonPred: A Tool for Predicting Plant-MirTRons. *Genomics.* **2012**, *99*, 370–375.

85. Chan, S. P.; Slack, F. J. And now Introducing Mammalian MirTRons. *Dev. Cell.* **2007**, *13*(5), 605–607.

86. Westholm, J. O.; Lai, E. C. Mirtrons: MicroRNA Biogenesis via Splicing. *Biochemie.* **2011**, *93*, 1897–1904.

87. Pratt, A. J.; MacRae, I. J. The RNA-Induced Silencing Complex: A Versatile Gene-Silencing Machine. *J. Biol. Chem.* **2009**, *284*, 17897–17901.

88. Tolia, N. H.; Joshua-Tor, L. Slicer and the Argonautes. *Nat. Chem. Biol.* **2006**, *3*, 36–43.

89. Bohmert, K.; Camus, I.; Bellini, C.; Bouchez, D.; Caboche, M.; Benning; C. Ago1 Defines a Novel Locusof *Arabidopsis* Conrtolling Leaf Development. *EMBO J.* **1998**, *17*, 170–180.

90. Aravin, A.; Gaidatzis, D.; Pfeffer, S.; Lagos-Quintana, M.; Landgraf, P.; Iovino, N.; Morris, P.; Brownstein, M. J.; Kuramochi-Miyagaw, a S.; Nakano, T.; Chien, M.; Russo, J. J.; Ju, J.; Sheridan, R.; Sander, C.; Zavolan, M.;Tuschl, T. A Novel Class of Small RNAs Bind to MILI Protein in Mouse Testes. *Nature.* **2006**, *442*(7099), 203–207.

91. Yigit, E.; Batista, P. J.; Bei, Y.; Pang, K. M.; Chen, C. C.; Tolia, N. H.; Joshua-Tor, L.; Mitani, S.; Simard, M. J.; Mello, C. C. Analysis of the *C. elegans Argonaute* Family Reveals that Distinct Argonautes Act Sequentially during RNAi. *Cell.* **2006**, *127*(4), 747–757.

92. Lee, Y.; Kim, M.; Han, J.; Yeom, K. H.; Lee, S.; Baek, S. H.; Kim, V. N. MicroRNA Genes are Transcribed by RNA Polymerase II. *EMBO J.* **2004**, *23*, 4051–4060.

93. Blevins, T.; Rajeswaran, R.; Shivaprasad, P. V.; Beknazariants, D.; Si-Ammour, A., et al. Four Plant Dicers Mediate Viral Small RNA Biogenesis and DNA Virus Induced Silencing. *Nucleic Acids Res.* **2006**, *34*, 6233–6246.

94. Collins, R. E.; Cheng, X. Structural Domains in RNAi. *FEBS Lett.* **2005**, *579*, 5841–5849.

95. Pratt, J. A.; MacRae, I. J. The RNA-Induced Silencing Complex: A Versatile Gene-Silencing Machine. *JBC Papers.* **2009**, *284*(27), 12897–12901.

96. Bouche, N.; Lauressergues, D.; Gasciolli, V.; Vaucheret, H. An Antagonistic Function for *Arabidopsis* DCL2 in Development and a New Function for DCL4 in Generating Viral SiRNAs. *EMBO J.* **2006**, *25*, 3347–3356.

97. Montgomery, T. A.; Yoo, S. J.; Fahlgren, N.; Gilbert, S. D.; Howell, M. D.; Sullivan, C. M.; Alexander, A.; Nguyen, G.; Allen, E.; Ahn, J. H.; Carrington, J. C. AGO1-iR173 Complex Initiates Phased SiRNA Formation in Plants. *Proc. Natl. Acad. Sci.* **2008**, *105*, 20055–20062.

98. Voinnet, O. Origin, Biogenesis, and Activity of Plant MicroRNAs. *Cell.* **2009**, *136*, 669–687.

99. Eun, C.; Lorkovic, Z. J.; Naumann, U.; Long, Q.; Havecker, E. R.; Simon, S. A.; Meyers, B. C.; Matzke, A. J. M.; Matzke, M. AGO6 Functions in RNA-Mediated Transcriptional Gene Silencing in Shoot and Root Meristems in *Arabidopsis thaliana. PLoS ONE.* **2011,** *6* (10), e25730.

100. Jouannet, V.; Moreno, A. B.; Elmayan, T.; Vaucheret, H.; Crespi, M. D.;Maizel, A. Cytoplasmic *Arabidopsis* AGO7 Accumulates in Membrane-Associated SiRNA Bodies and is Required for Ta-SiRNA Biogenesis. *EMBO J.* **2012,** *31,* 1704–1713.

101. Ross, J. F.; Carlson, J. A.; Brock; G. *MiRNA:* The New Gene Silencer. *Am. J. Clin. Pathol.* **2007,** *128*(5), 830–836.

102. Chryssa, K.; Stefan, A. M.; Andrew, L. K.; Shridar, G.; Ronny, D.; Thomas, J.; David, M. L.; Klaus, R. Dicer-Deficient Mouse Embryonic Stem Cells are Defective in Differentiation and Centromeric Silencing. *Genes Dev.* **2004,** *19,* 489–501.

103. Robert, J. S.; Lewis, H.; Kathleen, E. F.; Richard, M. A. DICER-LIKE 1 and DICER-LIKE 3 Redundantly Act to Promote Flowering via Repression of FLOWERING LOCUS C in *Arabidopsis thaliana. Genetics.* **2007,** *176,* 1359–1362.

104. Gasciolli, V.; Mallory, A. C.; Bartel, D. P.; Vaucheret, H. Partially Redundant Functions of Arabidopsis DICER-like Enzymes and a Role for DCL4 in Producing Trans-Acting SiRNAs. *Curr. Biol.* **2005,** *15,* 1494–1500.

105. Chen, X.; Liu, J.; Cheng, Y.; Jia, D. HEN1 Functions Pleiotropically in *Arabidopsis* Development and Acts in C Function in the Flower. *Development.* **2002,** *129,* 1085–1094.

106. Kidner, C. A.; Martienssen, R. A. The Developmental Role of MicroRNA in Plants. *Curr. Opin. Plant Biol.* **2005,** *8*(1), 38–44.

107. Carmell, M. A.; Xuan Z.; Zhang, M. Q.; Hannon, G. J. The Argonaute Family: Tentacles that Reach into RNAi, Developmental Control, Stem Cell Maintenance, and Tumorigenesis. *Genes Dev.* **2002,** *16,* 2733–2742.

108. Kim, J.; Jung, J. H.; Reyes, J. L.; Kim, Y. S.; Kim, S. Y.; Chung, K. S.; Kim, J. A.; Lee, M.; Lee, Y.; Narry Kim, V. MicroRNA-Directed Cleavage of ATHB15 mRNA Regulates Vascular Development in Arabidopsis Inflorescence Stems. *Plant J.* **2005,** *42,* 84–94.

109. Axtell, M. J. Classification and Comparison of Small RNAs from Plants. *Annu. Rev. Plant Biol.* **2013,** *64,* 137–159.

110. Marin, E.; Jouannet, V.; Herz, A.; Lokerse, A. S.; Weijers, D.; Vaucheret, H.; Nussaume, L.; Crespi, M. D.; Maizel, A. MiR390,*Arabidopsis* TAS3 Tasi RNAs, and their AUXIN RESPONSE FACTOR Targets Define an Autoregulatory Network Quantitatively Regulating Lateral Root Growth. *Plant Cell.* **2010,** *22,* 1104–1117.

111. Yu, B.; Yang, Z.; Li, J.; Minakhina, S.; Yang, M.; Padgett, R. W.; Steward, R.; Chen, X. Methylation as a Crucial Step in Plant MicroRNA Biogenesis. *Science.* **2005,** *307,* 932–935.

112. Donner, T. J.; Sherr, I.; Scarpelia, E. Regulation of Preprocambial Cell State Acquisition by Auxin Signaling in *Arabidopsis* Leaves. *Development.* **2009,** *136,* 3235–3246.

113. Palatnik, J. F.; Allen, E.; Wu, X.; Schommer, C.; Schwab, R.; Carrington, J. C.; Weigel, D. Control of Leaf Morphogenesis by MicroRNAs. *Nature.* **2003,** *425,* 257–263.

114. Chuck, G.; Meeley, R.; Hake, S. Floral Meristem Initiation and Meristem Cell Fate are Regulated by the Maize AP2 Genes Ids1 and Sid1. *Development.* **2008,** *135,* 3013–3019.

115. Yang, T.; Xue, L.; An, L. Functional Diversity of MiRNA in Plants. *Plant Sci.* **2007,** *172,* 423–432.

116. Yant, L.; Mathieu, J.; Dinh, T. T.; Ott, F.; Lanz, C.; Wollmann, H.; Chen, X.; Schmid, M Orchestration of the Floral Transition and Floral Development in *Arabidopsis* by the Bifunctional Transcription Factor APETALA2. *Plant Cell.* **2010,** *22,* 2156–2170.

117. Eldem, V.; Okay, S.; Unver, T. Plant MicroRNAs: New Players in Functional Genomics. *Turk. J. Agric. For.* **2013,** *37,* 1–21.

118. Dugas, D. V.; Bartel, B. MicroRNA Regulation of Gene Expression in Plants. *Curr. Opin. Plant Biol.* **2004,** *7,* 512–520.

119. Zhang, B.; Pan, X.; Cobb, G. P.; Anderson, T. A. Plant MicroRNA: A Small Regulatory Molecule with Big Impact. *Dev. Biol.* **2006,** *289,* 3–16.

120. Park, M. Y.; Wu, G.; Gonzalez-Sulser, A.; Vaucheret, H.; Poethig, R. S. Nuclear Processing and Export of Micrornas in *Arabidopsis*. *Proc. Natl. Acad. Sci. USA.* **2005,** *102,* 3691–3696.

121. Bollman, K. M.; Aukerman, M. J.; Park, M. Y.; Hunter, C.; Berardini, T. Z.;Poethig, R. S. HASTY, the Arabidopsis Ortholog of Exportin 5/MSN5, Regulates Phase Change and Morphogenesis. *Development.* **2003,** *130,* 1493–1504.

122. Laufs, P.; Peaucelle, A.; Morin, H.; Traas, J. MicroRNA Regulation of the CUC Genes is Required for Boundary Size Control in Arabidopsis Meristems. *Development.* **2004,** *131,* 4311–4322.

123. Wang, W.; Gaffney, B.; Hunt, A.; Tang, G. MicroRNAs (MiRNAs) in Plant Development. *ELS.* **2007,** 1–9.

124. Wang, J. W.; Schwab, R.; Czech, B.; Mica, E.; Weigel, D. Dual Effects of Mir156-Targeted SPL Genes and CYP78A5/KLUH on Plastochron Length and Organ Size in *Arabidopsis thaliana*. *Plant Cell.* **2008,** *20,* 1231–1243.

125. Wu, G.; Poethig, R. S. Temporal Regulation of Shoot Development in *Arabidopsis thaliana* by MiR156 and its Target SPL3. *Development.* **2006,** *133,* 3539–3547.

126. Shikata, M.; Koyama, T.; Mitsuda, N.; Ohme-Takagi, M. *Arabidopsis* SBP-Box Genes SPL10, SPL11 and SPL2 Control Morphological Change in Association with Shoot Maturation in the Reproductive Phase. *Plant Cell Physiol.* **2009,** *50,* 2133–2145.

127. Xie, Z.; Allen, E.; Fahlgren, N.; Calamar, A.; Givan, S. A.; Carrington, J. C. Expression of *Arabidopsis* MiRNA Genes. *Plant Physiol.* **2005,** *138,* 2145–2154.

128. Wang, J. W.; Park, M. Y.; Wang, L. J.; Koo, Y.; Chen, X. Y.; Weigel, D.; Poethig, R. S. MiRNA Control of Vegetative Phase Change in Trees. *PLoS Genet.* **2011,** *7,* e1002012.

129. Kim, J. J.; Lee, J. H.; Kim, W.; Jung, H. S.; Huijser, P.; Ahn, J. H. The MicroRNA156–SQUAMOSA PROMOTER BINDING PROTEINLIKE3 Module Regulates Ambient Temperature-Responsive Flowering via FLOWERING LOCUS T in *Arabidopsis*. *Plant Physiol.* **2012,** *159,* 461–478.

130. Martin, R. C.; Asahina, M.; Liu, P. P.; Kristof, J. R.; Coppersmith, J. L.; Pluskota, W. E.; Bassel, G. W.; Goloviznina, N. A.; Nguyen, T. T.; Martinez- Andujar, C.; Kumar, A. M. B.; Pupel, P.; Nonogaki, H. The MicroRNA156 and MicroRNA172 Gene Regulation Cascades at Post-Germinative Stages in *Arabidopsis*. *Seed Sci. Res.* **2010,** *20,* 79–87.

131. Wang, J. W.; Czech, B.; Weigel, D. MiR156-Regulated SPL Transcription Factors Define an Endogenous Flowering Pathway in *Arabidopsis thaliana*. *Cell.* **2009,** *138,* 738–749.

132. Yamaguchi, A.; Wu, M.; Yang, L.; Wu, G.; Poethig, R. S.; Wagner, D. The MicroRNA-Regulated SBP-box Transcription Factor SPL3 is a Direct Upstream Activator of LEAFY, FRUITFULL and APETALA1. *Dev. Cell.* **2009,** *17,* 268–278.

133. Zhou, C. M.; Wang, J. W. Regulation of Flowering Time by MicroRNAs. *J. Genet. Genomics.* **2013,** *40,* 211–215.

134. Mathieu, J; Yant, L. J.; Mürdter, F.; Küttner, F.; Schmid, M. Repression of Flowering by the MiR172 Target SMZ. *PLoS Biol.* **2009**, *7,* e1000148.

135. Wu, G.; Park, Y.; Conway, S. R. Wang, J. W.; Weigel, D.; Poethig, R. S. The Sequential Action of Mir156 and Mir172 Regulates Developmental Timing in *Arabidopsis. Cell.* **2009**, *138,* 750–759.

136. Zhu, Q. H.; Helliwell, C. A. Regulation of Flowering Time and Floral Patterning by MiR172. *J. Exp. Bot.* **2010**, *62,* 487–495.

137. Spanudakis, E.; Jackson, S. The Role of MicroRNA in the Control of Flowering Time. *J. Exp. Bot.* **2014**, *65* (2), 365–380.

138. Lauter, N.; Kampani, A.; Carlson, S.; Goebel, M.; Moose, S. P. MicroRNA172 Down-Regulates Glossy15 to Promote Vegetative Phase Change in Maize. *Proc. Natl. Acad. Sci. USA.* **2005**, *102,* 9412–9417.

139. Chuck, G.; Meeley, R.; Irish, E.; Sakai, H.; Hake, S. The Maize Tasselseed4 MicroRNA Controls Sex Determination and Meristem Cell Fate by Targeting Tasselseed6/in Determinate Spikelet1. *Nat. Genet.* **2007**, *39,* 1517–1521.

140. Jung, J.; Lee, S.; Yun, J.; Lee, M.; Park, C. TheMiR172 Target TOE3 Represses AGAMOUS Expression during Arabidopsis Floral Patterning. *Plant Sci.* **2014**, *215–216,* 29–38.

141. Achard, P.; Herr, A.; Baulcombe, D. C.; Harberd, N. P. Modulation of Floral Development by a Gibberellin-Regulated MicroRNA. *Development.* **2004**, *131,* 3357–3365.

142. Allen, R. S.; Li, J. Y.; Stahle, M. I.; Dubroue, A, Gubler, F.; Millar, A. A. Genetic Analysis Reveals Functional Redundancy and the Major Targets of the *Arabidopsis* MiR159 family. *Proc. Natl. Acad. Sci. USA.* **2007**, *104,* 16371–16376.

143. Terzi, L. C.; Simpson, G. G. Regulation of Flowering Time by RNA Processing. *Curr. Top. Microbiol. Immunol.* **2008**, *326,* 201–218.

144. Yamaguchi, A.; Abe, M. Regulation of Reproductive Development by Non-Coding RNA in Arabidopsis: To Flower or notto Flower. *J. Plant Res.* **2012**, *125,* 693–704.

145. Schwab, R.; Palatnik, J. F.; Riester, M.; Schommer, C.; Schmid, M.; Weigel, D. Specific Effects of MicroRNAs on the Plant Transcriptome. *Dev. Cell.* **2005**, *8,* 517–527.

146. Schommer, C.; Bresso, E. G.; Spinelli, S. V.; Palatnik, J. F. Role of MicroRNA MiR319 in Plant Development. *Signal. Commun. Plants.* **2012**, *15,* 29–47.

147. Rubio-Somoza, I, Weigel, D. Coordination of Flower Maturation by a Regulatory Circuit of Three MicroRNAs. *PLoS Genet.* **2013**, *9,* e1003374.

148. Sarvepalli, K.; Nath, U. Hyper-Activation of the TCP4 Transcription Factor in *Arabidopsis thaliana* Accelerates Multiple Aspects of Plant Maturation. *Plant J.* **2011**, *67,* 595–607.

149. Bowman, J. L.; Floyd, S. K. Patterning and Polarity in Seed Plant Shoots. *Annu. Rev. Plant Biol.* **2008**, *59,* 67–88.

150. Willmann, M. R.; Mehalick, A. J.; Packer, R. L.; Jenik, P. D. *MicroRNAs* Regulate the Timing of Embryo Maturation in *Arabidopsis. Plant Physiol.* **2011**, *155,* 1871–1884.

151. Williams, L.; Grigg, S. P.; Xie, M.; Christensen, S.; Fletcher, J. C. Regulation of *Arabidopsis* Shoot Apical Meristem and Lateral Organ Formation by Microrna MiR166g and Its AtHD-ZIP Target Genes. *Development.* **2005**, *132,* 3657–3668.

152. Itaya, A.; Bundschuh, R.; Archual, A. J.; Joung J. G.; Fei, Z.; Dai, X.; Zhao, P. X.; Tang, Y.; Nelson, R. S.; Ding, B. Small RNAs in Tomato Fruit and Leaf Development. *BBA.* **2008**, *1779,* 99–107.

153. Seefried, W.; Willmann, M. R.; Clausen, R. L.; Jenik, P. D. Global Regulation of Embryonic Patterning in Arabidopsis by MicroRNAs. *Plant Physiol.* **2014**, *165,* 67–687.

154. Moxon, S.; Jing R.; Szittya G.; Schwach F.; Rusholme Pilcher R. L.; Moulton V.; Dalmay T. Deep Sequencing of Tomato Short RNAs Identifies MicroRNAs Targeting Genes Involved in Fruit Ripening. *Genome Res.* **2008,** *18,* 1602–1609.

155. Karlova, R.; Rosin, F. M.; Busscher-Lange, J.; Parapunova, V.; Do P. T.; Fernie, A. R.; Fraser, P. D.; Baxter,C.; Angenent, G. C, de Maagd, R. A. Transcriptome and Metabolite Profiling Show that APETALA2a is a Major Regulator of Tomato Fruit Ripening. *Plant Cell.* **2011,** *23,* 923–941.

156. Zhang, X.; Zou, Z.; Zhang, J.; Zhang, Y.; Han, Q.; Hu, T.; Xu, X.; Liu, H.; Li, H.; Ye, Z. Over-Expression of Sly-Mir156a in Tomato Results in Multiple Vegetative and Reproductive Trait Alterations and Partial Phenocopy of the Sft Mutant. *FEBS Lett.* **2011,** *585,* 435–439.

157. Wang, C.; Leng, X.; Zhang, Y.; Kayesh, E.; Zhang, Y.; Sun, X.; Fang, J. Transcriptome-Wide Analysis of Dynamic Variations in Regulation of Grapevine MicroRNAs on their Target Genes during Grapevine Development. *Plant Mol. Biol.* **2014,** *84,* 269–285.

158. Houston, K.; McKim, S. M.; Comadran, J.; Bonar, N.; Druka, I.; Uzrek, N.; Cirillo, E.; Guzy-Wrobelska, J.; Collins, N. C.; Halpin, C. Variation in the Interaction between Alleles of HvAPETALA2 and MicroRNA172 Determines the Density of Grains on the Barley Inflorescence. *Proc. Natl. Acad. Sci. USA.* **2013,** *110,* 16675–16680.

159. Kang, M.; Zhao, Q.; Zhu, D.; Yu, J. Characterisation of MicroRNA Expression during Maize Seed Development. *BMC Genomics.* **2012,** *13,* 360–371.

160. Song, Q.; Liu, Y.; Hu, X.; Zhang, W.; Ma, B.; Chen, S.; Zhang, J. Identification of MiRNAs and their Target Genes in Developing Soybean Seeds by Deep Sequencing. *BMC Genomics.* **2011,** *11,* 5–21.

161. Korbes, A. P.; Machado, R. D.; Guzman, F.; Almerão, M. P.; de Oliveira, L, F. V.; Loss-Morais, G.; Turchetto-Zolet, A. C.; Cagliari, A.; Maraschin, F. S.; Margis-Pinheiro, M.; Margis, R. Identifying Conserved and Novel MicroRNA in Developing Seeds of *Brassica napus* Using Deep Sequencing. *PLoS ONE.* **2012,** *7,* e50663.

162. Song, J. B.; Shu, X. X.; Shen, Q.; Li, B. W.; Song, J.; Yang, Z. M. Altered Fruit and Seed Development of Transgenic Rapeseed (*Brassica napus*) Overexpressing MicroRNA394. *PLoS ONE.* **2015,** *10*(5), e0125427.

163. Rubio-Somoza, I.; Cuperus, J. T.; Weigel, D.; Carrington, J. C. Regulation and Functional Specialization of Small RNA-Target Nodes during Plant Development. *Curr. Opin. Plant. Biol.* **2009,** *12,* 622–627.

164. Guo, H. S.; Xie, Q.; Fei, J. F.; Chua, N. H. MicroRNA Directs mRNA Cleavage of the Transcription Factor NAC1 to Downregulate Auxin Signals for *Arabidopsis* Lateral Root Development. *Plant Cell.* **2005,** *17,* 1376–1386.

165. Xue, L. J.; Zhang, J. J.; Xue, H. W. Characterization and Expression Profiles of MiRNAs in Rice Seeds. *Nucleic Acids Res.* **2009,** *37,* 916–930.

166. Yoon, E. K.; Yang, J. H.; Lim, J.; Kim, S. H.; Kim, S. K.; Lee, W. S. Auxin Regulation of the MicroRNA390-Dependent Transacting Small Interfering RNA Pathway in *Arabidopsis* Lateral Root Development. *Nucleic Acids Res.* **2010,** *38,* 1382–1391.

167. Vidal, E. A.; Araus, V.; Lu, C.; Parry, G.; Green, P. J.; Coruzzi, G. M.; Gutierrez, R. A. Nitrate Responsive MiR393/AFB3 Regulatory Module Controls Root System Architecture in *Arabidopsis thaliana. Proc. Natl. Acad. Sci. USA.* **2010,** *107,* 4477–4482.

168. Mallory, A. C.; Bartel, D. P.; Bartel, B. MicroRNA-Directed Regulation of Arabidopsis AUXIN RESPONSE FACTOR17 is Essential for Proper Development and Modulates Expression of Early Auxin Response Genes. *Plant Cell.* **2005,** *17,* 1360–1375.

169. Wang, J. W.; Wang, L. J.; Mao, Y. B.; Cai, W. J.; Xue, H. W.; Chen, X. Y. Control of Root Cap Formation by Microrna-Targeted Auxin Response Factors in *Arabidopsis*. *Plant Cell.* **2005,** *17,* 2204–2216.

170. Gutierrez, L, Bussell, J. D.; Pacurar, D. I.; Schwambach, J.; Pacurar, M.; Bellini, C. Phenotypic Plasticity of Adventitious Rooting in Arabidopsis is Controlled by Complex Regulation of AUXIN RESPONSE FACTOR Transcripts and MicroRNA Abundance. *Plant Cell.* **2009,** *21,* 3119–3132.

171. Carlsbecker, A.; Lee, J. Y.; Roberts, C. J.; Dettmer, J.; Lehesranta, S.; Zhou, J.; Lindgren, O.; Moreno-Risueno, M. A.; Vaten, A.; Thitamadee, S. Cell Signalling by MicroRNA165/6 Directs Gene dose-Dependent Root Cell Fate. *Nature.* **2010,** *465,* 316–321.

172. Lima, J. S.; Loss-Morais, G.; Margis, R. MicroRNAs Play Critical Roles during Plant Development and in Response to Abiotic Stresses. *Genet. Mol. Biol.* **2012,** *35* (4), 1069–1077.

173. Rhoades, J. M. W.; Bartel, D. P. Computational Identification of Plant MicroRNAs and their Targets, Including a Stress-Induced MiRNA. *Mol. Cell.* **2004,** *14,* 787–799.

174. Liu, D.; Song, Y.; Chen, Z.; Yu, D. Ectopic Expression of Mir396 Suppresses GRF Target Gene Expression and Alters Leaf Growth in *Arabidopsis. Physiol. Plant.* **2009,** *136,* 223–236.

175. Rodriguez, R. E.; Mecchia, M. A.; Debernardi, J. M.; Schommer, C.; Weigel, D.; Palatnik, J. F. Control of Cell Proliferation in *Arabidopsis thaliana* by MicroRNA Mir396. *Development.* **2010,** *137,* 103–112.

176. Schommer, C.; Palatnik, J. F.; Aggarwal, P.; Chetelat, A.; Cubas, P.; Farmer, E. E.; Nath, U.; Weigel, D. Control of Jasmonate Biosynthesis and Senescence by Mir319 Targets. *PLoS Biol.* **2008,** *6*(9), e230.

177. Efroni, I.; Blum, E.; Goldshmidt, A.; Eshed, Y. A Protracted and Dynamic Maturation Schedule Underlies *Arabidopsis* Leaf Development. *Plant Cell.* **2008,** *20,* 2293–2306.

178. Nath, U.; Crawford, B. C.; Carpenter, R.; Coen, E. Genetic Control of Surface Curvature. *Science.* **2003,** *299,* 1404–1407.

179. Ori, N.; Cohen, A. R.; Etzioni, A.; Brand, A.; Yanai, O.; Shleizer, S.; Menda, N.; Amsellem, Z.; Efroni, I.; Pekker, I.; Alvarez, J. P.; Blum, E.; Zamir, D.; Eshed, Y. Regulation of LANCEOLATE by Mir319 is Required for Compound-Leaf Development Ii Tomato. *Nat. Genet.* **2007,** *39,* 787–791.

180. Chi, X.; Yang, Q.; Chen, X.; Wang, J.; Pan, L.; Chen, M.; Yang, Z.; He, Y.; Liang, X.; Yu, S. Identification and Characterization of MicroRNAs from Peanut (*Arachis hypogaea L.*) by High Throughput Sequencing. *PLoS ONE.* **2011,** *6,* e27530.

181. Yu, N.; Cai, W. J.; Wang, S. C.; Shan, C. M.; Wang, L. J.; Chen, X. Y. Temporal Control of Trichome Distribution by MicroRNA156-Targeted SPL Genes in *Arabidopsis thaliana*. *Plant Cell.* **2010,** *22,* 2322–2335.

182. Lal, S.; Pacis, L. B.; Smith, H. M. S. Regulation of the SQUAMOSA PROMOTER-BINDING PROTEIN-LIKE Genes/MicroRNA156 Module by the Homeodomain Proteins PENNYWISE and POUND-FOOLISH in *Arabidopsis. Mol. Plant.* **2011,** *4,* 1123–1132.

183. Fu, C, Sunkar, R.; Zhou, C.; Shen, H.; Zhang, J. Y.; Matts, J.; Wolf, J.; Mann, D. G. J.; Stewart, C. N.; Tang, Y.; Wang, Z. Y. Overexpression of MiR156 in Switchgrass (Panicum virgatum L.) Results in Various Morphological Alterations and Leads to Improved Biomass Production. *Plant Biotechnol. J.* **2012,** *10* (4), 443–452. DOI: 10.1111/j.1467-7652.

184. Xie, K. B.; Shen, J. Q.; Hou, X.; Yao, J. L.; Li, X. H.; Xiao, J. H.; Xiong, L. Z. Gradual Increase of Mir156 Regulates Temporal Expression Changes of Numerous Genes during Leaf Development in Rice. *Plant Physiol.* **2012,** *158,* 1382–1394.

185. Jung, J. H.; Seo, P. J.; Park, C. M. MicroRNA Biogenesis and Function in Higher Plants. *Plant Biotechnol. Rep.* **2009,** *3,* 111–126.

186. Liu, N.; Wu, S.; Van Houten, J.; Wang, Y.; Ding, B., et al. Down-Regulation of AUXIN RESPONSE FACTORS 6 and 8 by MicroRNA 167 Leads to Floral Development Defects and Female Sterility in Tomato. *J. Exp. Bot.* **2014,** *65,* 2507–2520.

187. Unver, T.; Parmaksiz, I.; Dündar, E. Identification of Conserved Micro-RNAs and their Target Transcripts in Opium Poppy (*Papaver somniferum L.*). *Plant Cell Rep.* **2010,** *29,* 757–769.

188. Puranik, S.; Sahu, P. P.; Srivastava, P. S.; Prasad, M. NAC Proteins: Regulation and Role in Stress Tolerance. *Trends Plant Sci.* **2012,** *17,* 369–381.

189. Aida, M.; Tasaka, M. Genetic Control of Shoot Organ Boundaries. *Curr. Opin. Plant. Biol.* **2006,** *9,* 72–77.

190. Baker, C. C.; Sieber, P.; Wellmer, F.; Meyerowitz, E. M. The Early Extra Petals1 Mutant Uncovers a Role for MicroRNA Mir164c in Regulating Petal Number in *Arabidopsis. Curr. Biol.* **2005,** *15,* 303–315.

191. Nikovics, K.; Blein, T.; Peaucelle, A.; Ishida, T.; Morin, H.; Aida, M.; Laufs, P. The Balance between the MIR164A and CUC2 Genes Controls Leaf Margin Serration in *Arabidopsis. Plant Cell.* **2006,** *18,* 2929–2945.

192. Peaucelle, A.; Morin, H.; Traas, J.; Laufs, P. Plants Expressing a MiR164-resistant CUC2 Gene Reveal the Importance of Post-Meristematic Maintenance of Phyllotaxy in *Arabidopsis. Development.* **2007,** *134,* 1045–1050.

193. Sieber, P.; Wellmer, F.; Gheyselinck, J.; Riechmann, J. L.; Meyerowitz, E. M. Redundancy and Specialization among Plant MicroRNAs: Role of the MIR164 Family in Developmental Robustness. *Development.* **2007,** *134,* 1051–1060.

194. Larue, C. T.; Wen, J. Q.; Walker, J. C. A *MicroRNA*-Transcription Factor Module Regulates Lateral Organ Size and Patterning in *Arabidopsis. Plant J.* **2009,** *58,* 450–463.

195. Kusumanjali, K.; Kumari, G.; Srivastava, P. S.; Das, S. Sequence Conservation and Divergence in Mir164c1 and Its Target, CUC1, in *Brassica* Species. *Plant Biotechnol. Rep.* **2012,** *6,* 149–163.

196. Zhang, B.; Wang, Q. MicroRNA-Based Biotechnology for Plant Improvement. *J. Cell. Physiol.* **2015,** *230,* 1–15.

197. Thiebaut, F.; Rojas, C. A.; Almeida, K. L.; Grativol, C.; Domiciano, G. C.; Lamb, C. R.; Engler, J. D. E. A.; Hemerly, A. S.; Ferreira, P. C. Regulation of MiR319 during Cold Stress in Sugarcane. *Plant Cell Environ.* **2012,** *35,* 502–512.

198. Jin, H.; Vacic, V.; Girke, T.; Lonardi, S.; Zhu, J. K. Small RNAs and the Regulation of Cis-Natural Antisense Transcripts in *Arabidopsis. BMC Mol. Biol.* **2008,** *9,* 6.

199. Jung, J. H.; Seo, Y. H.; Seo, P. J.; Reyes, J. L.; Yun, J.; Chua, N. H.; Park, C. M. The GIGANTEA-Regulated MicroRNA172 Mediates Photoperiodic Flowering Independent of CONSTANS in *Arabidopsis. Plant Cell.* **2007,** *19,* 2736–2748.

200. Gupta, O. P.; Permar, V.; Koundal, V.; Singh, U. D.; Praveen, S. MicroRNA Regulated Defense Responses in *Triticum aestivum L.* during *Puccinia graminis f.sp. tritici* Infection. *Mol. Biol. Rep.* **2012,** *39,* 817–824.

201. Kantar, M.; Unver, T.; Budak, H. Regulation of Barley MiRNAs upon Dehydration Stress Correlated with Target Gene Expression. *Funct. Integr. Genomics.* **2010,** *10,* 493–507.

202. Sunkar, R.; Li, Y. F.; Jagadeeswaran, G. Functions of MicroRNAs in Plant Stress Responses. *Trends Plant Sci.* **2012,** *17,* 196–203.

203. Sunkar, R, Zhu, J. Novel and Stress-Regulated MicroRNAs and other Small RNAs from *Arabidopsis*. *Plant Cell.* **2004,** *16,* 2001–2019.

204. Zhao, B.; Liang, R.; Ge, L.; Li, W.; Xiao, H.; Lin, H.; Ruan, K.; Jin, Y. Identification of Drought-Induced MicroRNAs in Rice. *Biochem. Biophys. Res. Commun.* **2007,** *354,* 585–590.

205. Kulcheski, F. R.; de Oliveira, L. F.; Molina, L. G.; Almerao, M. P.; Rodrigues, F. A.; Marcolino, J.; Barbosa, J. F.; Stolf-Moreira, R.; Nepomuceno, A. L.; Marcelino-Guimaraes, F. C.; Abdelnoor, R. V.; Nascimento, L. C.; Carazzolle, M. F.; Pereira, G. A. G.; Margis, R. Identification of Novel Soybean MicroRNAs Involved in Abiotic and Biotic Stresses. *BMC Genomics.* **2011,** *12,* 307.

206. Yin, Z, Li, Y, Han, X, Shen, F. Genome-Wide Profiling of MiRNAs and other Small Non-Coding RNAs in the *Verticillium dahliae* Inoculated Cotton Roots. *PLoSONE.* **2012,** *7,* e35765.

207. Yang, X.; Wang, L.; Yuan, D.; Lindsey, K.; Zhang, X. Small RNA and Degradome Sequencing Reveal Complex MiRNA Regulation during Cotton Somatic Embryogenesis. *J Exp. Bot.* **2013,** *64,* 1521–1536.

208. Ferdous, J.; Husain, S. S.; Shi, B. Role of MicroRNAs in Plant Drought Stress. *Plant Biotechnol. J.* **2015,** *13,* 293–305.

209. Xu, M. Y.; Dong, Y.; Zhang, Q. X.; Zhang, L.; Luo, Y. Z.; Sun, J.; Fan, Y. L.; Wang, L. Identification of MiRNAs and their Targets from *Brassica napus by* High-Throughput Sequencing and Degradome Analysis. *BMC Genomics.* **2012,** *13,* 421.

210. Dryanova, A.; Zakharov, A.; Gulick, P. J. Data Mining for MiRNAs and their Targets in the *Triticeae. Genome.* **2008,** *51,* 433–443.

211. Lucas, S. J.; Budak, H. Sorting the Wheat from the Chaff: Identifying MiRNAs in Genomic Survey Sequences of *Triticum aestivum* Chromosome 1AL. *PLoS ONE.* **2012,** *7,* e40859.

212. Kantar, M.; Lucas, S.; Budak, H. MiRNA Expression Patterns of *Triticum dicoccoides* in Response to Shock Drought Stress. *Planta.* **2011,** *233,* 471–484.

213. Pandey, R.; Joshi, G.; Bhardwa, A. R.; Agarwal, M.; Katiyar-Agarwal, S. A Comprehensive Genome-Wide Study on Tissue-Specific and Abiotic Stress-Specific MiRNAs in *Triticum aestivum*. *PLoS ONE.* **2014,** *9,* e95800.

214. Xin, M.; Wang, Y.; Yao, Y.; Xie, C.; Peng, H.; Ni, Z.; Sun, Q. Diverse Set of MicroRNAs are Responsive to Powdery Mildew Infection and Heat Stress in Wheat (*Triticum aestivum L.*). *BMC Plant Biol.* **2010,** *10,* 123. doi:10.1186/1471-2229-10-123.

215. Wei, L.; Zhang, D.; Xiang, F.; Zhang, Z. Differentially Expressed MiRNAs Potentially Involved in the Regulation of Defense Mechanism to Drought Stress in Maize Seedlings. *Int. J. Plant Sci.* **2009,** *170,* 979–989.

216. Xie, F.; Stewart, C. N. Jr.; Taki, F. A.; He, Q.; Liu, H.; Zhang, B. High Throughput Deep Sequencing Shows that MicroRNAs Play Important Roles in Switchgrass Responses to Drought and Salinity. *Plant Biotechnol. J.* **2014,** *12,* 354–366.

217. Ni, F. T.; Chu, L. Y.; Shao, H. B.; Liu, Z. H. Gene Expression and Regulation of Higher Plants Under Soil Water Stress. *Curr. Genomics.* **2009,** *10,* 269–280.

218. Li, W. X.; Oono, Y.; Zhu, J.; He, X. J.; Wu, J. M.; Iida, K.; Lu, X. Y.; Cui, X.; Jin, H, Zhu, J. K. The Arabidopsis NFYA5 Transcription Factor is Regulated Transcriptionally and Posttranscriptionally to Promote Drought Resistance. *Plant Cell.* **2008,** *20,* 2238–2251.

219. Trindade, I.; Capitão, C.; Dalmay, T.; Fevereiro, M. P.; Santos, D. M. MiR398 and MiR408 are Up-Regulated in Response to Water Deficit in *Medicago truncatula*. *Planta.* **2010**, *231,* 705–716.
220. Li, Y. F.; Zheng, Y.; Addo-Quaye, C.; Zhang, L.; Saini, A.; Jagadeeswaran, G, Axtell, M. J.; Zhang, W.; Sunkar, R. Transcriptome-Wide Identification of MicroRNA Targets in Rice. *Plant J.* **2010**, *62,* 742–759.
221. Liu, H. H.; Tian, X.; Li, Y. J.; Wu, C. A.; Zheng, C. C. Microarray-Based Analysis of Stress-Regulated MicroRNAs in *Arabidopsis thaliana*. *RNA.* **2008**, *14,* 836–843.
222. Reyes, J. L.; Chua, N. H. ABA Induction of Mir159 Controls Transcript Levels of Two MYB Factors during *Arabidopsis* Seed Germination. *Plant J.* **2007**, *49,* 592–606.
223. Khraiwesh, B.; Zhu, J. K.; Zhu, J. Role of MiRNAs and SiRNAs in Biotic and Abiotic Stress Responses of Plants. *Biochim. Biophys. Acta.* **2012**, *1819,* 137–148.
224. Cui, H.; Hao, Y.; Kong, D. SCARECROW has a SHORT-ROOT Independent Role in Modulating the Sugar Response. *Plant Physiol.* **2012**, *158,* 1769–1778.
225. Arenas-Huertero, C.; Pere, B.; Rabanal, F.; Blanco-Melo, D.; De la Rosa, C.; Estrada-Navarrete, G.; Sanchez, F.; Alicia Covarrubias, A.; Luis Reyes, J. Conserved and Novel MiRNAs in the Legume *Phaseolus vulgaris* in Response to Stress. *Plant Mol. Biol.* **2009**, *70,* 385–401.
226. Li, B.; Qin, Y.; Duan, H.; Yin, W.; Xia, X. Genome-Wide Characterization of New and Drought Stress Responsive MicroRNAs in *Populus euphratica*. *J. Exp. Bot.* **2011**, *62,* 3765–3779.
227. Ni, Z.; Hu, Z.; Jiang, Q.; Zhang, H. Overexpression of Gma-MIR394a Confers Tolerance to Drought in Transgenic *Arabidopsis thaliana*. *Biochem. Biophys. Res. Commun.* **2012**, *427,* 330–335.
228. Song, J. B.; Huang, S. Q.; Dalmay, T.; Yang, Z. M. Regulation of Leaf Morphology by MicroRNA394 and its Target Leaf Curling Responsiveness. *Plant Cell Physiol.* **2012**, *53,* 1283–1294.
229. Jagadeeswaran, G.; Zheng, Y.; Li, Y. F.; Shukla, L. I.; Matts, J.; Hoyt, P.; Macmil, S. L.; Wiley, G. B.; Roe, B. A.; Zhang, W.; Sunkar, R. Cloning and Characterization of Small RNAs from *Medicago truncatula* Reveals Four Novel Legume-Specific MicroRNA Families. *New Phytol.* **2009**, *184,* 85–98.
230. Liang, G.; Yang, F.; Yu, D. MicroRNA395 Mediates Regulation of Sulfate Accumulation and Allocation in *Arabidopsis thaliana*. *Plant J.* **2010**, *62,* 1046–1057.
231. Lu, S.; Sun, Y. H.; Chiang, V. L. Stress-Responsive MicroRNAs in Populus. *Plant J.* **2008**, *55,* 131–151.
232. Sun, X.; Xu, L.; Wang, Y.; Yu, R.; Zhu, X.; Luo, X.; Gong, Y.; Wang, R.; Limera, C.; Zhang, K.; Liu L. Identification of Novel and Salt- Responsive MiRNAto Explore Mirna-Mediated Regulatory Network of Salt Stress Response in Radish (*Raphanus sativus L.*). *BMC Genomics.* **2015**, *16,* 197.
233. Pant, B. D.; Buhtz, A.; Kehr, J.; Scheible, W. R. MicroRNA399is a Long-Distance Signal for the Regulation of Plant Phosphate Homeostasis. *Plant J.* **2008**, *53,* 731–738.
234. Paul, S.; Datta, S. K.; Datta, K. MiRNARegulation of Nutrient Homeostasis in Plants. Front. *Plant Sci.* **2015**, *6,* 232.
235. Rausch, T.; Wachter, A. Sulfur Metabolism: A Versatile Platform for Launching Defence Operations. *Trends Plant Sci.* **2005**, *10,*503–509.
236. Yamasaki, H.; Abdel-Ghany, S. E, Cohu, C. M.; Kobayashi, Y.; Shikanai, T.; Pilon, M. Regulation of Copper Homeostasis by Microrna in *Arabidopsis*. *J. Biol. Chem.* **2007**, *282,* 16369–16378.

237. Abdel-Ghany, S. E.; Pilon, M. MicroRNA-Mediated Systemic Down-Regulation of Copper Protein Expression in Response to Low Copper Availability in *Arabidopsis. J. Biol. Chem.* **2008,** *283,* 15932–15945.

238. Shukla, L. L.; Chinnusamy, V.; Sunkar, R. The Role of MicroRNAs and other Endogenous Small RNAs in Plant Stress Responses. *Biochim. Biophys. Acta.* **2008,** *1779,* 743–748.

239. Combier, J.; Frugier, F.; de Billy, F.; Boualem, A.; El-Yahyaoui, F.; Moreau, S.; Vernie, T.; Ott, T.; Gamas, P.; Crespi, M.; Niebel, A. MtHAP2-1 is a Key Transcriptional Regulator of Symbiotic Nodule Development Regulated by Microrna169 in *Medicago truncatula. Genes Dev.* **2006,** *20,* 3084–3088.

240. Lv, D. K.; Bai, X.; Li, Y.; Ding, X. D.; Ge, Y.; Cai, H.; Ji, W.; Wu, N.; Zhu, Y. M. Profiling of Cold-Stress-Responsive MiRNAs in Rice by Microarrays. *Gene.* **2010,** *459,* 39–47.

241. Zhou, X.; Wang, G.; Sutoh, K.; Zhu, J. K.; Zhang, W. Identification of Cold-Inducible Micrornas in Plants by Transcriptome Analysis. *Biochim. Biophys. Acta.* **2008,** *1779,* 780–788.

242. Zhang, J.; Xu, Y.; Huan, Q.; Chong, K. Deep Sequencing of Brachypodium Small RNAs at the Global Genome Level Identifies MicroRNAs Involved in Cold Stress Response. *BMC Genomics.* **2009,** *10,* 449.

243. Sunkar, R.; Kapoor, A.; Zhu, J. K. Post-Transcriptional Induction of Two Cu/Zn Superoxide Dismutase Genes in Arabidopsis is Mediated by Downregulation of Mir398 and Important for Oxidative Stress Tolerance. *Plant Cell.* **2006,** *18,* 2051–2065.

244. Li, T.; Li, H.; Zhang, Y. X.; Liu, J. Y. Identification and Analysis of Seven H2O2-Responsive MiRNAs and 32 New MiRNAs in the Seedlings of Rice (*Oryza sativa L. ssp. indica*). *Nucleic Acids Res.* **2011,** *39,* 2821–2833.

245. Navarro, L.; Dunoyer, P.; Jay, F.; Arnold, B.; Dharmasiri, N.; Estelle, M.; Voinnet, O.; Jones, J. D. A plant MiRNA Contributes to Antibacterial Resistance by Repressing Auxin Signaling. *Science.* **2006,** *312,* 436–439.

246. Fahlgren, N.; Howell, M. D.; Kasschau, K. D.; Chapman, E. J.; Sullivan, C. M.; Cumbie, J. S.; Givan, S. A.; Law, T. F.; Grant, S. R.; Dangl, J. L.; Carrington, J. C. High-Throughput Sequencing of *Arabidopsis* MicroRNAs: Evidence for Frequent Birth and Death of MIRNA Genes. *PLoS ONE.* **2007,** *2,* e219.

247. Katiyar-Agarwal, S.; Jin, H. Role of Small RNAs in Host-Microbe Interactions. *Annu. Rev. Phytopathol.* **2010,** *4,* 225–246.

248. Zhang, W.; Gao, S.; Zhou, X.; Chellappan, P.; Chen, Z.; Zhou, X.; Zhang, X.; Fromuth, N.; Coutino, G.; Coffey, M.; Jin, H. Bacteriaresponsive MicroRNAs Regulate Plant Innate Immunity by Modulating Plant Hormone Networks. *Plant Mol. Biol.* **2011a,** *75,* 93–105.

249. Zhang, X.; Zhao, H.; Gao, S.; Wang, W. C.; Katiyar-Agarwal, S.; Huang, H. D.; Raikhel, N.; Jin, H. *Arabidopsis* Argonaute 2 Regulates Innate Immunity via MiRNA393*-Mediated Silencing of a Golgi-Localized SNARE Gene, MEMB12. *Cell.* **2011b,** *42,* 356–366.

250. Zhou, X.; Zhu, Q.; Eicken, C.; Sheng, N.; Zhang, X.; Yang, L.; Gao, X. MicroRNA Profiling Using µParaflo Microfluidic Array Technology. *Methods Mol. Biol.* **2012,** *822,* 153–82.

251. Navarro, L.; Bari, R.; Achard, P.; Lison, P.; Nemri, A.; Harberd, N. P., et al. DELLAs Control Plant Immune Responses by Modulating the Balance of Jasmonic Acid and Salicylic Acid Signalling. *Curr. Biol.* **2008,** *18,* 650–655.

252. Bednarek, P.; Kwon, C.; Schulze-Lefert, P. Not a Peripheral Issue: Secretion in Plant Microbe Interactions. *Curr. Opin. Plant. Biol.* **2010,** *13,* 378–387.

253. Shivaprasad, P. V.; Chen, H. M.; Patel, K..; Bond, D. M.; Santos, B. A.; Baulcombe, D. C. A MicroRNA Superfamily Regulate Nucleotide Binding Site-Leucine-Rich Repeats and other MRNAs. *Plant Cell.* **2012,** *24,* 859–874.

254. Zhou, M.; Gu, L. F.; Li, P. C.; Song, X. W.; Wei, L. Y.; Chen, Z. Y.; Cao, X. F. Degradome Sequencing Reveals Endogenous Small RNA Targets in Rice (*Oryza sativa L. ssp. indica*). *Front Biol.* **2010,** *5,* 67–90.

255. Ding, D.; Zhang, L.; Wang, H.; Liu, Z.; Zhang, Z.; Zheng, Y. Differential Expression of MiRNAs in Response to Salt Stress in Maize Roots. *Ann. Bot.* **2009,** *103,* 29–38.

256. Li, Y.; Zhang, Q.; Zhang, J.; Wu, L.; Qi, Y.; Zhou, J. M. Identification of MicroRNAs Involved in Pathogen-Associated Molecular Pattern-Triggered Plant Innate Immunity. *Plant Physiol.* **2010a,** *152,* 2222–2223.

CHAPTER 6

SMALL RNAs-II: MODE OF ACTION AND POTENTIAL APPLICATIONS IN PLANT IMPROVEMENT

NEER K. SINGH, SAURABH ANAND, and SANDIP DAS*

Department of Botany, University of Delhi, New Delhi 110007, India

Corresponding author. E-mail: sdas@botany.du.ac.in; sandipdas04@ gmail.com

CONTENTS

ABSTRACT

Small non-protein coding RNAs, generated as short double-stranded duplexes, consisting of "guide" and "passenger" strands play a pivotal role in regulating the expression of target genes. The guide strand is loaded onto a multi-protein complex called RNA-induced silencing complex (RISC) and RNA induced transcriptional silencing (RITS). RISC is a multi-protein complex consisting of an *Argonaute* protein with Piwi/Argonaute/Zwille (PAZ) domain, which binds to small RNA, and Piwi domain that is responsible for endonucleolytic activity causing the cleavage of small RNA-messenger RNA (mRNA) complex and mediate regulation of target sequences. RITS is RNAi effector complex that leads to the heterochromatin assembly by histone modification. RITS consists of the chromodomain protein, Chp1, and the other components including Ago1 and Tas3. Short interfering RNA (siRNA) loaded in this complex interacts with dg-dh repeats of centromere and cause methylation of lysine 9 of histone H3 via Clr4 methyltransferase. Several types of small RNAs like micro RNAs (miRNAs), siRNAs, tasi-RNAs, and Piwi-interacting RNAs (piRNAs), differ from each other in their biogenesis, mode of action and function. Among them miRNA and siRNA are important riboregulators and have been extensively studied in eukaryotes. miRNAs act post-transcriptionally in mRNA degradation and translational repression via partial or full complementary base pairing at spatial and temporal level. siRNAs can act in a similar manner as miRNAs, but in addition have a role in DNA methylation, histone modification and mRNA degradation via specific base pairing allowing no mismatches. The mechanism of action of siRNA and miRNA are termed as auto-silencing and hetero-silencing, respectively. Understanding of the biogenesis and mode of action including composition of RISC complex have led to the development of various gene silencing strategies such as virus induced gene silencing (VIGS), RNAi and artificial miRNAs with their inherent advantages and disadvantages. VIGS is a recent approach in which the function of a gene in question can be characterized exploiting RNA-mediated antiviral defense machinery in plants. This technique has been studied well in *Arabidopsis thaliana*, *Nicotiana benthamiana*, tomato, and barley. It has proved to be efficient in gaining knowledge about entire gene families. VIGS, as compared to other functional genomics approaches, is rapid, overcomes functional redundancy of genes, avoids plant transformation, is less expensive and can be used to test gene function in multiple genetic backgrounds. Apart from targeted silencing for genetic manipulation for crop improvement; gene silencing has also been applied to understand plant development and adaptation through generation of loss-of-function, knock-out/knock-down mutants. RNAi technology is being employed to manipulate important agronomic traits such as height (miR167 in tomato), root (miR171 in *A. thaliana*), leaf morphology (miR319, miR164), flower development (miR167 in tomato, miR159 in *A. thaliana,* etc.), seed productivity (OsmiR156), seed development (miR397 and miR328 in Rice), oil quality (miR156, miR167, and miR6029in *Brassica napus*), viral and fungal (miR156) tolerance, and many others such as fruit shelf life (miR156/miR157), metabolite content (miR393), and anthocyanin content (miR156)in major crop improvement programs. This chapter will summarize how endogenous/native and artificially synthesized small RNAs are being used to manipulate traits of agricultural significance.

6.1   INTRODUCTION TO SMALL RNAS

The role of RNA as an important ribo-regulator molecule that has numerous roles in the development of an organism is being increasingly deciphered. It is not only a genomic messenger that passes heritable information from DNA to protein but also has a regulatory role in gene expression. Early 90's was seen as an era of discovery of RNA as a regulatory molecule which participates in various developmental processes of eukaryotes, and this includes various classes of small RNAs. In eukaryotes, it was observed that only a small fraction of genome account for the coding genome and a major transcriptional output involving ~25% of the genome, is non-protein coding RNA.[1] Besides transfer RNA (tRNA) and ribosomal RNA (rRNA), a class of non-coding RNAs, which are generally 20–40 nt in length and carry out many fundamental processes like gene expression, genome stability, and adaptive response to biotic and abiotic stresses have now emerged.[2]

Small RNAs are double stranded, non-coding and are differentiated on basis of their origin, mode of action and nature of occurrence. They are classified in two broad classes short interfering RNA (siRNA) and microRNA (miRNA). Other small RNAs include Piwi interacting RNA (piRNA), transacting RNAs (tasiRNA), and repeat associated RNA (rasiRNA). The first small RNA discovered was miRNA from *Caenorhabditis elegans* where lin-4 and let-7 regulate the expression of Lin-14, which in turn controls the developmental timing and patterning in the worm.[3] In plants small RNAs have now been reported extensively from *A. thaliana* and in several other species also.[4–6]

On the basis of their origin small RNAs that originate from inter-molecular interaction between two single stranded complementary RNAs results in siRNAs while those that occur from a precursor which on intra-molecular hybridization forms hairpin RNAs (hpRNAs) constitutes miRNAs.[7] The biochemical and molecular mechanisms governing biogenesis of both siRNA and miRNA shows lot of variation. Micro RNAs are transcribed from the discrete genomic loci by RNA pol II to generate primary miRNA transcripts.[8] In an example of a rare instance, Borchert et al. have demonstrated the involvement of RNA pol III in transcribing human miRNAs.[9] The primary miRNA is then processed by RNase III like enzyme Dicer which specifically cleaves double stranded RNAs (dsRNAs) to form precursor miRNA. DCL 1 further cleaves this precursor miRNA into a mature miRNA/miRNA* duplex (designated as 5-p-miRNA/3-p-miRNA) and their 3' ends are methylated by HEN1. Processing of precursor miRNA to produce the mature duplex proceed via the "stem-to-loop" or "loop-to-stem" mechanism.[10] This duplex is then transported to the nucleus via the nuclear export protein HASTY. The miRNA/miRNA* duplex then disassemble where one of the strand, the guide strand (the strand with 5' instability) is loaded into RNA-induced silencing complex (RISC), a protein complex consisting of *Argonaute*protein (core component of RISC). Here, the miRNA and its cognate messenger RNA (mRNA) transcript interact, causing the cleavage of the target mRNA.[11] The double stranded small interfering RNAs (siRNAs) are produced by RNA dependent RNA polymerases (RDRs/RdRP). A unique feature of plant siRNAs is the involvement of RNA pol IV which is found exclusively in plants.[12] Trans acting siRNAs (ta-siRNAs) forms another class of small RNAs that are generated from the cleaved fragments of noncoding TAS transcripts that has been subjected to miRNA-mediated cleavage. The cleaved fragments are processed into dsRNAs by RDRP6 which are then processed by DCL4 into tasiRNA. They regulate many non-identical genes and miRNAs.[13]

There is one more class of small RNA, the natural antisense siRNA (Nat-siRNA) which are expressed during stress.[14] These are processed by RDR6, HYL1, and DCL proteins into siRNAs. Nat-siRNAs can be found in two main forms, cis-Nat-siRNAs which are obtained from the same region as their sense transcript but in the opposite polarity while trans-Nat-siRNAs that are derived from the different genomic location with complementary transcripts to their sense transcript.[15] They are known to be involved in regulatory pathways as diverse as RNAi, DNA methylation, RNA editing, genomic imprinting, and X-chromosome inactivation but their roles have been studied in only a few species till date.[16] Their biogenesis requires DCL1, RDR6, HEN1, HYL1, SGS3, and Pol IV.[17,18]

In addition, there are reports of the occurrence of long siRNAs (lsiRNA), which unlike other small RNAs are 30–40 nt in length. They are either pathogen induced or express under specific growth conditions. RDR6, DCL-1, DCL-4, SGS3, HEN1, and HYL1 are involved in the biogenesis of lsiRNA.[19]

The loading of the guide strand is mediated by two major proteins—the TAR-RNA-binding protein (dsRNA binding domain protein /TRBP in humans) or by its homologs such as R2D2 as in *Drosophila* and *Argonaute*. As mentioned earlier, thermodynamic differences between the two ends of the double-stranded siRNA or miRNA determines which strand is to be loaded onto the RISC complex. The strand with 5' instability or lower thermodynamic stability acts as guide strand whereas the other strand called the passenger strand is destroyed. This strand selection is facilitated by R2D2 or TRBP protein. Upon loading of the small RNA into the RISC with the help of RISC loading complex (RLC)[20] the miRNA and its cognate mRNA transcript interact, causing the cleavage of the target gene.

6.2   COMPARATIVE ACCOUNT OF MODE OF ACTION

Small RNAs can induce gene silencing via three mechanisms:

  i.   Transcript cleavage
  ii. Translational repression
  iii. Chromatin remodeling/hetero-chromatinization

Small RNAs (miRNAs and siRNAs) are processed as 20–24 nt duplex molecules with two-nucleotide overhangs at their 3'end, a characteristic feature of DICER products. One of the strands of the duplex molecule (guide strand) is loaded onto the RISC and is complementary to the target sequence of the target gene which they regulate. The entry of the guide strand into the RISC is via interaction of the guide strand with the PAZ, middle and Piwi domain of the Ago proteins. Analysis of the crystal structure of Ago/guide DNA showed that middle and the Piwi domains are responsible for holding the 5'-P of the guide strand and may act as an anchor. As a result the 5'-nucleotide may not be available during target recognition. The 3' end of the guide strand on the other hand is bound to the hydrophobic pocket made by the PAZ domain. The entire binding pocket of Ago comprising of the PAZ, middle and Piwi domain interacts with the guide strand via the phosphodiester backbone.[21] Comparison of analysis of the solvent structure of an

Ago-silencing complex from *Thermus thermophilus*[22] and target prediction of miRNAs[23,24] revealed that bases 2–6 of the guide strand are exposed from the binding pocket of the Ago and are therefore the most critical component of target recognition of the RISC. The guide strand interacts with the Ago proteins to mediate the silencing of gene they target by cleavage of the phosphodiester bond of the backbone and degradation through RISC. The fate of the target mRNA depends on the degree of complementarity between the small RNA and the target which is highest in 2–7 nt region (known as seed region) from the 5' end of small RNA. If the degree of base pairing is high (perfect or near-perfect complementarity), then it causes the cleavage of mRNA at 10–11 nt position relative to 5' end of small RNA; if the base pairing is imperfect with unpaired bases, then it causes the translational repression. Figure 6.1 shows the crystal structure of human Ago2 bound with a synthetic target and miRNA. A study by Schirle et al.[25] has provided structural insights and evidence of nucleotide position 2–7 acting at the seed sequence. Transcript cleavage activity resides in the RNase H fold of the Piwi domain of Ago protein with the catalytic site being a conserved dyad of aspartic acid followed by either an aspartic acid or histidine (DD[D/H]). This is in contrasts to DD (D/E; glutamate) that other canonical RNAse H containing enzymes such as retroviral integrases or transposases employ.[26] The other key difference being that canonical RNaseH are involved in cleaving the RNA strand from a RNA:DNA hybrid unlike Ago which slices RNA from a RNA:RNA hybrid. Two metal ions are necessary to catalyze the cleavage reaction. The "slicer" activity however, is present only in select set of Ago proteins and not all Ago proteins have this property. For example, among the eight human Ago, only Ago2 has slicer activity; in *A. thaliana*, among 10 Ago, Ago1, and Ago4 have slicer property; as *Schizosaccharomyces pombe* has only one Ago, it is endowed with slicer property in order to perform post transcriptional gene silencing (PTGS). Members of Ago that lack this property are categorized as non-slicers.[27]

The second mode of small RNA mediated gene regulation is via preventing the mRNA from being translated. Translational repression requires assembly of the miRNA onto polysomes which is Ago dependent. In *A. thaliana*, Ago1 and Ago10 together with the functional miRNA have been shown to be associated with polysomes.[28–30] Translational repression has been proposed to occur via two primary mechanisms involving stearic hindrances. Iwakawa and Tomari[30] proposed that in case of high similarity in target site and when the target site resides toward the 5'end of ORF or 5'UTR, Ago/RISC may hinder recruitment of ribosomes. In case of target sites residing toward 3'end of ORF, RISC causes stearic hindrances that prevent assembly and

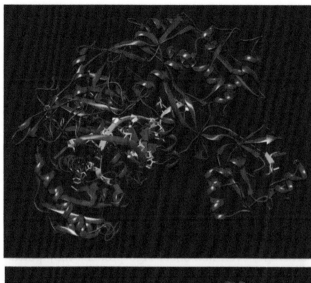

Lateralview of Human AGO2 (blue)
bound to 12 nt synthetic RNA (green)
and synthetic miRNA (red).
Crystal structure PDB ID- 4W5T
Schirle et al. 2014
Gene regulation. structural basis for
microrna targeting.
Schirle NT, Sheu-Gruttadauria J, Macrae IJ
Science (2014) 346 p.608

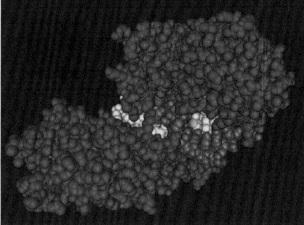

Top/bottom view of Human AGO2 (purple)
bound to 12 nt synthetic RNA (yellow)
and synthetic miRNA (brown).

Space filling model depicts that the two
RNA molecules are deep-seated within
the catalytic pocket;
Seed pairing occurs using position 2-7.

Crystal structure PDB ID- 4W5T
Schirle et al. 2014
Gene regulation. structural basis for
microrna targeting.
Schirle NT, Sheu-Gruttadauria J, Macrae IJ
Science (2014) 346 p.608

FIGURE 6.1   Crustal structure of AGO2 bound to RNA duplex. The PDB structure 4W5T was retrieved from NCBI structure database and viewed using Chimera and Cn3D.

movement of ribosome.[30] A component of RISC, GW182 protein is known to interact with the Poly A Binding protein (PABP; in the 3' poly A tail of mRNA) and this may prevent the closed loop formation involving PABP and e-IFs (e-IF4G) that are necessary for assembly of ribosome. GW182-PABP interaction also is known to promote deadenylation or shortening/removal of poly A tail via action of cytoplasmic deadenylases (PAN2-PAN3). Once the poly-A tail is acted upon by deadenylases recruited by interaction of GW182 (component of RISC)-PABP, the mRNA is marked for 5' de-capping followed by 5'-to-3' degradation via XRN1 or is stored in a "de-adenylated, repressed and non-functional" state.[31] In the eukaryotic system, cap-dependent translation of mRNA is the norm. Ago-RISC can also recruit decapping factors of mRNA such as DCP1, Me31B, and HPat to enhance the degradation of mRNA.[32,33] Therefore, translational repression can act at pre-initiation stage by promoting decapping and degradation of mRNA or by stalling ribosome recruitment and assembly (initiation), or at post-initiation stage by hindering ribosome movement (elongation). The mechanism of decapping and deadenylation is limited to caped mRNA because cap-independent mRNAs are found to be resistant to miRNA mediated translational repression.[34] Ago1, which is endowed with RNase H like slicer activity, can therefore perform gene regulation via both transcript cleavage and translational repression. However, as translational repression does not entail high degree of sequence complementarity and cleavage, even non-slicer Ago proteins may potentially be involved in such RISC. These processes occur once mRNA has formed and therefore this mechanism of regulation is known as PTGS.

The third mechanism by which small RNA regulates its target is by histone and DNA methylation of H3 histone at lys-9 which causes its silencing at transcript level.[35] The formation of repressive heterochromatin guided by small RNA is now well established and the mode of action is termed as transcriptional gene silencing (TGS). TGS causes gene silencing as it hinders the synthesis of mRNA. TGS can be differentiated biochemically from PTGS as it involves formation of what is termed as the RNA induced transcriptional silencing (RITS) complex.

Studies on transgenes revealed when multiple copies are inserted at a particular locus, it results in its silencing. This may be because of one of the two important processes occurring in the genome, TGS or PTGS. One of the major mechanisms by which TGS operates is through DNA methylation, a mechanism important for normal development of an organism as well as to maintain genome stability. For example TGS causes methylation of promoter, hinders mRNA synthesis, resulting in no or very low level of transcription and thus is a powerful mechanism to regulate transcriptional activity. Such modification of chromatin state is guided by either cytosine methylation in the sequence context of CG, CHG, or CHH (where H = A, C, or T), or via histone modifications. Small dsRNAs can act as a trigger for RNA-directed DNA Methylation (RdDM), characterized by *de novo* methylation of cytosine.[36,37] In an early study, tobacco plants infected with the transgene carrying PSTdV resulted in its methylation as the virus particles replicates.[36] TGS is heritable and corresponds to modification of the DNA or methylation of the promoter. In this process, DNA Methyl-transferases (DNMTs) such as DNMT1 and DNMT3 are known to methylate cytosine residues *denovo*or when the target is in a hemi-methylated condition; Met1 methyltransferase is important for the maintenance of DNA methylation in subsequent generations.[38] Factors required for RdDM to occur involve RNA polymerase-IV(Pol IV) that is involved in producing the siRNA trigger for RdDM, and Pol V that facilitates the methylation by producing scaffold transcript that interacts with siRNA trigger and recruits it to the target DNA,[39] chromatin remodeling proteins, transcription factors and DNMT-DRM2.[40] It must be borne in mind that both RNA Pol IV and Pol V are plant specific RNA polymerases.[41] RNA Pol IV has been shown to be responsible for biogenesis of precursor of 24 nt siRNA primarily from transposon and repeat DNA in the genome with the association of a protein *Sawadee homeodomain homolog* 1 (SHH1), but the entire mechanism of initial target recognition is not fully understood. It is known that SHH1 can recognize H3K9me and unmethylated H3K4 implying a direct correlation between histone modification (methylation/de-methylation) and DNA methylation. Pol IV transcribes a single stranded RNA (ssRNA) which is processed into a dsRNA by RdRP/RDR. The dsRNA is further acted upon by DCL3 to produce 24 nt siRNA, exported to cytoplasm, methylated at 3'-OH by HEN1 and loaded onto RITS complex with Ago4 or Ago9 in reproductive tissues.[42] The RITS complex with Ago4 and siRNA is then re-imported back into the nucleus. RNA Pol V is also known to be helped by members of SU (VAR) 3–9 family of histone methyltransferase enzymes such as SUVH2, SUVH9, and SUVR2 to associate with target DNA that contains some amount of methylation. Pol V also recruits the RITS complex with Ago4 and 24 nt siRNA and initiates transcription. The nascent transcript of Pol V pairs with the resident siRNA and further recruits *domain rearranged methyltransferase* 2 (DRM2) which catalyzes *denovo* DNA methylation at the genomic region homologous to siRNA. Further, a protein complex termed IDN2-IDP complex may be involved in stabilizing the interaction between siRNA and Pol V scaffold transcript, interact with SWI-SNF factor to alter the nucleosome position and facilitate RdDM.[37] Chromatin modification that involves both histone modification and DNA methylation can act in a feed-back loop to re-enforce each other. As mentioned earlier, recruitment of RNA Pol IV and Pol V is guided by modification of H3K9 and H3K4 residues, and also involves SU (VAR) 3–9 histone methyl transferase family of proteins. Additionally, removal of acetylation, H3K4me and H2B-ubiquitinylation needs to be removed to maintain DNA/promoter methylation and H3K9me. Similarly, HDAC6-histone deacetylase and MET1-DNMT act together to maintain C-methylation at 5'-CG-3'.

In a non-canonical pathway, RNA Pol II transcripts from transposons are converted to dsRNA by RdRP/RDR6 which is then acted upon by DCL2 and DCl4 to produce 21 nt siRNA. Such 21 nt siRNAs can cause PTGS of Pol II generated transcripts from transposons; the dsRNA generated by the non-canonical pathway can also cause Ago2, Pol V transcript scaffold and DRM2 dependent *denovo* DNA methylation. This *denovo* methylation can trigger the canonical Pol IV driven siRNA biogenesis and RdDM.

In *A. thaliana* Ago6 is shown to be involved in Pol V mediated RdDM and hence cause transcriptional gene silencing in shoot and root meristems.[43] Small interfering RNA bound to Ago4 can induce *denovo* DNA methylation by DRM2 methyltransferase. RDM1 encodes a small protein that associates with this effector complex also RDM1 and Pol V function together in the nucleoplasm in RdDM.[44] A DNMT DMS3 encodes a hinge dimerization protein similar to hinge domain region of structural maintenance of chromosome (SMC) protein which might be linked to the siRNA-based epigenetic modification and RdDM.[45] *Arabidopsis* proteins GMI1 and DMS11 have ATPase domain and work in sync with DMS3 toward siRNA mediated epigenetic silencing.[46] This DMS3-DMS11 complex form a V-shaped heterodimers to form a ring-like structure around the DNA/RNA strands so as to make them accessible to Pol V transcription and to small RNAs mediated DNA methylation.[47] Although TGS involve methylation of promoter region and PTGS the suppression of coding mRNA, both are found to be mechanistically related. In *Arabidopsis*, study on mutant lines of *ddm1* and *met1* revealed that DNA methylation and chromatin structure regulate TGS and PTGS, any mutation at DNA level can impair both TGS and PTGS.[48,49]

Studies on model systems in plants (*Arabidopsis*), animals (*C. elegans*) and fission yeast *S. pombe* have led to the discovery of components that are part of the RITS complex. RITS complex consists of a chromodomain protein Chp1 that binds to centromere, Tas3 and Ago1 protein. The Chp1-Tas3 association is important for maintaining the centromeric heterochromatin along with H3K9 methyltransferase which binds Chp1. Another complex that consists of RDP1, DCR1, and RDRC is recruited into RITS to initiate small RdDM.[50,51] RdDM is a complex process that involves multiple components to work in a co-ordinated manner for TGS to happen. RDM3 encodes a protein KTF1 that has similarity to a transcription elongation factor SPT5 of Pol II enzyme. KTF1 contains a putative NLS in the N-terminus and WG/GW rich repeats in the C-terminus region that has an Ago binding motif. It presumably binds to Pol V generated transcripts and recruits Ago4 to form a silencing complex to mediate RdDM.[52]

6.3   TRAITMODIFICATION USING VARIOUS GENE SILENCING TECHNOLOGIES BASED ON SiRNAs AND MiRNAs

6.3.1   RNA INTERFERENCE

RNA interference (RNAi) is a post-transcriptional gene silencing phenomenon based on sequence-specific degradation of target mRNA. It is capable of silencing one or multiple members of a gene family. The major components of RNAi are Dicer, *Argonaute*, RDRs and RISC. RNAi occurs when a dsRNA is processed by Dicer into small RNAs (21 nt in length), these small RNAs interact with *Argonaute*-Piwi complex to act as a trigger which bind to the endogenous homologous mRNA transcripts and mediate its cleavage through RISC causing transcriptional and post transcriptional gene silencing. A number of diverse roles have been associated with RNAi like PTGS, epigenetic modification, transposon regulation, and genome stability, and so forth.[53] The mechanism of RNAi was first observed in plants by Napoli et al. in 1990 where the term homology dependent gene silencing (HDGS) or co-suppression was coined; later Romano and Macino introduced a homologous RNA sequence in fungi *Neurospora crassa* which causes the suppression of gene of the carotenoid pigment biosynthesis pathway al-1, al-2, and al-3and they termed the mechanism as quelling.[54] In animals, this phenomenon was observed by Fire and Mello in 1998[55] while working on *C. elegans*; Fire and Mello employed genetic mapping using reciprocal crosses of WT and mutants of development timing (heterochronic mutants; Lin-4, Lin 14, let-7, and Lin 41) and provided the first evidence of involvement of small RNA, microRNA in the process of RNAi.

The initial work on RNAi in plants was done in *Petunia* when Napoli et al. introduced a chimeric CHS gene to increase the production of anthocyanin. This resulted in reduction in the levels of both endogenous as well as introduced CHS gene that give rise to either colorless flowers or flowers with patterned veins. They called this phenomenon as co-suppression, which co-ordinately silences both the transgene as well as the endogenous gene.[56] Also, Krol et al. showed that there is a reduction in the flower floral pigment of *Petunia* upon the introduction of a transgene overexpressing DFR gene which resulted in the reduced level of DFR mRNA in the flowers with reduced floral pigment. The sense inhibition of gene expression by RNA that interfere with the transcription process occur either by its interaction with the double stranded DNA or single stranded DNA[57] or because of the introduction of multiple copies of transgenes that resulted in promoter methylation.[58]

RNAi in plants can be achieved by delivering sense and antisense transgene constructs through *Agrobacterium* mediated transformation, through viruses and or through particle bombardment.[59] Sense and antisense transgene constructs were used to deliver RNAi in plants but for the efficient expression of silencing molecule, hpRNA constructs are used where sense and antisense sequence is separated by a spacer (a splicable intron) and is expressed under the control of strong constitutive promoter. Plant systems in which stable transformations is difficult to achieve, hpRNAs can be expressed transiently through particle bombardment.[60,61] Few specifically designed vector constructs are also available for inducing RNAi, such as pHANNIBAL and pKANNIBAL, these vectors possess a strong constitutive promoter, a selectable marker, and an insertion site for the target gene separated by an intron. Gateway vectors like pHELLSGATE, pWATERGATE, and pSTARGATE have also been used to mediate gene silencing in both dicots and monocots.[62] The following sections discusses a few examples where silencing has been achieved using RNAi strategy.

6.3.1.1   RNAi FOR METABOLIC ENGINEERING

Silencing strategies based on RNAi has been employed in variety of plants to alter their metabolic pathways. For instance, in *Nicotiana sylvestris* lysine production was altered by targeting the DHPS gene thus regulating its level in seeds as well as vegetative tissue.[63] In cotton, δ-cadinene synthase gene was targeted using RNAi to restrict the content of gossypol in cottonseed thereby producing seeds with high nutritional value.[64] The carotenoid and flavonoid content in tomato was enhanced by silencing the photomorphogenic gene DET1 using fruit specific promoters.[65] In 2007, Van Eck et al. reported that RNAi mediated blocking of the β-carotene hydroxylase enzyme to prevent its conversion into zeaxanthin resulted in increased beta-carotene content in potatoes.[66] In mustard, the reduced expression of εLCY increases the accumulation of β-carotene zeaxanthin, violaxanthin, and lutein in the seeds of *B. napus*.[67] Lignans in *Forsythia* cell suspension[68] and phenylpropene aroma compounds in cultivated strawberry[69] were elevated via RNAi. Similarly, the nornicotine levels were reduced in tobacco;[70] in Brassica species, the unwanted glucosinolateprogoitrin was reduced, while the glucosinolate glucoraphanin was enhanced.[71]

Starch is the key source of carbohydrate in human diet as well as a raw material for industries. It consists of two polymers amylose (linear) and amylopectin (branched) that consists of α, 1, 4-glucosidic bonds and α, 1, 6-glucosidic linkages. The content of amylose to amylopectin in any food governs its quality. RNAi has been used in various commercial food varieties to alter their levels and make them beneficial for human consumption. In wheat, silencing of SBEIIa and SBEIIb gene through RNAi led to an increase in amylose content of wheat by 70% thereby increasing the resistant starch (RS) fiber which has shown to improve the bowel health in rats and can have positive health effects in humans.[72,73] A similar approach was used in maize where the suppression of ZmSBEIIa and ZmSBEIIb through RNAi led to an increase in its amylose content[74] and in potatoes to produce high amylose plants.[75] Sweet potatoes have also been genetically manipulated using RNAi to produce amylose free starch by silencing GBSSI gene and high amylose starch by silencing SBEII gene.[76]

6.3.1.2   RNAi FOR MALE STERILITY IN PLANTS

Male sterility is an important agronomic trait especially for developing hybrid varieties for crop improvement. Many crop plants have been manipulated using RNAi to alter male fertility for such trait. The reduction in the expression of TA29 in *Nicotiana* and BCP1 in *A. thaliana* lead to pollen abortion and male sterility.[77] Genes involved in flavonoid biosynthesis pathway, CHS has been targeted for RNAi silencing to block the flavonoid synthesis pathway to elucidate its role in reproduction. Plants with reduced CHS activity have altered fruit color, hampered pollen development, affecting seed set and thereby producing parthenocarpic tomato fruit.[78] RNAi has also been used as a powerful reverse genetics tool for functionally characterizing the gene of interest. The *Oryza sativa* gene, OsGEN-L was cloned and found to belong to RAD2/XPG nuclease family and its role in male sterility has been elucidated using RNAi mediated gene silencing. Silencing of OsGEN-L led to defective microspore development.[79] UDP-glucose phosphorylase (UGPase) functions in carbohydrate metabolism, cell wall formation, starch synthesis, and many other developmental processes. Silencing of UGP1 using RNAi in rice resulted in many developmental aberrations including male sterility as it is shown to be essential for callose deposition during pollen development.[80,81] RNAi mediated silencing of the meiosis related genes resulted in sterile phenotype in *A. thaliana*.[82] SAMDC gene involved in polyamine synthesis pathway has three homologs in tomato. RNAi mediated silencing of these homologs produced plants with distorted pollen morphology hence affecting the pollen viability. There is reduction in the transcript level of SAMDC along with the PA levels.[83]

6.3.1.3   RNAi FOR DISEASE CONTROL

Several groups have attempted to employ RNAi to provide resistance against insect pests that cause huge crop loss and reduction in the yield. A few examples are described which aid the uptake of dsRNA in the cells; (a) SID (a transmembrane protein) assisted uptake through membrane channels, (b) the association of dsRNA with the vesicles indicate that the uptake might occur through receptor-mediated endocytosis, and (c) when the dsRNA released from the virus infected cells it is taken up readily by the uninfected cells so as to acquire immunity against virus.[84]

Cotton plants engineered to silence CYP6AE14, a p450 monooxygenase gene in *Helicoverpa armigera*, acquired protection against the bollworm.[85] Plants were engineered with RNAi constructs to target the genes expressed in the midgut of the insect larvae vacuolar H+ATPase (v-ATPase) and coatomer β which proved fatal for insect larvae.[86] Tobacco plants engineered with RNAi construct directed against the v-ATPase gene of whiteflies *Bemisia tabaci* conferred resistance against these whiteflies.[87] A gap gene, hunchback (hp), essential for the axial patterning in insects during zygotic stage when targeted for dsRNA mediated silencing in tobacco against *Myzus persicae* resulted in impaired reproduction.[88] Delivery of dsRNA via chloroplast genome is an evolving area of research as it has advantage over the nuclear delivery system of RNAi signal; the prokaryotic nature of chloroplast enables multigene engineering, high level of transgene expression and transgene containment.[89] In one such study, plastids when engineered to carry long dsRNA against the genes pivotal to development, CPB, ACT, and SHR render maximum protection against the Colorado potato beetle in potato. The silencing signal remains far longer in plastids as they lack any cellular machinery for RNAi. The expressed long RNA is therefore prevented from being cleaved by Dicer into short interfering RNA and insects that feed on these plants are targeted more effectively.[90] DsRNA expressed in chloroplast to target gut-specific genes hamper the larvae development and pupation of *H. armigera* in cotton plant.[91] RNAi-based methods have also been implemented extensively for pest management and insect toxicology.[92]

6.3.1.4 RNAi FOR MODIFICATION OF FLOWER COLOR AND FRAGRANCE

Flower color can be altered by genetic manipulation of genes involved in flavonoid biosynthesis pathway. Alteration of flower color is one of the primary objectives of horticultural practices to increase their commercial value. Chalcone isomerase (CHI) enzyme is a crucial enzyme that mediates the production of flavonols in plants. Suppression of CHI by RNAi led to reduction in the level of flavonoid components in pollens and petals.[93] The flower color of gentian plants were altered by RNAi mediated gene knock-down. Down-regulation of anthocyanin biosynthesis enzyme 5/3'-aromatic acyltransferase (5/3'AT) produces pale blue flowers while the suppression of both 5/3'AT and flavonoid 3', 5'-hydroxylase (F3'5'H) generated lilac-colored flowers.[94] In *Chrysanthemum morifolium*, carotenoid cleavage dioxygenase 4 (CCD4) gene is responsible for white petal development, RNAi mediated silencing of this gene leads to the production of yellow colored flower over the dominant white petal flowers. Thus, CmCCD4 gene is an important factor in determining the petal color.[95] Ethylene also regulates many physiological post-pollination events like pollen tube growth and petal wilting. Flowers emit volatile organic compounds (VOCs) to attract pollinators for reproduction. Ethylene regulates the emission of these compounds and also maintains the timings and cyclic clocks for their release. *Petunia* hybrid BSMT gene (PhBSMT) involved in the production of methylbenzoate, a VOC, when suppressed using RNAi leads to 75–99% decrease in methylbenzoate emission.[96]

6.3.1.5 RNAi FORIMPROVEMENT OF INDUSTRIAL TRAITS

The product of *SDP1* (patatin-domain triacylglycerol lipase) is responsible for catalysis of first step of TAG degradation during seed germination. RNAi lines of JcSDP1 were generated in *Jatropha* using a seed specific promoter, and these lines shows accumulation of TAGs in mature seeds, increase in seed size, altered proportion of fatty acids leading to an increase in seed oil content. These JcSDP1 deprived bioengineered plants of *Jatropha* accumulate 30% more seed lipid, hence increasing its value as an important biodiesel crop.[97] Rice starch branching enzyme (RBE) is involved in amylopectin synthesis. When RBE3 is targeted with RNAi, an increase in the content of amylopectin and growth rate was observed.[98] In maize, an important gene in lignin biosynthetic pathway caffeoyl-CoA O-methyltransferase

(CCoAOMT) encodes an enzyme that regulates the production of lignin in plant tissues. Maize straw, the post-harvest residue, is rich in protein and high in fiber content and can be used as livestock feed but owing to its high lignin content (~22%) the usage is limited. RNAi technology appears to be an excellent approach to suppress the expression of this gene.[99]

Advantages: Multiple gene family members can be knocked down with a single RNAi inducing constructs and can also silence homologous sequences in a polyploid genome. Gene knockdowns through RNAi are dominant compared to other loss of function approaches. *Disadvantages:* siRNA-based gene silencing suffers from lack of specificity as multiple small RNA is produced from a single dsRNA molecule thus leading to off-target silencing.

6.3.2   VIRUS INDUCED GENE SILENCING (VIGS)

The term virus induced gene silencing (VIGS) was coined by van Kammen[100] and refers to a strategy when a plant virus is engineered with a target gene that has to be silenced. Once the plant is infected with such a modified virus the silencing signal spreads through it causing systemic silencing of the target gene. VIGS technology utilizes viruses as vectors to introduce dsRNA in the host genome. dsRNA may originate as a replication intermediate of a ssRNA virus. The dsRNAs are then processed by Dicer and loaded into RISC in a mechanism similar to the synthesis of miRNA to mediate gene silencing. Thus VIGS can also be utilized as a reverse genetic tool to identify the function of a gene and can be applied to wide range of species. Much of the preliminary work on VIGS has been done on *N. benthamiana*, as it is easy to maintain also it is susceptible to broad range of viruses. It was first observed that a plant infected with a particular virus develops "loss-of-function" phenotype of the gene of interest. In one such report of pathogen-derived resistance it was shown that a plant infected with tobacco etch virus (TEV) develops symptoms of the infection but after 3–4 weeks, the new formed tissue develop resistance to that virus. However, the plant does not show any resistance to other viruses.[101] This mechanism was shown to occur in the cytoplasm and can be correlated with the phenomenon of sense suppression.[102] One more study showed that *Nicotiana clevelandii* develops resistance to the TBRV strain W22 after being inoculated with its culture. New emerging leaves show resistance to this virus but are susceptible to potato virus X (PVX) which indicates that the resistance is strain specific.[103]

Plant viruses are commonly ssRNA viruses or double stranded DNA viruses. RNA viruses use plant machinery to incorporate the small RNA in the genome which targets the sequence against which it has been developed. VIGS vectors have been manipulated to induce silencing of the endogenous gene of the host plant. First VIGS vector was developed using tobaccomosaic virus (TMV) harboring a fragment of the host PDS gene involved in carotenoid biosynthesis. Plants when infected with this virus, shows photobleaching.[104] Similarly other genes that are involved in the plant metabolic pathway such as chlorophyll synthesis and cell wall formation were silenced using this approach. PVX was also studied as a possible candidate of delivering a silencing signal into the host genome.[105] PVX-based vectors are stable but the virus (PVX) has narrow host range than TMV. These vectors had been exploited for VIGS for several years but the problem with these vectors is that upon infection they produce disease symptoms in the host plant which make the evaluation of the phenotype due to "loss-of-function" difficult.[106] Bean pod mottle virus(BPMV) has been used to silence PDS gene in *Glycine max*.[107]

Although this technique was developed for functional genomics studies with few alterations, other possibilities can be explored. A DNA-based VIGS vector has been altered with 35S promoter and fewer modifications so that it mediates RNA silencing via simultaneous expression of foreign gene for marker assisted gene silencing, also the insertion of noncoding sequences allows the silencing of cDNA, promoter and untranslated regions.[107] It also enables mechanical inoculation which provides an alternative to agroinoculation and biolistic methods and thus a "one step" approach for high throughput application.[108] Tobacco rattle virus (TRV)-based vectors has been exploited in number of plant species as it produces mild disease symptoms, wide host range and can infect meristem and adjacent tissues. Many RNA and DNA viruses are used for virus mediated gene silencing.

The transgene through VIGS can be delivered in the host system through variable approaches like *in vitro* transcription; *Agrobacterium* mediated transformation, mechanical inoculation, and agroinoculation. VIGS vectors are employed not only to express a transcribed segment of a gene which undergoes posttranscriptional silencing but there are few reports where these vectors are made to suppress the promoter activity post transcriptionally through via methylation.

Advantages: The first advantage is that VIGS is an easy and rapid method of gene silencing as plant develops symptoms within few weeks. Second, no need to establish stable plant transformation protocol. Third, partial sequence information is sufficient to silence a gene. Fourth, VIGS can be used for both forward and reverse genetics. Fifth, VIGS can be used to silence genes with multiple copies (e.g., genes from polyploid plants such as wheat (*Triticum aestivum*)) or multiple family members. Sixth, VIGS can be used to assess the function of genes whose mutation (or antisense-mediated knockdown) is lethal in sexually propagated plants.

Disadvantage: As VIGS does not require stable transformation system, some of the effects are transient and the silencing level is variable. VIGS being based on the principle of siRNA mediated can also have effects on non-target genes. VIGS is not amenable for all species or only those species which are infected by viruses can be targeted.

6.3.3 ARTIFICIAL MiRNA (AMiRNA) FOR GENE SILENCING

Before the development of amiRNA mediated silencing, various RNAi-based strategies were exploited for gene silencing to produce gene knockdowns and subsequently analyzing gene function. It was done by over expressing the target gene in an antisense orientation or by using RNAi vectors that has been developed with sense antisense sequence separated by an intron to produce hpRNA to silence the target gene. This often led to lack of specificity and off-target silencing and also transgene silencing. Another technique for gene silencing utilizes VIGS vectors but they need to be developed with viruses that infect the plant naturally and mediate effective gene silencing without producing much disease symptoms in the plant. As miRNAs are known to be a highly specific gene silencing molecule it can be an excellent strategy to silence the target in a highly specific manner. These short ~22 nt RNA molecules binds almost perfectly to their target mRNAs to mediate gene silencing through *Argonaute*-RISC interaction and causes suppression of gene function. Studies showed that fewer mismatches does not affect the efficacy of miRNA, taking this into consideration artificial miRNA vectors have been engineered that uses one of the endogenous precursor miRNA as a backbone to engineer synthetic miRNA whose sequence binds with our gene of interest in a nearly perfect complementarity hence and generate sequence specific silencing signal. Once in the system, it behaves exactly as the endogenous miRNA by using the biogenesis machinery. amiRNA can be designed to target a specific mRNA transcript or homologous genes.[109]

Artificial miRNA can be designed using an alignment algorithm web microRNA designer or WMD, to develop amiRNA against the any given target transcript[110,111] WMD designs amiRNAs using the target mRNA sequence and optimize it such to avoid any silencing of the off-targets. It then generates a list of amiRNAs sorted on the basis of their sensitivity and specificity. It also provides a tool for designing oligos that has to be used relative to the amiRNA sequence. Overlapping PCR is employed to engineer the artificial miRNA sequence in the native MIR319 backbone for *A. thaliana*.[110] There are several other precursor backbones that have been also been tested and developed for amiRNA construction in monocots and dicots species, ath-miR169d, ath-miRNA172a, ath-miR319a, ath-miR167b, ath-miR171a, ath-miR159a, ath-miR164b, osa-miR528, and osa-miR395. All of them have been reported to work equally effectively. In *A. thaliana*; however, artificial miRNA embedded in the backbone of ath-miR319a precursor was shown to produce more effectual silencing effects than ath-miR172a precursor.[110]

This method of incorporating the sequence into the endogenous precursor carrying a sequence against the target transcript can be time consuming and can cause PCR errors in the miRNA sequence. To avoid such errors, amiRNA constructs can also be generated through restriction sites using MIR319a and MIR395a backbone to incorporate the amiR/amiR* sequence.[112] Another method has been developed to over express amiRNA by TA cloning into pGREEN vectors.[113] A few examples of use of native and artificial miRNA for modifying traits are discussed below.

6.3.3.1 AMiRNA FOR METABOLIC ENGINEERING

In *Lemna minor*, miR166 backbone was used for expressing an amiRNA against the *A. thaliana* homologue of CHLORINA42 (CH42) which encodes a magnesium chelatase subunit required for chlorophyll production. MiR166 is found to be conserved among the mosses and the seed plants. Transgenic *L. minor* fronds showed reduced expression of CH42 and hence reduction in chlorophyll.[114] amiRNAs were generated against the two *Physcomitrella patens* genes PpFtsZ2–1 and PpGNT1 involved in chloroplast division and N-glycosylation of proteins, respectively. PpFtsZ2–1 null mutant shows macrochloroplast impeded defective plastid division whereas the PpGNT1 knockdown plants did not show any significant defect.[115] Due to ever-increasing fuel demand green alga *Chlamydomonas reinhardtii* has become the most studied alternative in the biofuel industry. A photosystem II related OEE2 gene involved in photosynthetic oxygen evolution targeted using amiRNA resulted in increased H_2 production, an excellent biofuel alternative.[116]

6.3.3.2 AMiRNA TO INDUCE PATHOGEN RESISTANCE

Effective resistance against viruses can be achieved by targeting the virus specific coat protein transcripts as has been reported in several plant species like *Nicotiana* sp.[117] and grapevine.[118] Also, genes involved in viral replication and mobility have been targeted for suppression. For generating efficient plant resistance against viral infection it is important to take into consideration certain parameters that determine the amiRNA design.[119] A DNA virus, tobacco leaf curl virus, requires coat proteins AV1 and AV2 along with other important genes for its replication and maintenance in host. amiRNA was synthesized to target these coat proteins in tomato. Plants expressing the amiRNA become tolerant to infection with geminivirus ToLCNDV.[120] Aphid resistance in *N. benthamiana* was acquired using amiRNA expressed against the key regulatory gene MpAChE2 (encodes acetylcholinesterase) of a hemiptera aphid *M. persicae*.[121] Cotton plants resistance to CLCuBuV virus were generated using amiRNA against the V2 gene important for pathogenesis in virus mobility in the host.[122] P25 is a movement protein in PVX whereas HC-pro regulates the virus mobility in Potato Virus Y (PVY) and prevents PTGS signaling and impairs the silencing response in the host. Tobacco plants highly resistant to PVX and PVY were generated using amiRNA against movement proteins.[117] Artificial miRNA against HC-pro of WSMV in *Triticum* developed resistance against this virus.[123] Broad range resistance to cucumber mosaic virus (CMV) and other non-related viruses can be obtained in tomato using amiRNA against the 2a/2b gene and 3'UTR region of cucumovirus.[17]

6.3.3.3 AMiRNA TO STUDY PLANT DEVELOPMENT AND FUNCTIONAL GENOMICS

In *O. sativa*, amiRNA was expressed to suppress Eui1 gene that controls internode elongation, in the cytoplasmic male sterile A lines whereas a target mimic MIM gene was expressed in the R line which suppresses the amiRNA-eui1 in the hybrids to produce semi-dwarf plants with panicle exertion and improved yield.[124]

amiRNA technology can be used as a reverse genetic tool to elucidate function of various genes regulating major developmental processes. The floral organ identity genes from *Arabidopsis* AP1 and CAL1 were targeted using amiRNA with slight modification in its design, having 5' mismatch to its amiRNA* counterpart and perfect complementary to the target, showed highly effective gene silencing. Both the genes were also targeted simultaneously by expressing heteromeric amiRNAs in tandem.[125]

6.3.3.4 AMiRNA TO ENGINEER MALE STERILITY

In *Solanum melongena*, amiRNA expressed under the anther specific promoter cause the suppression of TAF genes which render plant male sterile. SmTAf10 and SmTAF13 mainly expressed in anthers and ovary. The expression of these genes in anther occurs early at the time of tapetum development and remains till the release of microspores whereas in ovary the expression remains throughout the placenta formation and ovule development. For anther specific silencing, the amiRNA constructs for SmTAF10 and SmTAF13 were expressed under the control of tobacco pTA29 and pNTM19 promoters specific for tapetum and microspore, respectively. Male sterility can be reversed by ethanol induction of SlTAF10 and SlTAF13 genes placed under the control of ethanol inducible system.[126]

Advantages: This is a breakthrough technology in the era of RNA-based silencing as it has several advantages over other RNAi methods as it is precise and accurate with little off-target silencing and single amiRNA can be designed to target several homologous loci in a genome at a time.[110]

Disadvantages: The availability of transcriptome and genome sequence information of a plant is a major pre-requisite before an accurate prediction and design of artificial miRNA and target can be made.

6.3.4 AMiRNAs CONTROL IMPORTANT AGRONOMIC TRAITS

O. sativa miRNA 397 (OsmiR 397) regulate the expression of OsLAC gene. OsmiR397 overexpression lines down regulates the expression of this OsLAC thereby affecting the sensitivity of plants to brassinosteroids and increases the grain yield, and affects panicle branching. It was further shown that OsLAC and OsmiR397 are involved in brassinosteroid-mediated regulation of plant development.[127] Therefore OsmiR397 is a putative candidate toward increasing the grain yield in rice. *Auxin response factor* (ARF) transcription factors are known to have an important role in plant development. They bind to the promoter region to activate many auxin-induced genes. In *A. thaliana* it has shown that ARF6 and ARF8 regulate both the vegetative and reproductive growth, also they are found to have overlapping functions. Overexpression of miR167 in tomato lead to severe developmental defects like short internodes and smaller leaves. Plants overexpressing miR167 showed significant reduction in plant fertility due to defects in pollen recognition on stigmatic surface and shorter styles, thus making mir167/ARFs a possible candidate for regulating plant fertility.[128] Small RNA sequencing in *B. napus* low seed oil content (L) cultivar and high seed oil content (H) cultivar was done to generate small RNA expression profile in the developing siliques. Several miRNAs namely, miR390, miR167, miR156, miR2111, miR6028, and miR6029 have been identified to be involved in early embryonic development and regulation of seed oil content.[129] Plants have been genetically engineered to make them resistant to pathogen attack. Secondary metabolites are known to protect plant from pathogen attack, miR393 targets T1R1/AFB auxin receptors and thereby regulates the auxin signaling. Auxin represses the salicylic acid pathway whereas overexpression of miR393 in *A. thaliana* suppresses auxin and makes plants resistant biotrophs while susceptible to necrotrophs. This indicates that redirection of the plant secondary metabolic flow toward glucosinolates and decrease camalexin levels can be achieved by regulating ARF1 and ARF9.[130] SPL genes are known to regulate a large number of processes in plant development including regulation of phase change and flowering; they in turn are regulated by miR156. Constitutive expression of miR156 elevates the level of anthocyanin accumulation in plant tissue and phase change.[131] Members of GA-MYB transcription factors namely MYB33, MYB65, and MYB101 are regulated by miR159 and it has been shown that overexpression of miR159 induces male sterility in plants. MiR159 therefore acts as molecular switch to control the expression of GAMYB like transcription factors that is involved in various aspects of plant development.[132]

Several genes are not regulated via miRNA machinery and yet are important targets for crop modification. In order to expand the repertoire of genes that can be targeted via miRNA machinery, the concept of artificial miRNA was developed.[110] The following section presents a few examples where artificial miRNA has been used for trait modification in plants. Table 6.1 is a representative collection of traits that have been altered using various forms of gene silencing.

6.4   CONCLUSION

Small RNA silencing has matured as a promising technique for functional genomics studies as several of these are involved as endogenous regulatory molecules that control the expression of genes involved in many fundamental

TABLE 6.1 Trait Modification Employing Various Gene Silencing Strategies. RNAi:

Trait	System used	Technology	Gene targeted	Effect on plant	Reference
Secondary metabolites	Papaver somniferum		Codeinone reductase	Increased Reticuline	[133]
	Panax ginseng		Dammarenediol synthase (DDS)	reduced ginsenoside	[134]
	Artemisia annua	RNAi	Squalene synthase (SQS)	Increased artemisinin	[135]
Male sterility	Nicotiana tabacum		TA29	Pollen abortion	[77]
	Arabidopsis Thaliana	RNAi	Bcp1	Pollen abortion	[136]
	Nicotiana tabacum		Chalcone isomerase (CHI)	Reduced pigmentation and flavanoid composition	[93]
	Gentiana sp.		F30 50 H and 5/30 AT gene	Reduced flower colour	[94]
Flower colour and fragrance	Chrysanthemum	RNAi	CmCCD4a	Change in flower colour	[95]
	Petunia hybrida		PhBSMT	Reduction in methylbenzoate production	[96]
	Rosa hybrida		DFR gene	Overexpress Iris DFR gene to produce violet colour	[137]
	Torenia hybrida		CHS gene	Flower colour changed to white	[138]
	Jatropha curcas		JcSDP1	Increased lipid content in seeds	[97]
Industrial traits	Oryza sativa	RNAi	RBE3	Increase in amylose content	[98]
	Zea mays		CCoAOMT	Decrease in lignin content in maize straw	[99]

TABLE 6.1 *(Continued)*
VIGS:

Virus source	Genus	Genome	Host sp.	Reference
Tobacco mosaic virus (TMV)	Tobamovirus	ssRNA	*Nicotiana* sp.	[104,139]
Potato virus X (PVX)	Potexvirus	ssRNA	*N. benthamiana*, *Solanum* sp.	[105,140]
Tobacco rattle virus (TRV)	Tobravirus	ssRNA	*Arabidopsis, Solanum* sps., *Eschscholzia californica*	[141–143]
Bean pod mottle virus (BPMV)	Comovirus	RNA	*Glycine max*	[107,144]
Pea early browning virus (PEBV)	Tobravirus	ssRNA	*Pisum sativum, Medicago truncatula, Lathyrus odoratus*	[145,146]
Satellite tobacco mosaic virus (STMV)	N. A.	RNA satellite virus	*Nicotiana tabacum*	[147]
Poplar mosaic virus (PopMV)	Carlavirus	ssRNA	*N. benthamiana* (transgene silencing)	[148]
Sunnhemp mosaic virus (SHMV)	Tobamovirus	ssRNA	*Medicago truncatula*	[149]
Tomato bushy shunt virus (TBSV) CaLCuV	Tombusvirus	ssRNA	*N. benthamiana*	[150]
Apple latent spherical virus (ALSV)	Cheravirus	ssRNA	*A. thaliana, N. benthamiana, Glycine max*	[151,152]
Grapewine virus A (GVA)	Vitivirus	ssRNA	*N. benthamiana, Vitis vinifera*	[153]
Cucumber mosaic virus (CMV)	Cucumovirus	ssRNA	*Antirrhinum majus, Glycine max*	[154], [155]
Turnip yellow mosaic virus (TYMV)	Tymovirus	ssRNA	*A. thaliana*	[156]
Cotton leaf curl virus (CLCV)	Begomovirus	ssDNA	*Gossypium hirsutum*	[157]
Tomato golden mosaic virus (TGMV)	Begomovirus	ssDNA	*N. benthamiana*	[158]

TABLE 6.1   (Continued)

Virus source	Genus	Genome	Host sp.	Reference
African cassava mosaic virus (ACMV)	Begomovirus	ssDNA	*N. benthamiana*	[159]
Barley strip mosaic virus (BSMV)	Hordeivirus	ssRNA	*Triticum aestivum*	[160]
Brome mosaic virus (BMV)	Bromovirus	ssRNA	*Zea mays*	[161]
Rice tungro bacilliform virus (RTBV)	Tungrovirus	dsDNA	*Oryza sativa*	[162]
Cymbidium mosaic virus (CymMV)	Potexvirus	ssRNA	*Phalaenopsis*	[163]

Artificial miRNA:

Trait	Plant species	amiRNA backbone used	Gene targeted	Phenotypic effect	Reference
Male sterility	*Solanum melongena*	ath-miR319a	SmTAF10 and SmTAF13	Male sterile transgenic lines	[126]
Pathogen resistance	*Vitis vinifera*	ath-miR319a	Coat protein genes	Provide resistance against *Grapevine fanleaf virus* (GFLV)	[118]
	Nicotiana sp.	ath-miR171a	MpAChE2	Resistance against hemipteran aphid	[121]
	Nicotiana benthamiana	cotton miR169a	V2 gene	Resistance against Begomovirus *Cotton Leaf Curl Burewala Virus* (CLCuBuV)	[122]
	Nicotiana tabacum	ath-miR159a,-167b, -177a	P25 and HC-pro	Resistance against PVX and PVY viruses	[117]
	Solanum lycopersicum	ath-miR159a	2a/2b gene and 3'UTR region	Broad range resistance to CMV and other viruses like TMV and TYLCV	[17]
	Triticum sp.	Rice-miR395	HC-pro	Affects replication and mobility wheat streak mosaic virus	[123]

TABLE 6.1 (Continued)

Trait	Plant species	amiRNA backbone used	Gene targeted	Phenotypic effect	Reference
	Solanum lycopersicum	ath-miR319a	Coat proteins AV1 and AV2	Plant become resistant to ToLCNDV	[120]
Plant development	*Arabidopsis*	ath-miR172a, ath-miR319a	LFY, GUN4, FT, MADS1, MADS2, YABBY1, YABBY2, ML1	Defective growth resulted from the silencing of various MADS box genes as plant produces bleached seedlings due to the suppression of chlorophyll producing factor. Reduced leaf number, floral defects polarity defect, and trichome patterning.	[110]
	Oryza sativa	osa-MIR528	Spl11, Pds, Eui1	Spotted leaves, elongated internodal length, bleached phenotype	[164]
	Oryza sativa	osa-MIR528	Eui1	Semi dwarf plants with improved yield	[124]
	Lemna minor	*Lemna gibba*-miR166	CH42	Bleached fronds due to reduction in chlorophyll	[114]
Metabolic engineering	*Physcomitrella patens*	ath-miR319a	PpFtsZ2-1 and PpGNT1	Null mutants of PpFtsZ2-1 form macrochloroplast	[115]
	Chlamydomonas Reinhardtii	cre-miR1157	COX90, PSY and DCL-1	Light dependent growth, aldino phenotype, and less accumulation of transposon specific siRNA	[165]
	Chlamydomonas reinhardtii	cre-miR1162	OEE2	Increased H2 production	[116]

developmental processes. Our understanding of miRNA and siRNA-based gene silencing has gained in the last few years with rapid advances in a multitude of disciplines such as small RNA profiling strategies, biochemical and genetic analysis of mutants, and solving of crystal structure of several components such as RISC or miR-Ago complexes. Based on our current understanding of their mode of action, small RNAs especially, microRNA-based gene silencing technologies are now been pursued world over toward trait modification with very precise targeting. Indeed, one of the off-shoots of the knowledge gained has been to also to target those genes which are not otherwise regulated by endogenous small RNAs through artificial microRNA. Various agronomically and commercially important traits in plants such as pest control, metabolite content, and pathogen resistance are currently been manipulated using RNAi, ViGS, and through artificial miRNAs. TGS and PTGS which were earlier considered to be two discrete phenomena are now known to be controlled by common regulators; hence a detailed insight is required to further utilize them in plant genetic modification. An area that has not been discussed is the discovery and application of miRNAs as biomarkers in various human disorders, and this field is likely to see explosive advances in future.

6.5   ACKNOWLEDGMENTS

NKS gratefully acknowledges financial support from CSIR as JRF/SRF; SA would like to thank DBT (grant number BT/PR628/AGR/36/674/2011) for award of JRF. SD would like to acknowledge DBT (grant numbers BT/PR10071/AGR/36/31/2007 and BT/PR628/AGR/36/674/2011) and Delhi University R&D grant for financial support of research on miRNA and gene silencing.

KEYWORDS

- **post-transcriptional gene silencing**
- **translational repression**
- **hetero-chromatinization**
- **RNAi**
- **artificial miRNA**
- **trait modification**

REFERENCES

1. Szymañski, M.; Barciszewska, M. Z.; Ywicki, M.; Barciszewski, J. Noncoding RNA Transcripts. *J. Appl. Genet.* **2003,** *44* (1),1–9.
2. Bonnet, E.; van de Peer, Y.; Rouzze, P. The Small RNA World of Plants. *New Phytologist.* **2006,** *171,* 451–468.
3. Lee, R. C.; Feinbaum, R. L.; Ambrose, V. *The* C. Elegans Heterochronic Gene Lin-4 Encodes Small RNAs with Antisense Complementarity to Lin-14. *Cell.*1993, *75,* 843–854.
4. Llave, C.; Kasschau, K. D.; Rector, M. A. Carrington, J. C. Endogenous and Silencing Associated Small RNAs in Plants. *Plant Cell.* **2002,** *14* (7), 1605–1619.
5. Llave, C.; Xie, Z.; Kasschau, K. D.; Carrington, J. C. Cleavage of Scarecrow-like mRNA Targets Directed by a Class of *Arabidopsis* miRNA. *Science.* **2002,** *297,* 2053–2056.
6. Reinhart, B. J.; Weinstein. E. G.; Rhoades, M. W.; Bartel, B.; Bartel, D. P. MicroRNAs in Plants. *Genes Dev.* **2002,** *16,* 1616–1626.
7. Axtell, M. J. Classification and Comparison of Small RNAs from Plants. *Annu. Rev. Plant Biol.* **2013,** *64,*137–159.
8. Fang, X.; Cui, Y.; Li, Y.; Qi, Y. Transcription and Processing of Primary MicroRNAs are Coupled by ElongatorComplex in *Arabidopsis. Nat. Plants.* **2015,** *1* (6), 15075. doi:10.1038/nplants.2015.75
9. Borchert, G. M.; Lanier, W.; Davidson, B. L. RNA Polymerase III Transcribes Human MicroRNAs. *Nat. Struct. Mol. Biol.* **2006,** *13* (12),1097–1101.
10. Czech, B.; Hannon, G. J. Small RNA Sorting: Matchmaking for Argonautes. *Nat. Rev. Genet.* **2011,** *12* (1), 19–31.
11. Hibio, N.; Hino, K.; Shimizu, E.; Nagata, Y.; Ui-Tei, K. Stability of miRNA 5'Terminal and Seed Regions is Correlated with Experimentally Observed miRNA-Mediated Silencing Efficacy. *Sci. Rep.* **2012,** *2,* 996.
12. Zhang, X.; Henderson, I. R.; Lu, C.; Green, P. J.; Jacobsen, S. E. Role of RNA Polymerase IV in Plant Small RNA Metabolism. *PNAS.* **2007,** *104,* 4536–4541.
13. Yoshikawa, M.; Peragine, A.; Park, M. Y.; Poethig R. S. A Pathway for the Biogenesis of Trans-Acting siRNAs in *Arabidopsis. Genes Dev.* **2005,** *19* (18), 2164–2175.
14. Katiyar-Agarwal, S.; Morgan, R.; Dahlbeck, D.; Borsani, O.; Villegas, A.; Zhu, J. K.; Staskawicz, B. J.; Jin, H. A Pathogen-Inducible Endogenous siRNA in Plant Immunity. *Proc. Nat. Acad. Sci. USA.* **2006,** *103* (47), 18002–18007.
15. Britto-Kido, S. D. A.; Ferreira-Neto, J. R. C.; Pandolfi, V.; Marcelino-Guimarães, F. C.; Nepomuceno, A. L.; Vilela Abdelnoor, R.; Benko-Issepon, A. M.; Kido, E. A. Natural Antisense Transcripts in Plants: A Review and Identification in Soybean Infected with *Phakopsora pachyrhizi* SuperSAGE Library. *Sci. World J.* **2013,** *26,* 219798.
16. Wang, X. J.; Gaasterland, T.; Chua, N. H. Genome-Wide Prediction and Identification of Cis-Natural Antisense Transcripts in *Arabidopsis thaliana. Genome Biol.* **2005,** *6* (4), R30.
17. Zhang, X.; Li, H.; Zhang, J.; Zhang, C.; Gong, P.; Ziaf, K.; Xiao, F.; Ye, Z. Expression of Artificial MicroRNAs in Tomato Confers Efficient and Stable Virus Resistance in a Cell-Autonomous Manner. *Transgenic Res.* **2011,** *20,* 569–581.
18. Zhang, X.; Xia, J.; Lii, Y. E.; Barrera-Figueroa, B. E.; Zhou, X.; Gao, S.; Lu, L.; Niu, D.; Chen, Z.; Leung, C.; Wong, T.; Zhang, H.; Guo, J.; Li, Y.; Liu, R.; Liang, W.; Zhu, J. K.; Zhang, W. Jin, H. Genome-Wide Analysis of Plant Nat-siRNAs Reveals Insights into Their Distribution, Biogenesis and Function. *Genome Biol.* **2012,** *13* (3), R20.

19. Katiyar-Agarwal, S.; Gao, S.; Vivian-Smith, A.; Jin, H. A Novel Class of Bacteria-Induced Small RNAs in *Arabidopsis. Genes Dev.* **2007,** *21* (23), 3123–3134.
20. Preall, J. B.; Sontheimer, E. J. RNAi: RISC Gets Loaded. *Cell.* **2005,** *123* (4), 543–545.
21. Pratt, A. J.; MacRae, I. J. The RNA-Induced Silencing Complex: A Versatile Gene-Silencing Machine. *J. Biol. Chem.* **2009,** *284* (27), 17897–17901.
22. Wang, Y.; Sheng, G.; Juranek, S.; Tuschl, T.; Patel, D. J. Structure of the Guide-Strand-Containing ARGONAUTE Silencing Complex. *Nature.* **2008,** *456* (7219), 209–213.
23. Lewis, B. P.; Shih, I. H.; Jones-Rhoades, M. W.; Bartel, D. P.; Burge, C. B. Prediction of Mammalian MicroRNA Targets. *Cell.* **2003,** *115* (7), 787–798.
24. Lewis, B. P.; Burge, C. B.; Bartel, D. P. Conserved Seed Pairing, Often Flanked by Adenosines, Indicates that Thousands of Human Genes are MicroRNA Targets. *Cell.* **2005,** *120* (1), 15–20.
25. Schirle, N. T.; Sheu-Gruttadauria, J.; Macrae, I. J. Gene Regulation: Structural Basis for MicroRNA Targeting. *Science.* **2014,** *346,* 608.
26. Tolia, N. H.; Joshua-Tor, L. Slicer and the ARGONAUTEs. *Nat. Chem. Biol.* **2007,** *3* (1), 36–43.
27. Mallick, B.; Ghosh, Z. Probing Evolutionary Biography of MicroRNAs and Associated Factors. *Curr. Genomics.* **2012,** *13* (2), 144–152.
28. Lanet, E.; Delannoy, E.; Sormani, R.; Floris, M.; Brodersen, P.; Crété, P.; Voinett, O.; Robaglia, C. Biochemical Evidence for Translational Repression by *Arabidopsis* MicroRNAs. *Plant Cell.* **2009,** *21* (6), 1762–1768.
29. Reynoso, M. A.; Blanco, F. A.; Bailey-Serres, J.; Crespi, M.; Zanetti, M. E. Selective Recruitment of mRNAs and miRNAs to Polyribosomes in Response to Rhizobia Infection in *Medicago truncatula. Plant J.* **2013,** *73* (2), 289–301.
30. Iwakawa, H. O.; Tomari, Y. Molecular Insights into MicroRNA-Mediated Translational Repression in Plants. *Mol. Cell.* **2013,** *52* (4), 591–601.
31. Braun, J. E.; Truffault, V.; Boland, A.; Huntzinger, E.; Chang, C. T.; Haas, G.; Weichenreider, O.; Coles, M.; Izaurralde, E. A Direct Interaction between DCP1 and XRN1 Couples mRNA Decapping to 5'Exonucleolytic Degradation. *Nat. Struct. Mol. Biol.* **2012,** *19* (12), 1324–1331.
32. Nishihara, T.; Zekri, L.; Braun, J. E.; Izaurralde, E. MiRISC Recruits Decapping Factors to miRNA Targets to Enhance Their Degradation. *Nucleic Acids Res.* **2013,** *41* (18), 8692–8705.
33. Fukaya, T.; Tomari, Y. MicroRNAs Mediate Gene Silencing via Multiple Different Pathways in *Drosophila. Mol. Cell.* **2012,** *48* (6), 825–836.
34. Fabian, M. R.; Sonenberg, N.; Filipowicz, W. Regulation of mRNA Translation and Stability by MicroRNAs. *Annu. Rev. Biochem.* **2010,** *79,* 351–379.
35. Ambros, V. The Functions of Animal MicroRNAs. *Nature.* **2004,** *431,* 350–355.
36. Wassenegger, M.; Heimes, S.; Riedel, L.; Sanger, H. L. RNA-Directed *De Novo* Methylation of Genomic Sequences in Plants. *Cell.* **1994,** *76,* 567–576.
37. Matzke, M. A.; Mosher, R. A. RNA-Directed DNA Methylation: An Epigenetic Pathway of Increasing Complexity. *Nat. Rev. Genet.* **2014,** *15,* 394–408.
38. Jones, L.; Ratcliff, F. Baulcombe, D. C. RNA-Directed Transcriptional Gene Silencing in Plants can be Inherited Independently of the RNA Trigger and Requires Met1 for Maintenance. *Curr. Biol.* **2001,** *11,* 747–757.
39. Wierzbicki, A.; Ream, T. S.; Haag, J. R.; Pikaard, C. S. RNA Polymerase V Transcription Guides ARGONAUTE4 to Chromatin. *Nat. Genet.* **2009,** *41,* 630–634.

40. Naumann, U.; Daxinger, L.; Kanno, T.; Eun, C.; Long, Q.; Lorkovic, Z. J.; Matzke, M.; Matzke, A. J. Genetic Evidence that DNA Methyltransferase DRM2 has a Direct Catalytic Role in RNA-Directed DNA Methylation in *Arabidopsis thaliana*. *Genetics*. **2011,** *187,* 977–979.

41. Haag, J. R.; Pikaard, C. S. Multisubunit RNA Polymerases IV and V: Purveyors of Non-Coding RNA for Plant Gene Silencing. *Nat. Rev. Mol. Cell Biol.* **2011,** *12,* 483–492.

42. Olmedo-Monfil, V.; Duran-Figueroa, N.; Arteaga-Vázquez, M.; Demesa-Arévalo, E.; Autran, D.; Grimanelli, D.; Slotkin, R. K.; Martienssen, R. A.; Vielle-Calzada, J. P. Control of Female Gamete Formation by a Small RNA Pathway in Arabidopsis. *Nature.* **2010,** *464,* 628–632.

43. Eun, C.; Lorkovic, Z. J.; Naumann, U.; Long, Q.; Havecker, E. R.; Simon, S. A.; Meyers, B. C.; Matzke, A. J. M.; Matzke, M. AGO6 Functions in RNA-Mediated Transcriptional Gene Silencing in Shoot and Root Meristems In *Arabidopsis thaliana. PLoS ONE.* **2011,** *6,* e25730.

44. Gao, Z.; Liu, H. L.; Daxinger, L.; Pontes, O.; He, X.; Qian, W.; Lin, H.; Xie, M.; Lorkovic, Z. J.; Zhang, S.; Miki, D.; Zhan, X.; Pontier, D.; Thierry, L.; Jin, H.; Matzke, A. J. M.; Matzke, M.; Pikaard, C. S.; Zhu, J. K. An RNA Polymerase II-and AGO4-Associated Protein Acts in RNA-Directed DNA Methylation. *Nature.* **2010,** *465* (7294), 106–109.

45. Kanno, T.; Bucher, E.; Daxinger, L.; Huettel, B.; Bohmdorfer, G.; Gregor, W.; Kreil, D. P.; Matzke, M.; and Matzke, A. J. A Structural-Maintenance-of-Chromosomes Hinge Domain-Containing Protein is Required for RNA-Directed DNA Methylation. *Nat. Genet.* **2008,** *40,* 670–675.

46. Lorkovic, Z. J.; Naumann, U.; Matzke, A. J. M.; Matzke, M. Involvement of a GHKLATPase in RNA-Directed DNA Methylation in *Arabidopsis thaliana. Curr. Biol.* **2012,** *22,* 933–938.

47. Bender, J. RNA-Directed DNA Methylation: Getting a Grip on Mechanism. Curr. Biol. **2012,** *22* (10), 933–938.

48. Morel, J. B.; Mourrain, P.; Béclin, C.; Vaucheret, H. DNA Methylation and Chromatin Structure Affect Transcriptional and Post-Transcriptional Transgene Silencing in *Arabidopsis. Curr. Biol.* **2000,** *10* (24), 1591–1594.

49. Sijen, T.; Vijn, I.; Rebocho, A.; van Blokland, R.; Roelofs, D.; Mol, J. N.; Kooter, J. M. Transcriptional and Posttranscriptional Gene Silencing are Mechanistically Related. *Curr. Biol.* **2001,** *11* (6), 436–440.

50. Ekwall, K. The RITS Complex—a Direct Link between Small RNA and Heterochromatin. *Mol. Cell.* **2004,** *13* (3), 304–305.

51. DeBeauchamp, J. L.; Moses, A.; Noffsinger, V. J.; Ulrich, D. L.; Job, G.; Kosinski, A. M.; Partridge, J. F. Chp1-Tas3 Interaction is Required to Recruit RITS to Fission Yeast Centromeres and for Maintenance of Centromeric Heterochromatin. *Mol. Cell. Biol.* **2008,** *28* (7), 2154–2166.

52. He, X. J.; Hsu, Y. F.; Zhu, S.; Wierzbicki, A. T.; Pontes, O.; Pikaard, C. S.; Liu, H. L.; Wang, C. S.; Jin, H.; Zhu, J. K. An Effector of RNA-Directed DNA Methylation in *Arabidopsis* is an ARGONAUTE 4-and RNA-Binding Protein. *Cell.* **2009,** *137* (3), 498–508.

53. Castel, S. E.; Martienssen, R. A. RNA Interference in the Nucleus: Roles for Small RNAs in Transcription, Epigenetics and Beyond. *Nat. Rev. Genet.* **2013,** *14,* 100–112.

54. Romano, N.; Macino, G. Quelling: Transient Inactivation of Gene Expression in *Neurospora crassa* by Transformation with Homologous Sequences. *Mol. Microbiol.* **1992,** *6,* 3343–3353.

55. Fire, A.; Xu, S. Q.; Montgomery, M. K.; Kostas, S. A.; Driver, S. E.; Mello, C. C. Potent and Specific Genetic Interference by Double-Stranded RNA in *Caenorhabditis elegans*. *Nature.* **1998,** *391,* 806–811.

56. Napoli, C.; Lemieux, C.; Jorgensen, R. Introduction of a Chimeric Chalcone Synthase Gene into Petunia Results in Reversible Co-Suppression of Homologous Genes in Trans. *Plant Cell.* **1990,** *2,* 279–289.

57. Van der Krol, A. R.; Mur, L. A.; Beld, M.; Mol, J. N.; Stuitje, A. R. Flavonoid Genes in *Petunia*: Addition of a Limited Number of Gene Copies May Lead to a Suppression of Gene Expression. *Plant Cell.* **1990,** *2,* 291–299.

58. Matzke, M. A.; Primig, M.; Trnovsky, J.; Matzke, A. J. M. Reversible Methylation and Inactivation of Marker Genes in Sequentially Transformed Tobacco Plants. *EMBO J.* **1989,** *8,* 643–649.

59. Watson, J. M.; Fusaro, A. F.; Wang, M. B.; Waterhouse, P. M. RNA Silencing Platforms in Plants. *FEBS Lett.* **2005,** *579,* 5982–5987.

60. Kohli, A.; Gahakwa, D.; Vain, P.; Laurie, D.; Christou, P. Transgene Expression in Rice Engineered through Particle Bombardment: Molecular Factors Controlling Stable Expression and Transgene Silencing. *Planta.* **1999,** *208,* 88–97.

61. Travella, S.; Klimm, T.; Keller, B. RNA Interference-Based Gene Silencing as an Efficient Tool for Functional Genomics in Hexaploid Bread Wheat. *Plant Physiol.* **2006,** *142,* 6–20.

62. Rukavtsova, E. B.; Alekseeva, V. V.; and Buryanov, Y. I. The Use of RNA Interference for the Metabolic Engineering of Plants (Review). *Russ. J. Bioorg. Chem.* **2010,** *36* (2), 146–156.

63. Tang, G.; Galili, G.; Zhuang, X. RNAi and MicroRNA: Breakthrough Technologies for the Improvement of Plant Nutritional Value and Metabolic Engineering. *Metabolomics.* **2007,** *3,* 357–369.

64. Sunilkumar, G.; Campbell, L. M.; Puckhaber, L.; Stipanovic, R. D.; Rathore, K. S. Engineering Cottonseed for Use in Human Nutrition by Tissue-Specific Reduction of Toxic Gossypol. *Proc. Nat. Acad. Sci. USA.* **2006,** *103,* 18054–18059.

65. Davuluri, G. R.; van Tuinen, A.; Fraser, P. D.; Manfredonia, A.; Newman, R.; Burgess, D.; Brummell, D. A.; King, S. R.; Palys, J.; Uhlig, J.; Bramley, P. M.; Pennings, H. M. J.; Bowler, C. Fruit-Specific RNAi-Mediated Suppression of DET1 Enhances Carotenoid and Flavonoid Content in Tomatoes. *Nat. Biotechnol.* **2005,** *23,* 890–895.

66. Van Eck, J.; Conlin, B.; Garvin, D. F.; Mason, H.; Navarre, D. A.; Brown, C. R. Enhancing Beta-Carotene Content in Potato by RNAi-Mediated Silencing of the Betacarotene Hydroxylase Gene. *Am. J. Potato Res.* **2007,** *84,* 331–342.

67. Yu, B.; Lydiate, D. J.; Young, L. W.; Schafer, U. A.; Hannoufa, A. Enhancing the Carotenoid Content of Brassica Napus Seeds by Down Regulating Lycopene Epsolon Cyclase. *Transgenic Res.* **2008,** *17,* 573–585.

68. Kim, H. J.; Ono, E.; Morimoto, K.; Yamagaki, T.; Okazawa, A.; Kobayashi, A.; Satake, H. Metabolic Engineering of Lignan Biosynthesis in *Forsythia* Cell Culture. *Plant Cell Physiol.* **2009,** *50* (12), 2200–2209.

69. Hoffmann, T.; Kurtzer, R.; Skowranek, K.; Kiessling, P.; Fridman, E.; Pichersky, E.; Schwab, W. Metabolic Engineering in Strawberry Fruit Uncovers a Dormant Biosynthetic Pathway. *Metabolic. Eng.* **2011,** *13* (5), 527–531.

70. Gavilano, L. B.; Coleman, N. P.; Burnley, L. E.; Bowman, M. L.; Kalengamaliro, N. E.; Hayes, A.; Bush, L.; Siminszky, B. Genetic Engineering *of Nicotiana tabacum* for Reduced Nornicotine Content. *J. Agr. Food Chem.* **2006,** *54* (24), 9071–9078.

71. Liu, Z.; Hirani, A. H.; McVetty, P. B.; Daayf, F.; Quiros, C. F.; Li, G. Reducing Progoi-trin and Enriching Glucoraphanin in *Brassica napus* seeds through Silencing of the GSL-ALK Gene Family. *Plant Mol. Biol.* **2012**, *79* (1–2), 179–189.

72. Regina, A.; Bird, A.; Topping, D.; Bowden, S.; Freeman, J.; Barsby, T.; Kosar-Hashemi, B.; Li, Z.; Rahman, S.; Morell, M. High-Amylose Wheat Generated by RNA Interfer-ence Improves Indices of Large-Bowel Health in Rats. *Proc. Nat. Acad. Sci. USA.* **2006**, *103,* 3546–3551.

73. Sestili, F.; Janni, M.; Doherty, A.; Botticella, E.; D'Ovidio, R.; Masci, S.; Jones, H. D.; Lafiandra, D. Increasing the Amylose Content of Durum Wheat Through Silencing of the SBEIIa Genes. *BMC Plant Biol.* **2010**, *10,* 144.

74. Zhao, Y.; Li, N.; Li, B.; Li, Z.; Xie, G.; Zhang, J. Reduced Expression of Starch Branching Enzyme IIa and IIb in Maize Endosperm by RNAi Constructs Greatly Increases the Amylose Content in Kernel with Nearly Normal Morphology. *Planta.* **2015**, *241* (2), 449–461.

75. Andersson, M.; Melander, M.; Pojmark, P.; Larsson, H.; Bulow, L.; Hofvander, P. Targeted Gene Suppression by RNA Interference: An Efficient Method for Production of High-Amylose Potato Lines. *J. Biotechnol.* **2006**, *123,* 137–148.

76. Kitahara, K.; Hamasuna, K.; Nozuma, K.; Otani, M.; Hamada, T.; Shimada, T.; Fujita, K.; Suganuma, T. Physicochemical Properties of Amylose-Free and High-Amylose Starches from Transgenic Sweetpotatoes Modified by RNA Interference. *Carbohydr. Polym.* **2007**, *69,* 233–240.

77. Nawaz-ul-Rehman, M. S.; Mansoor, S.; Khan, A. A.; Zafar, Y.; Briddon, R. W. RNAi-Mediated Male Sterility of Tobacco by Silencing TA29. *Mol. Biotechnol.* **2007**, *36,* 159–165.

78. Schijlen, E. G. W. M.; Ric de Vos, C. H.; Martens, S.; Jonker, H. H.; Rosin, F. M.; Molthoff, J. W.; Tikunov, Y. M.; Angenent, G. C.; van Tunen, A. J.; Bovy, A. G. RNA Interference Silencing of Chalcone Synthase, the First Step in the Flavonoid Biosyn-thesis Pathway, Leads to Parthenocarpic Tomato Fruits. *Plant Physiol.* **2007**, *144,* 1520–1530.

79. Moritoh, S.; Miki, D.; Akiyama, M.; Kawahara, M.; Izawa, T.; Maki, H.; Shimamoto, K. RNAi-Mediated Silencing of OsGEN-L (OsGEN-like), a New Member of the RAD2/XPG Nuclease Family, Causes Male Sterility by Defect of Microspore Development in Rice. *Plant Cell Physiol.* **2005**, *46* (5), 699–715.

80. Chen, R.; Zhao, X.; Shao, Z.; Wei, Z.; Wang, Y.; Zhu, L.; Zhao, J.; Sun, M.; He, R.; He, G. Rice UDP-Glucose Pyrophosphorylase1 is Essential for Pollen Callose Deposition and Its Cosuppression Results in a New Type of Thermosensitive Genic Male Sterility. *Plant Cell.* **2007**, *19* (3), 847–861.

81. Woo, M. O.; Ham, T. H.; Ji, H. S.; Choi, M. S.; Jiang, W.; Chu, S. H.; Piao, R.; Chin, J. H.; Kim, J. A.; Park, B. S.; Seo, H. S.; Jwa, N. S.; Couch, S. M.; Koh, H. J. Inactivation of the UGPase1 Gene Causes Genic Male Sterility and Endosperm Chalkiness in Rice (*Oryza sativa* L.). *Plant J.* **2008**, *54* (2), 190–204.

82. Wang, X.; Singer, S. D.; Liu, Z. Silencing of Meiosis-Critical Genes for Engineering Male Sterility in Plants. *Plant Cell Rep.* **2012**, *31,* 747–756.

83. Sinha, R.; Rajam, M. V. Rnai Silencing of Three Homologues of S-Adenosylmethionine Decarboxylase Gene in Tapetal Tissue of Tomato Results in Male Sterility. *Plant Mol. Biol.* **2013**, *82* (1–2), 169–180.

84. Huvenne, H.; Smagghe, G. Mechanisms of dsRNA Uptake in Insects and Potential of RNAi for Pest Control: A Review. *J. Insect. Physiol.* **2010**, *56* (3), 227–235.

85. Mao, Y. B.; Tao, X. Y.; Xue, X. Y.; Wang, L. J.; Chen, X. Y. Cotton Plants Expressing CYP6AE14 Double-Stranded RNA Show Enhanced Resistance to Bollworms. *Transgenic Res.* **2011,** *20* (3), 665–673.

86. Mao, J.; Zhang, P.; Liu, C.; Zeng, F. Co-Silence of the Coatomer β and v-ATPase A Genes by siRNA Feeding Reduces Larval Survival Rate and Weight Gain of Cotton Bollworm, *Helicoverpa armigera. Pest. Biochem. Physiol.* **2015,** *118,* 71–76.

87. Thakur, N.; Upadhyay, S. K.; Verma, P. C.; Chandrashekar, K.; Tuli, R.; Singh, P. K. Enhanced Whitefly Resistance in Transgenic Tobacco Plants Expressing Double Stranded RNA of V-ATPase A Gene. *PLoS ONE.* **2014,** *9* (3), e87235.

88. Mao, J.; Zeng, F. Plant-Mediated RNAi of a Gap Gene-Enhanced Tobacco Tolerance against the *Myzus persicae. Transgenic Res.* **2014,** *23* (1), 145–152.

89. Daniell, H.; Kumar, S.; Dufourmantel, N. Breakthrough in Chloroplast Genetic Engineering of Agronomically Important Crops. *Trends Biotechnol.* **2005,** *23* (5), 238–245.

90. Zhang, J.; Khan, S. A.; Hasse, C.; Ruf, S.; Heckel, D. G.; Bock, R. Full Crop Protection From an Insect Pest by Expression of Long Double-Stranded RNAs in Plastids. *Science.* **2015,** *347* (6225), 991–994.

91. Jin, S.; Singh, N. D.; Li, L.; Zhang, X.; Daniell, H. Engineered Chloroplast dsRNA Silences Cytochrome p450 Monooxygenase, V-ATPase and Chitin Synthase Genes in the Insect Gut and Disrupts*Helicoverpa armigera*Larval Development and Pupation. *Plant Biotechnol. J.* **2015,** *13* (3), 435–446.

92. Kim, Y. H.; Issa, M. S.; Cooper, A. M.; Zhu, K. Y. RNA Interference: Applications and Advances in Insect Toxicology and Insect Pest Management. *Pest. Biochem. Physiol.* **2015,** *120,* 109–117.

93. Nishihara, M.; Nakatsuka, T.; Yamamura, S. Flavonoid Components and Flower Color Change in Transgenic Tobacco Plants by Suppression of Chalcone Isomerase Gene. *FEBS Lett.* **2005,** *579,* 6074–6078.

94. Nakatsuka, T.; Mishiba, K. I.; Kubota, A.; Abe, Y.; Yamamura, S.; Nakamura, N.; Tanaka, Y.; Nishihara, M. Genetic Engineering of Novel Flower Colour by Suppression of Anthocyanin Modification Genes in Gentian. *J. Plant Physiol.* **2010,** *167,* 231–237.

95. Ohmiya, A.; Kishimoto, S.; Aida, R.; Yoshioka, S.; Sumitomo, K. Carotenoid Cleavage Dioxygenase (CmCCD4a) Contributes to White Color Formation in *Chrysanthemum* petals. *Plant Physiol.* **2006,** *142* (3),1193–1201.

96. Underwood, B. A.; Tieman, D. M.; Shibuya, K.; Dexter, R. J.; Loucas, H. M.; Simkin, A. J.; Sims, C. A.; Schmelz, E. A.; Klee, H. J.; Clark, D. G. Ethylene-Regulated Floral Volatile Synthesis in Petunia Corollas. *Plant Physiol.* **2005,** *138,* 255–266.

97. Kim, M. J.; Yang, S. W.; Mao, H. Z.; Veena, S. P.; Yin, J. L.; Chua, N. H. Gene Silencing of Sugar-Dependent 1 (JcSDP1), Encoding a Patatin-Domain Triacylglycerol Lipase, Enhances Seed Oil Accumulation in *Jatropha curcas. Biotechnol. Biofuels.* **2014,** *7* (1), 36.

98. Jiang, H.; Zhang, J.; Wang, J.; Xia, M.; Zhu, S.; Cheng, B. RNA Interference-Mediated Silencing of the Starch Branching Enzyme Gene Improves Amylose Content in Rice. *Genet. Mol. Res.* **2013,** *2* (3), 2800–2808.

99. Li, X.; Chen, W.; Zhao, Y.; Xiang, Y.; Jiang, H.; Zhu, S.; Cheng, B. Down Regulation of Caffeoyl-CoA O-Methyltransferase (CCoAOMT) by RNA Interference Leads to Reduced Lignin Production in Maize Straw. *Genet. Mol. Biol.* **2013,** *36* (4), 540–546.

100. vanKammen, A. Virus-induced Gene Silencing in Infected and Transgenic Plants. *Trends Plant Sci.* **1997,** *2,* 409–411.

101. Lindbo, J. A.; Silva-Rosales, L.; Proebsting, W. M.; Dougherty, W. G. Induction of a Highly Specific Antiviral State in Transgenic Plants: Implications for Regulation of Gene Expression and Virus Resistance. *Plant Cell.* **1993,** *5,* 1749–1759.

102. Smith, H. A.; Swaney, S. L.; Parks, T. D.; Wernsman, E. A.; Dougherty, W. G. Transgenic Plant Virus Resistance Mediated by Untranslatable Sense RNAs: Expression, Regulation, and Fate of Nonessential RNAs. *Plant Cell.* **1994,** *6,* 1441–1453.

103. Ratcliff, F.; Harrison, B. D.; Baulcombe, D. C. A Similarity between Viral Defence and Gene Silencing in Plants. *Science.* **1997,** *276,* 1558–1560.

104. Kumagai, M. H.; Donson, J.; Della-Ciopa, G.; Harvey, D.; Hanley, K.; Grill, L. K. Cytoplasmic Inhibition of Carotenoid Biosynthesis with Virus-Derived RNA. *Proc. Natl. Acad. Sci. USA.* **1995,** *92,* 1679–1683.

105. Ruiz, M. T.; Voinnet, O.; Baulcombe, D. C. Initiation and Maintenance of Virus-Induced Gene Silencing. *Plant Cell.* **1998,** *10,* 937–946.

106. Ratcliff, F.; Martin-Hernandez, A. M.; Baulcombe, D. C. Tobacco Rattle Virus as a Vector for Analysis of Gene Function by Silencing. *Plant J.* **2001,** *25,* 237–245.

107. Zhang, C.; Ghabrial, S. A. Development of Bean Pod Mottle Virus-Based Vectors for Stable Protein Expression and Sequence-Specific Virus-Induced Gene Silencing in Soybean. *Virology.* **2006,** *344,* 401–411.

108. Zhang, C.; Bradshaw, J. D.; Whitham, S. A.; Hill, J. H. The Development of an Efficient Multipurpose Bean Pod Mottle Virus Viral Vector Set for Foreign Gene Expression and RNA Silencing. *Plant Physiol.* **2010,** *153,* 52–65.

109. Eamens, A. L.; Agius, C.; Smith, N. A.; Waterhouse, P. M.; Wang, M. B. Efficient Silencing of Endogenous MicroRNAs Using Artificial MicroRNAs in *Arabidopsis thaliana. Mol. Plant.* **2011,** *4* (1), 157–170.

110. Schwab, R.; Ossowski, S.; Riester, M.; Warthmann, N.; Weigel, D. Highly Specific Gene Silencing by Artificial MicroRNAs in *Arabidopsis. Plant Cell.* **2006,** *18,* 1121–1133.

111. Ossowski, S.; Schwab, R.; Weigel, D. Gene Silencing in Plants Using Artificial MicroRNAs and Other Small RNAs. *Plant J.* **2008,** *53,* 674–690.

112. Liang, G.; He, H.; Li, Y.; Yu, D. A New Strategy for Construction of Artificial miRNA Vectors in *Arabidopsis. Planta.* **2012,** *235* (6), 1421–1429.

113. Wang, C.; Yin, X.; Kong, X.; Li, W.; Ma, L.; Sun, X.; Guan, Y.; Todd, C. D.; Yang, Y.; Hu, X. A Series of TA-Based and Zero-Background Vectors for Plant Functional Genomics. *PLoS ONE.* **2013,** *8* (3), e59576.

114. Cantó-Pastor, A.; Mollá-Morales, A.; Ernst, E.; Dahl, W.; Zhai, J.; Yan, Y.; Meyers, B. C.; Shanklin, J.; Martienssen, R. Efficient Transformation and Artificial miRNA Gene Silencing in *Lemna minor. Plant Biol.* **2015,** *17,* 59–65.

115. Khraiwesh, B.; Ossowski, S.; Weigel, D.; Reski, R.; Frank, W. Specific Gene Silencing by Artificial MicroRNAs in *Physcomitrella patens*: An Alternative to Targeted Gene Knockouts. *Plant Physiol.* **2008,** *148,* 684–693.

116. Li, H.; Zhang, L.; Shu, L.; Zhuang, X.; Liu, Y.; Chen, J.; Hu, Z. Sustainable Photosynthetic H2-Production Mediated by Artificial miRNA Silencing of OEE2 Gene in Green Alga *Chlamydomonas reinhardtii. Int. J. Hydrogen Energy.* **2015,** *40,* 5609–5616.

117. Ai, T.; Zhang, L.; Gao, Z.; Zhu, C. X.; Guo, X. Highly Efficient Virus Resistance Mediated by Artificial MicroRNAs that Target the Suppressor of PVX and PVY in Plants. *Plant Biol.* **2011,** *13,* 304–316.

118. Jelly, N. S.; Schellenbaum, P.; Walter, B.; Maillot, P. Transient Expression of Artificial MicroRNAs Targeting *Grapevine fan leaf virus* and Evidence for RNA Silencing in Grapevine Somatic Embryos. *Transgenic Res.* **2012,** *21,* 1319–1327.

119. Duan, C. G.; Wang, C. H.; Fang, R. X.; Guo, H. S. Artificial MicroRNAs Highly Accessible to Targets Confer Efficient Virus Resistance in Plants. *J. Virol.* **2008,** *82,* 11084–11095.

120. Vu, T. V.; Roy Choudhury, N.; Mukherjee, S. K. Transgenic Tomato Plants Expressing Artificial MicroRNAs for Silencing the Pre-Coat and Coat Proteins of a Begomovirus, *Tomato leaf curlNew Delhi virus,* Show Tolerance to Virus Infection. *Virus Res.* **2013,** *172,* 35–45.

121. Guo, H.; Song, X.; Wang, G.; Yang, K.; Wang, Y.; Niu, L.; Chen, X.; Fang, R. Plant-Generated Artificial Small RNAs Mediated Aphid Resistance. *PLoS ONE.* **2014,** *9* (5), e97410.

122. Ali, I.; Amin, I.; Briddon, R. W.; Mansoor, S. Artificial MicroRNA-Mediated Resistance against the Monopartite Begomovirus *Cotton leaf curl Burewala virus. Virol. J.* **2013,** *10,* 231.

123. Fahim, M.; Millar, A. A.; Wood, C. C.; Larkin, P. J. Resistance to *Wheat streak mosaic virus* Generated by Expression of an Artificial Polycistronic MicroRNA in Wheat. *Plant Biotechnol. J.* **2012,** *10* (2), 150–163.

124. Chen, H.; Jiang, S.; Zheng, J.; Lin, Y. Improving Panicle Exsertion of Rice Cytoplasmic Male Sterile Line by Combination of Artificial MicroRNA and Artificial Target Mimic. *Plant Biotechnol. J.* **2013,** *11* (3), 336–343.

125. Park, W.; Zhai, J.; Lee, J. Y. Highly Efficient Gene Silencing Using Perfect Complementary Artificial miRNA Targeting AP1 or Heteromeric Artificial miRNA Targeting AP1 and CAL Genes. *Plant Cell Reports.* **2009,** *28* (3), 469–480.

126. Toppino, L.; Kooiker, M.; Lindner, M.; Dreni, L.; Rotino, G. L.; Kater, M. M. Reversible Male Sterility in Eggplant *(Solanum melongena L.)* by Artificial MicroRNA-Mediated Silencing of General Transcription Factor Genes. *Plant Biotechnol. J.* **2011,** *9* (6), 684–692.

127. Zhang, Y. C.; Yu, Y.; Wang, C. Y.; Li, Z. Y.; Liu, Q.; Xu, J.; Liao, J-Y.; Wang, X. J.; Qu, L. H,. Chen, H.; Xin, P.; Yan, C.; Chu, J.; Li, H. Q.; Chen, Y. Q. Overexpression of MicroRNA OsmiR397 Improves Rice Yield by Increasing Grain Size and Promoting Panicle Branching. Nat. Biotechnol. **2013,** *31* (9), 848–852.

128. Liu, N.; Wu, S.; Van Houten, J.; Wang, Y.; Ding, B.; Fei, Z.; Clarke, T. H.; Reed, J. W.; van der Knaap, E. Down-Regulation of AUXIN RESPONSE FACTORS 6 and 8 by MicroRNA 167 Leads to Floral Development Defects and Female Sterility in Tomato. J. Exp. Bot. **2014,** *65* (9), 2507–2520.

129. Zhao, Y. T.; Wang, M.; Fu, S. X.; Yang, W. C.; Qi, C. K.; Wang, X. J. Small RNA Profiling in Two *Brassica napus* Cultivars Identifies MicroRNAs with Oil Production- and Development-Correlated Expression and New Small RNA Classes. *Plant Physiol.* **2012,** *158* (2), 813–823.

130. Robert-Seilaniantz, A.; MacLean, D.; Jikumaru, Y.; Hill, L.; Yamaguchi, S.; Kamiya, Y.; Jones, J. D. The MicroRNA miR393 Re-Directs Secondary Metabolite Biosynthesis Away from Camalexin and Towards Glucosinolates. *Plant J.* **2011,** *67* (2), 218–231.

131. Gou, J. Y.; Felippes, F. F.; Liu, C. J.; Weigel, D.; Wang, J. W. Negative Regulation of Anthocyanin Biosynthesis in *Arabidopsis* by a Mir156-Targeted SPL Transcription Factor. *Plant Cell.* **2011,** *23* (4), 1512–1522.

132. Alonso-Peral, M. M.; Li, J.; Li, Y.; Allen, R. S.; Schnippenkoetter, W.; Ohms, S.; White R. G.; Millar, A. A. The MicroRNA159-Regulated GAMYB-Like Genes Inhibit Growth and Promote Programmed Cell Death in *Arabidopsis. Plant Physiol.* **2010,** *154* (2), 757–771.

133. Allen, R. S.; Millgate, A. G.; Chitty, J. A.; Thisleton, J.; Miller, J. A. C.; Fist, A. G.; Gerlach, W. A.; Larkin, P. J. RNAi-Mediated Replacement of Morphine with the Non Narcotic Alkaloid Reticuline in *Opium poppy*. *Nat. Biotechnol.* **2004,** *22,* 1559–1566.

134. Han, J. Y.; Kwon, Y. S.; Yang, D. C.; Jung, Y. R.; Choi, Y. E. Expression and RNA Interference-Induced Silencing of The Dammarenediol Synthase Gene in *Panax ginseng.* *Plant Cell Physiol.* **2006,** *47,* 1653–1662.

135. Zhang, L.; Jing, F.; Li, F.; Li, M.; Wang, Y.; Wang, G.; Sun, X.; Tang, K. Development of Transgenic *Artemisia annua* (Chinese wormwood) Plants with an Enhanced Content of Artemisinin, an Effective Antimalarial Drug, by Hairpin-RNA Mediated Gene Silencing. *Biotechnol. Appl. Biochem.* **2009,** *52,* 199–207.

136. Tehseen, M.; Imran, M.; Hussain, M.; Irum, S.; Ali, S.; Mansoor, S.; Zafar, Y. Development of Male Sterility by Silencing Bcp1 Gene of *Arabidopsis* through RNA Interference. *Afr. J. Biotechnol.* **2010,** *9,* 2736–2741.

137. Katsumoto, Y.; Mizutani, M.; Fukui, Y.; Brugliera, F.; Holton, T. A.; Karan, M.; Nakamura, N.; Yonekura-Sakakibara, K.; Togami, J.; Pigeaire, A.; Tao, G. Q.; Nehra, N. S.; Lu, C. Y.; Dyson, B. K.; Tsuda, S.; Ashikari, T.; Kusumi, T.; Mason, J. G.; Tanaka, Y. Engineering of the Rose Flavonoid Biosynthetic Pathway Successfully Generated Blue-Hued Flowers Accumulating Delphinidin. *Plant Cell Physiol.* **2007,** *48,* 1589–1600.

138. Fukusaki, E.; Kawasaki, K.; Kajiyama, S.; An, C. I.; Suzuki, K.; Tanaka, Y.; Kobayashi, A. Flower Color Modulations of *Torenia hybrida* by Downregulation of Chalcone Synthase Genes with RNA Interference. *J. Biotechnol.* **2004,** *111,* 229–240.

139. Hiriart, J. B.; Aro, E. M.; Lehto, K. Dynamics of the VIGS-Mediated Chimeric Silencing of the *Nicotiana benthamiana* ChlH Gene and of the *Tobacco mosaic virus* Vector. *Mol. Plant Microbe. Interact.* **2003,** *16,* 99–106.

140. Faivre-Rampant, O.; Gilroy, E. M.; Hrubikova, K.; Hein, I.; Millam, S.; Loake, G. J.; Birch, P.; Taylor, M.; Lacomme, C. Potato VirusX-Induced Gene Silencing in Leaves and Tubers of Potato. *Plant Physiol.* **2004,** *134,* 1308–1316.

141. Burch-Smith, T. M.; Schiff, M.; Liu, Y.; Dinesh-Kumar, S. P. Efficient Virus-Induced Gene Silencing in *Arabidopsis. Plant Physiol.* **2006,** *142,* 21–27.

142. Brigneti, G.; Martin-Hernandez, A. M.; Jin, H.; Chen, J.; Baulcombe, D. C.; Baker, B.; Jones, J. D. Virus-Induced Gene Silencing in *Solanum* Species. *Plant J.* **2004,** *39,* 264–272.

143. Wege, S.; Scholz, A.; Gleissberg, S.; Becker, A. Highly Efficient Virusinduced Gene Silencing (VIGS) in *California poppy* (*Eschscholzia californica*): An Evaluation of VIGS as a Strategy to Obtain Functional Data from Non-Model Plants. *Ann. Bot. (Lond.).* **2007,** *100,* 641–649.

144. Juvale, P. S.; Hewezi, T.; Zhang, C.; Kandoth, P. K.; Mitchum, M. G.; Hill, J. H.; Witham, H. A.; Baum, T. J. Temporal and Spatial *Bean pod mottle virus*-Induced Gene Silencing in Soybean. *Mol. Plant Pathol.* **2012,** *13* (9), 1140–1148.

145. Constantin, G. D.; Krath, B. N.; MacFarlane, S. A.; Nicolaisen, M.; Johansen, I. E.; Lund, O. S. Virus-Induced Gene Silencing as a Tool for Functional Genomics in a Legume Species. *Plant J.* **2004,** *40,* 622–631.

146. Gronlund, M.; Constantin, G.; Piednoir, E.; Kovacev, J.; Johansen, I. E.; Lund, O. S. Virus-Induced Gene Silencing in *Medicago truncatula* and *Lathyrus odorata. Virus Res.* **2008,** *135,* 345–349.

147. Gossele, V.; Fache, I.; Meulewaeter, F.; Cornelissen, M.; Metzlaff, M. SVISS—a Novel Transient Gene Silencing System for Gene Function Discovery and Validation in Tobacco Plants. *Plant J.* **2002,** *32,* 859–866.

148. Naylor, M.; Reeves, J.; Cooper, J. I.; Edwards, M. L.; Wang, H. Construction and Properties of a Gene-Silencing Vector Based on *Poplar mosaic virus (genus Carlavirus). J. Virol. Methods.* **2005,** *124* (1–2), 27–36.

149. Varallyay, E.; Lichner, Z.; Safrany, J.; Havelda, Z.; Salamon, P.; Bisztray, G.; Burgyan, J. Development of a Virus Induced Gene Silencing Vector from a Legumes Infecting Tobamovirus. *Acta. Biol. Hung.* **2010,** *61,* 457–469.

150. Pignatta, D.; Kumar, P.; Turina, M.; Dandekar, A.; Falk, B. W. Quantitative Analysis of Efficient Endogenous Gene Silencing in *Nicotiana benthamiana* Plants Using *Tomato bushy stunt virus* Vectors that Retain the Capsid Protein Gene. *Mol. Plant Microbe. Interact.* **2007,** *20,* 609–618.

151. Igarashi, A.; Yamagata, K.; Sugai, T.; Takahashi, Y.; Sugawara, E.; Tamura, A.; Yaegashi, H.; Yamagishi, N.; Takahashi, T.; Isogai, M.; Takahashi, H.; Yoshikawa, N. Apple Latent Spherical Virus Vectors for Reliable and Effective Virus-Induced Gene Silencing among a Broad Range of Plants Including Tobacco, Tomato, *Arabidopsis thaliana,* Cucurbits, and Legumes. *Virology.* **2009,** *386,* 407–416.

152. Yamagishi, N.; Yoshikawa, N. Virus-Induced Gene Silencing of Endogenous Genes and Promotion of Flowering in Soybean by *Apple latent spherical virus*-Based Vectors. *In Soybean – Molecular Aspects of Breeding;* Sudaric, A., Ed.; InTech Publishers: Rijeka, Croatia, 2011; pp 43–56.

153. Muruganantham, M.; Moskovitz, Y.; Haviv, S.; Horesh, T.; Fenigstein, A.; Preez J. D.; Stephan, D.; Burger, J. T.; Mawassi, M. Grapevine Virus A-Mediated Gene Silencing in *Nicotiana benthamiana* and *Vitis vinifera. J. Virol. Methods.* **2009,** *155* (2), 167–174.

154. Kim, B. M.; Inaba, J. I.; Masuta, C. Virus Induced Gene Silencing in *Antirrhinum majus* Using the *Cucumber Mosaic Virus* Vector: Functional Analysis of the AINTEGU-MENTA (Am-ANT) Gene of *A. majus. Hortic. Environ. Biotechnol.* **2011,** *52,* 176–182.

155. Nagamatsu, A.; Masuta, C.; Senda, M.; Matsuura, H.; Kasai, A.; Hong, J. S.; Kitamura, K.; Abe, J.; Kanazawa, A. Functional Analysis of Soybean Genes Involved in Flavonoid Biosynthesis by Virus-Induced Gene Silencing. *Plant Biotechnol. J.* **2007,** *5,* 778–790.

156. Pflieger, S.; Blanchet, S.; Camborde, L.; Drugeon, G.; Rousseau, A.; Noizet, M.; Planchais, S.; Jupin, I. Efficient Virus-Induced Gene Silencing in *Arabidopsis* Using a 'One-Step' TYMV-Derived Vector. *Plant J.* **2008,** *56,* 678–690.

157. Tuttle, J. R.; Idris, A. M.; Brown, J. K.; Haigler, C. H.; Robertson, D. Geminivirus-Mediated Gene Silencing from *Cotton leaf crumple virus* is Enhanced by Low Temperature in Cotton. *Plant Physiol.* **2008,** *148,* 41–50.

158. Carrillo-Tripp, J.; Shimada-Beltran, H.; Rivera-Bustamante, R. Use of Geminiviral Vectors for Functional Genomics. *Curr. Opin. Plant Biol.* **2006,** *9,* 209–215.

159. Fofana, I. B.; Sangare, A.; Collier, R.; Taylor, C.; Fauquet, C. M. A Geminivirus-Induced Gene Silencing System for Gene Function Validation in Cassava. *Plant Mol. Biol.* **2004,** *56,* 613–624.

160. Lee, W. S.; Rudd, J. J.; Kanyuka, K. Virus Induced Gene Silencing (VIGS) for Functional Analysis of Wheat Genes Involved in *Zymoseptoria tritici* Susceptibility and Resistance. *Fungal Genet. Biol.* **2015,** *79,* 84–88.

161. Benavente, L. M.; Ding, X. S.; Redinbaugh, M. G.; Nelson, R.; Balint-Kurti, P. Virus-Induced Gene Silencing in Diverse Maize Lines Using the *Brome mosaic virus*-Based Silencing Vector. *Maydica.* **2012,** *57* (3), 206–214.

162. Kant, R.; Sharma, S.; Dasgupta, I. Virus-Induced Gene Silencing (VIGS) for Functional Genomics in Rice Using Rice *Tungro bacilliform virus* (RTBV) as a Vector. In *Plant*

Gene Silencing: Methods and Protocols; Springer: New York, 2015; Series 1287, pp 201–217.

163. Hsieh, M. H.; Lu, H. C.; Pan, Z. J.; Yeh, H. H.; Wang, S. S.; Chen, W. H.; Chen, H. H. Optimizing Virus-Induced Gene Silencing Efficiency with *Cymbidium mosaic virus* in Phalaenopsis Flower. *Plant Sci.* **2013,** *201,* 25–41.

164. Warthmann, N.; Chen, H.; Ossowski, S.; Weigel, D.; Hervé, P. Highly Specific Gene Silencing by Artificial miRNAs in Rice. *PLoS ONE.* **2008,** *3* (3), e1829.

165. Molnar, A.; Bassett, A.; Thuenemann, E.; Schwach, F.; Karkare, S.; Ossowski, S.; Weigel, D.; Baulcombe, D. Highly Specific Gene Silencing by Artificial MicroRNAs in the Unicellular Alga *Chlamydomonas reinhardtii. Plant J.* **2009,** *58,* 165–174.

CHAPTER 7

PLANT MOLECULAR BREEDING: WAY FORWARD THROUGH NEXT-GENERATION SEQUENCING

SAIMA KHAN[1], PANKAJ PANDOTRA[2], AJAI PRAKASH GUPTA[3],
R. K. SALGOTRA[2], MALIK MUZAFAR MANZOOR[1],
SAJAD AHMAD LONE[1], and SUPHLA GUPTA[1,4*]

[1]*Plant Biotechnology Division, Indian Institute of Integrative Medicine, Jammu 180001, Jammu and Kashmir, India*

[2]*School of Biotechnology, Sher-e-Kashmir University of Agricultural Science and Technology of Jammu, Chatha, Jammu 180009, Jammu and Kashmir, India*

[3]*Quality Control and Assurance Division, Indian Institute of Integrative Medicine, Jammu 180001, Jammu and Kashmir, India*

[4]*Faculty at Academic of Scientific and Innovative Science, Indian Institute of Integrative Medicine, Jammu 180001, Jammu and Kashmir, India*

*Corresponding author. E-mail: suphlabg@gmail.com

CONTENTS

ABSTRACT

The classical approach for molecular breeding is heavily dependent on marker-assisted selection (MAS) and the trait linked DNA markers as an alternative to support phenotypic screening. Literature has demonstrated its ability as powerful genomic tools to enhance the effectiveness and accuracy of breeding practices for crop improvement. The major worries, however, are the specificity, cost, and the sequence knowledge requirement for each species in their development. Recent technological advancement has shown that the next-generation sequencing (NGS) can decode a whole genome or transcriptome in a very short time and is the most powerful tool to identify and develop DNA/RNA molecular markers for MAS. The amalgamation of NGS technique with molecular breeding for crop improvement has been successfully exploited in many crops including *Setaria italica* Beauv, *Lupinus angustifolius* L., *Mimulus, Cajanus cajan* L., and *Linum usitatissimum* L., to name few, but many more are being experimented. Still more studies are required to make it routine. This chapter showcases some of the advanced NGS techniques and their utility in the large-scale development of molecular markers which is expected to enhance our understanding toward population genomics, mapping studies in crop improvement, and reference genome sequence assembly for non-model plant, specially.

7.1   INTRODUCTION

Plant breeding is a mission of continuously discovering and pyramiding desirable genes of agronomical and economical interest into elite breeding lines to produce superior cultivars. Three decades ago Alec Jeffreys[1] published a pioneering paper on the use of mini-satellite probes for DNA fingerprinting of humans that escorted to change the course of diagnostic approaches. The novelty of the method, for the first time, in discrimination between humans, animals, plants, and fungi was possible on the individual level using DNA markers. The new technology was soon extended to many other organisms including plants. Discreet plant genetic stocks, including bi-parental and multi-parent mapping populations, mutant populations, and perpetuate collections of recombinant lines, have been generated to facilitate mapping and gene function analysis via association studies and quantitative trait loci (QTLs) mapping in several crop species using this technique. In the last three decades enough DNA-based comprehensive data has been generated using fingerprinting methods, such as restriction fragment length polymorphism (RFLP),[2] random amplified polymorphic DNA (RAPD),[3,4] simple sequence repeat (SSR),[5] diversity arrays technology (DArT),[6] amplified fragment length polymorphism (AFLP),[7,8] and microsatellite-anchored fragment length polymorphism (MFLP).[9] The data have helped in identification and map location of agriculturally important genes and QTL which forms the basis for parental selection and marker-assisted selection (MAS) in plant breeding. But due to labor-intensive, time-consuming, and highly intimidating task of identifying polymorphic markers which are tightly linked to genes of interest for molecular MAS,[10–12] this approach seems slow for crop improvement program. The development of improved breeding lines for commercial crop cultivation by the current breeding approaches has proved time consuming and more expensive. To meet the demands of the human population growth rate and uncertainty in climate changes, we must accelerate the pace of our current breeding practices and focused on alternative approaches. Plant research has transformed dramatically by the genomics-based approach providing breeders with a new set of tools and techniques, to allow the study of the whole genome and its relationship with the phenotype[13] thereby leading to a new revolution in plant research at the beginning of the 21st century.

Molecular biology advancement is highly linked to technological evolution and progress in the sequencing technology is a perfect example to showcase that. Based on chemistry and process used in sequencing, till date three generations of sequencing processes have been developed. First generation, also called "Sanger sequencing" has been the predominant sequencing method for the past 30 years. In Sanger sequencing, a *polymerase chain reaction* (PCR) in the presence of differentially dyed dideoxynucleotide triphosphates (ddNTPs) is performed. Sequences are deduced on the basis of color. The sequencing of whole genome of the model plant *Arabidopsis thaliana*[14] and selected genotypes such as rice,[15] maize,[16] sorghum,[17] populous,[18] grapevine,[19] papaya,[20] or soybean[21] have been sequenced which has helped in building base to many of the modern molecular biology techniques. However, on the contrary, it has certain limitations like high run time. The complete sequencing of large and complex genomes of important cultivated species such as wheat, sugarcane, or coffee, is very challenging and may take several years on the basis of the traditional Sanger technology. Second-generation sequencing (SGS) techniques had largely overcome this restriction. Here, enzymes such as cleavage enzyme and polymerase along with fluorescent dyed deoxynucleotide triphosphate (dNTP) and highly automated system operate for detection and assembly of sequences. Although, each step is highly sequential and slow; however, it can sequence millions of DNA fragments at once, a technique like Illumina is based on the same principle. This technique is also not perfect as read length produced is less (100 bp) and has high error rate (1–2%/base). Techniques like SGS require a lot of reagents and are not cost-effective. Since then, methodologies have undergone rapid evolution and diversification and thus many techniques were developed, optimized, eventually novel, more efficient and reliable methods took over. In plants, it provide breeders a genome-wide molecular tools such as large collections of markers, high-throughput genotyping strategies, high density genetic maps, new experimental populations, and so forth, that if incorporated into existing breeding methods[22] can develop more efficient plant cultivars in comparatively less time, which are needed to feed the world's growing population while preserving natural resources.

The salient features of these technologies have lowest input cost of sequencing reduced by more than one thousand times compared to Sanger technology, less time consumption, generate sequence data in very large volumes, no need to perform tedious traditional cloning steps, and making possible to perform millions of sequencing reactions in parallel.[23] New techniques are on the horizon that intend to overcome some of the constraints, including the second generation ion torrent, and the so-called third-generation sequencers, Pacific Biosciences, Oxford Nanopore and Life Sciences Qdot technology, all of which sequence single molecules of DNA in real-time.

According to Nybom et al.,[24] different technological advances could be divided into three phases (a) Non-PCR-based, the first phase of DNA finger-printing, which was dominated by restriction fragment analysis in conjunction with Southern blot hybridization, (b) PCR-based, single- or multi-locus profiling techniques, and (c) Next Gen Sequencing-based high-throughput approach. Eventually, the first phase was almost completely taken over by the second phase, after the invent of PCR, and many routine applications of plant DNA fingerprinting still rely on PCR-based markers—the DNA finger-printing, of the present age. The beginning of the third phase is very recent, almost a decade before, where several novel, highly parallel DNA sequencing strategies are being developed and employed to increase the efficiency and throughput over the current Sanger sequencing technology almost to 1000-fold and more in facilitated marker development, or directly in the sense of "genotyping-by-sequencing." The heritable polymorphism displayed by genetic markers lies at the heart of present day modern genetics that had enabled solutions to many important questions in population and ecological genetics and evolution. Way back, Luikart et al.[25] could foresee the ideal molecular approach for population genomics, to be high throughput, uncovering hundreds of polymorphic markers covering the entire genome in a single, simple, and reliable experiment. Now, with the advent of next-generation sequencing (NGS), there are several such approaches, that are capable of discovering, sequencing, and genotyping not hundreds but thousands of markers across almost any genome of interest in a single step,[26] even in populations in which little or no genetic information is available. The technological advancements in the sequencing methods, referred to as NGS, are extremely high throughput producing thousands or millions of sequences at a fraction of the cost of traditional Sanger methods.[27,28] Very quick handsome results are on the display in molecular breeding approaches that qualify not only comprehensive genome-wide association studies for any organism, but also genome-wide studies on wild populations, with substantial benefits for conservation genetics and ecology.[29] Genomics-assisted breeding approaches have greatly advanced with the increasing availability of genome and transcriptome sequence data for several model plant and crop species. Complete and/or draft genome sequences have become available for several plant species such as rice, sorghum (http://www.phytozome.net/sorghum), poplar (http://www.phytozome.net/poplar.php), grape (http://www.phytozome.net/grape.php), papaya, *Medicago* (http://www.medicago.org/ genome), and soybean (http://www.phytozome.net/soybean). Whole genome or gene-space sequencing is in progress for several other crops such as maize (http://www.maizegenome.org), wheat (http://www.wheatgenome.org), barley (http://www.public.iastate.edu/imagefpc /IBSC %20Webpage/IBSC%20Template-home.html), tomato (http://sgn.cornell.edu/about/tomato_sequencing/), and foxtail millet (http:// www.jgi.doe.gov/sequencing/why/99178.html. Complementary to genome sequencing is the widespread application of transcriptome sampling strategies, which has resulted in large collections of expressed sequence tags (ESTs) for nearly all economically important plant species (http://www.ncbi.nlm.nih.gov/dbEST/dbEST_ summary. html). Previously, most genome and transcriptome sequencing projects used Sanger sequencing methodology. However, owing to growing interest in human genome re-sequencing, a new generation of sequencing technologies has emerged. These NGS technologies are able to generate DNA sequence data inexpensively and at a rate that is several orders of magnitude faster than that of traditional technologies. Advances in sequencing technologies are driving down sequencing costs and increasing sequence capacity at an unprecedented rate, making whole-genome re-sequencing by individual laboratories possible. As a result, genomics-assisted breeding should gain momentum, with the potential for significant improvements in the precision and efficiency for predicting phenotypes from genotypes.

Notably, one of the current and emerging applications of NGS is large-scale genetic polymorphism discovery.[27,28] Different molecular marker systems such as RFLPs, AFLPs, sequence tagged sites (STS), SSRs, and single nucleotide polymorphism (SNP) have been greatly benefitted by NGS methods. A specific application of this new technology in plant molecular breeding is the possibility of specific, rapid, and cost-effective discovery of SSR or microsatellite loci. These widely used, hyper variable markers constitute an important genomic resource in DNA polymorphism. Microsatellite markers have numerous uses, including linkage map development, QTLs mapping, MAS, parentage analysis, cultivar fingerprinting, genetic diversity studies, gene flow, and evolutionary studies.[30,31] Many biological questions can now be responded with high precision, for example, genetics involved in adaptation and ecological interactions in natural populations, comparative genome-wide insights into natural variation by sequencing thousands of individual genomes from a species, species composition and gene content in complex communities, identifying recombination breakpoints for linkage mapping or QTL mapping, MAS, and phylogeography of wild populations in large scale. However, experimental design and data analysis for these new approaches can be complex.[32]

In this chapter, we have focused on the use of NGS for genome-wide marker discovery, discuss their advantages and limitations, and recommending the advancement for the use in future studies of molecular breeding.

7.2   REVOLUTION IN MARKER DISCOVERY: THE IMPACT

Traditional way of marker development is expensive, laborious involving computational process, cloning, optimization, and primer designing steps that could not be adopted for all individuals alike. The salient constraint of the traditional system of marker development being the specificity of the markers to the original population enabling biased genotyping of new populations toward alleles present in the original survey, creating serious problem for studies of wild or highly divergent populations. Contrastingly, the NGS-based techniques enable high-throughput discovery, sequencing, and genotyping of markers directly on genomic DNA in a single sequencing step and with mostly parallelized library preparation. Through a massive parallelization, NGS provides an enormous number of reads that permits sequencing of entire genomes at a fraction as compared to the costs of Sanger sequencing. Hence, it has become feasible to obtain the complete genomic sequence for a large number of individuals. Genotyping of the same markers in other populations can be achieved with further sequencing runs, accurately representing the new populations and avoiding bias toward the sequence of the markers in the originally surveyed population.[33] Although, the cost of SNP genotyping using sequence-based approaches remains higher than the existing SNP arrays, the high output of genome-wide SNP arrays using NGS is still more economical which may be extrapolated to many different populations. For several organisms, including humans, *Drosophila melanogaster,* and *A. thaliana,* large re-sequencing projects are well on their way. However, for small communities the cost of sequencing is likely to be far lower than that of array development. Nevertheless, despite the enormous cost reduction, genome sequencing on a population scale is still out of reach from the budget of most laboratories. Traditional genotyping using NGS can also be carried out, at a low cost, by developing microsatellite markers in small-scale panels' derived using NGS platform, extracting huge statistical information at a comparably lower cost. Current genome-wide re-sequencing projects collect the sequences on individual basis to obtain high coverage of the entire genome and to have high confidence for all heterozygous sites to be discovered. In order to trim down the cost without compromising with the sequence coverage, more effective sampling strategy should be devised.

7.3   ROLE OF RESTRICTION ENZYMES IN NGS

Restriction enzymes offer the first direct evidence for sequence variation at the specific site. The Restriction endonucleases have central position in marker discovery and genotyping for decades. These markers have been utilized in linking to many genes responsible for diseases in humans as in Huntington's disease,[34] cystic fibrosis,[35] and to construct the first complete linkage map of the human genome10. Restriction enzymes also remain core tool in the genome-wide NGS methodologies. The diversity in the restriction mechanism of available enzymes including length, symmetry, guanine-cytosine (GC)/AT biasness in their recognition sites, and methylation-sensitivity, allows flexibilities at experimental level making them an extremely versatile assay tool. The usage of enzyme can be customized at individual level by tailoring the approach to the genome of interest, the project goals, and the budget. For example, in plants, a methylation-sensitive enzyme can be employed to exclude repetitive regions that will avoid cutting most methylated repeat elements. Several methods have been developed for high-throughput genetic marker discovery and genotyping using restriction enzymes (REs). Interestingly, NGS technology executes comparable pattern for library preparation involving: shearing of DNA randomly, either by way of nebulization or by sonication; universal adapter ligation at both ends of the sheared DNA fragments; and lastly immobilization and amplification of the adapter-flanked fragments in order to generate clustered amplicons that serve as templates for the sequencing reactions.[27] Since restriction patterns are largely unaffected by the choice of sequencing platforms most innovations of particular methods can be broadly applied.

Majority of the methods in NGS, in conjunction with RE, involve the following key basic steps (a) the digestion of multiple samples of genomic DNA of the targeted individuals or populations with one or more REs, (b) selection or reduction of the resulting restriction fragments, and (c)NGS of the final set of fragments. The final set of sequence read length is generally<1 kb in size due to the read-length limits of most of the current NGS platforms. Nucleotide variation can be taken as polymorphisms in the resulting sequenced fragments and may be utilized as genetic marker. Since the discovery of genes, their location and function, as well as the development of large marker collections with high resolution is being possible through fully sequenced and well annotated genome thereby providing handy tools for the breeders. Moreover, gain in information concerning the identification and map location of agriculturally important genes and QTL grants the basis for parental selection and MAS accommodating plant-breeding program. Re-sequencing the genome of individual plants within the targeted genotype through existing SNP platforms[36] has become routine practice due to cost of genome sequencing, permitting the breeder to attain early selection of a trait on the basis of MAS in a breeding program. Depending on the application and the objectives, different standards of marker data are desirable. Several marker development methods utilizing NGS platforms were reported, including reduced-representation libraries (RRLs),[37,38] complexity reduction of polymorphic sequences (CRoPS),[39] restriction-site associated DNA sequencing (RAD-seq),[40] sequence based polymorphic marker technology (SBP),[41] low coverage multiplexed shotgun genotyping (MSG),[42] and genotyping by sequencing (GBS).[43]

In RRLs, reduced genome complexity is accomplished via the construction of an RRL by a restriction digestion followed by size select selection of the targeted samples. Further, similar selected subset of fragments (size-selected), obtained from different genotypes were deep sequenced for accurate SNP discovery. The use of RRLs was first proposed in humans to efficiently find SNPs using Sanger sequencing. The robustness of the high-throughput SNP discovery method was validated recently in cattle[44] and soybean.[45]

7.3.1 COMPLEXITY REDUCTION OF POLYMORPHIC SEQUENCES (CROPS)

CRoPS combine the power of reproducible genome complexity reduction of AFLP (of two or more samples) with the novel high-throughput sequencing by-synthesis technology. The obtained sequences are stored in a relational database and putative SNPs are extracted from this. The bioinformatics pipeline involves data processing, CRoPS analysis, and SNP mining. The technique has ability for efficient polymorphism discovery targeting unique genome regions in organisms lacking extensive genome sequence information.

SBP: In this technique the assembled database is mined for SNP using SHORE (Short Read) program (SHORE program is a mapping and analysis pipeline for short read data produced on the Illumina platform). The technology facilitates isolation of co-dominant molecular markers for targeted genomic regions whose genomic sequences have been assembled. SBP marker technology can convert SNPs to molecular markers for any genomic regions in a highly efficient manner resulting into development of high-density molecular marker maps. The information becomes essential for cloning genes based on their genetic map positions and identifying tightly linked molecular markers for selecting desirable genotypes. The reference genome sequence can be utilized to develop SBP markers for a specific genomic region with known physical location thereby marker-poor regions can be enriched with SBP markers. The technology was successfully employed by Sahu et al.[41] in Arabidopsis SBP marker development in marker-poor region.

Restriction-site associated DNA (RAD): Restriction-site associated DNA (RAD) is a category of DNA markers that carries out the detection of genetic variation being adjacent to RE cleavage sites, once NGS sequencing of a target genome is done. RAD was originally described by Miller et al. based on microarray platform which was later adapted on the massively parallel NGS platform by Baird et al.[46] In the RAD protocol, genomic DNA is digested with a 6–8 base-cutter RE and a barcoded adapter is ligated to compatible sticky ends. For multiplex sequencing, DNA samples, each with a different barcode, are pooled, randomly sheared, size selected (300–700 bp), and a second adapter is ligated after polishing and filling ends.[46] The methodology does not require any prior sequence knowledge for detecting DNA polymorphisms in the species under investigation. RAD sequencing produces two types of DNA markers (a) dominant markers related to DNA variations within the restriction sites and (b) co-dominant markers for sequence variation adjacent to the restriction sites.[47] RAD markers have been employed in genetic mapping on fungi,[48] fish,[46] insects,[49] and more recently on plants.[47,50,51] In *Lupinus angustifolius* L.,38 molecular markers linked to the disease resistance gene (R gene) Lanr1 were discovered using RAD approach. Five of them could be converted into cost effective, simple PCR-based markers which were linked to the R gene. These markers are now replacing those being previously developed by a traditional DNA fingerprinting method for MAS in the Australian national lupin breeding program.[52] With advent of RAD approach, plant breeders have developed the choice to select markers pertinent to a multitude of crosses in their breeding programs. Notably, NGS-based RAD sequencing offers one of the core advantage in marker development like ease in renovation of cost-effective, simple PCR-based candidate marker that is co-dominant in nature and enviable for high-throughput performance on modern SNP genotyping platforms for MAS in molecular plant breeding. The application of NGS based method of RAD-seq as DNA fingerprinting for rapid, cost-effective marker development tagging a disease R gene against anthracnose disease has been reported by Yang et al.[52] in *L. angustifolius* L. About total of 445 RAD markers were discovered across all seven barley chromosomes located that proved very valuable for the construction of linkage map in this crop.[50]

7.3.2   GENOTYPING BY SEQUENCING (GBS)

Technological advances in NGS have lowered down the costs of DNA sequencing to the point that GBS is now feasible and comparatively simpler for high diversity large genome species. In GBS methodology, DNA samples, barcode, and common adapter pairs are loaded and dried in multi-well plates and digested with appropriate RE (ApeKI), and adapters are ligated to the ends of genomic DNA fragments. The usage of RE reduces the genome complexity. Employing methylation-sensitive REs, repetitive regions of genomes can be avoided and lower copy regions could be targeted with 2–3-fold higher efficiency. This simplifies computationally challenging alignment problems in species. After the inactivation of the ligase by heating, aliquot of each sample is pooled and applied to a size exclusion column to remove unreacted adapters. Appropriate primers with binding sites on the ligated adapters are added and PCR is performed to increase the fragment pool. PCR products are cleaned up and fragment sizes of the resulting library are checked on a DNA analyzer. Libraries without adapter dimers are retained for DNA sequencing. In addition to the simplicity of the procedure (no fragment size selection and few enzymatic and purification steps), the protocol is time and cost efficient through its use of a single well for genomic DNA digestion and adapter ligation. These processes can be done in the same buffers so that no additional transfers are needed.[49]

7.3.3 MULTIPLEXED SHOTGUN GENOTYPING (MSG)

This is genotyping based on multiplexed shotgun sequencing that can identify recombination breakpoints in a large number of individuals simultaneously at a resolution sufficient for most mapping purposes. In this technique, which utilizes best aspects of RAD and whole genome sequencing (WSG), genomic DNA is fragmented with a frequent cutter RE resulting into "sticky ends." This initial step is followed by ligation of individual bar-coded adaptors to restriction fragments. Experimental samples are pooled, ligation products are size selected, PCR-amplified, and sequenced. The reads from the sequencing run are parsed based by barcode. Each read is mapped to each of two parental genomes. Genotypes and recombination breakpoints are used in downstream analyses, such as QTL mapping and ancestry assignment. The simplicity of the method is no shearing and repair of DNA prior to adapter ligation and small DNA (10 ng) requirement.[42]

Due to low cost and high-throughput screening, NGS has become a preferred choice of plant breeders. This has resulted in huge genetic information, high gene resolution, function-association studies, and QTL discoveries.

7.3.4 DESIGN OF MARKER DISCOVERY EXPERIMENTS

DNA sequencing technologies have yielded massive and accurate DNA sequence. However, several interacting factors affect the choice of an optimal NGS marker discovery method. Here, few factors and several technical variations and analytical challenges that apply to all methods are being outlined.

7.3.4.1 OBJECTIVE AND TARGET

The huge data produced by high-throughput sequencing can produce tens of thousands of markers with high genotyping accuracy, but this resolution and power may not be necessary to answer many biological questions like for crossing studies, high density markers are not required and low coverage sequencing may yield sufficient data to get the answer. Thus, how many markers are required to achieve the target and the depth of the genotyping are the critical questions to solve before the start of the experiment.

7.3.4.2   AVAILABILITY OF A REFERENCE GENOME

Availability of reference genome sequence decides the depth of the sequencing to be performed. The reads from the test sequence produced by any of the technologies, are mapped on physical map using assembled reference sequences. The coverage that is required to genotype each individual is highly influenced by the quality of the reference genome assembly as its high quality makes, easier to attribute missing genotypes. Further, reference genomes may also be employed to design *in silico* marker discovery experiments by simulating the number of markers produced by different enzymes. Challenges arise when a reference genome sequence is unavailable, inadequately assembled, come from a distantly related taxon, or are large and highly repetitive. Crude estimation of restriction sites is possible if the genome size and GC content are known, especially GC content, so it is worth making a high-quality estimate of this parameter. In practice, for the identification of unique loci and annotating sequence reads unambiguously, a lot of data are filtered out and discarded. This is essential to minimize spurious putative loci, unusually high read counts and repetitive sequences.[53]

7.3.4.3   POLYMORPHISM

As in conventional genetic polymorphism methodology, in NGS also, populations with low levels of polymorphism will require more markers, and therefore require an enzyme with a higher cutting frequency to produce sufficient polymorphic markers. Techniques like RRLs, MSG, and CRoPS, which sequence entire restriction fragments, may be more suitable for these populations. By contrast, high degree of polymorphism in a population will considerably alter the fragment distribution pattern as the probability of having variation at restriction sites will be higher. Alterations in the restriction site can vary the fragment size, resulting in exclusion of putative markers during size selection. Though more probable markers are filtered out in a highly polymorphic population, the remaining markers are more likely to be informative than in less polymorphic populations because these markers are more likely to contain a polymorphism. Unfortunately, these methods are not the perfect solution for genomes with a large repetitive fraction or high ploidy value. For example, using an RRL it was possible to validate 94% of a sample of putative swine SNPs14 but only 48% of a sample of SNPs in a rainbow trout19, largely owing to the ancestral whole-genome duplication in salmonid species.[33] However, repetitive sequences can be partially avoided by the careful choice of REs or by removal *in silico* by filtering putative loci, as described above.

7.3.4.4   RESTRICTION ENZYME

The choice of RE depends on (a) marker density and (b) genome complexity for the species under study. For example, Van et al.[39] put considerable effort into choosing an enzyme that avoids common repeats in cattle, and the methylation-sensitive ApeKI used by Elshire et al.[43] may not be appropriate for other methylated genomes. All methods except RRLs use endonucleases that produce overhangs. New adaptors are required for different overhangs, but the same adaptor set can be suitable for multiple enzymes producing the same overhang, such as the eight-base cutter SbfI (cut site CCTGCA^GG) and the six-base cutter PstI (cut site CTGCA^G).[33]

7.3.4.5   PREPARATION OF DNA SAMPLE

For the success of the experiment, a high-quality genomic DNA (free of contamination either with RNA or with DNA from other species) is crucial. As digestion, ligation and amplification of the DNA are the core of all the protocols. Any variations in these steps, due to the genomic DNA quality, can have significant effects on the final marker set. The requirement for the amount of genomic DNA depends on the technique being employed (RRL, 25 ng pooled;[54] CRoPS, 300 ng/sample;[55] RAD-seq, 300 ng/sample;[50] MSG, 10 ng/sample;[42] and GBS, 100 ng/sample[43]). Most importantly, the quantity of DNA from different samples should be evenly balanced before pooling to avoid losing markers from some individuals owing to lack of coverage.[33]

7.3.4.6   ADAPTOR DESIGN

Adaptors needs to be designed such that any barcode is at least three-base pairs distinct from all others, so that reads containing an error in the barcode sequence can be uniquely assigned to a sample.

7.3.4.7   PCR AMPLIFICATION

The PCR profile, in sample preparation, might favor sequencing toward GC-rich, short fragments and fragment pool. Some protocols for genome and transcriptome sequencing avoid PCR altogether.[56] Despite its demonstrable biased effect, it is difficult to avoid PCR in NGS marker methods, because PCR amplification is required to amplify the adaptor-ligated fragments to outcompete other fragments. To minimize its partialism, number of PCR cycles can be reduced and the amount of input DNA increased instead.

7.3.4.8   SEQUENCING

The depth of these quenching depends on the outcome based on the data produced, the sequencing platform can be broadly classified into (a) short reads and (b) long reads. For marker development, design and identification of polymorphisms, data having short reads solves the purpose so the platforms such as the Illumina Genome Analyzer and SOLiD are preferable to the Roche Genome Sequencer. Although, long reads from the Roche Genome Sequencer can be useful for assembling draft genomic sequence for unsequenced species.[54] The optimal mean coverage per locus varies widely across experimental goals and strategies, as it reflects adjustments with factors such as the number of individuals, the number of groups or populations, the genome size, the density of markers across the genome, and the total sequencing effort.

7.3.4.9   POOLING INDIVIDUALS

Pooling of genomic DNA requires a lot of cautiousness. Several factors influence the outcome in DNA pooling including both the quality as well as quantity of DNA being used. Many studies use one barcode for a pool of several individuals. This can be useful to avoid a whole-genome amplification step if the amount of DNA/individual is small.[57] Although, pooling improves SNP discovery and leads to better estimates of population allele frequencies,[58] but pooling averts genotyping of individuals after SNP discovery. So, simultaneous marker discovery and genotyping are not possible. If the sequencing resource permits, barcodes for individual samples provide greater flexibility for downstream analysis and this approach does not preclude ignoring the barcodes and pooling the samples bioinformatically. Pooling has the disadvantage of potentially missing rare variants and is highly sensitive to variation in the DNA concentration among individuals in a pool[59] (although individually barcoded sequencing also suffers from this problem). In the absence of a high-quality reference genome sequence, pooling also precludes filtering on the basis of observed heterozygosity.[33]

7.4   GENOME/TRANSCRIPTOME SEQUENCING

Sequence data generated through NGS methodology follows either of the two routes; a genomics (DNA sequencing (DNA-Seq)) or a transcriptome(RNA sequencing (RNA-Seq)) route, which has pave the way to study the genetic and functional information stored within any organism, at an extraordinary scale and speed. Transcriptome sequencing has been a cheaper preference for developing marker rather than costly WGS due to repeat content and those species with limited genetic or financial resources thereby equating the progress of functional genomics research significantly. These approaches helped to divulge applicable information regarding synchronized study of transcript structure (like alternative splicing), allelic information (SNPs), exon structure annotation and high resolution gene expression and large dynamic range such as different stage of development and environmental conditions;[60] however, no information about non-coding sequences is being provided.

Transcriptome or genome sequences could help to develop molecular markers within any organism on a large scale. Moreover, NGS generally allocates minimal cost and rapid identification of hundreds of microsatellite loci and thousands of SNPs, even while fraction of initial sequencing is used. Whatever sequence data sets are generated being by NGS platforms, microsatellite marker within them can be easily detected through various programs like msatcommander[61] and MSatFinder.[62] Additionally, SNPs positioned in the coding region or UTRs can also be revealed using varied approaches (http://seqanswers.com/wiki/Special:BrowseData/Bioinformatics _application).

7.5 IN SILICO ANALYSIS OF DENOVO TRANSCRIPTOME

In the direction of transcriptome studies, for model organisms, genome-guided advancement has become a standard approach to RNA-Seq analysis by utilizing various software tools.[63,64] However, organisms with no well-assembled genome or even if one is present, this approach cannot be either applied to or the results may vary across genome assembly versions. In such cases, transcriptome data can be assembled even when a genome sequence is not available, via a *denovo* transcriptome assembler based on computational approach. Numerous tools like Trans-ABySS,[65] Velvet-Oases,[66] SOAPde-novo-trans (http://soap.genomics.org.cn/SOAPdenovo-Trans.html), and Trinity[67] are nowadays offered for *de novo* assembly of RNA-Seq. On the other hand, programs like GigaBayes[68] and VarScan[69] statistically distinguishing true polymorphisms from sequencing errors, for SNP detection are brought in use. It brings about the probability to be calculated that a polymorphism represent either a true SNP or a sequencing error through Bayesian approach and once these markers get identified, while using a multitude of platforms for molecular level studies, they can be corroborated among large number of individuals[70] such as map construction, map saturation, and genome-wide diversity studies.

7.6 DISCOVERY OF SINGLE NUCLEOTIDE POLYMORPHISM (SNP) MARKER VIANGS

With recent advent of second- and third-generation DNA sequencing technology, various methods involving discovery and analysis of SNP marker have been enhanced.[71] The barrier put forward by previous inadequate application of high-throughput SNP analysis technology due to non-availability of DNA sequence information within the plant genome, has been conquered by NGS technology.[72] However, the supplementary variability read quality of SGS methods can be prevailed over by augmented depth of coverage, the important parameter during SNP marker discovery.[73,23] For semiautomated or fully automated SNP genotyping, a large number of commercial platforms are available,[74,75] the Illumina Golden Gate assays being the most widely used for mid-throughput applications. Under anthracnose infection condition, Narina et al.[76] studied the transcriptome sequence comparison of two resistant and one susceptible lines of water yam crop in order to detect SNPs in genes putatively involved in pathogen response. Additionally, to detect SNPs alfalfa for its improvement as a forage crop and cellulosic feedstock,

Yang et al.[77] sequenced two its genotypes contrasting for cellulose and lignin content.

7.7   ANALYTICAL CHALLENGES

NGS, although is fast, comprehensive highly sensitive, high-throughput process but the sequence reads obtained are from a large, heterogeneous pool of DNA fragments and lot of variation could be assumed to arise due to variation in steps. Several steps in the protocols may also worsen it further. Particularly pooling of DNA, PCR amplification, and size selection. A crucial feature of marker discovery using the techniques above is that they incorporate a multi-level sampling process. There is sampling deviation in the number of reads across individuals or barcodes within the pool, across loci within each individual, and across alternative alleles at polymorphic loci. Consequently, NGS approaches, although producing orders of magnitude more markers than previously possible, differ from traditional marker genotyping in important ways (a) there is unavoidable variance in the sample sizes of individuals across loci, (b) loci across individuals, and (c) uncertainty in genotype assignments across loci and individuals.

Library preparation of the samples regulates the success of the marker synthesis. Although, the protocols for the methods involve only standard molecular biology techniques (e.g., RE digestion, size selection by agarose gel extraction, shearing, ligation, and PCR) which forms the core of any competent molecular biologist lab. However, in practice, the steps need to be optimized for individual experiments. Also the length of the protocol for some methods can be challenging; therefore, it may take several attempts to produce a successful library. Alternatively, library preparation services available from several companies and sequencing facilities may be opted. A completed library can be sequenced on any NGS machine, either locally or at a sequencing facility. Molecular breeding applies molecular biology tools to accelerate the breeding process. One of the most important methods of molecular breeding is MAS, the use of DNA markers that are tightly linked to target loci as a substitute to assist phenotypic screening.

At the largest genomics center in the world, BGI-tech, facilities and new services have been developed which are available to advance the applications of NGS technology in the molecular breeding area. Here a breeding project can be completed in one year, which would have otherwise taken 6–8 years, using traditional strategies.

7.8   ADVANTAGES OF NGS APPLICATIONS IN MOLECULAR BREEDING: AN OVERVIEW

NGS technologies have several potential applications in plant genetics and breeding. The high-throughput sequencing have allowed rapid advances in genotyping that promises unparalleled resolution in studies of the genetic architecture of complex traits including the generation of genomic resources, marker development and QTL mapping, wide crosses and alien gene introgression, expression analysis, association genetics, and population biology. The huge data obtained from NGS technologies can provide genomic resources such as ESTs, gene space, and genome assembly. These resources have a direct impact on understanding the genome architecture for crop genetics. Another application of NGS is in parental genotyping of mapping populations or of wild relatives, which can accelerate the development of molecular markers, for example, SSR and SNP markers and enhance the reliability and utility of the markers. Functional or gene-based marker can be developed by performing NGS of complementary DNAs (cDNAs) of contrasting genotypes for the trait of interest that can be used to locate candidate genes involved in or associated with the trait. The expression mapping of the candidate genes, together with phenotyping of the segregating populations developed from the contrasting genotypes, will provide expression quantitative trait loci (eQTLs). Markers associated with these eQTLs thus would serve as the perfect markers for MAS in molecular breeding of the crop. Another important application of NGS is in association genetics or population biology, where either genomes or pools of PCR products of thousands of candidate genes can be sequenced in hundreds of individuals using barcodes. The sequence data obtained could then be used to identify SNPs or haplotypes across genes or genomes for use in association genetics and/or population biology.

Few examples associated with NGS-based marker breedings and their impact.

7.8.1   APPLICATION OF MOLECULAR MARKERS IN BREEDING FOR RESISTANCE TO BARLEY YELLOW MOSAIC VIRUS

One of the most important diseases of winter barley in Europe and East Asia, the barley yellow mosaic virus (BaYMV) disease is caused by Barley mild mosaic virus (BaMMV) and BaYMV. Chemical treatments to control the disease are neither efficient nor economical as the transmission of the disease is by the soil-borne fungus *Polymyxa graminis*. However, field selection for resistance genotypes is often difficult to perform because of unpredictable environmental conditions. The only feasible solution looks to be breeding for resistance to this disease. Consequently, the application of closely linked PCR-based markers for the transmission of R gene[s] against BaYMV is now successful and efficient.

7.8.2 MOLECULAR MARKERS IN BREEDING FOR RESISTANCE TO FUSARIUM HEAD BLIGHT IN WHEAT

Literature has revealed severe natural epidemics of *Fusarium* head blight (FHB) in *Triticum aestivum* L., once or twice in a decade, in Central Europe and humid and semi humid areas of the world, damaging the crop widely and severely thereby sharply reducing yield and quality of susceptible genotypes. Evaluation of FHB resistance is time consuming, laborious, and costly because the inheritance of resistance is complex and phenotypic expression is significantly affected by environmental factors. Molecular markers closely linked to the major QTL involved in FHB resistance have recently been found[78] and raise the possibility of using MAS for introducing resistance alleles into elite wheat varieties. However, due to the multifactorial nature of FHB resistance, the combination of MAS on the major QTL during seedling stage with phenotypic selection on the particular plants after flowering stage could be at the moment more sufficient and safety strategy in breeding of new varieties combining a high level of yield performance and high level of resistance to FHB.

7.8.3 APPLICATION OF MOLECULAR MARKERS FOR MALE FERTILITY RESTORATION IN PAMPA CMS IN BREEDING OF RYE

Hybrid rye breeding and seed production require a cytoplasmic male sterility (CMS) system as a hybridization mechanism. On the other hand, for the complete restoration of pollen fertility, effective, nuclear encoded restorer genes for CMS-inducing cytoplasm are indispensable. Partial restoration of male fertility causes a reduction in the amount of viable pollen thus encourages infection by the ergot fungus (*Clavicepspurpurea*). Ergot infection contaminates rye grains with sclerotia containing toxic alkaloids. To reduce or avoid this risk, rye hybrids need effective restorer genes. Applications of NGS technologies in plant genetics and breeding are briefly tabulated in Table 7.1.

TABLE 7.1 Some Applications of Next-Generation Sequencing in Plant Molecular Breeding.

S. no.	Plantname/system	Applications	References
1	Oryza sativa	Genome-wide target association, genome diversity, and evolution studies	[85]
2	Papaversomniferum	Increased sequence coverage, high number of opium poppy-specific SSR markers. Useful in diversity, identification, mapping, and breeding studies	[86]
3	Capsicum annuum L.	Transcriptome profiling for the identification of microsatellite (SSR) markers	[87]
4	Cicerarietinum L.	SSRs and ISSRs identified for developing molecular markers	[88]
5	Cucumismelo var. makawa	Facilitated chamoe breeding, discovery of SSR markers capable of discriminating chamoe varieties	[89]
6	Triticum durum	De novo single nucleotide polymorphisms (SNP) marker discovery genotyping of a bi-parental durum wheat population	[90]
7	Penstemon	Understanding of the phylogenetic structure within important drought tolerant genus	[91]
8	Vicia sativa	Assessing genetic diversity, population structure, and positional cloning, facilitating vetch breeding programs	[92]
9	Populusnigra	Identification defective variants of genes	[93]
10	Eragrostistef	Identification of potentially valuable mutations, improved lodging resistance	[94]
11	LupinusangustifoliusL.	Selection for phomopsis stem-blight disease (PSB) resistance	[95]
12	Viciafaba L.	Constructing genetic linkage maps, future QTL mapping, and marker-assisted trait for genetic gain	[96]
13	Exome sequencing	Advancement in crop improvement	[97]
14	Linumusitatissimum L.	Efficient discovery of a large number of microsatellite markers	[98]
15	Vacciniummacrocarpon Ait.	Accuracy of the sequence assembly and molecular markers for a genetic linkage map	[99]
16	Zea mays	Identification of key ancestors of modern germplasm	[100]
17	Brassica napus	High-density polymorphism detection	[101]

TABLE 7.1 (Continued)

S. no.	Plantname/system	Applications	References
18	Loliumperenne	Development of sequence-based markers linked to stem rust resistance	[102]
19	Sorghum bicolor L. Moench	Designing strategies for economical genotyping	[103]
20	Gossypiumhirsutum	Location of quantitative, economically important regions	[104]
21	Cymbidium ensifolium	Understanding the molecular mechanisms of floral development and flowering	[105]
22	Cucurbitapepo	Mapping and diversity studies	[106]
23	Lathyrussativus L.	Significant contribution to genomics enabled improvement of grasspea	[107]
24	Pisumsativum L.	Construction of genetic linkage map, marker-assisted pea breeding	[108]
25	Allium sativum L.	Genetic mapping, association mapping, genetic diversity, and comparison of the genomes of Allium species	[109]
26	Paeoniasuffruticosa	Genetic linkage map construction, QTL mapping, gene location and cloning, and molecular marker-assisted breeding	[110]
27	Ipomoea batatas	Discovery of novel genes associated with tuberous root formation and development and construct on high density microarrays for further characterization of gene expression profiles	[111]
28	SolanumlycopersicumL.	Marker-assisted breeding and genome-wide association studies	[112]
29	Saccharumofficinarum	Genome assembly and marker identification resulting in biotechnological improvements	[113]
30	Arabidopsis thaliana	Utility in crops and reference material for other completed genome sequence	[114]
31	Vitisvinifera	Play very significant roles in many research and non-research areas of plant virology	[115]
32	Carica papaya	Facilitated draft genome assembly, resource for comparative genomics and map-based cloning of agronomically and economically important genes, and for sex chromosome research	[116]
33	Medicagotruncatula	Powerful tool for forward genetic studies in identifying transposable element insertions causing phenotypes	[117]

TABLE 7.1 *(Continued)*

S. no.	Plantname/system	Applications	References
34	*Ricinus communis*	DNA fingerprinting and molecular diversity analysis of plant species	[118]
35	*Korean Glycine max*	Overall microbial community of Korean fermented soybean pastes	[119]
36	*Setariaitalica*	Reshuffling events were detected through collinearity identification between foxtail millet, rice, and sorghum. Rearrangements in the C4 photosynthesis pathway were also identified	[120]
37	*Hordeumvulgare L.*	Source of genetic introgressions for improving elite barley germplasm	[121]
38	*Helianthus*	Annotation of the sunflower genome sequence and for studying the genome evolution in dicotyledons	[122]
39	*Eucalyptus camaldulensis*	Unexplored diversity in the studied genes with potential applications for marker-trait associations	[123]
40	*Daucuscarota subsp. sativus*	Whole genome sequencing for *de novo* assembly of carrot mitochondrial genomes	[124]

7.9 FUTURE PROSPECTS

In order to attain more and longer sequence reads and also declined error rates in sequencing, the currently available technologies will further need to get improvised. This demand is usually met by the most imperative emerging NGS technique emerging as "third generation sequencing (TGS)" that is based on single-molecule sequencing, not only generating longer sequence reads within short span of time and at minimal costs but also lessen bias in detection of individual genes or allelic expression. However, even if the sequencing cost drops down further and the amount of generated data also increases, there will be innovative demands for novel analytical methods and infrastructure. By putting forth the genomic empire to study non-model organisms, NGS at present is drastically shifting our perception of conducting genetic research which will continue to do so in the predictable future. Though NGS-based strategies are undoubtedly escalating the efficiency of environment specific breeding crops, yet we oblige to provide farmers information regarding the availability of new varieties, crop management systems and marketing opportunities. However, it is crucially important that the efforts of the plant breeding community be fully integrated into the entire value chain so they can ultimately reach the people they are intended to benefit.

7.10 CONCLUSION

In recent times many DNA markers and NGS together have been developed and are powerful tools for successful breeding. The promise of MAS in crop breeding still remains but achieving practical benefits is taking longer than expected. The main reasons for this delay are the insufficient quality of markers regarding their predictive and/or diagnostic value, inadequate experimental design, high costs and complexity of quantitative traits. Rapidly increasing throughput will allow more individuals to be sequenced in a population, more markers to be sequenced per individual and each marker to be genotyped at greater depth and so with greater accuracy. We expect that it will be possible to sequence tens of thousands of markers in thousands of individuals in the near future. This will be far in excess of what is required for many studies in which a small number of markers are quite sufficient, and will be accessible. Here, NGS technology provides an opportunity to perform nucleotide variation profiling and simultaneously discovery of genetic markers on large scale, which in turn will aid in the quest of the genetic basis of ecologically important phenotypic variations either through the QTLs mapping or genome-wide association studies (GWAS). Extraordinarily raised number of markers will genuinely enable researchers to acquire superior exactitude in varied studies like population genetic studies,[79] QTL and linkage disequilibrium (LD) mapping projects,[70] kinship assignments,[80] historical demographic patterns, introgression and admixture.[81] However, the problems being associated with SGS could be rectified further through momentous improvised technology that comes forward as "Third-generation sequencing." These technologies interestingly are based upon methods involving direct scrutiny of single molecules without any requirement of repetitive wash and scan steps during DNA synthesis, synchronization of multiple reactions, or problems associated with PCR amplifications or phasing.[72] All promises such as lower cost per analysis, longer reads, direct RNA sequencing, and direct analysis of methylation are being taken care of by these technologies. Hence, they are being developed commercially and various sequencing projects such as PacBio RS (Pacific Biosciences, http://www.pacificbiosciences.com), Helicos (Helicos, http://www.helicosbio.com), or Ion Torrent (Life Technologies, http://www.iontorrent.com) are built-in and additionally they are expecting to replace nearly 47% of all SGS activity in the subsequent three years.[82] NGS generated sequences are usually deposited in the National Center for Biotechnology Information (NCBI) Sequence Read Archive (http://www.ncbi.nlm.nih.gov/unigene). Combinatorial use of newer technology applications based on high-throughput genotyping with that of newer marker systems (e.g., SNP), and new selection approaches such as AB-QTL, mapping-as-you-go, marker-assisted recurrent selection, and genome-wide selection[83,84] would pave way for the development of improvised complex polygenic traits associated with agronomic traits in the future.

KEYWORDS

- **next-generation sequencing**
- **plant molecular breeding**
- **molecular markers**
- **polymorphism**
- **transcriptome**
- **microsatellite**
- **restriction enzyme**

REFERENCES

1. Jeffreys, A. J.; Wilson, V.; Thein, S. L. Individual-Specific 'Fingerprints' of Human DNA. *Nature.* **1985,** *316,* 76–79.
2. Burr, B.; Burr, F. A.; Thompson, K. H.; Albertson, M. C.; Stuber, C. W. Gene Mapping with Recombinant Inbreds in Maize. *Genetics.* **1988,** *118,* 519–526.
3. Williams, J. G. K.; Kubelik, A. R.; Livak, K. J.; Rafalski, J. A.; Tingey, S. V. DNA Polymorphisms Amplified by Arbitrary Primers are Useful as Genetic Markers. *Nucleic Acids Res.* **1990,** *18,* 6531–6535.
4. Molnar, S. J.; James, L. E.; Kasha, K. J. Inheritance and RAPD Tagging of Multiple Genes for Resistance to Net Blotch in Barley. *Genome.* **2000,** *43,* 224–231.
5. Sundaram, R. M.; Naveenkumar, B.; Biradar, S. K.; Balachandran, S. M.; Mishra, B.; Ilyas, A. M.; Viraktamath, B. C.; Ramesha, M. S.; Sarma, N. P. Identification of Informative SSR Markers Capable of Distinguishing Hybrid Rice Parental Lines and their Utilization in Seed Purity Assessment. *Euphytica.* **2008,** *163,* 215–224.
6. Wittenberg, A. H.; Vander L. T.; Cayla, C.; Kilian, A.; Visser, R. G.; Schouten, H. J. Validation of the High-Throughput Maker Technology DArTusing the Model Plant *Arabidopsis thaliana. Mol. Genet. Genomics.* **2005,** *274,* 30–39.
7. Vos, P.; Hogers, R.; Bleeker, M.; Reijans, M.; Lee, T.; Hornes, M.; Frijters, A.; Peleman, J.; Kuiper, M.; Zabeau, M. AFLP: A New Technique for DNA Fingerprinting. *Nucleic Acids Res.* **1995,** *23,* 4407–4414.
8. Brugmans, B.; Van der, H. R. G. M.; Visser, R. G. F.; Lindhout, P.; Van, E. H. J. A New and Versatile Method for the Successful Conversion of AFLP Markers into Simple Single Locus Markers. *Nucleic Acids Res.* **2003,** *31,* e55.
9. Yang, H.; Shankar, M.; Buirchell, B. J.; Sweetingham, M. W.; Caminero, C.; Smith, P. M. C. Development of Molecular Markers Using MFLP Linked to a Gene Conferring Resistance to *Diaporthetoxica*in Narrow-Leafed Lupin (*Lupinusangustifolius* L.). *Theor. Appl. Genet.* **2002,** *105,* 265–270.
10. Kumar, L. S. DNA Markers in Plant Improvement: An Overview. *Biotechnol. Adv.* **1999,** *17,*143–182.
11. Koebner, R.; Summers, R. The Impact of Molecular Markers on the Wheat Breeding Paradigm. *Cell Mol. Biol. Lett.* **2002,** *7,* 695–702.
12. Moose, S. P.; Mumm, R. H. Molecular Plant Breeding as the Foundation for 21st Century Crop Improvement. *Plant Physiol.* **2008,** *147,* 969–977.
13. Tester, M.; Langridge, P. Breeding Technologies to Increase Crop Production in a Changing World. *Science.* **2010,** *327,* 818–822.
14. Altshuler, D., et al. An SNP Map of the Human Genome Generated by Reduced Representation Shotgun Sequencing. *Nature.* **2000,** *407,* 513–516.
15. Matsumoto, T.; Wu, J.; Kanamori, H.; Katayose, Y.; Fujisawa, M., et al. The Map-Based Sequence of the Rice Genome. *Nature.* **2005,** *436,* 793–800.
16. Schnable, P. S.; Ware, D.; Fulton, R. S.; Stein, J. C.; Wei, F.; Pasternak, S.; Liang, C.; Zhang, J.; Fulton, L.; Graves, T. A., et al. The B73 Maize Genome: Complexity, Diversity, and Dynamics. *Science.* **2009,** *326,* 1112–1115.
17. Paterson, A. H.; Bowers, J. E.; Bruggmann, R.; Dubchak, I.; Grimwood, J.; Gundlach, H.; Haberer, G.; Hellsten, U., et al. The Sorghum Bicolor Genome and the Diversification of Grasses. *Nature.* **2009,** *457,* 551–556.

18. Tuskan, G. A.; Difazio, S.; Jansson, S.; Bohlmann, J.; Grigoriev, I.; Hellsten, U.; Putnam, N.; Ralph, S.; Rombauts, S., et al. The Genome of Black Cottonwood, Populus Trichocarpa (Torr. &Gray). *Science.* **2006,** *313,* 1596–1604.

19. Jaillon, O.; Aury, J. M.; Noel, B.; Policriti, A.; Clepet, C.; Casagrande, A.; Choisne, N.; Aubourg, S.; Vitulo, N., et al. French-Italian Public Consortium for Grapevine Genome Characterization. The Grapevine Genome Sequence Suggests Ancestral Hexaploidizationin Major Angiosperm Phyla. *Nature.* **2007,** *449,* 463–467.

20. Ming, R.; Hou, S.; Feng, Y.; Yu, O.; Dionne-Laporte, A.; Saw, J. H.; Senin, P.; Wang, W.; Ly, B. V.; Lewis, K. L. T., et al. The Draft Genome of the Transgenic Tropical Fruit Tree Papaya *(Carica papaya Linnaeus). Nature.* **2008,** *452,* 991–996.

21. Schmutz, J.; Cannon, S. B.; Schlueter, J.; Ma, J.; Mitros, T.; Nelson, W.; Hyten, D. L.; Song, Q.; Thelen, J. J.; Cheng, J.; Xu, D.; Hellsten, U., et al. Genome Sequence of the Palaeopolyploid Soybean. *Nature.* **2010,** *463,* 178–183.

22. Lorenz, A. J.; Chao, S.; Asoro, F. G.; Heffner, E. L.; Hayashi, T.; Iwata, H.; Smith, K. P.; Sorrells, M. K.; Jannink, J. L. Genomic Selection in Plant Breeding: Knowledge and Prospects. *Adv. Agron.* **2011,** *110,* 77–123.

23. Varshney, R. V.; Glaszmann, J. C.; Leung, H. Ribaul, J. M. More Genomic Resources for Less Studies Crops. *Trends Biotechnol.* **2010,** *28,* 452–460.

24. Nybom, H.; Weising, K.; Rotter. B. DNA Fingerprinting in Botany: Past, Present, Future. *Invest. Genet.* **2014,** *5,* 1.

25. Luikart, G.; England, P. R.; Tallmon, D.; Jordan, S.; Taberlet, P. The Power and Promise of Population Genomics: From Genotyping to Genome Typing. *Nat. Rev. Genet.* **2003,** *4,* 981–994.

26. Stapley, J., et al. Adaptation Genomics: The Next Generation. *Trends Ecol. Evol.* **2010,** *25,* 705–712.

27. Shendure, J.; Ji, H. Next-Generation DNA Sequencing. *Nat. Biotechnol.* **2008,** *26,* 1135–1145.

28. Ekblom, R.; Galindo, J. Applications of Next Generation Sequencing in Molecular Ecology of Non-Model Organisms. *Heredity.* **2011,** *107,* 1–15.

29. Allendorf, F. W.; Luikart, G. Next-Generation RAD Sequencing Identifies Thousands of SNPs for Assessing Hybridization between Rainbow and Westslope Cutthroat *Trout. Mol. Ecol. Resour.* **2011,** *11,* 117–122.

30. Cavagnaro, P. F.; Senalik, D. A.; Yang, L.; Simon, P. W.; Harkins, T. T.; Kodria, C. D.; Huang, S.; Weng, Y. Genome-Wide Characterization of Simple Sequence Repeats in Cucumber *(Cucumissativus L.). BMC Genomics.* **2010,** *11,* 569–586.

31. Zhu, H.; Senalik, D.; Mccown, B. H.; Zeldin, E. L.; Speers, J.; Hyman, J.; Bassil, N., et al. 2Mining and Validation of Pyrosequenced Simple Sequence Repeats (SSRs) from American Cranberry *(VacciniummacrocarponAit.). Theor. Appl. Genet.* **2011,** *124,* 87–96.

32. Helyar, S. J., et al. Application of SNPs for Population Genetics of Non Model Organisms: New Opportunities and Challenges. *Mol. Ecol. Resour.* **2011,** *11,* 123–136.

33. Davey, J. W.; Blaxter, M. L. RADSeq: Next-Generation Population Genetics. *Brief Funct. Genomics.* **2010,** *9,* 416–423.

34. Gusella, J. F., et al. A Polymorphic DNA Marker Genetically Linked to Huntington's Disease. *Nature.* **1983,** *306,* 234–238.

35. Riordan, J., et al. Identification of the Cystic Fibrosis Gene: Cloning and Characterization of Complementary DNA. *Science.* **1989,** *245,* 1066–1073.

36. Imelfort, M.; Duran, C.; Batley, J.; Edwards, D. Discovering Genetic Polymorphisms in Next-Generation Sequencing Data. *Plant Biotechnol. J.* **2009,** *7,* 312–317.

37. Kraus, R. H. S., et al. Genome Wide SNP Discovery, Analysis and Evaluation in Mallard *(Anasplatyrhynchos). BMC Genomics.* **2011,** *12,* 150.

38. Margulies, M., et al. Genome Sequencing in Microfabricated High-Density Picolitre Reactors. *Nature.* **2005,** *437,* 376–380.

39. Van T. C. P., et al. SNP Discovery and Allele Frequency Estimation by Deep Sequencing of Reduced Representation Libraries. *Nat. Methods.* **2008,** *5,* 247–252.

40. Pfender, W. F.; Saha, M. C.; Johnson, E. A.; Slabaugh, M. B. Mapping with RAD (Restriction-Site Associated DNA) Markers to Rapidly Identify QTL for Stem Rust Resistance In *Loliumperenne. Theor. Appl. Genet.* **2011,** *122,* 1467–1480.

41. Sahu, B. B.; Sumit, R.; Srivastava, S. K.; Bhattacharyya, M. K. Sequence Based Polymorphic (SBP) Marker Technology for Targeted Genomic Regions: Its Application in Generating a Molecular Map of the *Arabidopsis thaliana* Genome. *BMC Genomics.* **2012,** *13,* 20.

42. Andolfatto, P., et al. Multiplexed Shotgun Genotyping for Rapid and Efficient Genetic Mapping. *Genome Res.* **2011,** *21,* 610–617.

43. Elshire, R. J., et al. A Robust, Simple Genotyping-by-Sequencing (GBS) Approach for High Diversity Species. *PLoS One.* **2011,** *6,* e19379.

44. Van, O. N. J., et al. Complexity Reduction of Polymorphic Sequences *(CRoPS):* A Novel Approach for Large-Scale Polymorphism Discovery in Complex Genomes. *PLoS One.* **2007,** *2,* e1172.

45. Hyten, D. L.; Cannon, S. B.; Song, Q.; Weeks, N., et al. High-Throughput SNP Discovery through Deep Resequencing of a Reduced Representation Library to Anchor and Orient Scaffolds in the Soybean Whole Genome Sequence. *BMC Genomics.* **2010,** *11,* 38.

46. Baird, N. A.; Etter, P. D.; Atwood, T. S.; Currey, M. C.; Shiver, A. L., et. al. Rapid SNP Discovery and Genetic Mapping Using Sequenced RAD Markers. *PLoS One.* **2008,** *3* (10), e63376.

47. Pfender. W. F.; Saha. M. C.; Johnson. E. A.; Slabaugh, M. B. Mapping with RAD (Restriction-Site Associated DNA) Markers to Rapidly Identify QTL for Stem Rust Resistance in *Loliumperenne. Theor. Appl. Genet.* **2011,** *122,* 1467–1480.

48. Lewis, Z. A.; Shiver, A. L.; Stiffler, N.; Miller, M. R; Johnson, E. A.; Selker, E. U. High Density Detection of Restriction-Site Associated DNA Markers for Rapid Mapping of Mutated Loci In *Neurospora. Genetics.* **2007,** *177,* 1163–1171.

49. Baxter, S. W.; Davey, J. W.; Johnston, J. S.; Shelton, A. M.; Heckel, D. G. Linkage Mapping and Comparative Genomics Using Next-Generation RAD Sequencing of a Non-Model Organism. *PLoS One.* **2011,** *6* (4), e19315.

50. Chutimanitsakun, Y.; Nipper, R. W.; Cuesta-Marcos, A.; Cistue, L.; Corey, A.; Filichkina, T.; Johnson, E. A.; Hayes, P. M. Construction and Application for QTL Analysis of a Restriction Site Associated DNA (RAD) Linkage Map in Barley. *BMC Genomics.* **2011,** *12,* 4.

51. Barchi, L.; Lanteri, S.; Portis, E.; Acquadro, A.; Vale, G.; Toppino, L.; Rotino, G. L. Identification of SNP and SSR Markers in Eggplant Using RAD Tag Sequencing. *BMC Genomics.* **2011,** *12,* 304.

52. Yang, T.; Bao, S.; Ford, R.; Jia, T.; Guan, J.; He, Y.; Sun, X.; Jiang, J.; Hao, J.; Zhang, X.; Zong, X. High-Throughput Novel Microsatellite Marker of Faba Bean via Next Generation Sequencing. *BMC Genomics.* **2012,** *13,* 602.

53. Hohenlohe, P. A.; Amish, S. J.; Catchen, J. M.; Allendorf, F. W.; Luikart, G. Next-Generation RAD Sequencing Identifies Thousands of SNPs for Assessing Hybridization between Rainbow and Westslope Cutthroat Trout. *Mol. Ecol. Resour.* **2011**, *1*, 117–122.

54. Ramos, A. M., et al. Design of a High Density SNP Genotyping Assay in the Pig Using SNPs Identified and Characterized by Next Generation Sequencing Technology. *PLoS One.* **2009**, *4*, e6524.

55. Gompert, Z., et al. Bayesian Analysis of Molecular Variance in Pyrosequences Quantifies Population Genetic Structure Across the Genome of *Lycaeides* Butterflies. *Mol. Ecol.* **2010**, *19*, 2455–2473.

56. Quail, M. A., et al. A Large Genome Center's Improvements to the Illumina Sequencing System. *Nat. Methods.* **2008**, *5*, 1005–1010.

57. Emerson, K. J., et al. Resolving Postglacial Phylogeography Using High-Throughput Sequencing. *Proc. Natl. Acad. Sci. USA.* **2010**, *107*, 16196–16200.

58. Futschik, A.; Schlotterer, C. The Next Generation of Molecular Markers from Massively Parallel Sequencing of Pooled DNA Samples. *Genetics.* **2010**, *186*, 207–218.

59. Cutler, D. J.; Jensen, J. D. To Pool, or not to Pool? *Genetics.* **2010**, *186*, 41–43.

60. Wang, Z.; Gerstein, M.; Snyder, M. RNA-Seq: A Revolutionary Tool for Transcriptomics. *Nat. Rev. Genet.* **2009**, *10*, 57–63.

61. Faircloth, B. C. Msatcommander: Detection of Microsatellite Repeat Arrays and Automated, Locus-Specific Primer Design. *Mol. Ecol. Resour.* **2008**, *8*, 92–94.

62. Thurston, M. I.; Field, D. *Msatfinder: Detection and Characterisation of Microsatellites;* Distributed by the authors athttp://www.genomics.ceh.ac.uk/msatfinder/. CEH Oxford: Oxford, 2005.

63. Trapnell, C., et al. Differential Gene and Transcript Expression Analysis of RNA-Seq Experiments with Top Hat and Cufflinks. *Nat. Protoc.* **2012**, *7*, 562–578.

64. Guttman, M., et al. AbInitio Reconstruction of Cell Type-Specific Transcriptomes in Mouse Reveals the Conserved Multi-Exonic Structure of LincRNAs. *Nat. Biotech.* **2010**, *28*, 503–510.

65. Robertson, G., et al. *De Novo* Assembly and Analysis of RNA-Seq Data. *Nat. Methods.* **2010**, *7*, 909–912.

66. Schliesky, S.; Gowik, U.; Weber, A. P.; Bräutigam, A. RNA-Seq Assembly–are We There Yet? *Front. Plant Sci.* **2012**, *3*, 220.

67. Schulz, M. H.; Zerbino, D. R.; Vingron, M.; Birney, E. Oases: Robust *De Novo* RNA-Seq Assembly across the Dynamic Range of Expression Levels. *Bioinformatics.* **2012**, *28*, 1086–1092.

68. Grabherr, M. G.; Hass, B. J.; Yassour, M.; Levin, J. Z.; Thompson, D. A., et al. Full-Length Transcriptome Assembly from RNA-Seq Data without a Reference Genome. *Nat. Biotech.* **2011**, *29*, 644–652.

69. Hillier, L. W.; Marth, G. T.; Quinlan, A. R.; Dooling, D. et al. Whole-Genome Sequencing and Variant Discovery in *C. elegans. Nat. Methods.* **2008**, *5*, 183–188.

70. Koboltd, D. C.; Chen, K.; Wylie, T.; Larson, D. E.; McLellan, M. D.; Mardis, E. R., et al. Varscan: Variant Detection in Massively Parallel Sequencing of Individual and Pooled Samples. *Bioinformatics.* **2009**, *25*, 2283–2285.

71. Slate, J.; Gratten, J.; Beraldi, D.; Stapley, J.; Hale, M.; Pemberton, J. Gene Mapping in the Wild with SNPs: Guidelines and Future Directions. *Genetica.* **2009**, *136*, 97–107.

72. Henry, R. J.; Edwards, K. New Tools for Single Nucleotide Polymorphism (SNP) Discovery and Analysis Accelerating Plant Biotechnology. *Plant Biotechnol. J.* **2009**, 3–11.

73. Thudi, M.; Li, Y.; Jackson, S. A.; May, G. D.; Varshney, R. K. Current State-of-the-Art Sequencing Technologies for Plant Genomics Research. *Brief Funct. Genomics.* **2012,** *11,* 3–11.

74. Deschamps, S.; Campbell, M. A. Utilization of Next Generation Sequencing Platforms in Plant Genomics and Genetic Variant Discovery. *Mol. Breed.* **2010,** *25,* 553–570.

75. Appleby, N.; Edwards, D.; Batley, J. New Technologies for Ultrahigh Throughput Genotyping in Plants. *Methods Mol. Biol.* **2009,** *513,* 19–39.

76. Lin, C. H.; Yeakley, J. M.; McDaniel, T. K.; Shen, R. Medium- to High-Throughput SNP Genotyping Using Vera Code Microbeads. *Methods Mol. Biol.* **2009,** *496,* 129–142.

77. Narina, S. S.; Buyyarapu, R.; Kottapalli, K. R.; Sartie, A. M.; Ali, M. I.; Robert, A.; Hodeba, M. J. D.; Sayre, B. L.; Scheffler, B. E. Generation and Analysis of Expressed Sequence Tags (Ests) for Marker Development in Yam (*Dioscorea alataL.*). *BMC Genomics.* **2011,** *12,* 100.

78. Yang, S. S.; Tu, Z. J.; Cheung, F.; Xu, W. W.; Lamb, J. F. S.; Jung, H. J. G.; Vance, C. P.; Gronwald, J. W. Using RNA-Seq for Gene Identification, Polymorphism Detection and Transcript Profiling in Two Alfalfa Genotypes with Divergent Dell Wall Composition in Stems. *BMC Genomics.* **2011,** *12,* 199.

79. Burr, B. F.; Burr, A.; Thompson, K. H.; Albertson, M. C.; Stuber, C. W. Gene Mapping with Recombinant Inbreds in Maize. *Genetics.* **1988,** *118* (5), 19–526.

80. Novembre, J.; Johnson, T.; Bryc, K.; Kutalik, Z.; Boyko, A. R., et al. Genes Mirror Geography within Europe. *Nature.* **2008,** *4,* 56–98.

81. Santure, A. W.; Stapley, J.; Ball, A. D.; Birkhead, T. R.; Burke, T.; Slate, J. On The Use of Large Marker Panels to Estimate Inbreeding and Relatedness: Empirical and Simulation Studies of a Pedigreed Zebra Finch Population Typed at 771 SNPs. *Mol. Ecol.* **2010,** *19,*1439–1451.

82. Jakobsson, M.; Scholz, S. W.; Scheet, P.; Gibbs, J. R.; VanLiere, J. M.; Fung, H. C., et al. Genotype, Haplotype and Copy-Number Variation in Worldwide Human Populations. *Nature.* **2008,** *451,*998–1003.

83. Peterson, T. W.; Nam, S. J.; Darby, A. Next-Generation Sequencing Survey. In *North America Equity Research;* JP Morgan Chase & Co.: New York, 2010.

84. Gupta, P. K; Langridge, P.; Mir, R. R. Marker-Assisted Wheat Breeding: Present Status and Future Possibilities. *Mol. Breed.* **2010,** *26,* 145–161.

85. Guo, L. B.; Ye, G. Y. Use of Major Quantitative Trait Loci to Improve Grain Yield of Rice. *Rice Sci.* **2014,** *21* (2), 65–82.

86. Celik, I.; Gultekin, V.; Allmer, J.; Doganlar, S.; Frary, A. Development of Genomic Simple Sequence Repeat Markers in Opium Poppy by Next-Generation Sequencing. *Mol. Breed.* **2014,** *34* (2), 323–334.

87. Ahn, Y. K.; Tripathi, S.; Kim, J. H., Cho, Y. I.; Lee, H. E.; Kim, D. S.; Yoon, M. K. Microsatellite Marker Information from High-Throughput Next-Generation Sequence Data of *Capsicum annuum* Varieties *Mandarin* and *Blackcluster.* *Sci. Hort.* **2014,** *170,* 123–130.

88. Kudapa, H.; Azam, S.; Sharpe, A. G.; Taran, B.; Li, R.; Deonovic, B.; Varshney, R. K. Comprehensive Transcriptome Assembly of Chickpea *(Cicerarietinum L.)* Using Sanger and Next Generation Sequencing Platforms: Development and Applications. *PLoS One.* **2014,** *9,* e86039.

89. Park, I.; Kim, J.; Lee, J.; Kim, S.; Cho, O.; Yang, K.; Kim, H. Development of SSR Markers by Next-Generation Sequencing of Korean Landraces of Chamoe *(Cucumismelo var. makuwa).* *Mol. Biol. Rep.* **2013,** *40* (12), 6855–6862.

90. Poecke, R. M.; Maccaferri, M.; Tang, J.; Truong, H. T.; Janssen, A.; Orsouw, N. J.; Vossen, E. A. Sequence-Based SNP Genotyping in Durum Wheat. *Plant Biotechnol. J.* **2013,** *11* (7), 809–817.

91. Dockter, R. B.; Elzinga, D. B.; Geary, B.; Maughan, P. J.; Johnson, L. A.; Tumbleson, D.; Stevens, M. R. Developing Molecular Tools and Insights into the *Penstemon genome* Using Genomic Reduction and Next-Generation Sequencing. *BMC Genetics.* **2013,** *14* (1), 66.

92. Chung, J. W.; Kim, T. S.; Suresh, S.; Lee, S. Y.; Cho, G. T. Development of 65 Novel Polymorphic cDNA-SSR Markers in Common Vetch *(Vicia sativa subsp. sativa)* Using Next Generation Sequencing. *Molecules.***2013,** *18* (7), 8376–8392.

93. Marroni, F.; Pinosio, S.; Di Centa, E.; Jurman, I.; Boerjan, W.; Felice, N.; Morgante, M. Large-Scale Detection of Rare Variants via Pooled Multiplexed Next-Generation Sequencing: Towards Next-Generation*Ecotilling. Plant J.* **2011,** *67* (4), 736–745.

94. Zhu, Q.; Smith, S. M.; Ayele, M.; Yang, L.; Jogi, A.; Chaluvadi, S. R.; Bennetzen, J. L. High-Throughput Discovery of Mutations in Tef Semi-Dwarfing Genes by Next-Generation Sequencing Analysis. *Genetics.* **2012,** *192* (3), 819–829.

95. Yang, H.; Tao, Y.; Zheng, Z.; Shao, D.; Li, Z.; Sweetingham, M. W.; Li, C. Rapid Development of Molecular Markers by Next-Generation Sequencing Linked to a Gene Conferring Phomopsis Stem Blight Disease Resistance for Marker-Assisted Selection in Lupin *(Lupinusangustifolius L.)* Breeding. *Theor. Appl. Genet.* **2013,** *126* (2), 511–522.

96. Yang, T.; Bao, S. Y.; Ford, R.; Jia, T. J.; Guan, J. P.; He, Y. H.; Zong, X. X. High-Throughput Novel Microsatellite Marker of Faba Bean via Next Generation Sequencing. *BMC Genomics.* **2012,** *13* (1), 602.

97. Singh, D.; Singh, P. K.; Chaudhary, S.; Mehla, K.; Kumar, S. Exome Sequencing and Advances in Crop Improvement. *Adv. Genet.* **2012,** *79,* 87–121.

98. Kale, S. M.; Pardeshi, V. C.; Kadoo, N. Y.; Ghorpade, P. B.; Jana, M. M.; Gupta, V. S. Development of Genomic Simple Sequence Repeat Markers for Linseed Using Next-Generation Sequencing Technology. *Mol. Breed.* **2012,** *30* (1), 597–606.

99. Polashock, J.; Zelzion, E.; Fajardo, D.; Zalapa, J.; Georgi, L.; Bhattacharya, D.; Vorsa, N. The American Cranberry: First Insights into the Whole Genome of a Species Adapted to Bog Habitat. *BMC Plant Biol.* **2014,** *14* (1), 165.

100. Adams, I. P.; Miano, D. W.; Kinyua, Z. M.; Wangai, A.; Kimani, E.; Phiri, N.; Boonham, N. Use of Next-Generation Sequencing for the Identification and Characterization of *Maize chlorotic mottle virus* and *Sugarcane mosaic virus*Causing Maize Lethal Necrosis in Kenya. *Plant Pathol.* **2013,** *62* (4), 741–749.

101. Bus, A.; Hecht, J.; Huettel, B.; Reinhardt, R.; Stich, B. High-Throughput Polymorphism Detection and Genotyping in *Brassica napus* Using Next-Generation RAD Sequencing. *BMC Genomics.* **2012,** *13* (1), 281.

102. Pfender, W. F.; Saha, M. C.; Johnson, E. A.; Slabaugh, M. B. Mapping with RAD (Restriction-Site Associated DNA) Markers to Rapidly Identify QTL for Stem Rust Resistance in *Loliumperenne. Theor. Appl. Genet.* **2011,** *122* (8), 1467–1480.

103. Nelson, J. C.; Wang, S.; Wu, Y.; Li, X.; Antony, G.; White, F. F.; Yu, J. Single-Nucleotide Polymorphism Discovery by High-Throughput Sequencing in Sorghum. *BMC Genomics.* **2011,** *12* (1), 352.

104. Byers, R. L.; Harker, D. B.; Yourstone, S. M.; Maughan, P. J.; Udall, J. A. Development and Mapping of SNP Assays in Allotetraploid Cotton. *Theor. Appl. Genet.* **2012,** *124* (7), 1201–1214.

105. Li, X.; Luo, J.; Yan, T.; Xiang, L.; Jin, F.; Qin, D.; Xie, M. Deep Sequencing-Based Analysis of the *Cymbidium ensifolium* Floral Transcriptome. *PLoS One.* **2013**, *8* (12), e85480.

106. Esteras, C.; Gómez, P.; Monforte, A. J.; Blanca, J.; Vicente-Dólera, N.; Roig, C.; Picó, B. High-Throughput SNP Genotyping in *Cucurbitapepo* for Map Construction and Quantitative Trait Loci Mapping. *BMC Genomics.* **2012**, *13* (1), 80.

107. Yang, T.; Jiang, J.; Burlyaeva, M.; Hu, J.; Coyne, C. J.; Kumar, S.; Zong, X. Large-Scale Microsatellite Development in Grasspea (*Lathyrussativus L.*), an Orphan Legume of the Arid Areas. *BMC Plant Biol.* **2014**, *14* (1), 65.

108. Sun, X.; Yang, T.; Hao, J.; Zhang, X.; Ford, R.; Jiang, J.; Zong, X. SSR Genetic Linkage Map Construction of Pea *(Pisumsativum L.)* Based on Chinese Native Varieties. *Crop J.* **2014**, *2* (2), 170–174.

109. Ipek, M.; Sahin, N.; Ipek, A.; Cansev, A.; Simon, P. W. Development and Validation of New SSR Markers from Expressed Regions in the Garlic Genome. *Sci. Agric.* **2015**, *72* (1), 41–46.

110. Gao, Z.; Wu, J.; Liu, Z. A.; Wang, L.; Ren, H.; Shu, Q. Rapid Microsatellite Development for Tree Peony and its Implications. BMC *Genomics.* **2013**, *14* (1), 886.

111. Wang, Z.; Fang, B.; Chen, J.; Zhang, X.; Luo, Z.; Huang, L.; Li, Y. *De Novo* Assembly and Characterization of Root Transcriptome Using Illumina Paired-End Sequencing and Development of cSSR Markers in Sweetpotato *(Ipomoea batatas)*. *BMC Genomics.* **2010**, *11* (1), 726.

112. Kim, J. E.; Oh, S. K.; Lee, J. H.; Lee, B. M.; Jo, S. H. Genome-Wide SNP Calling Using Next Generation Sequencing Data in Tomato. *Mol. Cells.* **2014**, *37* (1), 36.

113. Cardoso-Silva, C. B.; Costa, E. A.; Mancini, M. C.; Balsalobre, T. W. A.; Canesin, L. E. C.; Pinto, L. R.; Vicentini, R. *De Novo* Assembly and Transcriptome Analysis of Contrasting Sugarcane Varieties. *PLoS One.* **2014**, *9* (2), e88462.

114. Austin, R. S.; Vidaurre, D.; Stamatiou, G.; Breit, R.; Provart, N. J.; Bonetta, D.; Guttman, D. S. Next-Generation Mapping of *Arabidopsis* Genes. *Plant J.* **2011**, *67* (4), 715–725.

115. Barba, M.; Czosnek, H.; Hadidi, A. Historical Perspective, Development and Applications of Next-Generation Sequencing in Plant Virology. *Viruses.* **2014**, *6* (1), 106–136.

116. Yu, Q.; Tong, E.; Skelton, R. L.; Bowers, J. E.; Jones, M. R.; Murray, J. E.; Ming, R. A Physical Map of the Papaya Genome with Integrated Genetic Map and Genome Sequence. *BMC Genomics.* **2009**, *10* (1), 371.

117. Jiang, C.; Chen, C.; Huang, Z.; Liu, R.; Verdier, J. ITIS, A Bioinformatics Tool for Accurate Identification of Transposon Insertion Sites Using Next-Generation Sequencing Data. *BMC Bioinformatics.* **2015**, *16* (1), 72.

118. Arif, I. A.; Bakir, M. A.; Khan, H. A.; Al Farhan, A. H.; Al Homaidan, A. A.; Bahkali, A. H.; Shobrak, M. A Brief Review of Molecular Techniques to Assess Plant Diversity. *Int. J. Mol. Sci.* **2010**, *11* (5), 2079–2096.

119. Hong, K. J.; Lee, C. H.; Kim, S. W. *Aspergillus oryzae*GB-107 Fermentation Improves Nutritional Quality of Food Soybeans and Feed Soybean Meals. *J. Med. Food.* **2004**, *7* (4), 430–435.

120. Zhang, G.; Liu, X.; Quan, Z.; Cheng, S.; Xu, X.; Pan, S.; Wang, J. Genome Sequence of Foxtail Millet (*Setariaitalica*) Provides Insights into Grass Evolution and Biofuel Potential. *Nature Biotechnol.* **2012**, *30* (6), 549–554.

121. Wendler, N.; Mascher, M.; Nöh, C.; Himmelbach, A.; Scholz, U.; Ruge-Wehling, B.; Stein, N. Unlocking the Secondary Gene-Pool of Barley with Next-Generation Sequencing. *Plant Biotechnol. J.* **2014**, *12* (8), 1122–1131.

122. Natali, L.; Cossu, R. M.; Barghini, E.; Giordani, T., Buti, M.; Mascagni, F.; Cavallini, A. The Repetitive Component of the Sunflower Genome as Shown by Different Procedures for Assembling Next Generation Sequencing Reads. *BMC Genomics.* **2013,** *14* (1), 686.

123. Hendre, P. S.; Kamalakannan, R.; Varghese, M. High-Throughput and Parallel SNP Discovery in Selected Candidate Genes in *Eucalyptus camaldulensis* Using Illumina *NGS* Platform. *Plant Biotechnol. J.* **2012,** *10* (6), 646–656.

124. Iorizzo, M.; Senalik, D.; Szklarczyk, M.; Grzebelus, D.; Spooner, D.; Simon, P. De, Novo Assembly of the Carrot Mitochondrial Genome Using Next Generation Sequencing of Whole Genomic DNA Provides First Evidence of DNA Transfer into an Angiosperm Plastid Genome. *BMC Plant Biol.* **2012,** *12* (1), 61.

CHAPTER 8

INTEGRATION OF OMICS APPROACHES FOR LOW-PHOSPHORUS TOLERANCE IN MAIZE

MOHAMMED SHALIM UDDIN[1], A. B. M. KHALDUN[1], and M. TOFAZZAL ISLAM[2*]

[1]Bangladesh Agricultural Research Institute, Gazipur 1701, Bangladesh

[2]Department of Biotechnology, Bangabandhu Sheikh Mujibur Rahman Agricultural University, Gazipur 1706, Bangladesh

*Corresponding author. E-mail: tofazzalislam@gmail.com

CONTENTS

ABSTRACT

Phosphorus (P) is one of the most important macro-elements which is essential for growth and development of plant. Plants have evolved complex responsive and adaptive mechanisms for acquisition, remobilization, and recycling of phosphate (Pi) to maintain P homeostasis. Phosphate starvation responsive (PSR) genes are involved in acquisition, mobilization, and substitution of Pi, metabolic pathways, signal transduction, transcriptional regulation, and many processes related to growth and development. This chapter has integrated the phenomics, genomics, proteomics, breeding informatics, micromics, metabolomics, and transcriptomics of maize under low-phosphorus stress (LP), and also the opportunities for further understanding of regulation of maize in LP stress as summarized. Thus the integration of systems biology with high-throughput, high-dimensional, and precision phenotyping will contribute to the development of maize varieties for LP tolerance.

8.1 INTRODUCTION

Phosphorus (P) is essential for all living individuals because of its role in numerous key molecules, including DNA and RNA. However, the supply of P from soil is often limited to crop production,[1] and demand for P fertilizer is increasing worldwide.[2] Low-phosphorous (LP) stress is an important soil nutrient stress affecting more than 5.7 billion hectares (ha) of land throughout the world,[3] and most other agricultural soils have only achieved moderate or high P fertility. Plants have developed well-regulated systems for phosphate (Pi) scavenging, acquisition, and recycling for preventing P starvation, which are generally called as phosphate starvation responses (PSRs).[4] PSRs are general features related to morpho-physiological, biochemical, and molecular changes of plants in response to P starvation, including organic acids release and hydrolytic enzymes secretion, metabolic modifications, mobilization, and recycling of internal P resource to modify root architecture system (RAS).[5]

A large proportion of farming land worldwide is deficient in soluble Pi, the form of P most readily absorbed and utilized by plants.[6] In acidic and alkaline soils, huge amounts of P fertilizers are applied to maize fields for maximizing the yield.[4] The cost of P fertilizers is increasing due to decreasing natural reserves of rock Pi and an increasing cost of extraction from what remains. It is anticipated that the present production of P fertilizer is moved beyond the peak point[7] and it will be exhausted within the next 70–200 years.[8]

8.2   MECHANISM OF LOW PHOSPHORUS EFFICIENCY IN MAIZE

P efficiency generally describes the ability of genotypes to give higher yield under LP condition.[9] The P efficiency differs at both genotype and species levels.[10] The ability of a genotype to give higher yield under LP condition relates to the ability to adopt more P from the soil under LP condition or the ability to produce more dry matter per unit of P, or a combination of both.[11,12]

8.2.1   *ROOT ARCHITECTURE SYSTEM (RAS)*

RAS includes embryonic primary root, seminal roots, postembryonic shoot-borne and lateral roots, which play a vital role for plant P acquisition.[13] RAS is highly flexible in its developmental response to LP conditions. LP availability modifies RAS traits such as primary root length, number and length of lateral roots, root branching, and enhancement of root hairs and cluster root formation.[14–17] In particular, sugars, auxin, and ethylene play important roles in modulating RAS during P deficiency.[18]

Shallower root growth angles (RGA) of seminal and crown roots increased top soil scavenging and P acquisition.[13] RGA is regulated by multiple genes, which can be analyzed using recombinant inbred line (RIL) population. Variation in maize RIL population for RGA shows that these phenotypes have a dominant effect on P acquisition in LP soil.[19,20]

In the LP stress conditions, plants have developed complex mechanisms to integrate local and systemic senses and signals to maintain the cellular nutrient homeostasis at the whole plant level. Roots perceive fluctuations in extracellular nutrient levels and send signals to the shoot through the xylem, warning that inadequate supply of the particular nutrient impends. Shoots sense root-derived nutrient signals and send signals to both shoots and roots, through the phloem, for adjustment of developmental processes and nutrient uptake.[21–24] Local Pi sensing and signaling initiate adjustments of root system architecture (RSA) to enhance Pi foraging, whereas systemic or long-distance signaling pathways act to control Pi uptake, mobilization, and reallocation.[25–28]

8.2.2 ROOT–SHOOT RATIO

The mobility of P is low in the soil, thus some genotypes develop larger root systems for scavenging higher quantity of soil P reached to the root surface.[29] Higher root–shoot ratio was reported for maize lines in LP field as compared within normal-P (NP) field.[30–32] Leaf growth is severely reduced under LP stress, which leads to reduced leaf demand for assimilates, and subsequently leads to translocation of photosynthates to the root.[33] LP stress also modifies the photosynthesis, metabolism of sugar, and/or partitioning of carbohydrate between source and sink.[34] Therefore, special root growth helps the stressed plants to forage more P from the soil in response to LP stress conditions.

8.2.3 ROOT HAIRS

Root hairs are important for improving the acquisition of P.[35] Root hairs enhance the ability of roots to explore the rhizosphere by increasing surface area.[36] They cover up to 77% of the total root surface area and thus are the major points of contact between plants and the rhizosphere. A wide range of variability in root morphology and distribution exists between genotypes of many plants.[37] Maize genotypes modified their RSA in response to LP in the rooting medium,[19] and those performed better under P-deficient conditions develop shallow root systems to tap P accumulated in topsoil and have greater specific P absorption rate, tissue P contents, relative growth rate, and biomass accumulation than others.[38,39]

8.2.4 MORPHO-PHYSIOLOGICAL RESPONSES ASSOCIATED WITH LP IN MAIZE

LP stress causes numerous and often-adverse alterations of plant growth, morpho-physiological processes, and yield. The common consequence of LP stress is the accumulation of anthocyanin pigment in leaf tissue to protect chloroplasts and nucleic acids from the intense or ultraviolet light.[40] Although similar responses are observed under nitrogen (N) starvation, the meristem activity is gradually lost in P starvation, whereas it remains relatively active under N starvation conditions.[41,42] Sufficient P availability during early crop growth has significant effect on different crop species.[43] Some common morpho-physiological effects of LP tress on maize are summarized in Table 8.1.

TABLE 8.1 Genetic Variability and Effects of LP Stress on Morphological and Physiological Adaptations in Maize.

Sl. no	Treatment	Genotypes (no.)	Growth conditions	Growth stage	Major effect or findings	References
Genetic variability study in maize for LP stress						
1	Different P level	2 genotypes	Nutrition solution	Seedlings stage	Genetic variation was found	Clark and Brown [44]
2.	Two (LP 8–10 μM and HP 40–50 μM)	Inbreds (20)	sand–alumina culture	Seedlings	Variability found for accumulate dry matter.	DaSilva and Gabelman [45]
3.	Two level (LP & HP)	23 popcorn S_6 line	Pot experiment	Seedlings stage	Efficient and inefficient inbred lines identified	Mundim et al. [46]
4.	Two (LP (23 kg P_2O_5 ha^{-1}) and HP (97 kg P_2O_5 ha^{-1}))	Maize accessions (20)	Field	Mature	PUE is related to distribution of dry matter. Identified genetic variation among maize landraces	Bayuelo-Jiménez and Ochoa-Cadavid [47]
5.	Two (LP (23 kg P_2O_5 ha^{-1}) and HP (97 kg P_2O_5 ha^{-1}))	Land race (242)	Field	Seedling	P efficient accessions had greater biomass, root to shoot ratio, nodal rooting, nodal root laterals, and nodal root hair density.	Bayuelo-Jiménez et al. [48]
6.	Two level (LP & HP)	23 popcorn S_6 line	Pot experiment	Seedlings stage	Efficient and inefficient inbred lines were identified	Mundim et al. [46]
Effect of LP stress on maize root growth						
7.	LP	–	Solution culture	Seedlings	Root length increased in low-P conditions.	Anghinoni and Barber [49]
8.	Two (0 and 120 kg P_2O_5 ha^{-1})	Inbreds (116)	Field	Seedling and mature	LP decreased the rates of the root length, root surface area, and root volume.	Jiang [50]
9.	Two (LP, NP)	One	Hydroponics	Seedlings	The axial roots were increasing and the elongation of 1st order lateral roots was reduced.	Mollier and Pellerin [51]

TABLE 8.1 *(Continued)*

Sl. no	Treatment	Genotypes (no.)	Growth conditions	Growth stage	Major effect or findings	References
10.	Two (0 and 130 kg P_2O_5 ha^{-1})	Landrace (5)	Field (P 10.5 mg kg^{-1} and pH of 5.6)	Seedlings	Reduced root volume, increased root hairs density, and root shoot ratio.	Yao et al. [52]
11.	4 nutrition solutions	12 hybrid	Nutrition solution	Seedlings	The adaptation of maize genotypes to low levels of soil P is closely related to a better-developed root system.	Alves et al. [53]
Effect of LP stress on maize leaf area						
12.	Low-P soil	2 genotypes	Field condition	Mature stage	Reduced leaf appearance, and delayed anthesis and increased the anthesis-silking interval (ASI)	Chen et al. [54]
13.	Three	Maize cv Volga	Field	Seedling and mature	LAI and leaf elongation rate were significantly	Plenets [55]
14.						
Effect of LP stress on photosynthesis						
15.	Two (0.001 mm (low-P) or 0.5 mm Pi (control)	One	Nutrition solution	Seedlings	Photosynthates are differently distributed between shoots and roots.	Usuda and Shimogawara [56]

8.3 MOLECULAR BREEDING APPROACHES OF LP TOLERANCE IN MAIZE

8.3.1 QUANTITATIVE TRAIT LOCI (QTL) OF LP TOLERANCE IN MAIZE

QTL mapping has been a key tool to study genetic architecture of complex traits.[57] Agronomically important traits like yield, grain quality and resistance/ tolerance to biotic and/or abiotic stresses are complex traits. Genetic architecture refers to many genome locations with genes that affect the traits and their effects and relative contributions of additive, dominant, and epistatic effects.[58] Uncovering of QTL of agronomical importance and fundamental genes has greatly increased our understanding of complex traits.[59] Understanding and further identifying QTL that trigger the traits will significantly contribute to breeding with marker-assisted selection (MAS)[60] and pyramiding of multiple favorable alleles.[61]

QTL analysis of maize lateral root, seminal root, and root hair traits shows that these complex traits are genetically regulated by QTL with epistatic effects under high and LP conditions[19,62,63] (Table 8.2). Several studies have been performed to identify QTL for grain yield and for traits related to phosphorus utilization efficiency (PUE) and LP tolerance in maize under different P conditions[62,64–68] They identified QTL in all maize chromosomes, and these QTL explained from 1 to 22% of phenotypic variation. Other studies evaluated root traits with the aim of determining their importance in P acquisition from soils with LP availability. Studies have also been performed to understand the genetic basis of PUE-related traits and their contributions to grain yield in maize, and the result showed that dominance and epistatic effects were more important than additive effects for grain yield under LP.[69]

Genetic mapping is moving from traditional QTL mapping using several hundred individuals derived from biparental populations to joint linkage and linkage disequilibrium (LD) mapping using both biparental and nature populations,[70,71] the latter usually using more than 1000 lines and millions of single nucleotide polymorphisms (SNPs) or genotyping-by-sequencing (GBS) marker in order to precisely map the responsible loci.[71,72] Future genetic researches of crop abiotic stresses such as maize LP tolerance are expected to use a large number of samples, and high-throughput phenotyping combined with high-density genotyping will become a routine approach.

TABLE 8.2 QTL Mapping Reports on Low-Phosphorus Tolerance in Maize.

Sl. no.	Population type/size	Parent performance	Traits	Map length (cM)	Marker no./ density (cM)	QTL no.	References
1.	RIL/169	Mo17 had longer RHL and larger RHLP than B73 under LP treatment	RHL, RHLP	1703.0	196/8.73	2	Zhu et al. [63]
2.	RIL/160	Mo17 had longer LRL, more LRN, and higher LRNP than B73 under LP treatment	LRL, LRN, LRNP	1703.0	196/8.73	8	Zhu et al. [73]
3.	RIL/162	Mo17 had longer SRL, more SRN than B73 under LP treatment	SRL, SRN	1703.0	196/8.73	4	Zhu et al. [74]
4.	RIL/197	Mo17 had larger SW than B73 under LP treatment	SW, RV, CLN	?	?	5	Kaeppler and Parke [75]
5.	F2:3/241	082 was more tolerant to LP stress than Ye107, and had greater values for all tested traits in LP field	RPL, RBW, RRW, RLA, RTW, RFN, TPS, PAE, PE, RAP, RH	1681.3	375/3.84	75	Chen et al. [68]
6.	F2:3/241	082 had larger RST, and lower SPUE and WUPE than Ye107 in LP field	SPUE, WPUE, RSR	1681.3	375/3.84	5	Chen et al. [67]
7.	F2:3/241	All tested traits of 082 were larger than Ye107 in LP field	RW, TRL, FN, PAE	1681.3	375/3.84	11	Chen and Xu [76]
8.	F2:3/210	All tested traits of 178 were larger than 5003 in LP field	GY, HKW, EL, RN, KN, ED	1755.1	207/8.48	34	Li et al. [66]
9.	F8/218	Ye478 had higher PH, EH, GY, KN, and HKW under LP treatment than Wu312 under LP treatment	PH, EH, PH/EH, GY, KN, HKN	2084.1	184/11.3	26	Cai et al. [64]

TABLE 8.2 (Continued)

Sl. no.	Population type/size	Parent performance	Traits	Map length (cM)	Marker no./ density (cM)	QTL no.	References
10.	F8/218	Ye478 had higher GY, LA, and LL, similar LW and ChL, and shorter FT and ASI than Wu312 under LP treatment	GY, LA, LL, LW, ChL, FT, ASI	2084.1	184/11.3	21	Cai et al. [77]

Abbreviations: RIL: recombinant inbred lines; LP: low-P; CIM: composite interval mapping; ED: ear diameter; RN: row number; EL: ear length; HKW: one hundred kernel weight; GY: grain yield; KN: kernel number per row; PH: plant height; EH: ear height; PH/EH: the ratio of ear height to plant height; FT: flowering time; CHL: chlorophyll level; LL: leaf length; LW: leaf width; LA: leaf area; SRN: seminal root number; SRL: seminal root length; LRN: lateral root number; RCHL: relative chlorophyll level; RLA: relative leaf area; RLW: relative leaf width; LRNP: lateral root number plasticity; TRL: taproot length; FN: fibrous root number; SW: shoot weight; RW: root weight; RV: root volume; RFN: relative fibrous root number; RTW: relative topsoil root dry weight; RPH: relative plant height; RBW: relative above-ground biomass dry weight; RRW: relative root dry weight; TPS: tolerance index to low-phosphorus stress; RH: the rate of [H⁺] under low- and normal-P; RAP: the rate of acid phosphatase activity under low- and normal-P; CLN: Colonization refers to the percent root colonized based on detection of vesicles, hyphae, or arbuscules at 100 random intersects per root sample; RSR: root to shoot ratio; PAE: phosphorus absorption efficiency; PE: phosphorus efficiency; SPUE: shoot phosphorus utilization efficiency; WPUE: whole phosphorus utilization efficiency of plant.

8.3.2 MARKER-ASSISTED BREEDING AND TRANSPORTATION OF GENE FOR P EFFICIENCY

Phosphorus uptake 1 (Pup1) is a QTL located on chromosome 12 in rice that causes phosphorus deficiency tolerance. Pup1 was introgressed into several rice varieties through marker-assisted backcrossing approach[78] and these lines exhibited the dramatic increase in rice P uptake efficiency, especially on P-deficient soils. Moreover, overexpression of Pup1 or phosphorus starvation tolerance 1 (PSTOL1), the rice gene responsible for the Pup1 QTL, also enhanced grain yield on P-deficient soils [2], clearly confirming the significant potential for employing PSTOL1 in rice breeding for P use efficiency. It may have the potential to increase P uptake efficiency in other cereals. In maize, homolog of PSTOL1 on chromosome 8 was co-localized with a QTL for root traits. In sorghum, SNPs within homologs of PSTOL1 are associated with different root traits.[79] By using transgenic techniques, numerous genes have been successfully introduced into different crop species as usual in cereals with the aim of improving P efficiency (Table 8.3).

8.3.3 ASSOCIATION MAPPING

QTL mapping has moved from single marker-based, two flanking marker-based, to multiple marker-based approaches and finally to whole genome approaches. LD or association mapping can exploit to identify QTL using germplasm, cultivars, and all available genetic and breeding materials, by which molecular dissection of complex traits can be more closely linked up with plant breeding. Association mapping can be used in maize for dissecting the complex traits like plant height and flowering time in maize.[94] LP stress is a complex trait, dissection of this will depend on continuous research effort in applied QTL mapping and integrated utilization of various information and materials that has been accumulating, including genetic and breeding materials, molecular markers, and various phenotypic data collected across environments.

8.3.4 GENOME-WIDE ASSOCIATION STUDIES (GWAS)

GWAS is an approach that alleles at genotypic markers in a large population of unrelated individuals tested for significant association with a quantitative phenotype of interest.[95] *High-density SNP data allows GWAS to test all*

TABLE 8.3 List of Genes That Have Been Successfully Introduced into Different Cereal Crop for Improving P Efficiency.

Sl. no	Gene introduced	Main effect under P deficiency	References
Pi acquisition efficiency (PAE)			
Maize			
1.	Basic helix-loop-helix domain ZmPTF1 (Maize)	Increased P content and plant yield	Li et al. [80]
2.	H+-pyrophosphatase gene TsVP (*Thellungiella halophila*)	Increased Pi uptake and grain yield	Pei et al. [113]
3	Phosphate starvation response regulator TaPHR1	Increased Pi uptake and grain yield	Wang et al. [81]
Rice			
4.	Pi starvation response regulator OsPHR2	Involved in Pi starvation signaling and increased shoot P content	Zhou et al. [82]
5.	SPX (SYG/PHO81/XPR1) domain genes OsSPX1	Involved in Pi homeostasis and Pi starvation signaling	Wang et al. [83], [84]
6.	Pi transporters OsPht1;8	Involved in Pi homeostasis	Jia et al. [85]
7.	Leaf tip necrosis1 LTN1	Involved in Pi starvation signaling, Pi uptake and transport	Hu et al. [86]
8.	Pi transporters OsPht1;1	Pi uptake and translocation	Sun et al. [87]
9.	Pi transporters OsPht1;11	Involved in symbiotic Pi uptake	Yang et al. [88]
10.	Pi transcription factor OsPTF1	Increased P content and plant biomass	Yi et al. [89]
11.	High affinity Pi transporter NtPT1	Increased Pi acquisition and seed yield	Park et al. [90]
Pi utilization efficiency (PUE)			
14.	MYB transcription factor OsMYB2P-1	Involved in Pi starvation signaling, increased biomass under low P condition	Dai et al. [91]
15.	Acid phosphatases OsPAP10a	Improved ATP hydrolysis and utilization	Tian et al. [92]
16.	Type I H+ -pyrophosphatase AVP1	Improved shoot mass and higher yields	Yang et al. [93]

Sources: Modified from Zhang et al., 2014.

the genes in the genome for their association with selected traits. In maize, the nested association mapping (NAM) population has been used for analysis of leaf architecture[96] and quantitative resistance to southern corn leaf blight.[97] Thus, the next-generation genome sequencing (NGS) technology, in combination with the GWAS strategy, offer powerful tools for dissecting the complex traits.[98]

8.3.5   NEXT-GENERATION SEQUENCING (NGS)

NGS technology is a superior alternative technology for genome-wide measurements of messenger RNA (mRNA), small RNAs (sRNAs), DNA methylation, transcription-factor (TF) binding sites, chromatin structure, and structural variation.[99] One of the most widely adopted uses of NGS technology is RNA sequencing (RNA-Seq) which empowers the relative quantification of gene expression of various genotypes. RNA-Seq depends on the principle that read counts for each transcript from the NGS data reproduces relative transcript concentrations. The data are reproducible and highly accurate. RNA-Seq reads can also be source for DNA sequence polymorphisms such as SNPs, which can change into genetic markers.

GBS libraries based on dipping genome complexity with restriction enzymes (REs), which is simple, quick, very specific, highly reproducible, and can reach important regions of the genome that are inaccessible to sequence capture approaches. By using methylation-sensitive REs, monotonous regions of genomes can be avoided and lower copy regions can be targeted with higher efficiency. This GBS procedure is demonstrated with maize and barley RIL populations where 20K and 25K sequence tags were mapped, respectively.[100] Using the GBS system developed by Cornell University, large-scale genotyping is being employed by the CIMMYT Global Maize Program in MARS and genomic selection (GS) for improvement of complex traits.[98] It is anticipated that NGS technologies accelerate the application of basic research findings to breeding by re-sequencing several tropical and temperate maize germplasm, thus opening up the germplasm pool available for maize improvement.

8.3.6   GENOMIC SELECTION (GS)

GS, also known as genome-wide selection (GWS), is a strategy to predict phenotypes based on genotypic data from both phenotypic and genotypic information of a training population.[95] GS identifies the superior geno-types without phenotyping, thereby increasing the gains from selection and reducing the interval between generations. Prediction and selection are performed at very early plant stages in this method, thus accelerating the breeding process. Additionally, the prediction tends to be more accurate because it considers the actual genetic relatedness of the evaluated individuals, rather than the mathematically calculated expected average relatedness. GS is suitable for quantitative traits regulated by a large number of genes each with very small effect.[101] GS method is used in tropical maize breeding for root traits under conditions of nitrogen and phosphorus stress.[102]

8.3.7 TRANSCRIPTOMICS OF LP TOLERANCE IN MAIZE

Transcriptome analysis has shown that changes occur in cell growth, hormone biosynthesis, cellular transports, amino acid metabolism, signaling, TFs, and carbohydrate metabolism in response to LP stress. TFs frequently coordinate the expression of several genes in response to environmental signals.[103] Transcriptomics analysis in maize root for Pi starvation response was performed and it reported that a total of 1179 Pi responsive genes, among them 820 and 363 genes were found either up- or down-regulated, respectively.

ZmPTF1, a basic helix-loop-helix (bHLH) domain TF involved in tolerance to Pi starvation was cloned from maize.[104] Overexpression of PTF1 resulted in enhanced PUE and improved biomass production in both transgenic rice[105] and maize.[104] Overexpression of ZmPTF1 resulted in improved root development, increased tassel branch production, and larger kernel size in transgenic maize. Overexpression of ZmPHR1 led to the up-regulation of multiple genes that regulate metabolism during LP. Mostly, in *Arabidopsis* overexpressing ZmPHR1 showed better growth under LP conditions, which indicates that PUE could be improved through the use of the TF ZmPHR1 in maize.[103]

Under low Pi conditions, changes in RSA, PSI gene expression, anthocyanin accumulation, sugar/starch accumulation, Pi uptake, and subsequent xylem loading are transcriptionally and post-transcriptionally regulated by a combination of transcription factors, SIZ1, components of chromatin remodeling complexes and long non-coding RNAs (Fig. 8.1).

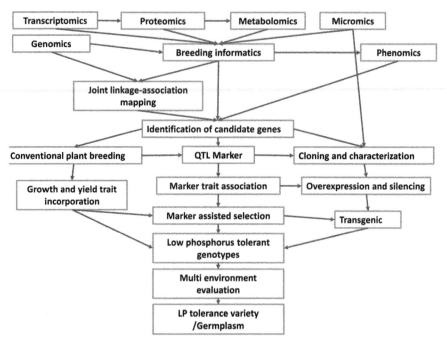

FIGURE 8.1   Low-P stress tolerance signaling and cross talk in plants.

8.3.8   MICRORNA OF LP TOLERANCE IN MAIZE

MicoRNAs (miRNAs) are endogenous, single stranded and non-coding sRNAs with a length of 19–24 nucleotides. It is biosynthesis from a single-stranded RNA precursor through a hairpin secondary structure[106,107] and which plays an important role in different biological process. miRNAs regulate gene expression by binding to transcripts of their target genes, which result into mRNA degradation, translational repression, or RNA-directed DNA methylation, and micromics will help for the better understanding of tolerance. Target genes of plant miRNAs are often transcription factors.[107–109]

In maize, a little is known about the regulatory mechanisms of complex processes as how phosphate starvation responsive (PSR) genes are regulated and how plants coordinate Pi uptake and allocate to maintain Pi homeostasis during Pi starvation. There are ten loci in maize for miRNA399 (zma-MIR399) family (www.mirbas.org), which was predicted by computational methods for PSR.[110]

From seedling stage of maize root, miRNAs were cloned and characterized under LP stress, and 12 maize miRNAs were discovered, including 10 newly cloned and two conserved (zma-miRNA156 and zma-miRNA399).[111]

8.3.9   OTHER MIRNAS RESPONSE TO PI DEFICIENCY

Using different high-throughput techniques, LP stress responsive miRNAs have been identified in *Arabidopsis*,[112–114] white lupin,[115] soybean,[116,117] common bean,[118] tomato,[119] wheat,[120] barley,[121] and maize.[122] miRNA families, miR156, miR159, miR166, miR169, miR319, miR395, miR398, miR399, miR447, and miR827 were found responsive to LP stress, indicating that miRNAs were coordinately involved in the conserved LP stress signaling pathways in plants.[123] Further research is needed to understand the possible morpho-physiological roles of P-responsive miRNAs during plant responses to LP stress.[124]

8.3.10   PROTEOMICS OF LP TOLERANCE IN MAIZE

Proteome studies have contributed to better understanding of the role of abscisic acid (ABA) regulation in the synthesis of chloroplast proteins and this has been one of the current developments.[125] Proteome analysis technique is used to study a variety of plant responses to environmental stress. Differential display of the soluble proteome provides a valuable overview of expression changes of the most abundant proteins in the cell, including post-translational modifications and degradation products.[126] Proteomics opens a wide scope of species-specific metabolic strategies.[126]

The proteome analysis of maize Qi-319 cultivar to 17 days of LP stress revealed differential accumulation of a number of carbohydrate metabolism enzymes in conditions of low Pi availability relative to nutrient-replete plants.[127] These included increased abundance of enzymes of the oxidative pentose phosphate pathway, which is involved in producing reducing equivalents in times of stress and increased accumulation of *tricarboxylic acid* (TCA) cycle enzymes. The abundance of pyruvate phosphate dikinase also increased in a typical Pi-stress response, which allows glycolysis to proceed by using Pi rather than the limiting Pi in the bypass reactions. A number of secondary metabolism enzymes involved in cell-wall lignification were also induced in Pi-stressed maize. In a subsequent proteomic analysis, the two cultivars have compared with a common genetic background but different root morphology.[128] In the Pi-efficient line 99038, carbohydrate metabolism during Pi stress appeared to be coordinated for higher synthesis and decreased degradation of citrate, as seen by the increase in malate dehydrogenase, pyruvate dehydrogenase, and citrate synthase, and the downregulation of aconitase and isocitrate dehydrogenase levels. This was confirmed by the higher activity of citrate synthase and malate dehydrogenase, and the amount of citrate that accumulated in the roots of 99038 compared to Qi-319.[129] The higher PUE of 99038 may also be complemented by the increase in pyruvate phosphate dikinase and UDP–glucose pyrophosphorylase. The increased abundance of 6-phosphogluconate dehydrogenase, the central regulatory enzyme of the oxidative pentose phosphate pathway, could also allow higher rates of citrate synthesis to be maintained. The PUE cultivar increased protein phosphatase 2A abundance, which is an important component of control of phosphorylated proteins, including auxin regulators, and is consistent with the increase in lateral root hair length and number, and the higher rate of meristematic cell division in this line.[128]

High-throughput array technologies have contributed to demonstrating the transcriptional regulation of abiotic stresses, including LP. Using the microarray platform, PSR in maize roots showed that the steady level of over 1100 transcripts is controlled by P-availability and among those, 33% do not have a significant match with an ortholog of *Arabidopsis*.[31]

Proteomics approach was used to identify proteins expression under LP conditions and among different maize inbred lines.[128,130] At seedling stage of maize, a total of 283 P responsive genes identified, of which 199 and 84 genes were found to be either up- or down-regulated, respectively.[131] The expression patterns of maize P transporters, acid phosphatase, 2-deoxymugineic acid synthase1, phytase, peroxidase (POD), and MYB TFs were validated in LP-efficient inbred line-178 roots in LP stress, of which two genes encoding acid phosphatase and phytase were significantly induced by P deficiency.[131] Furthermore, after long-term P starvation, changes in protein accumulation were identified,[127] indicating modifications in carbohydrate metabolism, synthesis, and degradation of nucleotide, amino acid, and in secondary metabolism.

8.3.11 HIGH-THROUGHPUT AND PRECISE PHENOTYPING (PHENOMICS)

High quality and large-scale phenotypic data are essential in modern plant breeding. The quality of marker–phenotype association is dependent on the quality of the phenotypic information.[132] Precise phenotypic characterization of germplasm will help breeders to make better selection decisions.

High-throughput methods can help breeders to quantify phenotypic traits with higher precision and in larger population sizes. Increasing population sizes allow improving the power of QTL. A high-throughput phenotyping tool should be popular if it satisfies the following criteria: Evaluates a large number of plants in a short time and traits can be measured with high accuracy. Improving the accuracy of phenotypic measurements and evaluating a larger number of plants will decrease the standard error, allowing us to be more confident on the quantity of the measured trait.

Root traits have not typically been used as selection criteria in traditional plant breeding programs, despite their central role in plant performance. In the last several decades, crop improvement has been focused on yield, and the benefits associated with this approach have been gradually decreasing.[133] More recently, the focus in crop breeding has shifted toward secondary traits that shape yield, including complex traits that could contribute to yield through greater physiological efficiency.[134]

The process of measuring traits should be reliable, consistent, and objective if genotypic differences were to be detected and selection optimized. Digital phenotyping has emerged as one method to accomplish these goals. Such methods are helping to drive the transition from categorical to quantitative phenotyping that links ontology terms and trait descriptors. Several root analyzer software have emerged as a tool to quantify the RAS in a fully automated or semi-automated approach.

Precision phenotyping mainly depends on how environmental errors and factors can be managed and controlled. This is particularly important for the traits such as abiotic stress tolerance to be phenotyped under controlled or stress environments. On the other hand, genotype–environment interaction (GEI) has been measured using phenotypic data collected under specific environments for various genotypes. However, environmental data *per se* have rarely been incorporated into the estimation of GEI or used to reveal the contribution of environmental factors to a specific phenotype. Xu et al.[135] proposed the concept of e-typing or environmental assay as a third dimension of the profile for modern plant breeding. E-typing provides a comprehensive set of environmental information that is required to describe a plant and complements genotypic and phenotypic information. MAS has changed plant breeding procedures by transitioning from phenotype-based selection to phenotype–genotype based selection, which represents a concept change from a "line" to a "plane." E-typing, together with genotyping and phenotyping, transforms the concept of plant breeding from a "plane," determined by genotype and phenotype, to a "space," measured by three dimensions, genotype, phenotype, and environment.[135]

8.3.12 INTEGRATION OF DIFFERENT MOLECULAR BREEDING APPROACHES FOR LP TOLERANCE IN MAIZE

A multidisciplinary approach is essential for breed LP tolerance in maize. A comprehensive understanding of the physiological basis of LP tolerance helps to identify the specific morphological and anatomical variations, and mechanisms involved. Based on this, germplasm collections can be thoroughly searched to identify appropriate LP tolerant donors that can be used in conventional and molecular breeding programs. The integration of various genomics technologies such as genomics, transcriptomics, proteomics, phenomics, metabolomics, and micromics, together with conventional breeding approaches, facilitates the identification and characterization of QTLs that regulate specific traits under LP (Fig. 8.2). A transgenic approach permits the exploitation of LP-tolerant genes from different sources. After gene incorporation and evaluation of transgenic plants, these genes can be transferred to other genetic backgrounds through breeding procedures. Functional genomics holds great promise for the future, but its effective integration is possible only when the appropriate technologies are made readily available to maize breeder.

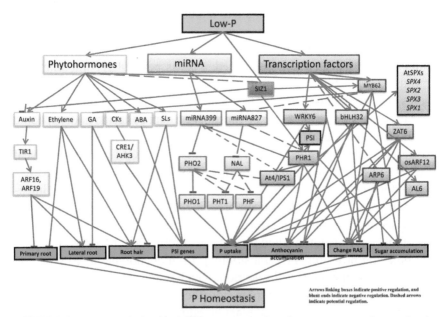

FIGURE 8.2 Integrated circuit of different approaches that are connected to each other associated with LP stress tolerance maize.

8.4   CONCLUSION

Phosphorous (P) is an important macronutrient for growth and development of plants. P availability in soil is often low and a large amount of P fertilizers is applied to attain high crop yields. P deficiency in maize is widespread and significantly reduces yield. The improvement of P efficiency in maize through breeding is important for sustainable crop production and world food security. The knowledge about how the plants sense P deficiency is growing rapidly, but a lot is still to be discovered in the near future. Phosphorus deficiency in plants triggers many transcriptional, biochemical, and morpho-physiological changes that ultimately help the plants absorb P from the soil or improve the PUE. Substantial genetic variation in P efficiency exists among the maize genotypes and a number of QTLs encoding traits for P efficiency have been identified in maize and rice. Dissection of QTLs and/ or the underlying genes now offers an important strategy to improve P efficiency. This could be achieved by conventional or marker-assisted breeding, genetic engineering with transformation, or a combination of these strategies. PSR mechanisms in plants are the complex coordination of root morph-physiological and biochemical responses that depend on P availability and plant sensing of heterogeneous P supply in soil. Future efforts should be focused on establishing a novel signaling mechanism and on understanding how different signaling pathways are integrated to give diverse and dynamic responses. With the progress of advanced technologies, a more comprehensive understanding of the Pi signaling network can be built. Ultimately, this knowledge will benefit future maize breeding aimed at improving the ability of plants to acquire and utilize P. Finally, it is expected that the integration of systems biology with high-throughput, high-dimensional, and precision phenotyping will contribute to the development of maize varieties with the ability to LP tolerance.

KEYWORDS

- **abiotic stress**
- **low phosphorus tolerant**
- **phosphate starvation responsive**
- **OMICS**
- **stress signaling**

REFERENCES

1. Elser, J. J. Phosphorus: A Limiting Nutrient for Humanity? *Curr. Opin. Biotechnol.* **2012,** *23,* 1–6.

2. Gamuyao, R.; Chin, J. H.; Pariasca-Tanaka, J.; Pesaresi, P.; Catausan, S.; Dalid, C.; Slamet-Loedin, I.; Tecson-Mendoza, E. M.; Wissuwa, M.; Heuer, S. The Protein Kinase PSTOL1 from Traditional Rice Confers Tolerance of Phosphorus Deficiency. *Nature.* **2012,** *488,* 535–539.

3. Rose, T. J.; Pariasca-Tanaka, J.; Rose, M. T.; Fukuta, Y.; Wissuwa, M. Genotypic Variation in Grain Phosphorus Concentration, and Opportunities to Improve P-use Efficiency in Rice. *Field Crop Res.* **2010,** *119,* 154–160.

4. Calderon-Vazquez, C.; Sawers, R. J.; Herrera-Estrella, L. Phosphate Deprivation in Maize: Genetics and Genomics. *Plant Physiol.* **2011,** *156,* 1067–1077.

5. Bucher, M. Functional Biology of Plant Phosphate Uptake at Root and Mycorrhiza Interfaces. *New Phytol.* **2007,** *173,* 11–26.

6. Yang, X. J.; Finnegan, P. M. Regulation of Phosphate Starvation Responses in Higher Plants. *Ann. Bot.* **2010,** *105,* 513–526.

7. Cordell, D. Towards Global Phosphorus Security: A Systems Framework for Phosphorus Recovery and Reuse Options. *Chemosphere.* **2011,** *84,* 747–758.

8. Lopez-Arredondo, D. L.; Herrera-Estrella, L. Engineering Phosphorus Metabolism in Plants to Produce a Dual Fertilization and Weed Control System. *Nat. Biotech.* **2012,** *30,* 889–893.

9. Graham, R. D. Breeding for Nutritional Characteristic in Cereals. In *Advances in Plant Nutrition;* Tinker, P. B., Lauchli, A., Eds.; Praeger: New York, 1984; pp 57–102.

10. Gunes, A.; Inal, A.; Alpaslan, M.; Cakmak, I. Genotypic Variation in Phosphorus Efficiency between Wheat Cultivars Grown Under Greenhouse and Field Conditions. *Soil Sci. Plant Nutr.* **2006,** *52,* 470–478.

11. Singh, G.; Nielsen, N. Root Traits as Tools for Creating Phosphorus Efficient Crop Varieties. *Plant Soil.* **2004,** *260,* 47–57.

12. Gahoonia, T.; Nielsen, N.; Lyshede, O. Phosphorus (P) Acquisition of Cereal Cultivars in the Field at Three Levels of P Fertilization. *Plant Soil.* **1999,** *211,* 269–281.

13. Lynch, J. P. Root Phenes for Enhanced Soil Exploration and Phosphorus Acquisition: Tools for Future Crops. *Plant Physiol.* **2011,** *156,* 1041–1049.

14. Kim, S. I.; Andaya, C. B.; Newman, J. W.; Goyal, S. S.; Tai, T. H. Isolation and Characterization of a Low Phytic Acid Rice Mutant Reveals a Mutation in the Rice Orthologue of Maize MIK. *Theor. Appl. Genet.* **2008,** *117,* 1291–1301.

15. Lambers, H.; Ahmedi, I.; Berkowitz, O.; Dunne, C.; Finnegan, P. M.; Hardy, G. E. S. J.; Jost, R.; Laliberte, E.; Pearse, S. J.; Teste, F. P. Phosphorus Nutrition of Phosphorus-Sensitive Australian Native Plants: Threats to Plant Communities in a Global Biodiversity Hotspot. *Conser. Physiol.* **2013,** *1,* 1–21.

16. Lambers, H.; Finnegan, P. M.; Laliberte, E.; Pearse, S. J.; Ryan, M. H.; Shane, M. W.; Veneklaas, E. J. Update on Phosphorus Nutrition in Proteaceae. Phosphorus Nutrition of Proteaceae in Severely Phosphorus-Impoverished Soils: Are There Lessons to be Learned for Future Crops? *Plant Physiol.* **2011,** *156,* 1058–1066.

17. Lambers, H.; Shane, M. W.; Cramer, M. D.; Pearse, S. J.; Veneklaas, E. J. Root Structure and Functioning for Efficient Acquisition of Phosphorus: Matching Morphological and Physiological Traits. *Ann. Bot.* **2006,** *98,* 693–713.

18. Niu, Y. F.; Chai, R. S.; Jin, G. L.; Wang, H.; Tang, C. X.; Zhang, Y. S. Responses of Root Architecture Development to Low Phosphorus Availability: A Review. *Ann. Bot.* **2013,** *112,* 391–408.

19. Zhu, J.; Kaeppler, S. M.; Lynch, J. P. Topsoil Foraging and Phosphorus Acquisition Efficiency in Maize (*Zea mays*). *Funct. Plant Biol.* **2005,** *32,* 749.

20. Lynch, J.; Brown, K. Topsoil Foraging – An Architectural Adaptation of Plants to Low Phosphorus Availability. *Plant Soil.* **2001,** *237,* 225–237.

21. Zhang, Z.; Liao, H.; Lucas, W. J. Molecular Mechanisms Underlying Phosphate Sensing, Signaling and Adaptation in Plants. *J. Integr. Plant Biol.* **2014,** *56,* 192–220.

22. Lucas, W. J.; Groover, A.; Lichtenberger, R.; Furuta, K.; Yadav, S. R.; Helariutta, Y.; He, X. Q.; Fukuda, H.; Kang, J.; Brady, S. M.; Patrick, J. W.; Sperry, J.; Yoshida, A.; López-Millán, A. F.; Grusak, M. A.; Kachroo, P. The Plant Vascular System: Evolution, Development and Functions. *J. Integr. Plant Biol.* **2013,** *55,* 294–388.

23. Liu, T. Y.; Chang, C. Y.; Chiou, T. J. The Long-Distance Signaling of Mineral Macronutrients. *Curr. Opin. Plant Biol.* **2009,** *12,* 312.

24. Lough, T. J.; Lucas, W. J. Integrative Plant Biology: Role of Phloem Long-Distance Macromolecular Trafficking. *Annu. Rev. Plant Biol.* **2006,** *57,* 203–232.

25. Nagarajan, V. K.; Smith, A. P. Ethylene's Role in Phosphate Starvation Signaling: More Than Just a Root Growth Regulator. *Plant Cell Physiol.* **2012,** *53,* 277–286.

26. Chiou, T. J.; Lin, S. I. Signaling Network in Sensing Phosphate Availability in Plants. *Annu. Rev. Plant Biol.* **2011,** *62,* 185–206.

27. Thibaud, M. C.; Arrighi, J. F.; Bayle, V.; Chiarenza, S.; Creff, A.; Bustos, R.; Paz-Ares, J.; Poirier, Y.; Nussaume, L. Dissection of Local and Systemic Transcriptional Responses to Phosphate Starvation in *Arabidopsis. Plant J.* **2010,** *64,* 775–789.

28. Lopez-Bucio, J.; Hernandez-Abreu, E.; Sanchez-Calderon, L.; Nieto-Jacobo, M. F.; Simpson, J.; Herrera-Estrella, L. Phosphate Availability Alters Architecture and Causes Changes in Hormone Sensitivity in the *Arabidopsis* Root System. *Plant Physiol.* **2002,** *129,* 244–256.

29. Jungk, A. Root Hairs and the Acquisition of Plant Nutrients from Soil. *J. Plant Nutr. Soil Sci.* **2001,** *164,* 121–129.

30. Li, Z.; Xu, C.; Li, K.; Yan, S.; Qu, X.; Zhang, J. Phosphate Starvation of Maize Inhibits Lateral Root Formation and Alters Gene Expression in the Lateral Root Primordium Zone. *BMC Plant Biol.* **2012,** *12,* 89.

31. Calderon-Vazquez, C.; Ibarra-Laclette, E.; Caballero-Perez, J.; Herrera-Estrella, L. Transcript Profiling of *Zea mays* roots Reveals Gene Responses to Phosphate Deficiency at the Plant- and Species-Specific Levels. *J. Exp. Bot.* **2008,** *59,* 2479–2497.

32. Gaume, A.; Mächler, F.; De León, C.; Narro, L.; Frossard, E. Low-P Tolerance by Maize (*Zea mays* L.) Genotypes: Significance of Root Growth, and Organic Acids and Acid Phosphatase Root Exudation. *Plant Soil.* **2001,** *228,* 253–264.

33. Cakmak, I.; Hengeler, C.; Marschner, H. Partitioning of Shoot and Root Dry Matter and Carbohydrates in Bean Plants Suffering from Phosphorus, Potassium and Magnesium Deficiency. *J. Exp. Bot.* **1994,** *45,* 1245–1250.

34. Sanchez-Calderon, L.; Lopez-Bucio, J.; Chacon-Lopez, A.; Gutierrez-Ortega, A.; Hernandez-Abreu, E.; Herrera-Estrella, L. Characterization of Low Phosphorus Insensitive Mutants Reveals a Crosstalk Between Low Phosphorus-Induced Determinate Root Development and the Activation of Genes Involved in the Adaptation of *Arabidopsis* to Phosphorus Deficiency. *Plant Physiol.* **2006,** *140,* 879.

35. Haling, R. E.; Brown, L. K.; Bengough, A. G.; Young, I. M.; Hallett, P. D.; White, P. J.; George, T. S. Root Hairs Improve Root Penetration, Root–Soil Contact, and Phosphorus Acquisition in Soils of Different Strength. *J. Exp. Bot.* **2013,** *64,* 3711–3721.

36. Zhu, J.; Brown, K. M.; Lynch, J. P. Root Cortical Aerenchyma Improves the Drought Tolerance of Maize (*Zea mays* L.). *Plant Cell Environ.* **2010,** *33,* 740–749.

37. Lynch, J. P.; Brown, K. M. New Roots for Agriculture: Exploiting the Root Phenome. *Philos. Trans. R Soc. Lond. B Biol. Sci.* **2012,** *367,* 1598–1604.

38. Zhu, J.; Zhang, C.; Lynch, J. P. The Utility of Phenotypic Plasticity of Root Hair Length for Phosphorus Acquisition. *Funct. Plant Biol.* **2010,** *37,* 313–322.

39. Zhu, J.; Shawn, M.; Kaeppler, Lynch, J. Mapping of QTL Controlling Root Hair Length in Maize (*Zea mays* L.) Under Phosphorus Deficiency. *Plant Soil.* **2005,** *270,* 299–310.

40. Bustos, R.; Castrillo, G.; Linhares, F.; Puga, M. I.; Rubio, V.; Perez-Perez, J.; Solano, R.; Leyva, A.; Paz-Ares, J. A Central Regulatory System Largely Controls Transcriptional Activation and Repression Responses to Phosphate Starvation in *Arabidopsis.* *PLoS Genet.* **2010,** *6,* e1001102.

41. Jain, A.; Poling, M. D.; Smith, A. P.; Nagarajan, V. K.; Lahner, B.; Meagher, R. B.; Raghothama, K. G. Variations in the Composition of Gelling Agents Affect Morphophysiological and Molecular Responses to Deficiencies of Phosphate and Other Nutrients. *Plant Physiol.* **2009,** *150,* 1033–1049.

42. Sanchez-Calderon, L.; Lopez-Bucio, J.; Chacon-Lopez, A.; Cruz-Ramirez, A.; Nieto-Jacobo, F.; Dubrovsky, J. G.; Herrera-Estrella, L. Phosphate Starvation Induces a Determinate Developmental Program in the Roots of *Arabidopsis thaliana.* *Plant Cell Physiol.* **2005,** *46,* 174–184.

43. Grant, C. B.; Bittman, S.; Montreal, M.; Plenchette, C.; Morel, C. Soil and Fertilizer Phosphorus: Effects on Plant P Supply and Mycorrhizal Development. *Can. J. Plant Sci.* **2005,** *85,* 3–14.

44. Clark, R. B.; Brown, J. C. Differential Phosphorus Uptake by Phosphorus-Stressed Corn Inbreds. *Crop Sci.* **1974,** *14,* 505–508.

45. Da Silva, A. E.; Gabelman, W. H. Screening Maize Inbred Lines for Tolerance to Low-P Stress Condition. *Plant Soil.* **1992,** *146,* 181–187.

46. Mundim, G. B.; Viana, J. M. S.; Maia, C.; Tuberosa, R. Early Evaluation of Popcorn Inbred Lines for Phosphorus Use Efficiency. *Plant Breeding.* **2013,** *132,* 613–619.

47. Bayuelo-Jiménez, J. S.; Ochoa-Cadavid, I. Phosphorus Acquisition and Internal Utilization Efficiency Among Maize Landraces from the Central Mexican Highlands. *Field Crop Res.* **2014,** *156,* 123–134.

48. Bayuelo-Jiménez, J. S.; Gallardo-Valdéz, M.; Pérez-Decelis, V. A.; Magdaleno-Armas, L.; Ochoa, I.; Lynch, J. P. Genotypic Variation for Root Traits of Maize (*Zea mays* L.) from the Purhepecha Plateau Under Contrasting Phosphorus Availability. *Field Crop Res.* **2011,** *121,* 350–362.

49. Anghinoni, I.; Barber, S. Phosphorus Influx and Growth Characteristics of Corn Roots as Influenced by Phosphorus Supply. *Agron J.* **1980,** *172,* 655– 668.

50. Jiang, H.; Yang, J.; Zhang, J. F.; Hou, Y. N. In *Screening of Tolerant Maize Genotypes in the Low Phosphorus Field Soil,* Proceedings of the 19th World Congress of Soil Science: Soil solutions for a Changing World, Symposium 3.1.2 Farm System and Environment Impacts 2010, Brisbane, Australia, August 1–6, 2010; Jiang, H., Yang, J., Zhang, J. F., Hou, Y. N., Eds.; International Union of Soil Sciences: Brisbane, 2010; pp. 214–217.

51. Mollier, A.; Pellerin, S. Maize Root System Growth and Development as Influenced by Phosphorus Deficiency. *J. Exp. Bot.* **1999**, *50*, 487–497.

52. Yao, Q. -l.; Yang, K. -c.; Pan, G. -t.; Rong, T. -z. The Effects of Low Phosphorus Stress on Morphological and Physiological Characteristics of Maize (*Zea mays* L.) Landraces. *Agr. Sci. China.* **2007**, *6*, 559–566.

53. Alves, V. M. C.; Parentoni, S. N.; Vasconcellos, C. A.; Bahia Filho, A. F. C.; Pitta, G. V. E. a.; Schaffert, R. E. Mechanisms of Phosphorus Efficiency in Maize. In *Plant Nutrition–Food Security and Sustainability of Agro-Ecosystems;* Walter Horst, Schenk, M. K., Bürkert, A., Claassen, N., Flessa, H., Frommer, W. B., Heiner Goldbach, E., Olfs, H. W., Römheld, V., Sattelmacher, B., Schmidhalter, U., Schubert, S., von Wirén, N., Wittenmayer, L., Eds.; Kluwer Academic Publishers: The Netherlands, 2001; pp 566–567.

54. Chen, F. -j.; Liu, X. -s.; Mi, G. -h. Varietal Differences in Plant Growth, Phosphorus Uptake and Yield Formation in Two Maize Inbred Lines Grown Under Field Conditions. *J. Integr. Agric.* **2012**, *11*, 1738–1743.

55. Plenets, D.; Mollier, A.; Pellerin, S. I. Growth Analysis of Maize Field Crops Under Phosphorus Deficiency. II. Radiation Use Efficiency, Biomass Accumulation and Yield Components. *Plant Soil.* **2000**, *224*, 259–272.

56. Usuda, H.; Shimogawara, K. Phosphate Deficiency in Maize III. Changes in Enzyme Activities During the Course of Phosphate Deprivation. *Plant Physiol.* **1992**, *99*, 1680–1685.

57. Kearsey, M. J.; Farquhar, A. G. L. QTL Analysis in Plants; Where Are We Now? *Heredity.* **1998**, *80*, 137–142.

58. Holland, J. B. Genetic Architecture of Complex Traits in Plants. *Curr. Opin. Plant Biol.* **2007**, *10*, 156–161.

59. Salvi, S.; Tuberosa, R. To Clone or Not to Clone Plant QTLs: Present and Future Challenges. *Trends Plant Sci.* **2005**, *10*, 297–304.

60. Collard, B. C.; Mackill, D. J. Marker-Assisted Selection: An Approach for Precision Plant Breeding in the Twenty-First Century. *Philos. Trans. R Soc. Lond. B Biol. Sci.* **2008**, *363*, 557–572.

61. Yang, Q.; Zhang, D.; Xu, M. A Sequential Quantitative Trait Locus Fine-Mapping Strategy Using Recombinant-Derived Progeny. *J. Integr. Plant Biol.* **2012**, *54*, 228–237.

62. Zhu, J.; Mickelson, S. M.; Kaeppler, S. M.; Lynch, J. P. Detection of Quantitative Trait Loci for Seminal Root Traits in Maize (*Zea mays* L.) Seedlings Grown Under Differential Phosphorus Levels. *Theor. Appl. Genet.* **2006**, *113*, 1–10.

63. Zhu, J.; Kaeppler, S. M.; Lynch, J. P. Mapping of QTL Controlling Root Hair Length in Maize (*Zea mays* L.) Under Phosphorus Deficiency. *Plant Soil.* **2005**, *270*, 299–310.

64. Cai, H.; Chu, Q.; Gu, R.; Yuan, L.; Liu, J.; Zhang, X.; Chen, F.; Mi, G.; Zhang, F. Identification of QTLs for Plant Height, Ear Height and Grain Yield in Maize (*Zea mays* L.) in Response to Nitrogen and Phosphorus Supply. *Plant Breeding.* **2012**, *131*, 502–510.

65. Cai, H.; Chen, F.; Mi, G.; Zhang, F.; Maurer, H. P.; Liu, W.; Reif, J. C.; Yuan, L. Mapping QTLs for Root System Architecture of Maize (*Zea mays* L.) in the Field at Different Developmental Stages. *Theor. Appl. Genet.* **2012**, *125*, 1313–1324.

66. Li, M.; Guo, X.; Zhang, M.; Wang, X.; Zhang, G.; Tian, Y.; Wang, Z. Mapping QTLs for Grain Yield and Yield Components Under High and Low Phosphorus Treatments in Maize (*Zea mays* L.). *Plant Sci.* **2010**, *178*, 454–462.

67. Chen, J.; Xu, L.; Cai, Y.; Xu, J. Identification of QTLs for Phosphorus Utilization Efficiency in Maize (*Zea mays* L.) Across P Levels. *Euphytica.* **2009**, *167*, 245–252.

68. Chen, J.; Xu, L.; Cai, Y.; Xu, J. QTL Mapping of Phosphorus Efficiency and Relative Biologic Characteristics in Maize (*Zea mays* L.) at Two Sites. *Plant Soil.* **2008**, *313,* 251–266.

69. Parentoni, S. N.; de Souza, C. L. Jr.; Alves, V. M. d. C.; Gama, E. E. G.; Coelho, A. M.; Oliveira, A. C. d.; Guimarães, P. E. O.; Guimarães, C. T.; Vasconcelos, M. J. V.; Pacheco, C. A. P.; Meirelles, W. F.; Magalhães, J. V. d.; Guimarães, L. J. M.; Silva, A. R. d.; Mendes, F. F.; Schaffert, R. E. Inheritance and Breeding Strategies for Phosphorus Efficiency in Tropical Maize (*Zea Mays* L.). *Maydica.* **2010**, *55,* 1–15.

70. Lu, Y.; Zhang, S.; Shah, T.; Xie, C.; Hao, Z.; Li, X.; Farkhari, M.; Ribaut, J. M.; Cao, M.; Rong, T.; Xu, Y. Joint Linkage-Linkage Disequilibrium Mapping is a Powerful Approach to Detecting Quantitative Trait Loci Underlying Drought Tolerance in Maize. *Proc. Natl. Acad. Sci. USA.* **2010**, *107,* 19585–19590.

71. Buckler, E. S.; Holland, J. B.; Bradbury, P. J.; Acharya, C. B.; Brown, P. J.; Browne, C.; Ersoz, E.; Flint-Garcia, S.; Garcia, A.; Glaubitz, J. C.; Goodman, M. M.; Harjes, C.; Guill, K.; Kroon, D. E.; Larsson, S.; Lepak, N. K.; Li, H.; Mitchell, S. E.; Pressoir, G.; Peiffer, J. A.; Rosas, M. O.; Rocheford, T. R.; Romay, M. C.; Romero, S.; Salvo, S.; Sanchez Villeda, H.; H. S. da Silva, Sun, Q.; Tian, F.; Upadyayula, N.; Ware, D.; Yates, H.; Yu, J.; Zhang, Z.; Kresovich, S.; McMullen, M. D. The Genetic Architecture of Maize Flowering Time. *Science.* **2009**, *325,* 714–718.

72. Li, X.; Zhu, C.; Yeh, C. T.; Wu, W.; Takacs, E. M.; Petsch, K. A.; Tian, F.; Bai, G.; Buckler, E. S.; Muehlbauer, G. J.; Timmermans, M. C.; Scanlon, M. J.; Schnable, P. S.; Yu, J. Genic and Nongenic Contributions to Natural Variation of Quantitative Traits in Maize. *Genome Res.* **2012**, *22,* 2436–2444.

73. Zhu, J.; Kaeppler, S. M.; Lynch, J. P. Mapping of QTLs for Lateral Root Branching and Length in Maize (*Zea mays* L.) Under Differential Phosphorus Supply. *Theor. Appl. Genet.* **2005**, *111,* 688–695.

74. Zhu, J.; Mickelson, S. M.; Kaeppler, S. M.; Lynch, J. P. Detection of Quantitative Trait Loci for Seminal Root Traits in Maize (*Zea mays* L.) Seedlings Grown Under Differential Phosphorus Levels. *Theor. Appl. Genet.* **2006**, *113,* 1–10.

75. Kaeppler, S. M.; Jennifer Parke, L.; Lynn Senior; Charles Stuber; William Tracy, F. Variation Among Maize Inbred Lines and Detection of Quantitative Trait Loci for Growth at Low Phosphorus and Responsiveness to Arbuscular Mycorrhizal Fungi. *Crop Sci.* **2000**, *40,* 358–364.

76. Chen, J.; Xu, L. Comparative Mapping of QTLs for H^+ Secretion of Root in Maize (*Zea mays* L.) and Cross Phosphorus Levels on Two Growth Stages. *Front. Agric. China.* **2011**, *5,* 284–290.

77. Cai, H.; Chu, Q.; Yuan, L.; Liu, J.; Chen, X.; Chen, F.; Mi, G.; Zhang, F. Identification of Quantitative Trait Loci for Leaf Area and Chlorophyll Content in Maize (*Zea mays*) Under Low Nitrogen and Low Phosphorus Supply. *Mol. Breed.* **2011**, *30,* 251–266.

78. Chin, J. H.; Gamuyao, R.; Dalid, C.; Bustamam, M.; Prasetiyono, J.; Moeljopawiro, S.; Wissuwa, M.; Heuer, S. Developing Rice with High Yield Under Phosphorus Deficiency: Pup1 Sequence to Application. *Plant Physiol.* **2011**, *156,* 1202–1216.

79. Magalhaes, J.; Guimaraes, C. T.; Azevedo, G. C.; Hufnagel, B.; Assis, L.; Negri, B.; Matos, F.; Sousa, S. M.; Guimaraes, L.; Guimaraes, F.; Parentoni, S. N.; Schaffert, R. E.; Gazaffi, R.; Garcia, A. A. F.; Viana, J. H.; Weltzien-Rattunde, E.; Kochian, L.; Heuer, S.; Wissuwa, M. In *Cloning and Validation of PSTOL1 Homologs in Maize and Sorghum,* Proceedings of the Plant & Animal Genome XXI Conference, San Diego, CA, USA, Jan 12–16, 2013.

80. Li, Z.; Gao, Q.; Liu, Y.; He, C.; Zhang, X.; Zhang, J. Overexpression of Transcription Factor ZmPTF1 Improves Low Phosphate Tolerance of Maize by Regulating Carbon Metabolism and Root Growth. *Planta.* **2011,** *233,* 1129–1143.

81. Wang, J.; Sun, J.; Miao, J.; Guo, J.; Shi, Z.; He, M.; Chen, Y.; Zhao, X.; Li, B.; Han, F.; Tong, Y.; Li, Z. A Phosphate Starvation Response Regulator Ta-PHR1 is Involved in Phosphate Signalling and Increases Grain Yield in Wheat. *Ann. Bot.* **2013,** *111,* 1139–1153.

82. Zhou, J.; Jiao, F.; Wu, Z.; Li, Y.; Wang, X.; He, X.; Zhong, W.; Wu, P. OsPHR2 is Involved in Phosphate-Starvation Signaling and Excessive Phosphate Accumulation in Shoots of Plants. *Plant Physiol.* **2008,** *146,* 1673–1686.

83. Wang, C.; Ying, S.; Huang, H.; Li, K.; Wu, P.; Shou, H. Involvement of OsSPX1 in Phosphate Homeostasis in Rice. *Plant J.* **2009,** *57,* 895–904.

84. Liu, F.; Wang, Z.; Ren, H.; Shen, C.; Li, Y.; Ling, H. Q.; Wu, C.; Lian, X.; Wu, P. OsSPX1 Suppresses the Function of OsPHR2 in the Regulation of Expression of OsPT2 and Phosphate Homeostasis in Shoots of Rice. *Plant J.* **2010,** *62,* 508–517.

85. Jia, H.; Ren, H.; Gu, M.; Zhao, J.; Sun, S.; Zhang, X.; Chen, J.; Wu, P.; Xu, G. The Phosphate Transporter Gene OsPht1;8 is Involved in Phosphate Homeostasis in Rice. *Plant Physiol.* **2011,** *156,* 1164–1175.

86. Hu, B.; Zhu, C.; Li, F.; Tang, J.; Wang, Y.; Lin, A.; Liu, L.; Che, R.; Chu, C. LEAF TIP NECROSIS1 Plays a Pivotal Role in the Regulation of Multiple Phosphate Starvation Responses in Rice. *Plant Physiol.* **2011,** *156,* 1101–1115.

87. Sun, S.; Gu, M.; Cao, Y.; Huang, X.; Zhang, X.; Ai, P.; Zhao, J.; Fan, X.; Xu, G. A Constitutive Expressed Phosphate Transporter, OsPht1;1, Modulates Phosphate Uptake and Translocation in Phosphate-Replete Rice. *Plant Physiol.* **2012,** *159,* 1571–1581.

88. Yang, S. Y.; Grønlund, M.; Jakobsen, I.; Grotemeyer, M. S.; Rentsch, D.; Miyao, A.; Hirochika, H.; Kumar, C. S.; Sundaresan, V.; Salamin, N.; Catausan, S.; Mattes, N.; Heuer, S.; Paszkowski, U.; Yang, S. Y.; Grønlund, M.; Jakobsen, I.; Grotemeyer, M. S.; Rentsch, D.; Miyao, A.; Hirochika, H.; Kumar, C. S.; Sundaresan, V.; Salamin, N.; Catausan, S.; Mattes, N.; Heuer, S.; Paszkowski, U. Nonredundant Regulation of Rice Arbuscular Mycorrhizal Symbiosis by Two Members of the Phosphate Transporter 1 Gene Family. *Plant Cell.* **2012,** *24,* 204236–204251.

89. Yi, K.; Wu, Z.; Zhou, J.; Du, L.; Guo, L.; Wu, Y.; Wu, P. OsPTF1, a Novel Transcription Factor Involved in Tolerance to Phosphate Starvation in Rice. *Plant Physiol.* **2005,** *138,* 2087–2096.

90. Park, M.; Baek, S. H.; Reyes, B.; Yun, S. Overexpression of a High-Affinity Phosphate Transporter Gene from Tobacco (NtPT1) Enhances Phosphate Uptake and Accumulation in Transgenic Rice Plants. *Plant Soil.* **2007,** *292,* 259–269.

91. Dai, X.; Wang, Y.; Yang, A.; Zhang, W. H. OsMYB2P-1, an R2R3 MYB Transcription Factor, is Involved in the Regulation of Phosphate-Starvation Responses and Root Architecture in Rice. *Plant Physiol.* **2012,** *159,* 169–183.

92. Tian, J.; Wang, C.; Zhang, Q.; He, X.; Whelan, J.; Shou, H. Overexpression of OSPAP10a, a Root-Associated Acid Phosphatase, Increased Extracellular Organic Phosphorus Utilization in Rice. *J. Integr. Plant Biol.* **2012,** *54,* 631–639.

93. Yang, H.; Knapp, J.; Koirala, P.; Rajagopal, D.; Peer, W. A.; Silbart, L. K.; Murphy, A.; Gaxiola, R. A. Enhanced Phosphorus Nutrition in Monocots and Dicots Over-Expressing a Phosphorus-Responsive Type I H+-Pyrophosphatase. *Plant Biotech. J.* **2007,** *5,* 735–745.

94. Peiffer, J. A.; Romay, M. C.; Gore, M. A.; Flint-Garcia, S. A.; Zhang, Z.; Millard, M. J.; Gardner, C. A. C.; McMullen, M. D.; Holland, J. B.; Bradbury, P. J.; Buckler, E. S. The Genetic Architecture of Maize Height. *Genetics.* **2014,** *196,* 1337–1356.

95. Hamblin, M. T.; Buckler, E. S.; Jannink, J. L. Population Genetics of Genomics-Based Crop Improvement Methods. *Trends Genet.* **2011,** *27,* 98–106.

96. Tian, F.; Bradbury, P. J.; Brown, P. J.; Hung, H.; Sun, Q.; Flint-Garcia, S.; Rocheford, T. R.; McMullen, M. D.; Holland, J. B.; Buckler, E. S. Genome-Wide Association Study of Leaf Architecture in the Maize Nested Association Mapping Population. *Nat. Genet.* **2011,** *43,* 159–162.

97. Kump, K. L.; Bradbury, P. J.; Wisser, R. J.; Buckler, E. S.; Belcher, A. R.; Oropeza-Rosas, M. A.; Zwonitzer, J. C.; Kresovich, S.; McMullen, M. D.; Ware, D.; Balint-Kurti, P. J.; Holland, J. B. Genome-Wide Association Study of Quantitative Resistance to Southern Leaf Blight in the Maize Nested Association Mapping Population. *Nat. Genet.* **2011,** *43,* 163–168.

98. Prasanna, B. M.; Cairns, J.; Xu, Y. Genomic Tools and Strategies for Breeding Climate Resilient Cereals. In *Genomics and Breeding for Climate-Resilient Crops;* Kole, C., Ed.; Springer: Berlin, Heidelberg, 2013; pp 213–239.

99. Liu, S.; Yeh, C. T.; Tang, H. M.; Nettleton, D.; Schnable, P. S. Gene Mapping Via Bulked Segregant RNA-Seq (BSR-Seq). *PLoS ONE.* **2012,** *7,* e36406.

100. Elshire, R. J.; Glaubitz, J. C.; Sun, Q.; Poland, J. A.; Kawamoto, K.; Buckler, E. S.; Mitchell, S. E. A Robust, Simple Genotyping-By-Sequencing (GBS) Approach for High Diversity Species. *PLoS ONE.* **2011,** *6,* e19379.

101. Xu, Y.; Xie, C.; Wan, J.; He, Z.; Prasanna, B. Marker-Assisted Selection in Cereals: Platforms, Strategies and Examples. In *Cereal Genomics II;* Gupta, P. K., Varshney, R. K., Eds.; Springer: Netherlands, 2013; pp 375–411.

102. Fritsche-Neto, R.; Do Vale, J. C.; Lanes, É. C. M. d.; Resende, M. D. V. d.; Miranda, G. V. Genome-Wide Selection for Tropical Maize Root Traits Under Conditions of Nitrogen and Phosphorus Stress. *Acta Sci. Agron.* **2012,** *34,* 389–395.

103. Wang, X.; Bai, J.; Liu, H.; Sun, Y.; Shi, X.; Ren, Z. Overexpression of a Maize Transcription Factor ZmPHR1 Improves Shoot Inorganic Phosphate Content and Growth of *Arabidopsis* Under Low-Phosphate Conditions. *Plant Mol. Biol. Rep.* **2013,** *31,* 665–677.

104. Li, Z.; Gao, Q.; Liu, Y.; He, C.; Zhang, X.; Zhang, J. Overexpression of Transcription Factor ZmPTF1 Improves Low Phosphate Tolerance of Maize by Regulating Carbon Metabolism and Root Growth. *Planta.* **2011,** *233,* 1129–1143.

105. Yi, K.; Wu, Z.; Zhou, J.; Du, L.; Guo, L.; Wu, Y.; Wu, P. OsPTF1, a Novel Transcription Factor Involved in Tolerance to Phosphate Starvation in Rice. *Plant Physiol.* **2005,** *138,* 2087–2096.

106. Voinnet, O. Origin, Biogenesis, and Activity of Plant microRNAs. *Cell.* **2009,** *136,* 669–687.

107. Jones-Rhoades, M.; Bartel, D.; Bartel, B. MicroRNAs and Their Regulatory Roles in Plants. *Annu. Rev. Plant Biol.* **2006,** *57,* 19–53.

108. Jones-Rhoades, M. Conservation and Divergence in Plant microRNAs. *Plant Mol Biol.* **2012,** *80,* 3–16.

109. Jones-Rhoades, M.; Bartel, D. Computational Identification of Plant microRNAs and Their Targets, Including a Stress-Induced miRNA. *Mol. Cell.* **2004,** *14,* 787–799.

110. Zhang, L.; Chia, J. M.; Kumari, S.; Stein, J. C.; Liu, Z.; Narechania, A.; Maher, C. A.; Guill, K.; McMullen, M. D.; Ware, D. A Genome-Wide Characterization of microRNA Genes in Maize. *PLoS Genet.* **2009,**, 5.

111. Zhang, Z.; Lin, H.; Shen, Y.; Gao, J.; Xiang, K.; Liu, L.; Ding, H.; Yuan, G.; Lan, H.; Zhou, S.; Zhao, M.; Gao, S.; Rong, T.; Pan, G. Cloning and Characterization of miRNAs from Maize Seedling Roots Under Low Phosphorus Stress. *Mol. Biol. Rep.* **2012**, *39,* 8137–8146.

112. Lundmark, M.; Kørner, C. J.; Nielsen, T. H. Global Analysis of microRNA in Arabidopsis in Response to Phosphate Starvation as Studied by Locked Nucleic Acid-Based Microarrays. *Physiol. Plant.* **2010**, *140,* 57–68.

113. Pant, B.; Musialak-Lange, M.; Nuc, P.; May, P.; Buhtz, A.; Kehr, J.; Walther, D.; Scheible, W. Identification of Nutrient-Responsive Arabidopsis and Rapeseed microRNAs by Comprehensive Real-Time Polymerase Chain Reaction Profiling and Small RNA Sequencing. *Plant Physiol.* **2009**, *150,* 1541–1555.

114. Hsieh, L.; Lin, S.; Shih, A.; Chen, J.; Lin, W.; Tseng, C.; Li, W.; Chiou, T. Uncovering Small RNA-Mediated Responses to Phosphate Deficiency in *Arabidopsis* by Deep Sequencing. *Plant Physiol.* **2009**, *151,* 2120–2132.

115. Zhu, Y.; Zeng, H.; Dong, C.; Yin, X.; Shen, Q.; Yang, Z. microRNA Expression Profiles Associated with Phosphorus Deficiency in White Lupin (*Lupinus albus* L.). *Plant Sci.* **2010**, *178,* 23.

116. Xu, F.; Liu, Q.; Chen, L.; Kuang, J.; Walk, T.; Wang, J.; Liao, H. Genome-Wide Identification of Soybean microRNAs and Their Targets Reveals Their Organ-Specificity and Responses to Phosphate Starvation. *BMC Genomics.* **2013**, *14,* 66.

117. Zeng, H. Q.; Zhu, Y. Y.; Huang, S. Q.; Yang, Z. M. Analysis of Phosphorus-Deficient Responsive miRNAs and Cis-Elements from Soybean (*Glycine max* L.). *J. Plant Physiol.* **2010**, *167,* 1289–1297.

118. Valdés-López, O.; Yang, S.; Aparicio-Fabre, R.; Graham, P.; Reyes, J. MicroRNA Expression Profile in Common Bean (*Phaseolus vulgaris*) Under Nutrient Deficiency Stresses and Manganese Toxicity. *New Phytol.* **2010**, *187,* 805.

119. Gu, M.; Xu, K.; Chen, A.; Zhu, Y.; Tang, G.; Xu, G. Expression Analysis Suggests Potential Roles of microRNAs for Phosphate and Arbuscular Mycorrhizal Signaling in *Solanum lycopersicum. Physiol. Plant.* **2010**, *138,* 226–237.

120. Zhao, X.; Liu, X.; Guo, C.; Gu, J.; Xiao, K. Identification and Characterization of microRNAs from Wheat (*Triticum aestivum* L.) Under Phosphorus Deprivation. *J. Plant Biochem. Biotechnol.* **2012**, *22,* 113–123.

121. Hackenberg, M.; Huang, P. J.; Huang, C. Y.; Shi, B. J.; Gustafson, P.; Langridge, P. A Comprehensive Expression Profile of MicroRNAs and Other Classes of Non-Coding Small RNAs in Barley Under Phosphorous-Deficient and -Sufficient Conditions. *DNA Res.* **2013**, *20,* 109–125.

122. Pei, L.; Jin, Z.; Li, K.; Yin, H.; Wang, J.; Yang, A. Identification and Comparative Analysis of Low Phosphate Tolerance-Associated microRNAs in Two Maize Genotypes. *Plant Physiol. Biochem.* **2013**, *70,* 221–234.

123. Kuo, H. F.; Chiou, T. J. The Role of microRNAs in Phosphorus Deficiency Signaling. *Plant Physiol.* **2011**, *156,* 1016–1024.

124. Zeng, H.; Wang, G.; Hu, X.; Wang, H.; Du, L.; Zhu, Y. Role of microRNAs in Plant Responses to Nutrient Stress. *Plant Soil.* **2014**, *374,* 1005–1021.

125. Hua, D.; Wang, C.; He, J.; Liao, H.; Duan, Y.; Zhu, Z.; Guo, Y.; Chen, Z.; Gong, Z. A Plasma Membrane Receptor Kinase, GHR1, Mediates Abscisic Acid- and Hydrogen Peroxide-Regulated Stomatal Movement in Arabidopsis. *Plant Cell.* **2012**, *24,* 2546–2561.

126. Alexova, R.; Millar, A. H. Proteomics of Phosphate Use and Deprivation in Plants. *Proteomics.* **2013,** *13,* 609–623.

127. Li, K.; Xu, C.; Zhang, K.; Yang, A.; Zhang, J. Proteomic Analysis of Roots Growth and Metabolic Changes Under Phosphorus Deficit in Maize (*Zea mays* L.) Plants. *Proteomics.* **2007,** *7,* 1501–1512.

128. Li, K.; Xu, Z.; Zhang, K.; Yang, A.; Zhang, J. Efficient Production and Characterization for Maize Inbred Lines with Low-Phosphorus Tolerance. *Plant Sci.* **2007,** *172,* 255–264.

129. Li, K.; Xu, C.; Li, Z.; Zhang, K.; Yang, A.; Zhang, J. Comparative Proteome Analyses of Phosphorus Responses in Maize (*Zea mays* L.) Roots of Wild-Type and a Low-P-Tolerant Mutant Reveal Root Characteristics Associated with Phosphorus Efficiency. *Plant J.* **2008,** *55,* 927–939.

130. Li, K.; Xu, C.; Li, Z.; Zhang, K.; Yang, A.; Zhang, J. Comparative Proteome Analyses of Phosphorus Responses in Maize (*Zea mays* L.) Roots of Wild-Type and a Low-P-Tolerant Mutant Reveal Root Characteristics Associated with Phosphorus Efficiency. *Plant J.* **2008,** *55,* 927.

131. Lin, H. J.; Gao, J.; Zhang, Z. M.; Shen, Y. O.; Lan, H.; Liu, L.; Xiang, K.; Zhao, M.; Zhou, S.; Zhang, Y. Z.; Gao, S. B.; Pan, G. T. Transcriptional Responses of Maize Seedling Root to Phosphorus Starvation. *Mol. Biol. Rep.* **2013,** *40,* 5359–5379.

132. Eathington, S. R.; Crosbie, T. M.; Edwards, M. D.; Reiter, R. S.; Bull, J. K. Molecular Markers in a Commercial Breeding Program. *Crop Sci.* **2007,** *47,* S-154-S-163.

133. Araus, J. L.; Slafer, G. A.; Royo, C.; Serret, M. D. Breeding for Yield Potential and Stress Adaptation in Cereals. *Crit. Rev. Plant Sci.* **2008,** *27,* 377–412.

134. Sinclair, T. R. Challenges in Breeding for Yield Increase for Drought. *Trends Plant Sci.* **2011,** *16,* 289–293.

135. Xu, Y.; Lu, Y.; Xie, C.; Gao, S.; Wan, J.; Prasanna, B. M. Whole-Genome Strategies for Marker-Assisted Plant Breeding. *Mol. Breed.* **2012,** *29,* 833–854.

CHAPTER 9

OMICS: MODERN TOOLS FOR PRECISE UNDERSTANDING OF DROUGHT ADAPTATION IN PLANTS

KARABA N. NATARAJA*, MADHURA BHAT G., and PARVATHI M. S.

Department of Crop Physiology, University of Agricultural Sciences, GKVK, Bangalore, India

Corresponding author. E-mail: nataraja_karaba@yahoo.com

CONTENTS

ABSTRACT

The global climate change reports and predictions of imminent deterioration in the conditions considered optimal for crop production, has formed the basis for currently identifying drought as one of the major challenges to be dealt with in the coming years. In order to achieve the requirements of food production under variable environmental conditions and increasing human population, understanding the various plants' stress responses is very relevant. There have been many technological advances in recent years, among which *omics* technologies have emerged as the forerunner in unraveling intricate plant response patterns. The various *omics* tools help in identifying the genetic factors controlling drought adaptation by elucidating functions of important genes, their regulators, as well as whole stress response pathways. Rapid advances are being noted in these techniques and novel tools are being developed at a considerable pace. Effective use of these discoveries in a timely manner would serve as the panacea for developing crop plants tolerant to drought as well as other stressful conditions. This chapter focuses on the role of the respective technologies categorized under genomics, transcriptomics, proteomics, and metabolomics approaches for prospecting genes, identification of key regulators/effectors and discovery of specific pathways linked to drought acclimation.

According to the FAO 2014 Hunger Report, achieving the Millennium Development Goal (MDG-1)—of halving the proportion of undernourished people in developing countries—seems possible (http://www.fao.org/hunger/en/). However, the latest National Aeronautics and Space Administration (NASA) Global Climate Change observations indicate elevated levels of atmospheric CO_2 and global atmospheric temperature, as well as rapidly diminishing forest cover. According to the report, the present CO_2 level in the air is highest in 650,000 years (400.57 ppm), while atmospheric temperature is 1.4 °F more since 1880, and nine out of 10 recorded warmest years on the planet occurred since 2000. Also, forest areas have decreased by 1.5 million km^2 (http://climate.nasa.gov/). These facts reflect foreseeing of the occurrence of more frequent and higher intensity stressful conditions for crop production. In most cases, drought stress would accompany high light and temperature stresses. In addition, world population is expected to increase by over one third (2.3 billion people) between 2009 and 2050. Mere growth of human population demands a nearly 70% increase in food production   (http://www.fao.org/fileadmin/templates/wsfs/docs/Issues_papers/ HLEF2050_Global_Agriculture.pdf). Taken together, these reports indicate a need for development of crop varieties with higher biomass and yield potential even under stressful conditions. Drought stress is projected to continue as one of the major constraints for improvement in agricultural productivity. In the dynamic scenario of global climatic conditions, it is crucial to identify and breed for drought tolerant crop varieties. Dissecting the tolerance conferring mechanism would enable crop scientists to use the relevant genes or genomic regions for targeted manipulation of traits in important crop species. Given that drought tolerance is governed by multiple traits, there is a need to use a comprehensive approach to study the adaptive responses in tolerant species.

From the agriculture point-of-view, in order to meet the penultimate goal of creating crops more adept for cultivation in the current scenario of climate change, applications of different "*Omics*" tools in understanding diverse aspects of plant physiology are numerous. One of the chief advantages of *omics* technologies is the direct, unbiased monitoring of the various factors governing plant growth and development. This would aid in investigation of the complex interplay between the plant and its environment at various levels; namely, whole genome, specific regions of the genome, transcriptome, proteome, and metabolome. Integration at system level of gene functions would further elucidate the intricacies of gene networks. Now known as "systems breeding," would lead to accelerated crop improvement strategies. The complex issues related to the analysis of trait architecture, discovery and detection of potential pathogens, understanding various disease epidemiologies, identification of resistance conferring genes, study of specific mutants, regulatory events, signaling and developmental aspects can be addressed effectively using *omics*. Eventually, it is expected that all the information generated would lead to a clearer elucidation of the pathways associated with plant growth and tolerance to different stressful conditions, and thus, serving as the panacea for crop scientists to achieve the said objective.

With this perspective, the issues discussed here are crop genomics and its relevance in dissecting the complexity of drought adaptive responses; transcriptomics, proteomics, and metabolomics-based approaches for prospecting genes/pathways, identification of key regulators/effectors, and discovery of specific pathways linked to drought adaptation processes.

9.1 CROP GENOMICS: UNRAVELING GENETIC CONTROL OF DROUGHT ADAPTIVE TRAITS

Genomics is the branch of molecular biology concerned with the structure, function, evolution, and mapping of genomes. It is the systematic examination of an organism's genome. Genomics has insofar broadened the vista of genetics remarkably by providing practicable and accurate information about ordered genes, epigenetic states, genetic variations, and gene expressions. Technological advancement in genomics has led to the advent of next generation sequencing (NGS) platform. These tools are generating gargantuan loads of information and deciphering the data has found a wide array of applications in various field, such as analyzing the cell lineage of different organisms, identification of genetic disorders and disease prognosis, drug testing, and other clinical applications; as well as in crop improvement.

Crop genomics and related tools offer one of the most promising paths through which some of these challenges can be addressed. Since plants are sessile organisms, they have evolved more complex stress adaptive mechanisms compared to animals. The traits that confer tolerance to stresses in plants are quantitative in nature and genomics hold the key to elucidation of relevant traits linked to stress adaptation. The genomics along with other *omics* tools as assist in consolidation of the vast amount of data generated from different studies on cellular and system level responses. It is, now possible to decipher genetic information from any given biological source, ever since the introduction of capillary electrophoresis (CE)-based Sanger sequencing, and exemplified after the arrival of NGS technologies. In principle, both CE-based Sanger sequencing and NGS are very similar. Initially, the objective of NGS technology was to study whole genomes. The NGS approaches helped us to sequence many organisms and the reference genome sequences of many important crops are now available (Table 9.1). The ability to perform high-throughput re-sequencing have provided opportunities for improving our understanding of crop plant domestication, and adaptive responses to diverse stresses which are essential to accelerate crop improvement. New generation experimental and computational approaches have transformed the comparative genomic research in crop plants.

Genomics has altered the scope of genetics by providing a landscape of ordered genes and their epigenetic states. Structural genomics that mainly deals with the genome sequencing has contributed for genome mapping and trait-based cloning of target genes. Traditionally, genes have been analyzed individually by cumbersome approaches, and modern NGS approaches have boosted this area of research rapidly. In the absence of plant genome sequences, several alternative approaches are followed to gather information on genes and their expression. There are many expressed sequence tags (ESTs) databases, cDNA libraries, microarrays, and serial analysis of gene expression (SAGE) datasets, which can be used to identify genes and predict their putative function. Crop genomic information would provide options for

TABLE 9.1 Important Crop Plants and Model Plants with Sequenced Genomes, Their Genome Size and Sequencing Approach.

Crop species	Genome size	Sequencing approach	Reference
Amaranthus hypochondriacus (grain amaranth)	466 Mb	Illumina	[119]
Arabidopsis thaliana (thale cress; model plant)	125 Mb	BAC end, TAC shotgun sequencing	[120]
Brassica rapa (field mustard)	485 Mb	WGS, Illumina, BAC end and sanger sequencing	[121]
Cajanus cajan (pigeon pea)	833 Mb	WGS, Illumina , 454 GS-FLX	[122]
Carica papaya (papaya)	372 Mb	WGS	[123]
Cicer arietinum (chickpea)	738 Mb	WGS, Illumina, BAC end sequence	[124]
Citrus sinensis (sweet orange)	367 Mb	Dihaploid WGS, Illumina	[125]
Cucumis melo (melon)	312 Mb	WGS, Roche 454, BAC end sequencing	[126]
Fragaria sp. (octoploid strawberry)	240 Mb	WGS, Roche 454, Illumina, SOLiD	[127]
Glycine max (soybean)	1115 Mb	WGS, Sanger sequencing	[128]
Gossypium raimondii (D genome cotton)	880 Mb	WGS, Illumina	[129]
Hordeum vulgare (barley)	5100 Mb	WGS, Illumina, BAC physical map, BAC sequence (Roche 454, Illumina)	[130]
Malus domestica (apple)	750 Mb	WGS, Sanger, Roche 454	[131]
Manihot esculenta (cassava)	770 Mb	WGS, Roche 454, BAC end Sanger sequencing	[132]
Medicago truncatula (barrel medic; model plant for legume genome research)	375 Mb	BAC physical map, Sanger, Illumina	[133]
Morus notabilis (tree mulberry)	330 Mb	WGS, Illumina HiSeq-2000	[134]
Musa acuminata (dwarf banana)	523 Mb	WGS, Roche 454, Sanger, Illumina	[135]
Nicotiana tabacum (model plant for gene expression experiments)	4.41 Gb	WGS	[136]
Oryza sativa (rice)	389 Mb	BAC physical map, Sanger sequencing	[137]

TABLE 9.1 *(Continued)*

Crop species	Genome size	Sequencing approach	Reference
Phoenix dactylifera (date palm)	658 Mb	WGS, Illumina	[138]
Phyllostachys heterocycla (bamboo)	2 Gb	WGS, Illumina BAC end sequence	[139]
Populus trichocarpa (black cottonwood)	550 Mb	BAC physical map, WGS, Sanger sequencing	[140]
Prunus persica (peach)	265 Mb	Sanger WGS	[141]
Setaria italica (foxtail millet; grass modelsystem)	500 Mb	WGS, Sanger, Illumina, BAC end sequence	[142]
Solanum lycopersicum (tomato)	900 Mb	WGS, Roche 454, Illumina and SOLiD, BAC end Sanger sequencing	[143]
Solanum tuberosum (potato)	844 Mb	Double monoploid DM and diploid RH, WGS, Illumina, Roche 454	[144]
Sorghum bicolor (sorghum)	700 Mb	WGS, Sanger sequencing	[145]
Theobroma cacao (cacao)	430 Mb	WGS, Sanger, Illumina, Roche 454	[146]
Triticum aestivum (bread wheat)	17,000 Mb	WGS, Roche 454	[147]
Vitis vinifera (wine grape)	475 Mb	WGS, Sanger Sequencing	[148]
Zea mays (maize)	2300 Mb	BAC physical map, BAC sequence 4-6 x deep	[149]

WGS: Whole Genome Sequencing; BAC: Bacterial Artificial Chromosome; TAC: Trasformation Complement Artificial Chromosome; SOLiD: Sequencing by Oligonucleotide Ligation and Detection.

Illumina, Roche 454, and SOLiD are DNA sequencing NGS platforms.

combining genetic mapping strategies and evolutionary analyses to direct and also to optimize the discovery and use of genetic variation. Analysis of the genome-wide sequence variation by re-sequencing improved the availability of information that can be used to develop markers, required for genetic mapping of agronomic and stress related traits. Molecular markers such as simple sequence repeat (SSR) and single nucleotide polymorphism (SNPs), identified from whole genome programs are being used in genetic analyses. The International Rice Genome Sequencing Project (IRGSP) that began in September 1997, at a workshop held in conjunction with the International Symposium on Plant Molecular Biology in Singapore helped us to unravel rice genome and identify genes contributing for different traits. For example, whole genome analysis of *Swarna* helped us to understand the genetic basis of low glycemic index in rice, which is a complex trait involving multiple factors.[1]

A major branch of crop genomics is functional genomics, which aims at the development and application of various approaches to study gene function with the aid of data generated from structural genomics ventures. Although a considerable number of plant genomes have been sequenced, an encyclopedia of DNA elements (ENCODE)-like project, to identify the functional elements for human genome, has not been carried out in plants. This would provide information about gene function and variations, and help predict phenotype from genotype. Similar to the prediction of genetic risk factors for human diseases, in order to predict agronomically important plant traits, like growth rate, drought tolerance, and yield, it is required to pinpoint specific loci that are responsible for phenotype and genetic architecture of that trait. Forward genetics has emerged as the tool of choice for addressing such questions. Gene functions in relation to phenotype/trait manifestation can be studied either by over expression or down regulation of gene of interest. The phenotypic differences identified either through mutagenesis or sampling from natural populations, can be traced back to their respective loci through suitable mapping approaches, such as quantitative trait loci (QTL) mapping. Recently, genome-wide association studies (GWAS) emerged as a means of overcoming some of the limitations of QTL mapping, which is considered as a powerful complementary tool for connecting genotype–phenotype maps. Well-studied model plant, *Arabidopsis thaliana* has emerged as an ideal model organism for GWAS as well, since it can be maintained as inbred lines by continuous self-fertilization. Crop genomics programs have given opportunities for diverse studies related to drought tolerance. For example, understanding role of abscisic acid (ABA), evaluation of the roles of compatible solutes, the role of stress related proteins like late embryogenesis abundant (LEA) proteins, and so forth, have been possible due to the availability of genomic information.

Identification of import regulatory genes linked to cellular tolerance and thus, overall plant tolerance to drought and other environmental stresses has been uplifted by the advances in genomics. Further validation and in-depth studies in transgenic model plants takes it closer toward the development of drought tolerant crops. In order to achieve "super-domestication"[2] to meet the demands of the human populace and the drastically worsening global climatic conditions, dynamic changes in plant breeding techniques and targeted trait manipulation by transgenic technology, which combine conventional breeding with crop genomics, is essential.

9.2 TRANSCRIPTOMICS: TOOL FOR PROSPECTING GENES LINKED TO DROUGHT ADAPTATION

Transcriptomics is the study of the transcriptome which comprises of the complete set of RNA transcripts that are transcribed by the genome, under specific circumstances or in a specific cell or tissue, using different tools and techniques. Prospecting genes from differentially expressed pool of transcripts can be done by the comparison of transcriptomes from distinct cell populations, or in response to different treatments (http://www.nature.com/subjects/transcriptomics). Transcriptome profiling has emerged as a promising strategy to assess the gene expression dynamics in response to a spectrum of drought stress at the global level even in the absence of genome information. Transcriptomics studies help the investigator to recognize the differentially expressed genes under an experimental condition when accurately compared to its control. This approach helps in a better understanding of the first level response to drought.

9.2.1 EMERGENCE OF TRANSCRIPTOMICS AND ITS GROWTH

Transcriptomics have evolved over the years with the advent of technology with respect to the techniques and strategies adopted to generate transcriptome data. The identification of differentially expressed genes was attempted using techniques like suppression subtractive hybridization (SSH), differential display (DDRT-PCR/DD-PCR) and traditional expression profiling in the initial years. The dawn of microarray technology marked the era of gene prospecting with more accuracy and specificity. Further in the *omics* era, global transcriptome studies have begun to emerge since 2005 and were taken to new heights with the progress in the field of NGS technologies.[3] With the advent of NGS technology, quantification of RNA at a much higher resolution is possible, in comparison to conventional microarray-based gene expression studies.

Subtractive hybridization method has been used to enrich differentially expressed tissue, developmental or treatment specific transcripts in desired cDNA libraries in various organisms.[4] Many used this approach to discover novel genes associated with drought tolerance.[5,6] The differential display (DD) concept using short arbitrary primers along with anchored oligo-dT primers led to tracking majority of the mRNA population in any tissue or cell. Subtractive hybridization was quickly overtaken by DD to clone differentially expressed genes (http://www.genhunter.com/support/). Microarrays prepared by high-speed robotic printing of complementary DNAs on glass were used for quantitative expression measurements of the corresponding genes.[7] Differential expression measurements of 45 *Arabidopsis* genes were made by means of simultaneous, two-color fluorescence hybridization way back in 1995 itself. Microarray technologies have become a common tool for transcript analysis in different organisms across various spheres in life sciences. Transcriptional profiling has been used for the analysis of responses to multiple stimuli.[8] The microarray analysis of *Arabidopsis* paved the way for identification of many key gene expression patterns,[9] which was further attempted in different plant species.[10] The SAGE approach that emerged during 1995, allowed the quantitative and simultaneous analysis of a large number of gene transcripts.[11]

Amidst all these approaches, conventional gene expression analyses have been adopted as a more common strategy worldwide. However, this approach itself saw a scale-up from northern blot and dot-blot analysis through reverse transcriptase- mediated PCR to the present day quantitative real time PCR (RT-qPCR) strategies.[12,13] Expression analysis of drought-responsive pathway linked genes was carried out in stress adapted crop species like peanut and finger millet.[14,15] However, of late, RT-qPCR strategies are now being used to validate many of the techniques listed above. For example, SSH library was validated through qRT-PCR in *Phaseolus vulgaris* roots.[16]

Global transcriptome analysis approaches through the use of various NGS technologies are the hot topics of research these days. Sufficient background information on the technical aspects and advances in the area of high throughput RNA sequencing has been reported.[17] The transcriptome analyses of *Porteresia* a wild relative of rice with capability of high salinity and submergence tolerance, using Illumina platform has led to the identification of candidate genes involved in salinity and submergence tolerance.[18] Many similar studies have been carried out to identify drought stress-responsive key players by RNA-seq approaches in different plant species like rice,[19,20] wheat,[21] maize,[22] red clover,[23] *Brassica juncea*,[24] barley[25] and even in trees species like Maxim,[26] *Haloxylon*,[27] mulberry (unpublished, 2015), and poplar.[28] A summary of diverse drought-responsive genes identified through transcriptomics analysis is depicted in the Figure 9.1.

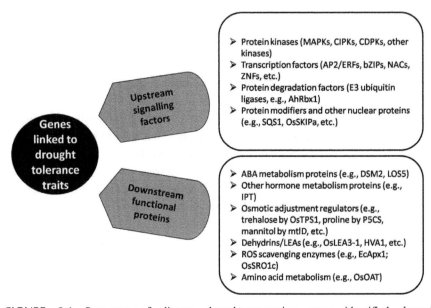

FIGURE 9.1 Summary of diverse drought-responsive genes identified through transcriptomics analyses.

9.2.2 TRANSCRIPTOMICS: NEW DIMENSIONS

Comparative expression profiling is yet another virtue which helps in getting an insight into differential responses of different crop species or contrasting genotypes or ecotypes of the same species. For example, droughts stress response analysis in upland and paddy rice. This approach helps in prospecting the candidate genes underlying the major drought resistant traits, which facilitates the dissection of different drought tolerant mechanisms.[29] Using a comparative transcriptome analysis, 16 common genes involved in heat stress response in plants have been identified while comparing between switch grass, rice, wheat, and maize.[30]

Plants often face a combination of stressful conditions and are forced to respond to each type of stress in a conservative manner, ultimately ensuring their survivability. These responses leading to adaptation of plants to various stresses may involve cross-talks among individual signaling pathways. The transcriptome response of *Arabidopsis* to simultaneous water deficit and biotic stress by infection with the plant-parasitic nematode *Heterodera schachtii*, when analyzed by microarray, yielded candidate genes with potential roles in controlling the response to multiple stresses.[31] An in-depth analysis of the potato transcriptome in response to multiple abiotic, biotic, and hormonal treatments using ribonucleic acid sequencing (RNA-seq) helped in unraveling multiple cross-talk pathways.[32] Many other similar strategies, including comparative microarray analysis in cotton, revealed the cross-talk of responsive genes or pathways to multiple abiotic or even biotic stresses.[33] Thus, the critical assessment of the regulatory networks and cross-talk between multiple stresses occurring simultaneously is required for crop improvement to stress tolerance, as done in *Arabidopsis* recently.[34] Cross-species comparison between diverse species will help in identifying the differentially expressed orthologs and define evolutionarily conserved genes associated with drought tolerance.

9.2.3 RECENT ADVANCES IN TRANSCRIPTOMICS

The normal procedure for analysis of gene expression involves the extraction of total RNA from homogenized tissue samples. This does not allow the identification of the location of gene expression. Novel advancement in techniques has given rise to *in situ* RNA-seq,[35] a methodology of sequencing RNA in live tissues and cells. It involves amplification and sequencing of RNAs by *in situ* ligation. Similarly, many more recent advances are made in the current era of transcriptomics. The recent focus is on "spatially resolved transcriptomics" which complements gene expression data with spatial information on cellular or tissue level expression patterns to achieve a better comprehension of the molecular basis for the manifestation of physiological processes.[36]

One of the critical phenomena that can improve understanding of the evolution of complex plant response patterns is alternative splicing (AS). AS is a phenomenon by which multiple transcripts are generated from a genomic region by virtue of different splicing combinations, thereby increasing the protein diversity in a cell. Unraveling the AS events in response to stress can aid in understanding the adaptive trait evolution in plant species. Recently, exon-skipped transcript abundance has been shown to be the dominant form of AS involved in the responses to heat and drought stress, utilizing RNA-seq.[37,38] Similarly, AS events in many plant developmental phenomena and environmental responses have been addressed through different algorithms associated with high throughput RNA-seq which can shed light on the origin, distribution and contrasting patterns of variation in protein-coding diversity in plant species.[39,40] More recently, RNA-seq was coupled with bioinformatics analysis to identify the extensive splicing network in animal system.[41] Similar approaches in plant system could yield results which help in a better understanding of stress regulatory responses. Novel methods need to be devised to address the genome wide transcriptome changes that occur owing to the transcriptome dynamics governed by mRNA production and degradation kinetics, which ultimately determine the capacity of a plant to modulate mRNA stability in response to stress.[42] A new platform, CytoSeq, detains single cells in picoliter wells to generate transcriptional profiles in large numbers.[43]

Many publicly available databases, such as EST databases, Reference mRNA (refseq_mrna), nucleotide collection (nr/nt), provide considerable information about relevant transcripts identified across different species (Table 9.3). Using suitable analytical tools, the identified differentially expressing transcript sequences are placed to proper pathways. The overall signaling pathways functional under drought adaptive stress conditions are thus identified. The genes recognized to be crucial for the adaptation to stress are further functionally validated using transgenic approaches in a suitable model plant system. Transcriptome analysis and functional characterization of individual genes involved in the multiple stress response can provide further opportunities for fully understanding the complex interactions that regulate multiple stress responses.

9.3 PROTEOMICS: AN APPROACH TO PROSPECT KEY REGULATORS/EFFECTORS OF DROUGHT ADAPTATION

Proteomics is defined as the large-scale study of proteins, including their structure and function, within a cell/system/organism. In other words, it is the study of proteomes and their functions. Proteome is the complete set of all expressed proteins in a cell, tissue, or organism. Study of protein profiles of particular plant species under specific stressful conditions is recognized as one of the finest and most accurate molecular methods in plant stress response studies. This is because proteins and other metabolites are the essential functional products of gene expression that ultimately determine the cellular phenotype. The pattern of protein profiles are altered by plants upon facing stress, in terms of up-regulation of proteins in an existing pool, *de novo* synthesis of proteins required in defense mechanisms and also degradation/turn-over of proteins that prove to be energy-expensive to be maintained under stress.

9.3.1 PLANT PROTEOMICS APPROACHES FOR DROUGHT STRESS STUDIES

Plant proteomics studies have to deal with several experimental challenges; however, advances in high-throughput proteomics helped to address complex biological query in various species. Improvements in protein extraction and protein separation contributed for rapid growth in plant proteomic research. Proteins that are induced or suppressed under drought conditions in plants can be identified by two-dimensional electrophoresis (2-DE)[44] or liquid chromatography coupled with tandem mass spectrometry (LC–MS/MS).[45] These protein separation strategies are being widely used. The gel-based approaches are used in most laboratories because of their simplicity, wide molecular weight coverage, and reproducibility. The protein spots obtained in 2-DE are trypsin digested into peptides for further identification. In most cases, in LC–MS/MS approach, the proteins are digested before separation, which can help to detect broad molecular weight range proteins and low abundant proteins.[46] The soft ionization methods such as matrix assisted laser desorption ionization (MALDI)[47] or electrospray ionization (ESI)[48] and collision-induced dissociation (CID) in tandem MS[49] are aiding in the detection of diverse proteins. The peptides detected through MS and MS/MS is searched against particular protein database to identify the protein of interest.

Since drought tolerance is governed by multiple traits and pathways, there is need to detect and quantify proteins at a global level. Different types of LC–MS-based tagging approaches such as isotope-coded affinity tags (ICAT),[50] stable isotope labeling by amino acids in cell culture (SILAC),[51–53] isobaric tags for relative and absolute quantitation (iTRAQ)[54–57] are useful for global analysis. Application of gel-free protein separation approaches and "second generation" proteomic techniques such as multidimensional protein identification technology (MudPIT), ICATs, targeted mass tags (TMTs), and iTRAQ, are being widely used in descriptive and comparative proteomic studies of plant development and abiotic stress adaptation. These types of global analysis are possible mainly because of the advancement in statistical tools and bioinformatic approaches. Some of the regularly used proteomic approaches in drought studies are mentioned in Table 9.2. However, several gaps still exist in the understanding of this field and further optimization is required for proteomics data analysis in plants.

9.3.2 DROUGHT INDUCED PROTEOME

Several studies have identified profiles of differentially up-regulated and down-regulated proteins under abiotic stress conditions. There are studies on the mechanisms involved in the drought response at the protein level using comparative proteomic approaches.[46] These proteomic studies identified different types of drought related functional proteins (Table 9.2). A majority of the drought-induced proteins seems to be associated with or involved in photosynthesis, redox regulation, oxidative stress response, protein folding and stabilization, signal transduction, and secondary metabolism[58] as depicted in Figure 9.2. Drought induces oxidative stress due to over production of reactive oxygen species (ROS). Plants under drought, to counteract the ill effects, induce the production of several ROS scavengers. Some of the proteins linked to ROS scavenging such as dehydrins, quinone reductase, dehydroascorbate reductase, γ-glutamylcysteine synthetase, and glutathione S-transferases have been detected from proteomic studies in soybean,[59–61] tomato,[62,63] sunflower,[64] and wild watermelon.[65] Transcriptome analysis indicated induction of several molecular chaperone genes under drought. Proteome analysis indicated increased levels of molecular chaperones, such as heat shock proteins in different plants like sugar beet,[66] wheat,[67] wild watermelon[68] and sugarcane[69] under drought stress. Drought affects primary metabolic processes such as carbon and nitrogen metabolism. The proteins related to these processes such as triosephosphate isomerase, UDP-glucose-6-phosphate dehydrogenase, malate dehydrogenase, NADP-malic enzyme, UDP-sugar pyrophosphorylase, α-mannosidase, and phosphoglucomutase were found to be in higher amounts in different plant roots under drought, probably due to increased cellular activities.[46] Although plant response to drought stress has not been very well documented at the proteomic level, recent finding highlights organ-specific proteome in soybean.[70]

TABLE 9.2 Overview of Comparative Proteomic Analyses in Different Crop Species in Response to Drought Stress.

Crop species	Proteomic approach	Drought-responsive protein classes	Stress-response pathways	References
Soybean	2-DE, IPG, SDS-PAGE, LC-MS/MS, nanoLC-MS/MS, IEF tube gel, MALDI-TOF MS, protein sequencing	ROS scavengers; enzymes related to metabolism, lignin biosynthesis, proteolysis; small G-protein family members; osmolytes; transmembrane water channels; chaperones; protease inhibitors; factors related to translation and proteosome	Cellular energy pathways (Glycolysis, TCA cycle, oxidative phosphorylation); photosynthesis; glycoprotein synthesis; amino acid biosynthesis; antioxidant defense mechanisms; flavonoid biosynthesis	[58–60,150]
Wild watermelon	2-DE			[65,68]
Rapeseed	2-DE			[58]
Wheat	2-DE			[67]
Sugarcane	1-DE, 2-DE			[69]
Cotton	2-DE and peptide mass fingerprinting (PMF) using MALDI-TOF-TOF.	Sixteen proteins were up-regulated while 6 were down-regulated	Cell cycle regulation, ATP synthesis, ion Transport, Energy metabolism (photosynthesis), ion transport, transcription, translation , signal transduction	[151]

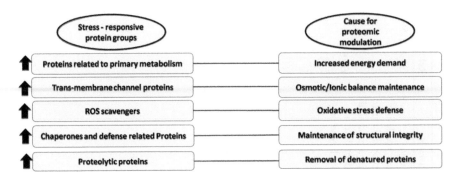

FIGURE 9.2   Common proteins identified by proteomic analyses in plants experiencing drought stress. Proteins belonging to diverse families have been identified and major classes are highlighted here.

9.4   METABOLOMICS: TOOL TO ELUCIDATE SPECIFIC PATHWAYS INVOLVED IN DROUGHT ADAPTATION

The primary objective of metabolomics technology is to better understand the phenotype from genotype information. Since the presence of metabolites is the proof of gene expression, protein–protein interactions and various regulatory cellular processes, metabolomics is a closer and clearer indication of the possible phenotype. The term metabolomics is defined as comprehensive and quantitative determination of all small molecules in a biological system. In general, plant kingdom may contain between 200,000 and 1,000,000 diverse metabolites.[71] Due to a large variety of chemical structure and properties of these metabolites, so far there is no single approach to detect and quantify these metabolites. The techniques such as GC–MS, LC–MS, and CE–MS, and nuclear magnetic resonance (NMR) spectroscopy are some of the commonly used techniques to detect metabolites. In modern metabolome research, these techniques are used in combination to detect and quantify diverse types of metabolites.

Metabolism in plant systems are perturbed by drought stress. Plants are known to adjust metabolic networks during the process of drought acclimation in order to maintain metabolic homeostasis. Stress perception activates stress-responsive metabolic pathways leading to the production of compounds that alleviate the stress effects. The modern metabolomics and system biology approaches enable us to gain a comprehensive knowledge of the acclimation responses essential for targeted manipulation of crops.

In recent years, *omics* analyses centered on metabolomics and hence it has become a powerful technique in functional genomics to identify the functions of genes involved in metabolic processes.[72] Metabolomics is becoming increasingly common in crop physiology and biochemistry especially to understand the complex regulatory network under stressful conditions. Considerable research has been made to understand the metabolic regulations, including regulation of photosynthesis, respiration, and other chemical regulators (hormones) in drought stress response. Initial reports on metabolomic changes under drought conditions were attempted in model system *Arabidopsis*.[73] Accumulation of diverse metabolites, including amino acids, raffinose family oligosaccharides, respiratory metabolites, have been shown to be responsive to drought stress in plants. Since the metabolomics changes are too many and complex, it would be desirable to dissect the metabolites into different components for better understanding of stress response in plants.

9.4.1 OXIDATIVE STRESS-RESPONSIVE METABOLOMICS

The effect of drought stress is enhanced by the ROS produced in the different cellular compartments, mainly, chloroplasts, peroxisomes, and mitochondria. Plants have developed mechanisms to mitigate the cellular damage caused by the ROS that are constantly generated in aerobic environments. Oxygen species like singlet oxygen, superoxide anion radicals, hydrogen peroxide (H_2O_2), and hydroxyl radicals are more reactive than O_2. These ROS can influence various cellular processes and cell components.[74] Some of the effects of ROS are mediated by cell signaling pathways,[75] although chemical nature and sub-cellular localization of ROS plays a crucial role in cellular responses.[76] Cellular metabolites can be affected ROS by different mechanisms as given below;

   a. *Direct effect*: The ROS can encourage the production of oxidized compounds by chemical reactions.[74,77]
   b. *Indirect effect*: The reactions of ROS with certain enzymes that can either hold back catalysis (e.g., respiratory protein and aconitase) or that post-translationally alter activities or decrease protein stability through carbonylation or thiol oxidation.[74,78–81]
   c. *Metabolic adjustments*: This mechanism may not involve direct oxidation, but can result from increased ROS-dependent engagement

of reductant-generating pathways to support increased ROS metabolism.[82–84]

The above mechanisms can act alone or in combination, and may direct secondary effects leading to altered metabolite contents, and modified enzyme activities guided through transcriptional or post-transcriptional mechanisms. ROS induce peroxidation of fatty acid chains and malondialdehyde (MDA) is breakdown products of fatty acid peroxides. Peroxidation of fatty acids by non-enzymatic means occurs through either addition of singlet oxygen or H abstraction followed by O_2 and H addition.[74] Complex types of compounds that are produced from lipid peroxidation have been reported, many of them have been termed "reactive electrophile species."[77] The activity of *electrophiles* seems to be regulated by conjugation to glutathione (GSH). Although the conjugation can occur spontaneously, the process is accelerated by the activity of glutathione S-transferase. This process can deplete the GSH pool.[85] Lipid peroxidation product, 4-hydroxy-2-nonenal (HNE, a reactive aldehyde) has been shown to be accumulated under oxidative stress.[86] Similar to HNE, the other reactive aldehydes, acrolein (2-propenal), has been detected under stress, which can decrease photosynthesis through inactivation of photosynthetic enzymes and the depletion of the GSH pool.[87]

In plants, there are mechanisms to prevent ROS formation and scavenge them both by enzymatic and non-enzymatic processes. Ascorbic acid (AsA), GSH, α-tocopherols, amino acids (e.g., proline), sugars, carotenoids, and quinic acid derivatives are some of the antioxidants produced under stress.[88] The tocopherols and tocotrienols, like carotenoids, are derived from geranylgeranyl pyrophosphate in the terpenoid pathway. Both tocopherols and tocotrienols can directly quench singlet oxygen or to interrupt peroxidation.[89] Since metabolic processes altered under drought are complex, critical analysis of key metabolites is still needed for clarity on regulatory network mediated by ROS.

9.4.2 OTHER METABOLITES LINKED TO DIVERSE CELLULAR PROCESSES

Drought stress induces as a general response leading to the accumulation of several amino acids such as valine (Val), leucine (Leu), isoleucine and agmatine (as a precursor of polyamines) along with and carbohydrate alcohols which, in combination with proline (Pro), play roles in osmoprotection.[90] Osmolytes such as soluble sugars and amino acids contribute to turgor maintenance by osmotic adjustment[91] and thus better water relations under drought. In perennial plant grapevine, N metabolism is altered in response to drought. Marked increase in amino acids, such as Pro, Val, Leu, threonine (Thr) and tryptophan (Trp), reductions of most organic acids, and changes in the phenylpropanoid pathway have been noticed in vine.[92] There are reports on the altered levels of polyamines, putrescine, spermidine, and spermine under stress. Exogenous application of polyamines to plants subjected to drought can alleviate stress, probably by reducing H_2O_2 and MDA levels.[93]

Along with the accumulation of metabolites exhibiting anti-oxidative activity, and osmoregulation, drought induces the production of various types of specialized metabolites. Chlorophyll breakdown increases under drought due to excessive ROS production. Chlorophyll breakdown involves loss of magnesium and the phytol chain, and export from chloroplast to the cytoplasm for further degradation,[94] and further removal from cytosol to the vacuole for breakdown. Metabolites such as ajmalicine, anthocyanins, flavonol glycosides, organic acids and sugar derivatives, glycyrrhizin, angustifolin, 13-α-hydroxylupanin, isoangustofolin, isolupanin, lupanin, phenylpropanoid glycosides and quinic acid derivatives, α-chaconine, and α-solanine have been linked to drought stress response in different plants types.[73,88,95–100] In maize (*Zea mays* L.) hybrids under drought, increased levels of various aromatic and branched chain amino acids have been detected, whereas several TCA cycle intermediates decreased.[101] There are genotypic differences in metabolites under stress, for example, in maize; significant difference in 26 of 40 total foliar metabolites during water stress has been reported.[101] Drought induced accumulation of stress hormone ABA has been reported and the accumulation of Pro and other amino acids are shown to be under ABA regulation.

Among the diverse types of metabolites, alterations in the primary metabolites are the most evident under drought. It can be concluded that changes in levels of sugars and sugar alcohols, aminoacids, and TCA cycle intermediates are common in different plant types under drought stress. Changes in the secondary metabolites are more specific in a given species. Comprehensive analysis of metabolome and fluxome along with transcriptome and proteome is needed for precise understanding of drought adaptation in plants (Fig. 9.3).

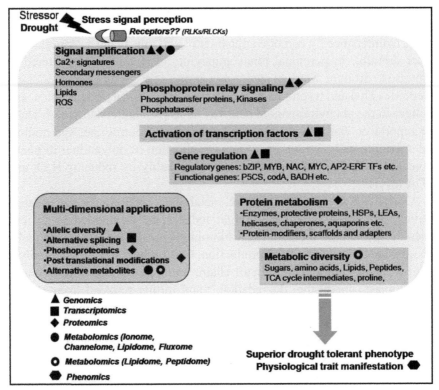

FIGURE 9.3   Schematic representation of the applications of different *omics* technologies in a drought stress response scenario. Plants respond to a drought stressor by signal perception, transduction of the signal and appropriate downstream physiological response which determines the phenotype manifestation. This entire cascade is under the control of multi- level regulations involving various gene products, protein complexes, and metabolites. *Omics* approaches aids in the identification of several of these components thereby fine-tuning critical stress adaptive pathways.

9.5   MULTI-OMICS-ANALYSES: INTEGRATED APPROACHES TO UNDERSTAND COMPLEX STRESS RESPONSES

The "OMES puzzle" was the name of an article that appeared in Nature in 2003. It then highlighted the importance of compiling the information generated from all the "omes" into a single basket for integrated analysis.[102] Systems analysis by combining the information fed in from different *omics* analyses will help in evaluating the plant performance under challenging situations. The magnitude of *omic* data generated ranging from whole-genome sequencing data, to extensive transcriptomic, proteomic, and metabolomic data invites the need for the development of effective models to predict the regulation of complex traits. There are several databases and repositories that provide multiple options for omics data compilation and analyses. A review of the "omics" approach to biomarkers of oxidative stress in *Oryza sativa* for identification of oxidative stress-related biomarkers and the integration of data from different disciplines shed light on the oxidative response pathways.[103] An integrated proteome and metabolome analysis was applied to understand the differential responses in rice and wheat coleoptiles under anoxic conditions.[104] An integrated transcriptome, proteome, and metabolome analysis was helped in unraveling the complexity of UV-B-mediated responses in maize.[105] With increasing value for and throughput from plant *omics* technologies, bioinformatics is emerging as the key to integrate the varied data generated.[106]

Numerous experiments conducted to evaluate plant behavior to different stresses have made phenomenal contributions to our understanding of plant stress responses. However, the lack of common platforms with respect to plant growth conditions, experiment profiles and data analysis has made the comparison of independent experiments and subsequent derivation of tangible information from such comparisons, tricky and challenging.[107,108] In recent years, integrated approaches, such as systems biology methods, have been evolving, providing promising tools for studying plant stress responses.[108–110]

The integration of different *omics* approaches would enable more comprehension on the complex interplay of events that lead to overall cellular level tolerance. A systems-based approach combined hormonome, metabolome, and transcriptome analyses in *Arabidopsis* transgenic lines exhibiting increased leaf growth that gave an insight into the molecular basis for leaf size control.[111] An overview of an integrated metabolomics approach has also been compiled very recently, which will ultimately give a deeper insight into the *in vivo* functions of well- known and novel unknown metabolites.[88]

Novel statistical and machine-learning approaches will have to be used to address the intricate interplay of different types of *omic* data; in the way machine learning approaches helped in distinguishing multiple stress conditions using stress-responsive genes and identify candidate genes for broad resistance in rice[112] Although there is a strong requirement for the development of new strategies to harvest all possible information under a particular stress scenario, there are hassles with respect to adopting the most effective

model for *multi-omics*-based analysis. Biologists have to collaborate efficiently with statisticians, mathematicians, computer scientists, and bioin-
formaticians for meta-dimensional analysis.[113] These integrated approaches
will give a concise insight into the complexity of plant stress tolerance trait
patterning in order to know about whole plant level reciprocation to any
stress, by resulting in the generation of a plant stress specific *integrome* (Fig.
9.4). A list of commonly used and publicly available set of databases used in
various *omics* studies is provided in Table 9.3.

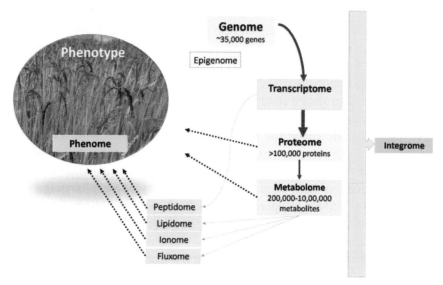

FIGURE 9.4 Overview of omics approaches contributing for the critical understanding of
traits governing plant phenotype.

TABLE 9.3 List of Commonly Used *Omics* Databases.

Common Name	Weblink
Genomics	
DNA Data Bank of Japan (DDBJ)	http://www.ddbj.nig.ac.jp/
European Molecular Biology Laboratory (EMBL)	http://www.embl.org/
GenBank	http://www.ncbi.nlm.nih.gov/genbank
PlantGDB	http://www.plantgdb.org/
Gramene	http://www.gramene.org/
Maize Genome Database	http://www.maizegdb.org/
Rice Genome Annotation Project	http://rice.plantbiology.msu.edu/

TABLE 9.3 *(Continued)*

Common Name	Weblink
Transcriptomics	
Array Express	http://www.ebi.ac.uk/arrayexpress/
Gene Expression Omnibus (GEO)	http://www.ncbi.nlm.nih.gov/geo/
RiceXPro	http://ricexpro.dna.affrc.go.jp/
Botany Array Resource	http://www.bar.utoronto.ca/
Stress- Genomics	http://www.stress-genomics.org/
PLEXdb	http://www.plexdb.org/
Proteomics	
Plant Proteome Database (PPBD)	http://ppdb.tc.cornell.edu/
Proteomics Identifications Database (PRIDE)	http://www.ebi.ac.uk/pride/archive/
Global Proteome Machine (GPMDB)	http://www.gpmdb.thegpm.org/
ProteomicsDB	http://www.proteomicsdb.org/
PeptideAtlas	http://www.peptideatlas.org/
PlaPID	http://www.plapid.net/
Metabolomics	
PlantMetabolomics.org	http://www.plantmetabolomics.org/
Plant Metabolome Database	http://www.sastra.edu/scbt/pmdb/
Plant Metabolic Network (PMN)	http://www.plantcyc.org/
MetNet online database	http://www.metnetonline.org/
MetNet	http://metnet.vrac.iastate.edu/
Phenomics	
Phenome Networks	http://phnserver.phenome-networks.com/
Integromics	
Plant Omics Data Center	http://www.bioinf.mind.meiji.ac.jp/podc/
Next- Gen Sequence Databases	http://mpss.udel.edu/
AtCAST	http://atpbsmd.yokohama-cu.ac.jp/cgi/ network/home.cgi

9.6 CONCLUSION AND PERSPECTIVES

Technological progression has always been the high plus for the rapid progress in crop improvement. Among the many breakthroughs, *omics* technologies have provided the major avenue for accelerating basic science research in a direction favorable for the development of applied strategies for abiotic stress tolerance targeted crop improvement. However, drought tolerance being a multigenic trait requires more rigorous and careful examination with the aid of multi-*omics* data generation and integration. Although many crop species have been sequenced, even in earlier sequenced model plants like *Arabidopsis* and rice only about half of the genes are functionally annotated. Strategic research needs to be carried out to direct crop improvement programs to evolve crop types with improved drought tolerance that ensures a sustainable future.

Omics have opened up the possibility to fill the gap between the genome and phenotype. The identification of pathway linked component trait underlying genes should be the current thrust for tangible levels of success in crop improvement.[15,114–116] Single and multi-gene transgenesis is the current strategy since the past one decade, which if aptly exercised alongside other molecular breeding strategies, can yield satisfactorily performing drought tolerant crop plants.[117,118] This eventually patterns the proteome and metabolome complements of a plant cell leading to the relevant *cellome* under abiotic stresses, which has to be carefully examined since there are certain setbacks in each case. However, a single *omics* approach can be misleading; there should be an option to corroborate between the different *omics* data to arrive at a consensus. High throughput *omics* data generation in association with centrally integrated databases can be the best facilitators for plant stress researchers for unraveling the complex regulatory pathway(s)/network(s) associated with that particular stress scenario. However, rationalization of the authenticity of *omics* output is required with respect to generation of meaningful information that can be correlated with the stress situation under scrutiny.

There are prospects for all sensible data generated from these innovative tools in crop improvement. In spite of the existence of various integration techniques devised for multi-dimensional analysis, evolution of more precise landmark multi-faceted approaches stands as a challenge.

KEYWORDS

- stressful condition
- transcriptome
- potential pathogen
- epidemiologies
- statistical tool

REFERENCES

1. Rathinasabapathi P., et al. Whole Genome Sequencing and Analysis of Swarna, a Widely Cultivated *Indica* Rice Variety with Low Glycemic Index. *Sci. Rep.* **2015**, *5,* 11303.doi:10.1038/srep11303.

2. Vaughan, D. A.; Balász, E.; Heslop-Harrison, J. S. From Crop Domestication to Super-Domestication. *Ann. Bot.* **2007**, *100,* 893–901.

3. Parvathi, M. S.; Nataraja, K. N. Emerging Tools, Concepts and Ideas to Track the Modulator Genes Underlying Plant Drought Adaptive Traits: An Overview. *Plant Signal. Behav.* **2015** (In Press).

4. Bui, L. C.; Leandri, R. D.; Renard, J. P.; Duranthon, V. SSH Adequacy to Pre-Implantation Mammalian Development: Scarce Specific Transcripts Cloning Despite Irregular Normalization. *BMC Genomics.* **2005**, *6,* 155–164.

5. Govind, G., et al. Identification and Functional Validation of a Unique Set of Drought Induced Genes Preferentially Expressed in Response to Gradual Water Stress in Peanut. *Mol. Genet. Genomics.* **2009**, *281* (6), 591–605.

6. Priyanka, B.; Sekhar, K.; Reddy, V. D.; Rao, K. V. Expression of Pigeon Pea High-Proline-Rich Protein Encoding Gene (*CcHyPRP*) in Yeast and *Arabidopsis* Affords Multiple Abiotic Stress Tolerance. *Plant Biotechnol.* **2010**, *8,* 76–87.

7. Schena, M., et al. Quantitative Monitoring of Gene Expression Patterns with a Complementary DNA Microarray. *Science.* **1995**, *270,* 467–470.

8. Kilian, J., et al. Prerequisites, Performance and Profits of Transcriptional Profiling the Abiotic Stress Response. *Biochim. Biophys. Acta.* **2012**, *1819,* 166–175.

9. Redman, J. C., et al. Development and Evaluation of an *Arabidopsis* Whole Genome Affymetrix Probe Array. *Plant J.* **2004**, *38,* 545–561.

10. Liu, Y., et al. Microarray Analysis of Transcriptional Responses to Abscisic Acid and Salt Stress in *Arabidopsis thaliana. Int. J. Mol. Sci.* **2013**, *14,* 9979–9998.

11. Velculescu, V. E., et al. Serial Analysis of Gene Expression. *Science.* **1995**, *270,* 484–487.

12. Guimarães-Dias, F., et al. Expression Analysis in Response to Drought Stress in Soybean: Shedding Light on the Regulation of Metabolic Pathway Genes. *Genet. Mol. Biol.* **2012**, *35,* 222–232.

13. Basu, S.; Roychoudhury, A. Expression Profiling of Abiotic Stress-Inducible Genes in Response to Multiple Stresses in Rice (*Oryza sativa* L.) Varieties with Contrasting Level of Stress Tolerance. *Bio. Med. Res. Int.* **2014**, *2014,* 12.

14. Pruthvi, V., et al. Expression Analysis of Drought Stress Specific Genes in Peanut (*Arachis hypogaea* L.). *Physiol. Mol. Biol. Plants.* **2013**, *19* (2), 277–281.

15. Parvathi, M. S., et al. Expression Analysis of Stress Responsive Pathway Genes Linked to Drought Hardiness in an Adapted Crop, Finger Millet (*Eleusine coracana*). *J. Plant Biochem. Biotechnol.* **2013**, *22* (2), 193–201.

16. Recchia, G. H., et al. Transcriptional Analysis of Drought-Induced Genes in the Roots of a Tolerant Genotype of the Common Bean (*Phaseolus vulgaris* L.). *Int. J. Mol. Sci.* **2013**, *14,* 7155–7179.

17. Wolf, J. B. W. Principles of Transcriptome Analysis and Gene Expression Quantification: An RNA-Seq Tutorial. *Mol. Ecol. Res.* **2013**, *13,* 559–572.

18. Garg, R., et al. Deep Transcriptome Sequencing of Wild Halophyte Rice, *Porteresia coarctata*, Provides Novel Insights into the Salinity and Submergence Tolerance Factors. *DNA Res.* **2013**, 1–16.

19. Zhai, R., et al. Transcriptome Analysis of Rice Root Heterosis by RNA-Seq. *BMC Genomics.* **2013,** *14,* 19.

20. Huang, L., et al. Comparative Transcriptome Sequencing of Tolerant Rice Introgression Line and its Parents in Response to Drought Stress. *BMC Genomics.* **2014,** *15,* 1026.

21. Okay, S.; Derelli, E.; Unver, T. Transcriptome-Wide Identification of Bread Wheat WRKY Transcription Factors in Response to Drought Stress. *Mol. Genet. Genomics.* **2014,** *289* (5), 765–781.

22. Kakumanu, A., et al. Effects of Drought on Gene Expression in Maize Reproductive and Leaf Meristem Tissue Revealed by RNA-Seq. *Plant Physiol.* **2012,** *160,* 846–867.

23. Yates, S. A., et al. *De Novo* Assembly of Red Clover Transcriptome Based on RNA-Seq Data Provides Insight into Drought Response, Gene Discovery and Marker Identification. *BMC Genomics.* 2014, *15,* 453.

24. Bhardwaj, A. R., et al. Global Insights into High Temperature and Drought Stress Regulated Genes by RNA-Seq in Economically Important Oilseed Crop *Brassica juncea. BMC Plant Biol.* **2015,** *15,* 9.

25. Hübner, S., et al. RNA-Seq Analysis Identifies Genes Associated with Differential Reproductive Success under Drought-Stress in Accessions of Wild Barley *Hordeum spontaneum. BMC Plant Biol.* **2015,** *15,* 134.

26. Gao, F., et al. Transcriptomic Analysis of Drought Stress Responses in *Ammopiptanthus mongolicus* Leaves Using the RNA-Seq Technique. *PLoS ONE.* **2015,** *10* (4), e0124382. [Online] doi:10.1371/journal.pone.0124382.

27. Long, Y., et al. *De Novo* Assembly of the Desert Tree *Haloxylon ammodendron* (C. A. Mey.) Based on RNA-Seq Data Provides Insight into Drought Response, Gene Discovery and Marker Identification. *BMC Genomics.* **2014,** *15,* 1111.

28. Barghini, E., et al. Transcriptome Analysis of Response to Drought in Poplar Interspecific Hybrids. *Genomics Data.* **2015,** *3,* 143–145.

29. Ding, X.; Li, X.; Xiong, L. Insight into Differential Responses of Upland and Paddy Rice to Drought Stress by Comparative Expression Profiling Analysis. *Int. J. Mol. Sci.* **2013,** *14,* 5214–5238.

30. Li et al. Transcriptome Analysis of Heat Stress Response in Switchgrass (*Panicum virgatum* L.). *BMC Plant Biol.* **2013,** *13,*153.

31. Atkinson, N. J., et al. Identification of Genes Involved in the Response of *Arabidopsis* to Simultaneous Biotic and Abiotic Stresses. *Plant Physiol.* **2013,** *162,* 2028–2041.

32. Massa, A. N.; Childs, K. L.; Buell, C. R. Abiotic and Biotic Stress Responses in *Solanum tuberosum* Group Phureja DM1-3516R44 as Measured through Whole Transcriptome Sequencing. *Plant Gen.* **2013,** *6* (3), 15. doi: 10.3835/plantgenome2013.05.0014.

33. Zhu, Y-N., et al. Transcriptome Analysis Reveals Crosstalk of Responsive Genes to Multiple Abiotic Stresses in Cotton (*Gossypium hirsutum* L.). *PLoS ONE.* **2013,** *8* (11), e80218. [Online] doi:10.1371/journal.pone.0080218.

34. Rasmussen, S., et al. Transcriptome Responses to Combinations of Stresses in *Arabidopsis. Plant Physiol.* **2013,** *161,* 1783–1794.

35. Ke, R., et al. *In Situ* Sequencing for RNA Analysis in Preserved Tissue and Cells. *Nat. Methods.* **2013,** *10,* 857–860.

36. Burgess, D. J. Putting Transcriptomics in its Place. *Nat. Rev. Genet.* **2015,** *16,* 319.

37. Gulledge, A. A., et al. Mining *Arabidopsis thaliana* RNA-Seq Data with Integrated Genome Browser Reveals Stress-Induced Alternative Splicing of the Putative Splicing Regulator SR45a. *Am. J. Bot.* **2012,** *99,* 219–231.

38. Sablok, G.; Harikrishna, J. A.; Min, X. J. Next Generation Sequencing for Better Under-standing Alternative Splicing: Way Ahead for Model and Non-Model Plants. *Transcrip-tomics.* **2013,** *1,* e103. doi:10.4172/2329-8936.1000e103.

39. Loraine, A., et al. RNA-Seq of *Arabidopsis* Pollen Uncovers Novel Transcription and Alternative Splicing. *Plant Physiol.* **2013,** *162,* 1092–1109.

40. Li, W., et al. Genome-Wide Detection of Condition-Sensitive Alternative Splicing in *Arabidopsis* Roots. *Plant Physiol.* **2013,** *162,* 1750–17563.

41. Zhou, X., et al. Transcriptome Analysis of Alternative Splicing Events Regulated by SRSF10 Reveals Position-Dependent Splicing Modulation. *Nucleic Acids Res.* **2014,** *42,* 1–12.

42. Shalem, O., et al. Transcriptome Kinetics is Governed by a Genome-Wide Coupling of mRNA Production and Degradation: A Role for RNA pol II. *PLoS Genet.* **2011,** *7* (9), e1002273. [Online] doi:10.1371/journal.pgen.1002273.

43. Fan, H. C., et al. Combinatorial Labeling of Single Cells for Gene Expression Cytom-etry. *Science.* **2015,** *347,* 1258367.

44. Wittmann-Liebold, B.; Graack, H. R.; Pohl, T. Two-Dimensional Gel Electrophoresis as Tool for Proteomics Studies in Combination with Protein Identification by Mass Spectrometry. *Proteomics.* **2006,** *6* (17), 4688–4703.

45. Fournier, M. L., et al. Multidimensional Separations-Based Shotgun Proteomics. *Chem. Rev.* **2007,** *107* (8), 3654–3686.

46. Ghosh, D.; Xu, J. Abiotic Stress Responses in Plant Roots: A Proteomics Perspective. *Front. Plant Sci.* **2014,** *5,* 6. [Online] doi: 10.3389/fpls.2014.00006.

47. Tanaka, K., et al. Protein and Polymer Analyses up to *m/z* 100 000 by Laser Ionization Time-of-Flight Mass Spectrometry. *Rapid Commun. Mass Spectrom.* **1988,** *2,* 151–153.

48. Yamashita, M.; Fenn, J. B. Electrospray Ion Source. Another Variation on the Free-Jet Theme. *J. Phys. Chem.* **1984,** *88* (20), 4451–4459.

49. McLuckey, S. A.; Stephenson, J. L. Jr. Ion/ion Chemistry of High-Mass Multiply Charged Ions. *Mass Spectrom. Rev.* **1998,** *17* (6), 369–407.

50. Gygi, S. P., et al. Quantitative Analysis of Complex Protein Mixtures Using Isotope-Coded Affinity Tags. *Nat. Biotechnol.* **1999,** *17* (10), 994–999.

51. Martinović, S., et al. Selective Incorporation of Isotopically Labeled Amino Acids for Identification of Intact Proteins on a Proteome-Wide Level. *J. Mass Spectrom.* **2002,** *37* (1), 99–107.

52. Ong, S. E., et al. Stable Isotope Labeling by Amino Acids in Cell Culture, SILAC, as a Simple and Accurate Approach to Expression Proteomics. *Mol. Cell Proteomics.* **2002,** *1*(5), 376–386.

53. Ibarrola, N., et al. A proteomic Approach for Quantitation of Phosphorylation Using Stable Isotope Labeling in Cell Culture. *Anal. Chem.* **2003,** *75* (22), 6043–6049.

54. Ross, P. L., et al. Multiplexed Protein Quantitation in *Saccharomyces cerevisiae* Using Amine-Reactive Isobaric Tagging Reagents. *Mol. Cell Proteomics.* **2004,** *3* (12), 1154–1169.

55. Choe, L., et al. 8-Plex Quantitation of Changes in Cerebrospinal Fluid Protein Expres-sion in Subjects Undergoing Intravenous Immunoglobulin Treatment for Alzheimer's Disease. *Proteomics.* **2007,** *7* (20), 3651–3660.

56. Ghosh, D., et al. Identification of Key Players for Colorectal Cancer Metastasis by iTRAQ Quantitative Proteomics Profiling of Isogenic SW480 and SW620 Cell Lines. *J. Proteome Res.* **2011,** *10* (10), 4373–4387.

57. Ghosh, D., et al. iTRAQ Based Quantitative Proteomics Approach Validated the Role of Calcyclin Binding Protein (CacyBP) in Promoting Colorectal Cancer Metastasis. *Mol. Cell Proteomics.* **2013,** *12* (7), 1865–1880.

58. Mohammadi, P. P.; Moieni, A.; Komatsu, S. Comparative Proteome Analysis of Drought-Sensitive and Drought-Tolerant Rapeseed Roots and their Hybrid F1 Line Under Drought Stress. *Amino Acids.* **2012,** *43,* 2137–2152.

59. Toorchi, M.; Yukawa, K.; Nouri, M. Z.; Komatsu, S. Proteomics Approach for Identifying Osmotic-Stress-Related Proteins in Soybean Roots. *Peptides.* **2009,** *30* (12), 2108–2117.

60. Alam, I., et al. Proteome Analysis of Soybean Roots Subjected to Short-Term Drought Stress. *Plant Soil.* **2010,** *333,* 491–505.

61. Mohammadi, P. P.; Moieni, A.; Hiraga, S.; Komatsu, S. Organ-Specific Proteomic Analysis of Drought-Stressed Soybean Seedlings. *J. Proteomics.* **2012,** *75* (6), 1906–1923.

62. Shalata, A., et al. Response of the Cultivated Tomato and its Wild Salt-Tolerant Relative *Lycopersicon pennellii* to Salt-Dependent Oxidative Stress: The Root Antioxidative System. *Physiol. Plant.* **2001,** *112* (4), 487–494.

63. Mittova, V., et al. Salinity Up-Regulates the Antioxidative System in Root Mitochondria and Peroxisomes of the Wild Salt-Tolerant Tomato Species *Lycopersicon. J. Exp. Bot.* **2004,** *55,* 1105–1113.

64. Di-Baccio, D.; Navari-Izzo, F.; Izzo, R. Seawater Irrigation: Antioxidant Defence Responses in Leaves and Roots of a Sunflower (*Helianthus annuus* L.) Ecotype. *J. Plant Physiol.* **2004,** *161* (12), 1359–1366.

65. Yoshimura, K., et al. Programmed Proteome Response for Drought Avoidance/Tolerance in the Root of a C(3) Xerophyte (Wild Watermelon) Under Water Deficits. *Plant Cell Physiol.* **2008,** *49* (2), 226–241.

66. Mohsen, H., et al. Proteome Analysis of Sugar Beet Leaves under Drought Stress. *Proteomics.* **2005,** *5* (4), 950–960.

67. Demirevska, K., et al. Drought Induced Leaf Protein Alterations in Wheat Varieties. *Gen. Appl. Plant Physiol.* **2008,** *34* (1–2), 79–102.

68. Yoshimura. K., et al. Expression of Spinach Ascorbate Peroxidase Isoenzymes in Response to Oxidative Stresses. *Plant Physiol.* **2000,** *123,* 223.

69. Jangpromma, N., et al. Rapid Assessment of Chlorophyll Content in Sugarcane Using a SPAD Chlorophyll Meter Across Different Water Stress Conditions. *Asian J. Plant Sci.* **2010,** *9* (6), 368–374.

70. Hossain, Z.; Khatoon, A.; Komatsu, S. Soybean Proteomics for Unraveling Abiotic Stress Response Mechanism. *J. Proteome Res.* **2013,** *12* (11), 4670–4684.

71. Obata, T.; Fernie, A. R. The Use of Metabolomics to Dissect Plant Responses to Abiotic Stresses. *Cell Mol. Life Sci.* **2012,** *69* (19), 3225–3243.

72. Saito, R., et al. Functional Analysis of the AKR4C Subfamily Family of *Arabidopsis thaliana*: Model Structures, Substrate Specificity, Acrolein Toxicity, and Responses to Light and [CO_2]. *Biosci. Biotechnol. Biochem.* **2013,** *77,* 2038–2045.

73. Urano, K., et al. Characterization of the ABA-Regulated Global Responses to Dehydration in *Arabidopsis* by Metabolomics. *Plant J.* **2009,** *57,* 1065–1078.

74. Møller, I. M.; Jensen, P. E.; Hansson, A. Oxidative Modifications to Cellular Components in Plants. *Annu. Rev. Plant Biol.* **2007,** *58,* 459–481.

75. Achard, P., et al. The Cold-Inducible CBF1 Factor–Dependent Signaling Pathway Modulates the Accumulation of the Growth-Repressing DELLA Proteins Via its Effect on Gibberellin Metabolism. *The Plant Cell.* **2008,** *20,* 2117–2129.

76. Noctor, G.; Lelarge-Trouverie, C.; Mhamdi, A. The Metabolomics of Oxidative Stress. *Phytochemistry*. **2015**, *112*, 33–53.
77. Farmer, E. E.; Mueller, M. J. ROS-Mediated Lipid Peroxidation and RES-Activated Signaling. *Annu. Rev. Plant. Biol.* **2013**, *64*, 429–450.
78. Davletova, S. Cytosolic Ascorbate Peroxidase 1 is a Central Component of the Reactive Oxygen Gene Network of Arabidopsis. *Plant Cell*. **2005**, *17* (1), 268–281.
79. Oracz, K., et al. ROS Production and Protein Oxidation as a Novel Mechanism for Seed Dormancy Alleviation. *Plant J.* **2007**, *50* (3), 452–465.
80. Holtgrefe, S., et al. Regulation of Plant Cytosolic Glyceraldehyde 3-Phosphate Dehydrogenase Isoforms by Thiol Modifications. *Physiol. Plant*. **2008**, *133*(2), 211–228.
81. Zaffagnini, M., et al. Glutathionylation in the Photosynthetic Model Organism *Chlamydomonas reinhardtii*: A Proteomic Survey. *Mol. Cell Proteomics*. **2012**, *11* (2), M111.014142. doi: 10.1074/mcp.M111.014142.
82. Noctor, G. Metabolic Signalling in Defence and Stress: The Central Roles of Soluble Redox Couples. *Plant Cell Environ*. **2006**, *29*(3), 409–425.
83. Valderrama, R., et al. The Dehydrogenase-Mediated Recycling of NADPH is a Key Antioxidant System against Salt-Induced Oxidative Stress in Olive Plants. *Plant Cell Environ*. **2006**, *29*, 1449–1459.
84. Dizengremel, P., et al. Metabolic-Dependent Changes in Plant Cell Redox Power after Ozone Exposure. *Plant Biol*. **2009**, *11*, 35–42.
85. Davoine, C., et al. Adducts of Oxylipin Electrophiles to Glutathione Reflect a 13 Specificity of the Downstream Lipoxygenase Pathway in the Tobacco Hypersensitive Response. *Plant Physiol*. **2006**, *140* (4), 1484–1493. DOI:10.1104/pp.105.074690.
86. Taylor, N. L., et al. Lipoic Acid-Dependent Oxidative Catabolism of Alpha-Keto Acids in Mitochondria Provides Evidence for Branched-Chain Amino Acid Catabolism in *Arabidopsis*. *Plant Physiol*. **2004**, *134* (2), 838–848.
87. Mano, S., et al. Seeing is Believing: On the Use of Image Databases for Visually Exploring Plant Organelle Dynamics. *Plant Cell Physiol*. **2009**, *50* (12), 2000–2014.
88. Nakabayashi, R.; Mori, T.; Saito K. Alternation of flavonoid accumulation under drought stress in *Arabidopsis thaliana*. *Plant Signal. Behav.* **2014**, 9.
89. DellaPenna, D.; Mène-Saffrané, L. Vitamin E in "Biosynthesis of Vitamins in Plants". *Adv. Botanical Res.* **2011**, *59*, 179–227.
90. Arbona, V., et al. Metabolomics as a Tool to Investigate Abiotic Stress Tolerance in Plants. *Int. J. Mol. Sci.* **2013**, *14*, 4885–4911.
91. Arbona, V., et al. Enzymatic and Non-Enzymatic Antioxidant Responses of Carrizo Citrange, a Salt-Sensitive Citrus Rootstock, to Different Levels of Salinity. *Plant Cell Physiol*. **2003**, *44*, 388–394.
92. Hochberg. U., et al. Metabolite Profiling and Network Analysis Reveal Coordinated Changes in Grapevine Water Stress Response. *BMC Plant Biol*. **2013**, *13*, 184.
93. Alcázar, R., et al. Putrescine Accumulation Confers Drought Tolerance in Transgenic *Arabidopsis* Plants Over-Expressing the Homologous Arginine Decarboxylase 2 Gene. *Plant Physiol. Biochem*. **2010**, *48*, 547–552.
94. Christ, B., et al. Cytochrome P450 CYP89A9 is Involved in the Formation of Major Chlorophyll Catabolites during Leaf Senescence in *Arabidopsis*. *Plant Cell*. **2013**, *25*, 1868–1880.
95. Bejarano, L., et al. Glycoalkaloids in Potato Tubers: The Effect of Variety and Drought Stress on the Alpha-Solanine and Alpha-Chaconine Contents of Potatoes. *J. Sci. Food Agr.* **2000**, *80*, 2096–2100.

96. Sanchez, D. H., et al. Comparative Metabolomics of Drought Acclimation in Model and Forage Legumes. *Plant Cell Environ.* **2012**, *35*, 136–149.

97. Jaleel, C. A., et al. Induction of Drought Stress Tolerance by Ketoconazole in *Catharanthus roseus* is Mediated by Enhanced Antioxidant Potentials and Secondary Metabolite Accumulation. *Colloid Surf. B.* **2007**, *60*, 201–206.

98. Nasrollahi, V., et al. The Effect of Drought Stress on the Expression of Key Genes Involved in the Biosynthesis of Triterpenoid Saponins in Liquorice (*Glycyrrhiza glabra*). *Phytochemistry.* **2014**, *103*, 32–37.

99. Christiansen, J. L., et al. Effect of Drought Stress on Content and Composition of Seed Alkaloids in Narrow-Leafed Lupin, *Lupinus angustifolius* L. *Eur. J. Agron.* **1997**, *7*, 307–314.

100. Torras-Claveria, L., et al. Analysis of Phenolic Compounds by High-Performance Liquid Chromatography Coupled to Electrospray Ionization Tandem Mass Spectrometry in Senescent and Water-Stressed Tobacco. *Plant Sci.* **2012**, *182*, 71–78. [Online]

101. Barnaby, J. Y.; Kim, M.; Bauchan, G.; Bunce, J.; Reddy, V.; Sicher, R. C. Drought Responses of Foliar Metabolites in Three Maize Hybrids Differing in Water Stress Tolerance. *PLoS ONE.* **2013**, *8*(10), e77145. doi:10.1371/journal.pone.0077145. [Online]

102. Baker, M. The 'OMES' Puzzle. *Nature.* **2013**, *494*, 416.

103. Ma, N. L., et al. A Review of the "Omics" Approach to Biomarkers of Oxidative Stress in *Oryza sativa. Int. J. Mol. Sci.* **2013**, *14*, 7515–7541.

104. Shingaki-Wells, R. N., et al. Differential Molecular Responses of Rice and Wheat Coleoptiles to Anoxia Reveal Novel Metabolic Adaptations in Amino Acid Metabolism for Tissue Tolerance. *Plant Physiol.* **2011**, *156*, 1706–1724.

105. Casati, P., et al. Transcriptomic, Proteomic and Metabolomic Analysis of UV-B Signaling in Maize. *BMC Genomics.* **2011**, *12*, 321.

106. Edwards, D.; Batley, J. Plant Bioinformatics: From Genome to Phenome. *Trends Biotech.* **2004**, *22* (5), 232–237.

107. Moreau, Y., et al. Comparison and Meta-Analysis of Microarray Data: From the Bench to the Computer Desk. *Trends Genet.* **2003**, *19*, 570–577.

108. Barah, P., et al. Molecular Signatures in *Arabidopsis thaliana* in Response to Insect Attack and Bacterial Infection. *PLoS ONE.* **2013**, *8* (3), e58987. [Online] doi:10.1371/journal.pone.0058987.

109. Konika Chawla, P. B., et al. Systems Biology: A Promising Tool to Study Abiotic Stress Responses. In *Omics and Plant Abiotic Stress Tolerance;* Bentham Publishers: Houston, TX, 2010.

110. Cramer, G. R., et al. Effects of Abiotic Stress on Plants: A Systems Biology Perspective. *BMC Plant Biol.* **2011**, *11*, 163.

111. Gonzalez, N., et al. Increased Leaf Size: Different Means to an End. *Plant Physiol.* **2010**, *153*, 1261–1279.

112. Shaik, R.; Ramakrishna, W. Machine Learning Approaches Distinguish Multiple Stress Conditions Using Stress-Responsive Genes and Identify Candidate Genes for Broad Resistance in Rice. *Plant Physiol.* **2014**, *164*, 481–495.

113. Ritchie, M. D., et al. Methods of Integrating Data to Uncover Genotype–Phenotype Interactions. *Nat. Rev. Genet.* **2015**, *16*, 85–97.

114. Liu, C., et al. *OsbZIP71*, a bZIP transcription factor, confers salinity and drought tolerance in rice. *Plant Mol. Biol.* **2014**, *84(1–2)*, 19–36.

115. Xiong, L., et al. ABA-Independent *OsPP18* Regulates Stress Tolerance in Rice. *Plant Physiol.* **2014**, DOI:10.1104/pp.114.251116.

116. Fuchs, S., et al. Abscisic Acid Sensor *RCAR7/PYL13*, Specific Regulator of Protein Phosphatase Coreceptors. *Proc. Natl. Acad. Sci. USA.* **2014,** *111,* 5741–5746.
117. Babitha, K. C., et al. Co- Expression of *AtbHLH17* and *AtWRKY28* Confers Resistance to Abiotic Stress in *Arabidopsis. Transgenic Res.* **2012,** *22,* 327–341.
118. Pruthvi, V.; Narasimhan, R.; Nataraja, K. N. Simultaneous Expression of Abiotic Stress Responsive Transcription Factors, *AtDREB2A, AtHB7* and *AtABF3* Improves Salinity and Drought Tolerance in Peanut (*Arachis hypogaea* L.). *PLoS ONE.* **2014,** *9* (12), e111152. [Online]
119. Sunil, M., et al. The Draft Genome and Transcriptome of *Amaranthus hypochondriacus*: A C4 Dicot Producing High-Lysine Edible Pseudo-Cereal. *DNA Res.* **2014,** *21,* 585–602.
120. The Arabidopsis Genome Initiative. Analysis of the Genome Sequence of the Flowering Plant *Arabidopsis thaliana. Nature.* **2000,** *408,* 796–815.
121. Wang, X., et al. The Genome of the Mesopolyploid Crop Species *Brassica rapa. Nat. Genet.* **2011,** *43,* 1035–1039.
122. Singh, N. K., et al. The First Draft of the Pigeonpea Genome Sequence. *J. Plant Biochem. Biotechnol.* **2012,** *21* (1), 98–112.
123. Ming, R., et al. The Draft Genome of the Transgenic Tropical Fruit Tree Papaya (*Carica papaya* Linnaeus). *Nature.* **2008,** *452* (7190), 991–996.
124. Jain, M., et al. A Draft Genome Sequence of the Pulse Crop Chickpea (*Cicer arietinum* L.). *Plant J.* **2013,** *74,* 715–729.
125. Xu, Q., et al. The Draft Genome of Sweet Orange (*Citrus sinensis*). *Nat. Genet.* **2013,** *45,* 59–67.
126. Garcia-Mas, J., et al. The Genome of Melon (*Cucumis melo* L.). *Proc. Natl. Acad. Sci. USA.* **2012,** *109,* 11872–11877.
127. Hirakawa, H., et al. Dissection of the Octoploid Strawberry Genome by Deep Sequencing of the Genomes of *Fragaria* Species. *DNA Res.* **2014,** *21,* 169–181.
128. Schmutz, J., et al. Genome Sequence of the Palaeopolyploid Soybean. *Nature.* **2010,** *463,* 178–183.
129. Wang, K., et al. The Draft Genome of a Diploid Cotton *Gossypium raimondii. Nat. Genet.* **2012,** *44,* 1098–1103.
130. Mayer, K. F. X., et al., A Physical, Genetic and Functional Sequence Assembly of the Barley Genome. *Nature.* **2012,** *491,* 711–716.
131. Velasco, R., et al. The Genome of the Domesticated Apple (*Malus* × *domestica* Borkh.). *Nat. Genet.* **2010,** *42,* 832 – 839.
132. Bancroft, I., et al. Dissecting the Genome of the Polyploid Crop Oilseed Rape by Transcriptome Sequencing. *Nat. Biotechnol.* **2011,** *29,* 762–766.
133. Young, N. D., et al. The *Medicago* Genome Provides Insight into the Evolution of Rhizobial Symbioses. *Nature.* *480* (7378), 520–524. doi:10.1038/nature10625.
134. He, N., et al. Draft Genome Sequence of the Mulberry Tree *Morus notabilis. Nat. Commun.* **2013,** *4,* 2445.
135. D'Hont, A., et al. The Banana (*Musa acuminata*) Genome and the Evolution of Monocotyledonous Plants. *Nature.* **2012,** *488,* 213–217.
136. Sierro, N., et al. The Tobacco Genome Sequence and its Comparison with those of Tomato and Potato. *Nat. Commun.* **2014,** *5,* 3833.
137. Takeshi, I., et al. The Rice Annotation Project. Curated Genome Annotation of *Oryza sativa* ssp. *japonica* and Comparative Genome Analysis with *Arabidopsis thaliana. Genome Res.* **2007,** *17,* 175–183.

138. Al-Dous, E., et al. *De Novo* Genome Sequencing and Comparative Genomics of Date Palm (*Phoenix dactylifera*). *Nat. Biotechnol.* **2011,** *29,* 521–527.

139. Peng, Z., et al. The Draft Genome of the Fast-Growing Non-Timber Forest Species Moso Bamboo (*Phyllostachys heterocycla*). *Nat. Genet.* **2013,** *45,* 456–461.

140. Tuskan, G. A., et al. The Genome of Black Cottonwood, *Populus trichocarpa* (Torr. & Gray). *Science.* **2006,** *313,* 1596–1604.

141. Ignazio, V., et al. The International Peach Genome Initiative. The High-Quality Draft Genome of Peach (*Prunus persica*) Identifies Unique Patterns of Genetic Diversity, Domestication and Genome Evolution. *Nat. Genet.* **2013,** *45,* 487–494.

142. Bennetzen, J. L., et al. Reference Genome Sequence of the Model Plant *Setaria. Nat. Biotechnol.* **2012,** *30,* 555–561.

143. Sato, S., et al. The Tomato Genome Sequence Provides Insights into Fleshy Fruit Evolution. *Nature.* **2012,** *485,* 635–641.

144. Amoros, W., et al. The Potato Genome Sequencing Consortium. Genome Sequence and Analysis of the Tuber Crop Potato. *Nature.* **2011,** *475,* 189–195.

145. Paterson, A. H., et al. The *Sorghum bicolor* Genome and the Diversifi Cation of Grasses. *Nature.* **2009,** *457,* 551–556.

146. Argout et al., The Genome of *Theobroma cacao. Nat. Genet.* **2011,** *43,* 101–109.

147. Brenchley, R., et al. Analysis of the Bread Wheat Genome Using Whole-Genome Shotgun Sequencing. *Nature.* **2012,** *491,* 705–710.

148. Jaillon, O., et al. The Grapevine Genome Sequence Suggests Ancestral Hexaploidization in Major Angiosperm Phyla. *Nature.* **2007,** *449,* 463–467.

149. Schnable, P. S., et al. The B73 Maize Genome: Complexity, Diversity, and Dynamics. *Science.* **2009,** *326,* 1112–1115.

150. Nouri, M. Z.; Komatsu, S. Comparative Analysis of Soybean Plasma Membrane Proteins Under Osmotic Stress Using Gel-Based And LC MS/MS-Based Proteomics Approaches. *Proteomics.* **2010,** *10* (10), 1930−1945.

151. Deeba, F., et al. Physiological and Proteomic Responses of Cotton (*Gossypium herbaceum* L.) to Drought Stress. *Plant Physiol. Biochem.* **2012,** *53,* 6–18.

CHAPTER 10

SALINITY STRESS: "OMICS" APPROACHES

NISHA KHATRI and YASHWANTI MUDGIL*

Department of Botany, University of Delhi, New Delhi, India

Corresponding author. E-mail: ymudgil@gmail.com

CONTENTS

ABSTRACT

Salinity stress is one of the major threats to plant performance which leads to the significant loss of the crop production worldwide. Being polygenic in nature, salinity stress induced changes are complex to understand. On exposure to salinity stress, plants undergo various molecular, physiological, and biochemical changes to combat the stress response. In general, salinity stress affects ion transport, synthesis of reactive oxygen species (ROS), solute accumulation and leads to the hormonal changes as well. These changes altogether have toxic and inhibitory effect on cell metabolism, photosynthetic apparatus, membrane structure, cell division, and growth. During the past two decades, various "omics" approaches have been used for the comprehensive understanding of the salinity stress signaling and tolerance mechanism in plants. The system biology approach using transcriptomics, proteomics, and metabolomics have discovered that the salinity overly sensitive (SOS) signaling is an early key pathway which helps to maintain ion homeostasis in plants during salinity stress. Large numbers of salinity-responsive genes and transcription factors have been identified and characterized using transcriptomic, genomic approaches and confirmed by proteomics analysis. This chapter highlights the recent advances made to understand salinity stress using various "omics" tools.

10.1 INTRODUCTION

According to Food and Agricultural Organization land and plant nutrition management services, out of the current 230 million hectare of irrigated land, near about 45 million hectares land is affected by salt stress worldwide and further exceeding due to improper management and drainage system. Salt stress affects every stage of plant development starting from the seed germination to reproductive development. Based on the ability of salt stress tolerance, plants can be divided into halophytes and glycophytes. Halophytes grows naturally in environments with high salt concentrations like maritime estuaries and salt lakes while glycophytes are salt stress susceptible and get damaged easily by high salinity. Unfortunately, most of the crops comes under the category of glycophytes and thus cannot tolerate the elevated salt concentrations leading to the decline in their productivity.[1,2] Therefore, understanding the science behind salinity stress is utmost important in order to develop the salt-tolerant crops. However, conventional breeding methods achieved limited success due to complex network of the process. In general, salt stress leads to various physiological, metabolic, and biochemical changes in the plant cell. Salt stress induced changes include alterations in the membrane permeability, nutrient imbalance, reduced photosynthesis, and decrease in the stomatal aperture.[3,4] In addition by unknown mechanisms salinity stress also leads to accumulation of reactive oxygen species (ROS). These ROS include singlet oxygen, superoxide, hydroxyl radical, and hydrogen peroxide, all are enhanced during stress.[5,6] Higher concentrations of ROS are lethal for the cell as it can lead to oxidative damage to various biomolecules including proteins, lipids, and DNA, resulting in an impairment of cellular processes. Salinity stress also results in ionic imbalance in the plants. During normal conditions, ion mobilization include ion uptake from the soil, their transportation from roots to shoots, and finally to the leaves[7-9]. However, during salt stress, this ion cycle is disturbed, leading to the accumulation of ions inside the cells. Therefore, maintenance of cellular Na^+/K^+ homeostasis is one of the primary tasks during salt stress.

Unfortunately, not much information is available on how Na^+ is sensed in any cellular system. Elevated salt concentrations are perceived by the receptors present on the cellular membranes.[10] In general, ion transporters are supposed to be the sensors for salt stress which activate the downstream signaling by generation of many secondary signal molecules.

Salinity overly sensitive (SOS) signaling pathway is the most studied pathway which is activated in response to salinity stress. This pathway includes SOS1 (putative plasma membrane Na^+/H^+ antiporter),[11] SOS2 (a serine/threonine kinase),[12] and SOS3 (myristoylated Ca^+ binding protein)[13,14] which mediate the cellular signaling under salt stress, and maintain the ion homeostasis.[15,16] Increased concentrations of Na^+ is lethal for the plants as it perturbs cytoplasmic K^+/Na^+ homeostasis. SOS signaling starts by sensing increased concentration of Na^+ thereby increasing intracellular Ca^{2+} level. SOS3 binds to Ca^{2+} and further activates SOS2 protein and forms a SOS2/SOS3 protein complex. This activated complex act as kinase and phosphorylates SOS1. Phosphorylated SOS1 reduces Na^+ toxicity by efflux. Other strategies utilized by plants to cope up with stress are reducing Na^+ influx and vacuolar compartmentation of Na^+.[17,18]

Plasma membrane localized transporters belong to the histidine kinase transporter (HKT) family, which are involved in transportation of Na^+ and K^+ across the membrane.[19] Other intracellular transporters of Na^+ and K^+ include NHX proteins which are Na^+, K^+/H^+ antiporters, localized at tonoplast, involved in K^+ homeostasis, endosomal pH regulation, and salt tolerance.[20]

Plants also encounter series of hormonal changes during salt stress. Expression of salt-responsive genes is modulated by abscisic acid (ABA). In *Hordeum vulgare* expression of HVP1 and HVP10, for vacuolar H^+-inorganic pyrophosphatase, and HvVHA-A, for catalytic subunit (subunit A) of vacuolar H^+-ATPase are regulated by ABA.[21] *Arabidopsis thaliana* seedlings pretreated with salicylic acid (SA) showed improved K^+ retention during salt stress due to up regulation of H^+-ATPase activity. SA also restores membrane potential via gated outwardly-rectifying K^+ (GORK) channel.[22] Biosynthesis of SA is stimulated by gibberellic acid (GA) via inducing the isochorismate synthase 1 (*ICS1*) also known as *SID2* (synthesize SA from chorismate) gene under high salinity. Involvement of SA is known in ROS mediated damage during stress, exogenous application of SA in *sid2* mutant seeds showed reduced levels of H_2O_2 depicting its involvement in ROS mediated damage during stress.[23]

In addition to hormonal changes, transporters, several other proteins and metabolites play crucial roles salt stress tolerance in plants. In order to identify those proteins and metabolites, several lines of studies have been conducted in past three decades using various, genomics, transcriptomics, proteomics, and metabolomics approach, which no-doubt increased our current understanding of how plant respond to the salinity stress. In this chapter, we are briefly discussing those studies in order to present the current scenario of salinity stress tolerance in plants.

10.2   OMICS FOR SALINITY STRESS

10.2.1   GENOMICS AND TRANSCRIPTOMICS

Genomic techniques such as high-throughput analyses of expressed sequence tags (ESTs), large-scale parallel analysis of gene expression; molecular markers have greatly increased our current understanding on plant responses to salinity stress. These genomics-based approaches aim to identify the salt responsive genes in order to provide a complete picture of salt signaling in plants.

ESTs are fragments of mRNA sequences derived through single sequencing reactions performed on randomly selected clones from cDNA libraries. In fields such as phylogenetics, transcript profiling and proteomics, ESTs have proven to be an efficient and rapid means to identify novel genes.[24-26] In recent example, suppression subtractive hybridization was used in sugar beet monosomic addition line M14 (unique germplasm that contains genetic materials from *Beta vulgaris* L. and *Beta corolliflora* Z.); cDNA libraries of differentially expressed genes were constructed and 58 unigenes including 14 singletons and 44 contigs were obtained under salt stress.[27]

Understanding of genetic diversity at the molecular level plays important role in crop improvement during stress. Various molecular markers were used to study genetic variability during salinity stress. Candidate gene based simple sequence repeats markers (cgSSRs) were developed in rice, which can be utilized as novel candidate for diversity analysis among rice genotypes differing in salinity response. Study showed that tri-nucleotide repeats (56.11%) are more abundant than di- (42.11%) and tetra- (2.8%) nucleotide repeats.[28] To test polymorphism between two contrasting genotypes of *Medicago truncatula* (Tru 131, tolerant genotype and Jemalong, sensitive genotype), two expressed sequence tags simple sequence repeats (EST–SSRs) markers (MTIC 044 and MTIC 124) were used.[29] There is a advantage of EST–SSR primers over genomic SSRs as EST–SSR are less polymorphic than genomic SSRs in crop plants because of greater DNA sequence conservation in transcribed regions.[30] The results obtained showed that these two EST–SSRs loci MTIC 124 and MTIC 044 encode cysteine proteinase inhibitor and GATA transcription factor, respectively. These data suggest that these two loci are involved in salt stress tolerance and are appropriate to use in salinity responses.[29]

Transcriptome profiles of *B. napus* cv. canola were explored using digital gene expression (DGE). DGE is a next generation sequencing (NGS) based technique, 163 genes were identified to be differentially expressed in response to salt stress. Gene ontology (GO) and Kyoto encyclopedia of genes and genomes (KEGG) enrichment analyses fished out genes involved in osmotic stress, oxidative stress, and cell wall damage caused by salt stress. These candidate genes can be used to enhance salt tolerance.[31]

In the absence of complete genome sequence availability for all the crop plants, RNA sequencing (RNA-Seq) is a very successful application of NGS technologies for quantitative analysis of gene expression.[32] RNA-Seq technology was implemented to get insights of salt stress transcriptome in a salt-tolerant variety CS52 of *Brassica juncea,* at seedling stage. RNA-Seq libraries were obtained under normal and in response to 24 h of salinity stress conditions. A total of 42,327 high-confidence unigenes were obtained. Only 47.4% of these are represented by *Brassica* ESTs in public databases indicating that 52.6% of the unigenes, predicted in this study, are novel compared to EST sequence available for *B. juncea* and constituent genomes (*Brassica rapa* and *Brassica nigra*).[33]

Similar RNA-Seq analysis was performed on *Hordeum spontaneum* (wild salt-tolerant barley) in order to discover novel genes and transcription factors. Transcriptome sequencing yielded 103–115 million reads for all replicates of each treatment, corresponding to over 10 billion nucleotides per sample. Of the total reads, between 74.8 and 80.3% could be mapped and 77.4–81.7% of the transcripts were found in the *H. vulgare* (cultivated barley with well-studied genome) unigene database (unigene-mapped). The unmapped wild barley reads for all treatments and replicates were assembled *de novo* and the resulting contigs were used as a new reference genome. This resulted in 94.3–95.3% of the unmapped reads mapping to the new reference. The number of differentially expressed transcripts was 9277, 3861 of which were unigene-mapped. The annotated unigene- and *de novo*-mapped transcripts (5100) were utilized to generate expression clusters across time of salt stress treatment.[34] These RNA-Seq methods generate high-resolution transcriptome maps sensitive enough to display transcripts with low-levels of expression.[35]

In *Oryza sativa* var. *japonica*, linkage mapping and association mapping were used to identify the quantitative trait locus (QTLs) for Na^+ and K^+ concentration in shoots and roots, which can be used for marker-assisted selection in rice breeding programs and may accelerate the development of salt-tolerant rice varieties.[36]

These transcriptomic datasets will help to elucidate the genetic basis of the plant species response to salt stress as in response to external stimuli plant continuously needs to rearrange their transcriptome profile. In-depth understanding of transcriptomic datasets will help to develop stress-tolerant crops based on favorable genetic resources.

10.2.2 PROTEOMICS

Analysis of plant proteome has led to the identification of several hundreds of salt responsive proteins in shoots, leaves, roots, seedlings, radicles, hypocotyls, grains, gametophytes, and unicells.[37] For analyzing the plant proteome in response to salt stress, mostly a gel-based methodology including two-dimensional gel electrophoresis (2DE) and mass spectrometry (MS) has been used. A two-dimensional difference gel electrophoresis (2D DIGE) technology was used to investigate the salt responsive proteins in the model plant *A. thaliana* (a glycophyte) and its wild relative, *Thellungiella salsuginea* (a halophyte), both contain remarkably similar genome, untreated plants showed notable differences in protein abundance as compared to salt treated, showing how species maintain cellular processes in spite of similar genetic makeup.[38] Using a 2DE-MS approach 59 and 104 salt responsive proteins were identified in rice leaves and roots, respectively.[39] In *H. vulgare*, root proteome of salt-sensitive (DH14) and tolerant (DH187) lines were investigated to understand the mechanism of salt tolerance. Matrix assisted laser desorption/ionization time-of-flight (MALDI-ToF/ToF) mass spectrometry showed that they respond differently in stress.[40]

More recently, a combination of both gel-based (2DE) and gel-free proteomics isobaric tags for relative and absolute quantitation (iTRAQ) approach was utilized to identify the salt responsive proteins from the salt tolerant sugar beet monosomic addition M14 line. Out of the total 274 identified proteins, 50 expressed differentially, among them 40 were up-regulated while 10 were down regulated following salinity stress.[41] In another similar study, DIGE and iTRAQ technologies were used to identify salt-responsive proteins while gas chromatography–mass spectrometry (GC–MS) analyses were carried out to identify differentially accumulated metabolites in rice suspension cultures.[42] Majorly carbohydrate and energy metabolism related proteins were identified as salt-responsive proteins while metabolome analysis led to the identification of 134 known metabolites. Out of these identified metabolites, 30 included amines and amides, 40 were organic acids, 40 were sugars, sugar acids, and sugar alcohols, 21 were fatty acids and sterols, and 3 were others. In particular, glucose, which might be acting as an osmoprotectant and a carbon source under salt stress, showed accumulation over time.

Comparative proteomic analyses of seedling leaves and roots were done in salt-sensitive genotype Jackson and salt-tolerant genotype Lee 68 of *Glycine max*. Out of the total 800 spots detected on 2-DE gels of leaf proteins, of which 78 differentially expressed proteins were successfully identified by MALDI-ToF-ToF.[43] In case of root, 68 differentially expressed proteins identified by 2-DE coupled with MS/MS. Functional categorization of the identified proteins showed that these were involved in diverse metabolic pathways and cellular processes.[44]

Capriotti et al. used a label-free shotgun proteomics approach to identify the salt responsive proteins in *Triticum durum* cv. Duilio (salt and drought tolerant). A total of 83 proteins showed significant change during high salt treatment, which are involved in primary metabolism, protein metabolism, and cellular defense mechanisms.[45]

Protein–protein interactions (PPIs) play an important role in stress responses. Since last decade, bimolecular fluorescence complementation (BiFC) has emerged as a key technique to visualize PPIs. Recently a nuclear-localized calcium-binding protein, short root in salt medium 1 (RSA1), required for salt tolerance is characterized and its interaction with RSA1 interacting transcription factor 1 (RITF1), a *basic helix-loop-helix* (bHLH) transcription factor is confirmed by BiFC and co-immunoprecipitation. Both of them regulate the *SOS1* and transcription of several genes involved in the detoxification of ROS generated by salt stress.[46] BiFC is also used in *B. napus* to confirm the interaction of *B. napus* CBL-interacting protein kinase (*BnCIPK6*) and calcineurin B-Like 1 (*BnCBL1*). Expressions of *BnCIPK6* and *BnCBL1* were significantly up-regulated by salt and transgenic plants over expressing *BnCIPK6* shows enhanced salt tolerance.[47]

Since proteins are the final product of central dogma and they are important components of various biochemical pathways. They play essential role in revealing molecular mechanisms behind any cellular process. The plant proteome is highly complex and dynamic and it needs information at genomic and metabolomic level to complete the picture.

10.2.3   METABOLOMICS

Metabolomics is a rapidly developing technology which involves quantitative estimation of all the low molecular weight compounds present in a cell or organism in a particular physiological or developmental state, as originally defined by Oliver et al.[48] Major approaches currently used in plant metabolomics research include nuclear magnetic resonance (NMR) spectroscopy, GC–MS, liquid chromatography–mass spectrometry (LC–MS), capillary electrophoresis–mass spectrometry (CE–MS) and Fourier transform ion cyclotron resonance mass spectrometry or Fourier transform infrared (FT-IR) spectroscopy.

Several metabolomics studies on salt-stressed plants have been reported. NMR is considered as a powerful and simple tool as it allows structure elucidation and detailed analysis of the biomolecular composition of a plant extract with relatively simple sample preparation.[49,50] NMR spectroscopy was performed separately on shoot and root extracts in *Zea mays* during salt stress. Partial least squares-discriminant analysis (PLS-DA) showed that salt stress initiates a clear and progressive metabolic response that was more evident in shoot as compared to root extracts. Moreover, metabolites like glycine-betaine, sucrose and asparagine (increased levels observed in shoot extracts), and *γ*-amino-*N*-butyric acid, malic acid, aspartic acid, and *trans*-aconitic acid (increased levels observed in root extracts) can be used as possible biomarkers for osmotic adjustment during high salinity environment.[51] However, due to poor sensitivity and poor dynamic range relative to

MS, NMR-based plant metabolomics can only detect and quantify the most-abundant metabolites; thereby, NMR is most widely used as a metabolic fingerprinting technique. Metabolic fingerprinting of salt stress in *Solanum lycopersicum* was used to identify metabolic changes in fruits under salinity stress.[52]

In T87 cell culture of *A. thaliana*, time-course metabolic profiling during salt-stress was performed. An analysis of primary metabolites was performed by GC/MS and LC/MS. The mass chromatographic data were converted into matrix data sets, which were subjected to data mining processes, including principal component analysis (PCA), a statistical analysis and batch learning self-organizing mapping analysis (BLSOM); which predicts functions on the basis of similarity in oligopeptide composition of proteins.[53] The mining results suggest that the phenylpropanoid pathway for lignin production, methylation cycle for the supply of methyl groups and glycine betaine biosynthesis are synergetically induced as a short-term response against salt-stress treatment.[54] GC–MS and microarrays were used to analyze *A. thaliana* and *T halophila*, a closely related halophytic species. Metabolome analyses revealed drastically different profiles between the two species. *T. halophila* maintained higher levels of metabolites like, such as fructose, sucrose, complex sugars, malate, and proline in both the absence and presence of salt stress as compared to *A. thaliana*.[55] In two cultivars of *H. vulgare* L. cv. Clipper and cv. Sahara, total of 72 compounds of known structure, comprising 23 amino acids, 20 organic acids, and 29 sugars, were identified via GC–MS in leaf extracts in salt stress. A number of metabolites (30) were also found that could neither be identified nor be classified using either commercial libraries. By contrast, in Sahara plants, the levels of the *tricarboxylic acid* (TCA) cycle intermediates, hexose phosphates, and metabolites involved in cellular protection increased in response to salt whereas these solutes remain unchanged in the more sensitive Clipper plants explaining why Sahara is more tolerant.[56]

In *Zea mays*, by comparing a salt-sensitive and a salt-resistant hybrid, it was identified that sugars such as glucose, fructose, and sucrose get accumulated in leaves as a salt-resistance adaption. GC-(TOF)-MS was used and interestingly a consistent reduction in the concentrations of the metabolites of the TCA cycle with increasing salt stress in both hybrids was observed.[57] TCA cycle intermediates should increase during stress in resistant hybrid to help plant to improve its growth as observed in barley.[56]

One major limitation of GC–MS technology is that it can only be used for analysis of volatile and thermally stable metabolites or metabolites that can be chemically modified to produce volatile derivatives. GC–MS is not suitable for analysis of thermolabile di- and triphosphates, or larger oligosaccharides due to their limited volatility, even after derivatization.[58,59] LC–MS is the most-important complementary technology to GC–MS, and different LC–MS-based methods have been used in current plant metabolomics studies to profile specific compound classes of the plant metabolome (also known as target metabolite analysis). LC–MS methods have an advantage over GC–MS methods; they can be used to analyze thermolabile, polar metabolites, and high-molecular weight compounds.

Metabolic changes of *G. max* L. Merr (cultivated soybean C08) and *Glycine soja* Sieb. et Zucc. (wild soybean W05) under salt stress were profiled using GC–MS and LC-FT/MS from extracts of soybean seedling leaves. It was observed that wild is more tolerant than cultivated contained higher amounts of disaccharides, sugar alcohols, and acetylated amino acids than cultivated soybean, but with lower amounts of monosaccharides, carboxylic acids, and unsaturated fatty acids.[60]

In *O. sativa* var. *japonica* cv. nipponbare, comparison of 2-DE protein profiles between the control and salt-stressed third leaves revealed 55 differentially expressed spots, where 47 spots were up regulated as compared to control. Among these spots, 33 protein spots (27 increased and 5 decreased) were determined by nano-electrospray ionization liquid chromatography-tandem mass spectrometry (nESI-LC–MS/MS). Most of the identified proteins falls into categories including major metabolic processes like photosynthetic carbon dioxide assimilation and photorespiration probably responsible for wilting and browning of leaf.[61]

Detoxification of ROS which is generated during stress is an important event. Ascorbate peroxidases (APXs) act as scavenger on H_2O_2 as it converts it into H_2O and O_2.[62] By using an in-gel APX enzyme assay strong induction of APX enzyme activity in the shoot of IR-29 after salt treatment was observed. The APX enzyme activity coincided with a significant decrease in H_2O_2 levels, as well as an increase in oxidized ascorbic acid content, in salt-stressed IR-29 seedlings.[63] Eight APX isozymes have been annotated in the rice genome (*O. sativa* L. cv. "Nipponbare").[64] LC–MS/MS identifies seven candidates in IR-29 (two APXs (Accession Nos. NP_001049769 and NP_001060741) and five antioxidant-related proteins).[65] These can be exploited to detoxify ROS in rice.

To understand the mechanism of salt tolerance in C_4 halophyte plant, *A. lagopoides*, combination of proteomic (2-DE) and metabolomic (CE–MS) approach revealed that TCA cycle related metabolites were down regulated whereas metabolism related proteins were up-regulated in salt-stressed shoots[66] to combat stress.

10.3 CONCLUSION AND FUTURE PERSPECTIVES

Due to advancements in the technology, various precise and more sensitive techniques and instruments have been developed for the analysis of genes, proteins, and metabolites. These newly developed techniques have advanced our understanding of how plants respond to salinity stress. Although genomics, transcriptomics, proteomics, and metabolomics studies results in the identification of several thousands of putative salt responsive candidates, but the information obtained from these studies are scattered and needs to be compiled and validated. Moreover, genomics, transcriptomic, proteomic, or metabolomics cannot provide us a global view of the salinity stress tolerance, so a focused system biology approach should provide a holistic view. Furthermore, more work needs to be done to identify and characterize the transmembrane ion transport mechanisms, sensors, and receptors in the signal transduction during stress. As recent advancements in "omic" technologies now offer opportunities to overcome the hurdle of polygenic nature of salt stress and to identify candidate genes which would help in improving the global food security. Figure 10.1 shows different omics approaches that can be used for analyzing effect of salt stress on plants.

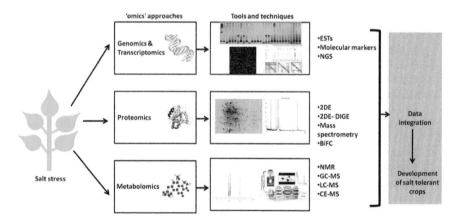

FIGURE 10.1 Schematic diagram showing different omics approaches used for analyzing effect of salt stress on plants. For genomics and transcriptomics, EST and NGS are being used while for analyzing the salt stress responsive proteins, 2DE, 2DE-DIGE, MS, and BiFC are used. For metabolites, NMR, GC–MS, LC–MS, and CE–MS are routinely used. Data integrated from all these omics approaches is integrated and used for understanding mechanism of salt stress tolerance.

KEYWORDS

- salinity
- signaling
- genomics
- proteomics
- metabolomics

REFERENCES

1. Greenway, H.; Munns, R. Mechanisms of Salt Tolerance in Nonhalophytes. *Annu. Rev. Plant Physiol.* **1980,** *31,* 149–190.
2. Flowers, T. J.; Colmer, T. D. Salinity Tolerance in Halophytes. *New Phytol.* **2008,** *179,* 945–963.
3. Munns, R.; Tester, M. Mechanisms of Salinity Tolerance. *Annu. Rev. Plant Biol.* **2008,** *59,* 651–681.
4. Rahnama, A.; James, R. A.; Poustini, K.; Munns, R. Stomatal Conductance as a Screen for Osmotic Stress Tolerance in Durum Wheat Growing in Saline Soil. *Funct. Plant Biol.* **2010,** *37* (3), 255–263.
5. Apel, K.; Hirt, H. Reactive Oxygen Species: Metabolism, Oxidative Stress, and Signal Transduction. *Annu. Rev. Plant Biol.* **2004,** *55,* 373–399.
6. Mittler, R. Oxidative Stress, Antioxidants and Stress Tolerance. *Trends Plant Sci.* **2002,** *7,* 405–410.
7. Munns, R.; Termaat, A. Whole-Plant Responses to Salinity. *Aust. J. Plant Physiol.* **1986,** *13,* 143.
8. Yeo, A. Molecular Biology of Salt Tolerance in the Context of Whole-Plant Physiology. *J. Exp. Bot.* **1998,** *49* (323), 915–929.
9. Tester, M.; Davenport, R. Na+ Tolerance and Na+ Transport in Higher Plants. *Ann. Bot.* **2003,** *91,* 503–527.
10. Xiong, L.; Schumaker, K. S.; Zhu, J. Cell Signaling during Cold, Drought, and Salt Stress. *Plant Cell.* **2002,** *14 supplement,* S165–S184.
11. Shi, H.; Ishitani, M.; Kim, C.; Zhu, J. K. The Arabidopsis Thaliana Salt Tolerance Gene SOS1 Encodes a Putative Na+/H+ Antiporter. *Proc. Natl. Acad. Sci. U. S. A.* **2000,** *97* (12), 6896–6901.
12. Guo, Y.; Qiu, Q.-S.; Quintero, F. J.; Pardo, J. M.; Ohta, M.; Zhang, C.; Schumaker, K. S.; Zhu, J.-K. Transgenic Evaluation of Activated Mutant Alleles of SOS2 Reveals a Critical Requirement for its Kinase Activity and C-Terminal Regulatory Domain for Ssalt Tolerance in Arabidopsis Thaliana. *Plant Cell.* **2004,** *16* (2), 435–449.
13. Ishitani, M.; Liu, J.; Halfter, U.; Kim, C. S.; Shi, W.; Zhu, J. K. SOS3 Function in Plant Salt Tolerance Requires N-myristoylation and Calcium Binding. *Plant Cell.* **2000,** *12* (9), 1667–1678.

14. Liu, J.; Zhu, J. K. A Calcium Sensor Homolog Required for Plant Salt Tolerance. *Science.* **1998,** *280* (5371), 1943–1945.
15. Zhu, J. K.; Liu, J.; Xiong, L. Genetic Analysis of Salt Tolerance in Arabidopsis. Evidence for a Critical Role of Potassium Nutrition. *Plant Cell.* **1998,** *10* (7), 1181–1191.
16. Halfter, U.; Ishitani, M.; Zhu, J.-K. The Arabidopsis SOS2 Protein Kinase Physically Interacts with and is Activated by the Calcium-Binding Protein SOS3. *Proc. Natl. Acad. Sci. U. S. A.* **2000,** *97* (7), 3735–3740.
17. Ward, J. M.; Hirschi, K. D.; Sze, H. Plants Pass the Salt. *Trends Plant Sci.* **2003,** *8,* 200–201.
18. Liu, J.; Zhu, J. K. An Arabidopsis Mutant that Requires Increased Calcium for Potassium Nutrition and Salt Tolerance. *Proc. Natl. Acad. Sci. U. S. A.* **1997,** *94* (26), 14960–14964.
19. Horie, T.; Hauser, F.; Schroeder, J. I. HKT Transporter-Mediated Salinity Resistance Mechanisms in Arabidopsis and Monocot Crop Plants. *Trends Plant Sci.* **2009,** *14,* 660–668.
20. Barragán, V.; Leidi, E. O.; Andrés, Z.; Rubio, L.; De Luca, A.; Fernández, J. A.; Cubero, B.; Pardo, J. M. Ion Exchangers NHX1 and NHX2 Mediate Active Potassium Uptake into Vacuoles to Regulate Cell Turgor and Stomatal Function in Arabidopsis. *Plant Cell.* **2012,** *24* (3), 1127–1142.
21. Fukuda, A.; Tanaka, Y. Effects of ABA, Auxin, and Gibberellin on the Expression of Genes for Vacuolar H+-Inorganic Pyrophosphatase, H+-ATPase Subunit A, and Na+/H+ Antiporter in Barley. *Plant Physiol. Biochem.* **2006,** *44* (5), 351–358.
22. Jayakannan, M.; Bose, J.; Babourina, O.; Rengel, Z.; Shabala, S. Salicylic Acid Improves Salinity Tolerance in Arabidopsis by Restoring Membrane Potential and Preventing Salt-Induced K+ Loss via a GORK Channel. *J. Exp. Bot.* **2013,** *64,* 2255–2268.
23. Lee, S.; Park, C.-M. Modulation of Reactive Oxygen Species by Salicylic Acid in Arabidopsis Seed Germination under High Salinity. *Plant Signal. Behav.* **2010,** *5* (12), 1534–1536.
24. Bouchez, D.; Höfte, H. Functional Genomics in Plants. *Plant Physiol.* **1998,** *118* (3), 725–732.
25. Alba, R.; Fei, Z.; Payton, P.; Liu, Y.; Moore, S. L.; Debbie, P.; Cohn, J.; D'Ascenzo, M.; Gordon, J. S.; Rose, J. K.; Martin, G.; Tanksley, S. D.; Bouzayen, M.; Jahn, M. M.; Giovannoni, J. ESTs, cDNA Microarrays, and Gene Expression Profiling: Tools for Dissecting Plant Physiology and Development. *Plant J.* **2004,** *39,* 697–714.
26. Parkinson, J.; Blaxter, M. Expressed Sequence Tags: An Overview. *Methods Mol. Biol.* **2009,** *533,* 1–12.
27. Yang, L.; Ma, C.; Wang, L.; Chen, S.; Li, H. Salt Stress Induced Proteome and Transcriptome Changes in Sugar Beet Monosomic Addition Line M14. *J. Plant Physiol.* **2012,** *169* (9), 839–850.
28. Molla, K. A.; Debnath, A. B.; Ganie, S. A.; Mondal, T. K. Identification and Analysis of Novel Salt Responsive Candidate Gene Based SSRs (cgSSRs) from Rice (*Oryza sativa* L.). *BMC Plant Biol.* **2015,** *15* (1), 122.
29. Amouri, A. A.; Udupa, S. M.; Lameche, F. Z. F.; Iraqi, D.; Henkrar, F.; El Hanafi, S. Molecular Investigation of Two Contrasting Genotypes of Medicago Truncatula to Salt Stress Using Two Expressed Sequence Tag-Simple Sequence Repeat (EST-SSRs) Markers. *Afr. J. Biotechnol.* **2014,** *13* (41), 4046–4051.
30. Scott, K. D.; Eggler, P.; Seaton, G.; Rossetto, M.; Ablett, E. M.; Lee, L. S.; Henry, R. J. Analysis of SSRs Derived from Grape ESTs. *Theor. Appl. Genet.* **2000,** *100,* 723–726.

31. Long, W.; Zou, X.; Zhang, X. Transcriptome Analysis of Canola (Brassica napus) under Salt Stress at the Germination Stage. *PLoS One.* **2015**, *10* (2), e0116217.

32. Wang, Z.; Gerstein, M.; Snyder, M. RNA-Seq: A Revolutionary Tool for Transcriptomics. *Nat. Rev. Genet.* **2009**, *10* (1), 57–63.

33. Sharma, R.; Mishra, M.; Gupta, B.; Parsania, C.; Singla-Pareek, S. L.; Pareek, A. De Novo Assembly and Characterization of Stress Transcriptome in a Salinity-Tolerant Variety CS52 of *Brassica juncea. PLoS One.* **2015**, *10* (5), e0126783.

34. Bahieldin, A.; Atef, A.; Sabir, J. S. M.; Gadalla, N. O.; Edris, S.; Alzohairy, A. M.; Radhwan, N. A.; Baeshen, M. N.; Ramadan, A. M.; Eissa, H. F.; Hassan, S. M.; Baeshen, N. A.; Abuzinadah, O.; Al-Kordy, M. A.; El-Domyati, F. M.; Jansen, R. K. RNA-Seq Analysis of the Wild Barley (*H. spontaneum*) Leaf Transcriptome under Salt Stress. *C. R. Biol.* **2015**, *338* (5), 285–297.

35. Marioni, J. C.; Mason, C. E.; Mane, S. M.; Stephens, M.; Gilad, Y. RNA-Seq: An Assessment of Technical Reproducibility and Comparison with Gene Expression Arrays. *Genome Res.* **2008**, *18* (9), 1509–1517.

36. Zheng, H.; Zhao, H.; Liu, H.; Wang, J.; Zou, D. QTL Analysis of Na+ and K+ Concentrations in Shoots and Roots under NaCl Stress Based on Linkage and Association Analysis in Japonica Rice. *Euphytica.* **2015**, *201* (1), 109–121.

37. Zhang, H.; Han, B.; Wang, T.; Chen, S.; Li, H.; Zhang, Y.; Dai, S. Mechanisms of Plant Salt Response: Insights from Proteomics. *J. Proteome Res.* **2012**, *11*, 49–67.

38. Vera-Estrella, R.; Barkla, B. J.; Pantoja, O. Comparative 2D-DIGE Analysis of Salinity Responsive Microsomal Proteins from Leaves of Salt-Sensitive Arabidopsis Thaliana and Salt-Tolerant Thellungiella Salsuginea. *J. Proteomics.* **2014**, *111*, 113–127.

39. Liu, C.; Chang, T.; Hsu, Y.; Wang, A. Z.; Yen, H.; Wu, Y.; Wang, C.; Lai, C. Comparative Proteomic Analysis of Early Salt Stress Responsive Proteins in Roots and Leaves of Rice. *Proteomics.* **2014**, *14* (15), 1759–1775.

40. Mostek, A.; Börner, A.; Badowiec, A.; Weidner, S. Alterations in Root Proteome of Salt-Sensitive and Tolerant Barley Lines under Salt Stress Conditions. *J. Plant Physiol.* **2015**, *174*, 166–176.

41. Li, H.; Pan, Y.; Zhang, Y.; Wu, C.; Ma, C.; Yu, B.; Zhu, N.; Koh, J.; Chen, S. Salt Stress Response of Membrane Proteome of Sugar Beet Monosomic Addition Line M14. *J. Proteomics.* **2015**, *127*, 18–33.

42. Liu, D.; Ford, K. L.; Roessner, U.; Natera, S.; Cassin, A. M.; Patterson, J. H.; Bacic, A. Rice Suspension Cultured Cells are Evaluated as a Model System to Study Salt Responsive Networks in Plants Using a Combined Proteomic and Metabolomic Profiling Approach. *Proteomics.* **2013**, *13* (12–13), 2046–2062.

43. Ma, H.; Song, L.; Shu, Y.; Wang, S.; Niu, J.; Wang, Z.; Yu, T.; Gu, W. Comparative Proteomic Analysis of Seedling Leaves of Different Salt Tolerant Soybean Genotypes. *J. Proteomics.* **2012**, *75* (5), 1529–1546.

44. Ma, H.; Song, L.; Huang, Z.; Yang, Y.; Wang, S.; Wang, Z.; Tong, J.; Gu, W.; Ma, H.; Xiao, L. Comparative Proteomic Analysis Reveals Molecular Mechanism of Seedling Roots of Different Salt Tolerant Soybean Genotypes in Responses to Salinity Stress. *EuPA Open Proteom.* **2014**, *4*, 40–57.

45. Capriotti, A. L.; Borrelli, G. M.; Colapicchioni, V.; Papa, R.; Piovesana, S.; Samperi, R.; Stampachiacchiere, S.; Laganà, A. Proteomic Study of a Tolerant Genotype of Durum Wheat under Salt-Stress Conditions. *Anal. Bioanal. Chem.* **2014**, *406* (5), 1423–1435.

46. Guan, Q.; Wu, J.; Yue, X.; Zhang, Y.; Zhu, J. A Nuclear Calcium-Sensing Pathway is Critical for Gene Regulation and Salt Stress Tolerance in Arabidopsis. *PLoS Genet.* **2013**, *9* (8), e1003755.

47. Chen, L.; Ren, F.; Zhou, L.; Wang, Q.-Q.; Zhong, H.; Li, X.-B. The Brassica Napus Calcineurin B-Like 1/CBL-Interacting Protein Kinase 6 (CBL1/CIPK6) Component is Involved in the Plant Response to Abiotic Stress and ABA Signalling. *J. Exp. Bot.* **2012**, *63* (17), 6211–6222.

48. Oliver, S. G.; Winson, M. K.; Kell, D. B.; Baganz, F. Systematic Functional Analysis of the Yeast Genome. *Trends Biotechnol.* **1998**, *16* (9), 373–378.

49. Kim, H. K.; Choi, Y. H.; Verpoorte, R. NMR-Based Metabolomic Analysis of Plants. *Nat. Protoc.* **2010**, *5* (3), 536–549.

50. Kim, H. K.; Choi, Y. H.; Verpoorte, R. NMR-Based Plant Metabolomics: Where do We Stand, Where do We Go? *Trends Biotechnol.* **2011**, *29* (6), 267–275.

51. Gavaghan, C. L.; Li, J. V.; Hadfield, S. T.; Hole, S.; Nicholson, J. K.; Wilson, I. D.; Howe, P. W. A.; Stanley, P. D.; Holmes, E. Application of NMR-Based Metabolomics to the Investigation of Salt Stress in Maize (Zea Mays). *Phytochem. Anal.* **2011**, *22* (3), 214–224.

52. Johnson, H. E.; Broadhurst, D.; Goodacre, R.; Smith, A. R. Metabolic Fingerprinting of Salt-Stressed Tomatoes. *Phytochemistry.* **2003**, *62* (6), 919–928.

53. Abe, T.; Kanaya, S.; Ikemura, T. In *Batch-Learning Self-Organizing Map for Predicting Functions of Poorly-Characterized Proteins Massively Accumulated.* Proceedings in WSOM '09 Proceedings of the 7th International Workshop on Advances in Self-Organizing Maps, St. Augustine, FL, June 08–10, 2009: Springer-Verlag: Berlin, Germany, pp 1–9.

54. Kim, J. K.; Bamba, T.; Harada, K.; Fukusaki, E.; Kobayashi, A. Time-Course Metabolic Profiling in Arabidopsis Thaliana Cell Cultures after Salt Stress Treatment. *J. Exp. Bot.* **2007**, *58* (3), 415–424.

55. Gong, Q.; Li, P.; Ma, S.; Indu Rupassara, S.; Bohnert, H. J. Salinity Stress Adaptation Competence in the Extremophile Thellungiella Halophila in Comparison with its Relative Arabidopsis Thaliana. *Plant J.* **2005**, *44* (5), 826–839.

56. Widodo; Patterson, J. H.; Newbigin, E.; Tester, M.; Bacic, A.; Roessner, U. Metabolic Responses to Salt Stress of Barley (*Hordeum vulgare* L.) Cultivars, Sahara and Clipper, Which Differ in Salinity Tolerance. *J. Exp. Bot.* **2009**, *60* (14), 4089–4103.

57. Richter, J. A.; Erban, A.; Kopka, J.; Zörb, C. Metabolic Contribution to Salt Stress in Two Maize Hybrids with Contrasting Resistance. *Plant Sci.* **2015**, *233,* 107–115.

58. Weckwerth, W. Metabolomics in Systems Biology. *Annu. Rev. Plant Biol.* **2003**, *54,* 669–689.

59. Kopka, J.; Fernie, A.; Weckwerth, W.; Gibon, Y.; Stitt, M. Metabolite Profiling in Plant Biology: Platforms and Destinations. *Genome Biol.* **2004**, *5* (6), 109.

60. Lu, Y.; Lam, H.; Pi, E.; Zhan, Q.; Tsai, S.; Wang, C.; Kwan, Y.; Ngai, S. Comparative Metabolomics in Glycine Max and Glycine Soja under Salt Stress to Reveal the Phenotypes of Their Offspring. *J. Agric. Food Chem.* **2013**, *61* (36), 8711–8721.

61. Kim, D. W.; Rakwal, R.; Agrawal, G. K.; Jung, Y. H.; Shibato, J.; Jwa, N. S.; Iwahashi, Y.; Iwahashi, H.; Kim, D. H.; Shim, I. S.; Usui, K. A Hydroponic Rice Seedling Culture Model System for Investigating Proteome of Salt Stress in Rice Leaf. *Electrophoresis.* **2005**, *26* (23), 4521–4539.

62. Shigeoka, S.; Ishikawa, T.; Tamoi, M.; Miyagawa, Y.; Takeda, T.; Yabuta, Y.; Yoshimura, K. Regulation and Function of Ascorbate Peroxidase Isoenzymes. *J. Exp. Bot.* **2002**, *53* (372), 1305–1319.

63. Lee, M. H.; Cho, E. J.; Wi, S. G.; Bae, H.; Kim, J. E.; Cho, J. Y.; Lee, S.; Kim, J. H.; Chung, B. Y. Divergences in Morphological Changes and Antioxidant Responses in Salt-Tolerant and Salt-Sensitive Rice Seedlings after Salt Stress. *Plant Physiol. Biochem.* **2013,** *70,* 325–335.

64. Teixeira, F. K.; Menezes-Benavente, L.; Margis, R.; Margis-Pinheiro, M. Analysis of the Molecular Evolutionary History of the Ascorbate Peroxidase Gene Family: Inferences from the Rice Genome. *J. Mol. Evol.* **2004,** *59* (6), 761–770.

65. Lee, S.; Chung, M.-S.; Kim, J. E.; Lee, G. W.; Jeong, Y. S.; Lee, M. H.; Hong, S. H.; Lee, S. S.; Kim, J.-H.; Chung, B. Y. Liquid Chromatography-Tandem Mass Spectrometry-Assisted Identification of Two Salinity-Inducible Ascorbate Peroxidases in a Salt-Sensitive Rice Cultivar (*Oryza sativa* L. cv."IR-29"). *Plant Growth Regul.* **2015,** *75* (1), 143–153.

66. Sobhanian, H.; Motamed, N.; Jazii, F. R.; Nakamura, T.; Komatsu, S. Salt Stress Induced Differential Proteome and Metabolome Response in the Shoots of Aeluropus Lagopoides (Poaceae), a Halophyte C4 Plant. *J. Proteome Res.* **2010,** *9* (6), 2882–2897.

CHAPTER 11

UNRAVELING THE ABIOTIC STRESS TOLERANCE IN COMMON BEAN THROUGH OMICS

REETIKA MAHAJAN[1], MUSLIMA NAZIR[2], NUSRAT SAYEED[3], VANDNA RAI[4], ROOHI MUSHTAQ[5], KHALID Z. MASOODI[2], SHAFIQ A. WANI[2], UNEEB URWAT[2], SHEEZAN RASOOL[2], R. K. SALGOTRA[1], and SAJAD MAJEED ZARGAR[2*]

[1]School of Biotechnology, S K University of Agricultural Sciences and Technology of Jammu, Chatha, Jammu, Jammu and Kashmir, India

[2]Centre for Plant Biotechnology, Division of Biotechnology, S K University of Agricultural Sciences and Technology of Kashmir, Shalimar, Srinagar, Jammu and Kashmir, India

[3]Depatment of Botany, Government College for Women, M A Road, Srinagar, Jammu and Kashmir, India

[4]National Research Centre on Plant Biotechnology, Indian Agricultural Research Institute, New Delhi, India

[5]Department of Biotechnology, S P College, Srinagar, Jammu and Kashmir, India

*Corresponding author. E-mail: smzargar@gmail.com

CONTENTS

ABSTRACT

Common bean (*Phaseolus vulgaris*) represents the most important food legume for direct human consumption with an adequate amount of proteins, carbohydrates, fibers, amino acids, micronutrients (Fe & Zn), and vitamins (e.g., B vitamins, folate, and vitamin K). Common bean also have potential disease preventing and health-promoting compounds. All these qualities makes common bean a food of choice for the poor's in developing countries. Due to climate change the yield of crops is constrained by various biotic and abiotic stresses. Drought, cold, high-salinity, and heat are major abiotic stresses that have reduced the yield of food crops worldwide. In addition to these problems, population explosion has also posed enormous threat to the sustainable agriculture. To address these problems, various strategies have been implemented by scientists across the world. These strategies cover the traditional as well as the modern technologies. Earlier scientists used breeding strategies (conventional and non-conventional) to develop improved varieties but as these strategies are highly influenced by the external environment, required 7–8 generations for development of new variety and due to the multigenic nature of stress tolerance, breeding practices have been limited to a particular area. With the advancement in the science, various new technologies have been invented which made it easier for the scientists to understand the multigenic nature of the stress tolerance that further helped the scientists to developed a number of new and improved varieties. Development of stress adapted common bean cultivars is a strategy to improve food security in bean producing areas under various stresses. Hence, in order to increase common bean production, there is a need to develop improved varieties that can withstand future challenges, such as climate change, sustainability, and food security. Here in this chapter, a brief discussion about the strategies (breeding and biotechnology) used for the improvement of common bean is given.

11.1 IMPACT OF ABIOTIC STRESS ON SUSTAINABLE AGRICULTURE

Agriculture is the world's largest industry. It is the source of livelihood for more than 1 billion people and generates food worth 1.3 trillion dollars annually. India's economy is mainly based on agriculture and presently it is one of the top two farm producers in the world. Indian agriculture is one of the largest sectors which provides employment to ~60% of population and contributes ~17% to Gross Domestic Product.[1] Besides generation of food for feeding the population agricultural products can be used for the production of fibers, fuels, raw materials, and biopharmaceuticals, the fundamental basis of human and animal sustenance.[2] Sustainable agriculture maintains balance among society, economy, and environment. Change in temperature, precipitation, altered soil conditions, and many more factors can result in decrease in food production which further deteriorates human health. Abiotic stresses affect plant at various stages like germination, flowering, time of pollination, fertilization, seed development which decreases crop yield and finally leads to decline in food production. One of the major challenges faced by the scientists around the globe is maintaining the balance between the ever-growing population and food production. The continuous growing population and increase in consumption of the available resources are constraining the world's economy and creating hurdles for human being to feed themselves. It is predicted that the world's population which was around 7.2 billion in 2013 will cross 9 billion by 2050 and there is need to increase the agriculture output by 70% to overcome the feeding problems.[3] The problem of feeding a billion of population became more serious due to the adverse effect of climate change as it mainly affects the agriculture sector. With the introduction of science, crop improvement has become possible. Therefore, combination of conventional and modern breeding approaches like, back crossing, foreground and background selection, phenotyping, gene pyramiding, marker-assisted selections (MASs), identification of quantitative trait loci (QTLs), and many more can be used to have new and improved varieties. Further, biotechnological approaches like identification of genes by using markers, genetic transformation, regulating signal transductions, and different omics approaches (genomics, transcriptomics, proteomics, and metabolomics) are choice of scientists to develop next generation crops to tackle the challenge of having sustainable agriculture, adverse effect of climate change, and to feed looming population.

11.2   ABIOTIC STRESS DUE TO CLIMATE CHANGE

Climate change poses threat to the world. Scientists across the globe are worried about the drastic climate change. Climate change is defined as the change in the state of the climate that can be identified (e.g., using statistical tests) by changes in the mean and/or the variability of its properties and that persists for an extended period, typically decades or longer.[4] Climate change includes parameters such as rise in atmospheric temperature, changes in rainfall pattern, sea level rise, salt-water intrusion, changes in precipitation patterns, excess UV radiation, and generation of floods and droughts are of great concern as it is a major risk for maintaining sustainable agriculture, environment, and health over the world.[5,6] According to International Panel for Climate Change, CO_2 concentration in the atmosphere has increased drastically from 280 to 370 ppm due to human activities and is likely to be doubled in 21st century.[7] This increase in CO_2 concentration resulted in global warming. With change in climate, drought stress would be escalated because of increased evapotranspiration. It is expected that due to global warming the frequency and intensity of drought will increase from 1 to 30% in extreme drought land area by 2100 which would neutralize the favorable effect from the elevated CO_2 concentration and restrict the structure and function of the terrestrial ecosystem.[8] Although with various negative impacts of CO_2, its higher concentration in atmosphere can enhance the crop production of rice, wheat, and soybean.[4] It is also reported that surface average temperature of earth would increase by 2.8 °C ranging from 1.8 to 4.0 °C by the end of 2050.[7] This increase in temperature will create problems like decline plant growth, terrestrial ecosystem productivity, in many regions all over the world, particularly in arid and semiarid area.[8–10] Moreover, it is predicted that by 2080 the overall production of cereal could decline by 2–4% which ultimately leads to increased food price by 13–45%. Further this increased price will increase risk of hunger. It is believed that by 2060 the changing climate will affect 40–300 million people of poor countries.[11] Over the last 100 years, the climate of India has undergone significant changes showing increasing trends in annual temperature with an average of 0.56 °C.[12,13]

As already discussed that abiotic stresses arise due to change in climate which ultimately affect the crop yield. Abiotic stress at any phase of plant growth and development can affect the normal growth, flowering, pollination, and fruit development and subsequently decrease the crop yield.[14] It is reported that altered temperature and precipitation can lead to the introduction of new combinations of pests and outbreak of secondary pests. Further, change in climate could also destroy the predator–prey relationships that normally keep pest populations in control and may increase pest population that could become one of the reasons for loss of biodiversity.[11] Thus, in order to preserve the biodiversity and reduce the adverse impact of climatic change on productivity and quality of crops there is need to develop new varieties that are tolerant to high temperature, moisture stress, salinity, and climate proofing through conventional and non-conventional, breeding techniques and various biotechnological approaches. Various abiotic stresses and their possible impact on plant are depicted in Figure11.1.

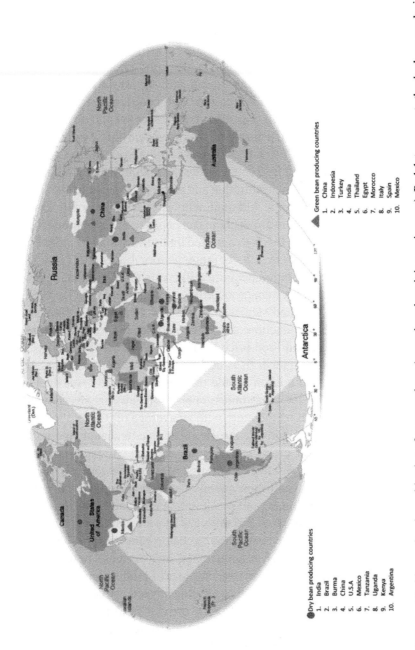

FIGURE 11.1 World map showing cultivation of common bean (countries where it is a major crop). Red dot represents the dry bean producing countries and green triangle represents green bean producing countries.

11.3   COMMON BEAN: A CROP THAT PROVIDES CALORIES

Common bean (*Phaseolus vulgaris* L.) is one of the five cultivated species from the genus *Phaseolus*. It is currently estimated to be one of the most important legumes worldwide.[15] Common bean is cultivated in many parts of world as dry beans and green beans. A map showing the countries where common bean (dry bean or green bean) is a major crop is shown in Figure 11.2. It is a major grain legume crop grown by the small-scale farmers of tropics, subtropics, and temperate regions.[16,17] In terms of importance it holds the third place after soybean and peanut whereas first for direct human consumption.[18] It is an important source of protein (20–28%), dietary fiber (56%), and complex carbohydrates for more than 300 million people in parts of Eastern Africa and Latin America, representing 65% of total protein consumed and 32% of energy.[18–20] It is also a major source of micronutrients, for example, iron (70 mg/kg) and zinc (33 mg/kg), vitamin B complex such as niacin, riboflavin, folic acid, and thiamine as well as being a good source of polyunsaturated fatty acids.[18,21–24] Being a good and cheap source of protein, fiber, carbohydrates, minerals, and vitamins it can solve malnutrition and hunger related problems for the poor of the world thus considered as the "poor man's meat."[25–27] Micronutrients like iron and zinc are important for metabolism of both plants and animals. In plants, both Fe and Zn play an important role in maintaining proper metabolic and physiological cellular processes. Fe is essential for chlorophyll biosynthesis, nitrogen fixation, DNA replication, reactive oxygen species (ROS) scavenging, and electron transport chain in both mitochondria and chloroplasts.[28,29] Chlorosis (low chlorophyll content) of young leaves is the most obvious visible symptom of Fe deficiency [30] and its deficiency also triggers the oxidative stress.[31,32] Like Fe, Zn, a divalent cation, also plays diverse roles in different cellular processes.[33] The metabolism of proteins, nucleic acids, carbohydrates, and lipids is dependent on Zn to a large extent.[34,35] While in human iron and zinc deficiencies affect mainly pregnant women, infants, and children. Iron deficiency increases maternal and infant morbidity and mortality, decreases immunity and impair mental and psychomotor development of children[36,37] Whereas, zinc deficiency debilitates growth, sexual maturation, and immune function, causes complications during pregnancy and retard fetal growth and gastro-enteric and respiratory infections, especially in children.[38,39] Iron and zinc are found commonly together in foods, so poor iron and zinc nutritional statuses are usually associated, particularly in developing countries where the diets are based on plant foods with low iron and zinc bioavailability and/or concentration.[40] Common bean also has medicinal value due to the presence of many[41,42] phytochemicals like polyphenolic compounds, fiber, lectins, trypsin inhibitors, and flavonoids. Beans are considered to have important role as antiplatelet, antiangiogenic, anti-cancer, anti-depressant, anti-leukemia, antimelanómico, antiprostático, apoptotic, cardioprotective, estrogenic, hepatoprotective, chemopreventive, hipocolesteronémico, hypotensive, lipolytic, lipotropic, plus antibacterial, antidiabetic, diuretic, antiviral and mutagenic, antipyretic, carminative, depurative, diaphoretic, fungicidal, and hypoglycemic.[43–45]

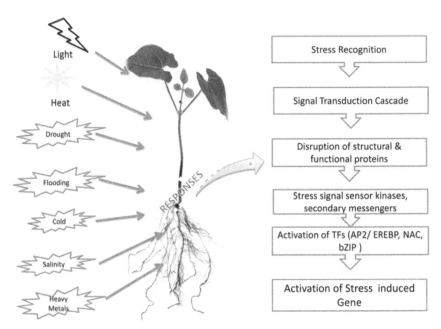

FIGURE 11.2   A pictorial representation of responses of various abiotic stresses on common bean *(Phaseolus vulgaris* L.).

11.4   IMPACT OF DROUGHT STRESS ON COMMON BEAN

Drought is one of the major abiotic stresses that have affected the agricultural systems and food production. It has introduced many physiological, biochemical, and molecular responses in several crop plants like bean.[46–50] Common bean is mainly grown in Latin America, Africa, Asia, China, Europe, the United States, and Canada.[18] Globally, bean production is ~18,943 million tons with 4340 million tons in India per annum.[51] It has been estimated about 60% of all crop production suffers from drought stress.[52,53] Various reports have shown that drought has decreased the common bean crop performance by damaging specific tissues and whole plant.[54-58] It has been reported that drought alone has reduced the yield of soybean by 40%.[59] Drought stress has adversely effected the total biomass production, pod number, seed number, seed weight and quality, seed yield per plant, days to maturity and root length, and mass[60-70] and it has reduced the average global production of beans (<900 kg/ha).[71] It can also cause flower abortion, pod drop, lower seed filling, photosynthate translocation and partitioning, reduces P uptake as well as N concentration, and fixation.[72-74] At biochemical level, drought stress causes inhibition of photosynthesis by altering pathway regulated by stomatal closure and decrease in flow of CO_2 into mesophyll tissues and also by altering the activity of ribulose-1,5-bisphosphate carboxylase/oxygenase.[75-77] A declined photosynthetic rate results in lesser amount of assimilates available for export to the sink organs.[78] Further drought stress disturbs respiration, translocation, ion uptake, carbohydrates, nutrient assimilation, and growth promoters.[79,80] Drought stress can affect the successful fruit and seed set in common bean by altering the metabolism of carbohydrates in plant reproductive organs while decreased water potential and increased accumulation of abscisic acid (ABA) in reproductive part of plants leads to loss of fruit and seed set.[81] Drought stress also decreases the symbiotic nitrogen fixation rate due to (a) the accumulation of ureides in both nodules and shoots,[61,82] (b) decline in shoot N demand, (c) lower xylem translocation rate due to a decreased transpiration rate, (d) decline of metabolic enzyme activity,[59] and (e) decline in photosynthetic rate.[59,83] Drought stress resulted in stimulation of sucrose and total sugars in nodules of legumes.[59,84-88] Various reports suggested that there is inhibition in nodule initiation, nodule growth and development, as well as nodule functions due to drought stress.[59,82,86] Drought stress leads to overproduction of ROS in plant which is directly responsible for oxidation of multicellular components like proteins, lipids, DNA, and RNA, resulting in death of cells.[89] Oxidative damage also affects the nodule performance and biological nitrogen fixation.[90] Moreover, drought stress has also affected the plant by inducing the expression of proteins which are related to stress whether directly or indirectly and assigned functions to some of the sequenced proteins.[91] Various stress related proteins are identified which involve both up regulated and down regulated proteins. Up regulated proteins involve proteins responsible for biosynthesis of osmolytes,[92] uptake and compartmentation of ions,[93] hydroxyl-radical scavenging,[94] and protection of cellular structure.[91] Whereas down regulated proteins are involved in photosynthesis-related function.[91] Furthermore, the onset of physiological, biochemical, and molecular changes in plant due to drought stress ultimately leads to death of plant.

11.4.1 MORPHOLOGICAL, PHYSIOLOGICAL, BIOCHEMICAL, AND MOLECULAR RESPONSES OF DROUGHT STRESS

A major challenge for the agricultural scientists and researchers is to face the consequences of abiotic and biotic stress. These stresses have affected every aspect of plant system drastically from physiological to molecular level. Out of the various abiotic stresses drought is of major concern for the scientists and researchers as it is reported that more than 60% of crop productivity losses occurred due to drought. Drought stress has raised the issue of feeding millions of individual in developing countries. During the course of evolution plants developed various mechanisms to cope up with the drought stress. These mechanisms vary in response from molecular expression, biochemical metabolism through plant physiological processes to ecosystem levels[10,95,96] and mainly include six approaches. (a) Plant can escape drought stress via escape mechanism. In this mechanism plant completes its life cycle before severe drought like early flowering in annuals species before the onset of severe drought.[97] (b) Plant can avoid drought stress via avoidance mechanism. In which plant enhance its water retaining capacity by modifying its root systems or conserve water by reducing the size of stomata and by decreasing leaf area/canopy cover.[98,99] (c) Plants can tolerate drought stress via tolerance mechanism. Tolerance in plant can be achieved by improving osmotic adjustment ability and increasing cell wall elasticity to maintain tissue turgidity.[100] (d) Plant can also resist drought stress via resistance mechanism. To resist drought stress plants alter their metabolic pathways like increased antioxidant metabolism,[101,102] increase in osmolytes (proline, glycine betaine, mannitol, trehalose, ononitol and ectoine) for maintaining cellular homeostasis under severe stress.[103-105] (e) Drought stress can be abandoned by removing a part like shedding of elder leaves.[10] (f) Plants can also cope up with the biochemical and physiological traits of drought stress via genetic mutation and genetic modification.[106-108] On the other hand at molecular level certain signal transduction pathways (ABA, Ca^{2+} mediated response) are activated to tolerate drought stress. Thus, the knowledge of these signaling pathways is of great importance to develop improved varieties. However, the response of plants to drought stress basically depends on two factors: the length and the severity of water deficit; whereas, plant species and genotypes, age and development stage, the organ, cell type, and type of sub cellular component also play an important role in determining the response.[109,110] The various responses of drought stress are briefly explained in Figure 11.3.

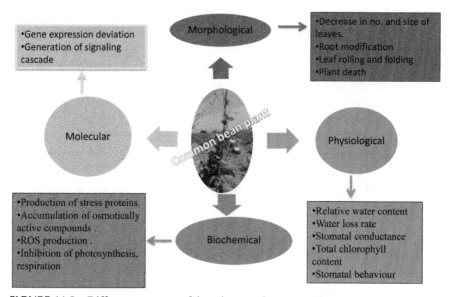

FIGURE 11.3   Different responses of drought stress in common bean.

11.5 APPROACHES TO TACKLE DROUGHT STRESS

Naturally plants have the tendency to overcome the drought stress by three mechanisms— avoidance, escape, and tolerance. Numerous approaches right from breeding to the currently available omics approaches which involve any of the three above-mentioned mechanisms can be used to tackle the drought stress. Some of the approaches are discussed here.

11.5.1 BREEDING APPROACHES

Genetic makeup of the crops has been altered by the farmers or the scientists since the agricultural practice began, some eight to ten thousand years ago. Earlier farmers used to save the seeds of the best-looking plant for next season. Later on with the introduction of science of genetics, plant breeders develop improved varieties by having the knowledge of specific and desirable genes for a trait. Therefore, the selection for best traits like faster growth, higher yields, pest and disease resistance, larger seeds, or sweeter fruits has dramatically increased the productivity and quality of the plants we grow for food, feed, and fiber. Generally, there are two breeding strategies to cope with drought stress.[111–113] (a) Development of early-maturing varieties to escape drought through early maturity. (b) Development of drought-tolerant varieties that can withstand or recover from drought stress during growth and produce reasonably high seed yield. Although, the introduction of new techniques in breeding for drought tolerance has produced high yielding and drought-tolerant genotypes which are of more interest to the farmers than their local varieties in these areas.[114] But these genotypes have not yet been exposed to farmer preferences.

With the advancement in science of plant breeding, plant breeders understood better how to select superior plants and breed them to create new and improved varieties of different crops by using various breeding approaches. Here we will be discussing few of the breeding approaches to overcome drought stress.

11.5.1.1 CONVENTIONAL BREEDING

Conventional plant breeding methods involve the improvement of plant by producing new individuals based on their genetic variation and by incorporating the selected better character in the progeny.[50] In this approach, two plants with desirable traits were selected and crossed to exchange genes, so that the offspring will have new and desirable genetic arrangements and then these individuals are tested for the expression of the desirable traits and are maintained in future generations.[115] Conventional breeding has been going on for hundreds of years and are still in use. As reported, earlier drought has affected each and every aspect of plant system. This created the need for development of drought resistant/tolerant variety. Drought tolerance is defined as the ability of plant to sustain reasonable yields under moderate water stress.[116] Drought stress occurs at different times in the growing season, and with different intensity, while its effects on crops are modified by soil type and fertility.[112] To survive under drought stress, various mechanisms play a role in providing drought tolerance, and their importance depends on growing stage at which the drought occurs.[113] Earlier in 1980's, common set of genotype studied in many countries reveals that local adaptation is one of the important factors of drought resistance.[117] Further in 1994, conventional genetic studies on common bean genotype showed that drought resistant genotypes behave differently at different environment, for example, genotypes which were selected from the Mexican highlands did not adapted well in Colombia. Moreover, it is further suggested that the value of drought resistance sources as parents and the yield of the parent in the given environment were closely associated. It was also reported that expression of the identified genes/QTLs could be masked if the component of local adaptation is more than drought tolerance.[118] A Mesoamerican race BAT 477 was identified as drought stress resistance in lowland tropical environments in a research at International Center for Tropical Agriculture (CIAT)-Colombia.[118] Later it was reported that Durango germplasm from the semiarid highlands of Mexico can be considered as best drought resistance in lowland tropical environment.[119] They also reported that the combination of the race Durango (Guanajuato 31) and race Mesoamerica (BAT 477) through double cross can result in much better line SEA 5. Another line L88–63, a black seeded variety of tropical environment selected from a cross having one of the parents derived from a combination of Durango and Mesoamerica races. This line performed better than SEA-5 in both stressed and unstressed conditions.[120] Moreover, superior drought resistance lines with better rooting trait improved seed filling and yielding around 50% more than SEA 5 were developed at CIAT.[121] Thus it can be concluded that improved drought resistance for lowland tropical environments can be developed by the combination of two races Durango and Mesoamerica. However, more effective introgression of Durango genes to lowland races can be done by increasing the breeding cycles. Whereas PintoVilla, combination of race Durango with Andean germplasm, is superior in highland environments of Mexico.[122] Drought tolerance is a genetically and physiologically complex trait with low heritability and limited genetic variation for which there is no suitable selection criteria. Moreover, drought tolerance is highly influenced by environment making it difficult for improvement through conventional breeding as it reduces the genotype–phenotype gap[67] and making the drought tolerance susceptible to genotype by environment (G × E) interaction.[113,123] Furthermore, drought stress is accompanied by other abiotic stresses like nutrient deficiencies, high temperature which further complicates the breeding efforts.[124] In spite of development of improved variety conventional breeding is losing its importance now a day as it is time consuming, it requires almost 7–8 years for development of improved varieties. Moreover, in this approach there is no surety that the desired gene will be inherited by the next generation. Sometimes the undesired traits also get transferred. To overcome this, breeders have to use a very large number of genotypes for screening and selection of desirable trait. Limitation of conventional breeding approaches created a platform for modern breeding approaches to tackle various biotic and abiotic stresses. With the better understanding of the modern techniques now it is easy to locate genes/QTLs on chromosomes and their individual effect. In common bean, identification of major genes/QTLs contributing for drought tolerance and their introgression in the backgrounds of our choice through MAS can be used for development of drought tolerant genotypes.[125]

11.5.1.2 PHYSIOLOGICAL BREEDING

This breeding approach requires optimization of the physiological processes involved in the plant response to tackle drought stress which leads to the increase in crop yield. Avoidance mechanism plays an important role in mitigating the adverse effect of drought stress by controlling the stomatal behavior. It controls the stomatal closure by reducing crop evapotranspiration during its vegetative phase, thus conserving moisture for the grain-filling period. This mechanism was studied in soybean.[126,127] In a grafting experiment conducted at two different locations (Palmira and Quilichao stations of CIAT) in which shoots and roots of susceptible and resistant genotypes were interchanged affirmed that drought resistance and increased yield were attributed to the roots. H6 Mulatinbo and BAT 1297 genotypes of common bean were found to be drought resistant.[128] Similar studies carried out in a greenhouse trial on common bean lines also showed a similar trait.[129] Another common escape mechanism is early maturity. Many farmers favored early maturity cultivars to minimize exposure to terminal drought. One of the early maturity common bean cultivar from Guatemala is "ICTA Ligero." However, each day of earliness reduces the yield potential by 74 kg/ha.[130] BAT 477 and San Cristobal are drought tolerant common bean varieties having avoidance mechanism that involves deeper rooting in response to soil drying.[131] The ability of segregating large amount of carbohydrates to the seeds under stress can be considered as an important trait for developing improved variety.[112] A black-seeded bean variety from Colombia (G 21212) with potential to set pods and fill seed and expresses unique genes under stress conditions fascinated the scientists at CIAT for conducting genetic studies.[132,133] Apart from avoidance and escape mechanisms, desiccation tolerance and water use efficiency (WUE) require different strategies to tackle drought stress. It has been suggested that an increase in transpiration efficiency (TE) or decline in soil evaporation will increase WUE,[134] but it is difficult to elevate the effective use of water for transpiration when there is scarcity of water.[135] TE can be used in many breeding programs as it is constant. It is suggested that traits related to TE must be studied to increase the effective use of available water through the growing season. However, there is decline in crop yield due to the increased TE which is obtained by partial stomatal closure, and consequently by a decrease of transpiration.[127] Genetic variation in the stomatal response to water stress in both controlled and field condition can be detected by using infrared thermograph, a new and feasible screening method.[136] Harvest Index (HI) is the measurement of crop yield and maintenance of HI is of great importance to develop drought resistance common bean variety and to enhance crop productivity through breeding program under drought stress.[58,69,137,138] In Middle American gene pool of common bean, most of the improvement is done by combining the small seeded race Mesoamerica with the Durango race originating from the dry highlands of Mexico. However, lines with superior drought yield in Colombia display better remobilization of photosynthate to pod formation and to grain during the grain-filling period.[58,114,139,140] Another important physiological parameter is Pod harvest index (PHI; seed weight/total pod weight × 100). It has been employed as an indicator of photosynthate remobilization capacity under stress.[140] PHI has higher heritability than yield under drought, and the selection of PHI over drought yield would result in greater genetic gain, and enhance non-stressed yield.[141] Combination of earliness, deep rooting, and photosynthate mobilization in an improved variety make it more resilient and can be used by farmers in smallholder farm conditions to minimize the risk from climate change and low soil fertility.[142] Tepary bean (*Phaseolus acutifolius*) is grown in an area where water is the limiting factor. Thus exhibiting several drought resistant traits like early maturity, excellent remobilization capacity under stress, deep rooting to avoid dehydration, small leaves for reduced water use, and stomatal control but not with osmotic adjustment.[143] Tepary bean can be considered as a paradigm for improvement of drought resistance in common bean as it exhibited the above-mentioned multiple traits.[140] Further, for the development of improved variety of common beans such traits can be found within common bean germplasm, or can be introduced from tepary bean through interspecific crosses. The genes contributing to drought resistance can be explored from a reserve of tepary bean which is obtained by common bean hybrid progenies.[144,145]

11.5.1.3 MOLECULAR BREEDING

As already discussed in the conventional breeding section, the limitation of conventional breeding has created platform for the modern breeding approaches. Here, in this section we will be discussing applications of molecular breeding in tackling stress. In Molecular breeding, two different approaches are used for the development of new and improved bean cultivars. In the first approach crop's yield is taken in consideration where as in second approach genes or QTLs related to a specific trait are identified directly for the development of stress resistance variety. Comparison of crop's yield in stressed and unstressed condition is the base for the quantification of stress response to identify QTL for resistance. Statistical tools like geometric means, regression deviation of stressed yields on unstressed yields, and percentage loss in yield are used to estimate resistance based on the stressed and unstressed treatments. This approach will estimate values of resistance directly for statistical analysis and identification of QTL.[69,146–148] Second approach is the identification of yield specific QTL by analyzing QTLs for both stressed and unstressed conditions yield independently. For candidate gene as markers, gene-based selection can be used as an alternative. However, authenticated physiological studies should be done to reveal the positive contribution of traits/genes to yield under stress conditions at different environments.[149] Various molecular breeding approaches which are used to tackle stresses are briefly explained in Figure 11.4.

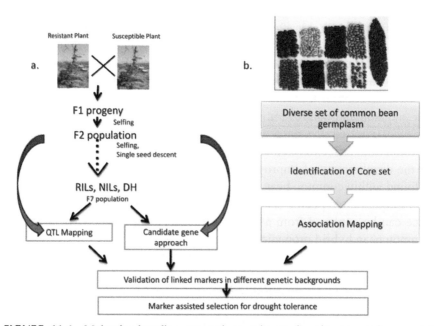

FIGURE 11.4 Molecular breeding approach to enhance drought stress tolerance. (a) Development of drought tolerant variety by QTL mapping and candidate gene approach. (b) Development of drought tolerant variety by association mapping approach.

The very first drought tolerant QTL was investigated in seven different environments in highland Mexico and Michigan. Random amplification of polymorphic DNAs (RAPDs) were used to identify four markers for QTL in one population and five were identified in second population.[150] It was noticed that MAS selection was more effective in Michigan than in Mexico under severe drought stress. It was also reported that G × E interaction affects the QTL expression, incomplete genome coverage and certain unrevealed QTL could determine yield in the Mexican environments. QTL for drought has been identified in BAT 477 under water deficit conditions at CIAT.[133] Gene-based MAS can be efficiently used for developing stress resistant/tolerant varieties as various stress-induced genes have been identified having drought resistance functions in common bean.[121] QTL for various traits like rooting pattern, grain yield, and photosynthate remobilization were identified in recombinant inbred line (RIL) population developed from BAT 477 and DOR 364 and the QTL for grain yield and photosynthate remobilization traits were independent of QTL for rooting depth or rooting pattern.[67-70]

11.5.2 BIOTECHNOLOGY APPROACH

Earlier, scientists across the globe used breeding approaches for development of improved variety of plants to overcome the food issues. But these approaches were time consuming and take 5–6 years to develop a new variety. With the introduction of plant biotechnology the development of new varieties become easier. This approach uses combination of few techniques like plant tissue culture and molecular biology, isolation and cloning of favorable traits, and the creation of transgenic crops, enhancing the expression of gene/gene silencing or gene knockout, and various newly introduced techniques to develop commercial important. This approach however enables us to select best genotype and has moved agriculture from a resource-based to a science-based industry.

11.5.2.1 IDENTIFICATION OF STRESS TOLERANCE GENE/TRANSCRIPTION FACTORS

Understanding of the stress responses and signal transduction pathways is an important step to provide resistance to plant during unfavorable environments. Drought stress has induced changes in gene expression profiling.[151] Transcription factors (TFs) are small molecules that attach to specific sites on a DNA molecule in order to activate or deactivate the expression of certain genes. TFs are basically proteins that are involved in gene regulation and play very important role in almost every stress response.[152] Various TFs which induce stress-responsive genes have been identified with the help of bioinformatics.[153–157] Genes which are induced by drought stress are induced by other abiotic stresses like salt, low and high temperature also suggesting that there exist similar mechanisms of stress responses: the production of stress proteins, up-regulation of antioxidants and accumulation of compatible solutes.[158–162] These genes not only protect the cells from water deficit by producing important metabolic proteins but also help in regulating signal transduction in plants under drought stress.[163–165] Stress-induced genes are categorized into following three major groups: (a) genes that directly protect plant cells against stresses such as heat stress proteins (HSPs) or chaperones, late embryogenesis abundant (LEA) proteins, anti-freeze proteins, detoxification enzymes, osmoprotectants, and free-radical scavengers,[166,167] (b) involved in signaling cascades and in transcriptional control, such as mitogen-activated protein kinase (MAPK), calcium-dependent protein kinase (CDPK), and salt overlay sensitive (SOS) kinase,[168,169] phospholipases,[170] and transcriptional factors,[151,171] and (c) involved in water and ion uptake and transport such as aquaporins and ion transporters.[172] In Arabidopsis, a model for understanding various processes about 7% of genome codes for more than 1800 TFs which belong to large multigene families.[173–175] ABA-responsive elements (ABREs) are a major cis-acting element in ABA-responsive gene expression which plays an important role in promoter activity.[176] These elements have a core ACGT-containing G-box motif and a protein containing basic region leucine zipper (bZIP) motifs binds to ABREs.[177–179] Dehydration response element (DRE) is another factor which is involved regulating gene expression in more than one stress, that is, drought, low-temperature, and salt stress.[180,181] Moreover, members of the DRE-binding protein (DREB) family, the ethylene-responsive element binding factor (ERF) family, the zinc-finger family, the basic helix-loop helix (bHLH) family, the basic-domain leucine zipper (bZIP) family, and the homeodomain TF family are mainly stress-inducible TFs. These could regulate various stress-inducible genes cooperatively or separately, and may constitute gene networks. The TF DREB1A is activated by drought, salinity, and low-temperature stress[164,182] whereas DREB2A and DREB2B are activated by osmotic stress and regulates osmotic stress by inducing specific stress-responsive genes.[180] Dehydrins are one of the group 2 LEA proteins[183] which are induced by ABA in response to drought, freezing, salt, and heavy metals.[184] These proteins play specific protective roles against drought in plant by preventing denaturation of macromolecules[185] and safeguarding membrane structure.[186] Plants also have some transcriptional factors with unique DNA-binding domains such as the AP2: EREBP domain. More than half of the drought-inducible genes are also induced by high-salinity and/or by ABA but only 10% are induced by cold stress.

Another class of TF no apical meristem (NAM), Arabidopsis transcription activation factor (ATAF), and cup-shaped cotyledon (CUC) (NAC) has been identified which are multifunctional proteins and have been implicated in abiotic and biotic stress responses, lateral root and plant development, flowering, secondary wall thickening, anther dehiscence, senescence, and seed quality.[187–203] NAC pathway along with several major stress signal transduction pathways that are activated during abiotic stress have been identified and well elucidated in *Arabidopsis thaliana* and rice (*Oryza sativa*).[204–206] Recently, NAC TFs are also identified in soybean, one of the important legume crops, it has been estimated that there are around 205 NAC or NAC-like TFs members in the GmNAC family, out of which 31 GmNAC members have been recently cloned and characterized in a comprehensive abiotic stress response manner.[207–209] The control of gene expression by TFs is being dissected using a genomic approach. There are various possibilities for decreasing TF expression: using knockout mutants or RNA interference (RNAi) constructs, or for increasing TF levels by over expression, either constitutively or with the use of an inducible promoter. Changing the expression of a number of TFs affects signaling, particularly the responses to disease and abiotic stress.[210] The stress-inducible TFs are promising candidates for generation of stress tolerant transgenic plants possessing traits best suited for survival or decreased yield loss under stressed conditions. Rodriguez-Uribe and Connell in 2006 conducted a study in common bean under drought stress and found a root-specific bZIP TF to be responsible for water deficit stress in tepary bean (*P. acutifolius*) and common bean (*P. vulgaris*).[211] In another study, AtNAP (a NAC family TF member) important for leaf senescence was characterized in Arabidopsis. They identified an orthologous NAC TF "PvNAP" from *P. vulgaris* that shared the same leaf senescence-specific expression pattern as AtNAP. In this study, *P. vulgaris* leaves were analyzed at five distinct developmental stages and "PvNAP" a transcript was detected in senescing leaves only. It was also reports that AtNAP homologues in *P. vulgaris* were able to restore the Arabidopsis AtNAP null mutant to wild type.[212] Recently, CA1 with sequence similarity to the ERF family AP2/EREBP was identified in *P. vulgaris* that was 10-fold induced by drought stress.[213] Some of the genes providing stress tolerance in various crops are given in Table 11.1.

TABLE 11.1 Gene Providing Stress Tolerance.

S.no.	Crop species	Gene	Function	References
1.	Rice	SNAC1	Tolerance to drought stress	[214]
		Os LEA-3–1	Increased growth under drought	[215]
		Os Cc1 & AP37(transcription factor)	Increased growth under drought	[216]
		OsDREB2A with 4ABRC/ rd29A	Enhanced tolerance to drought and salinity	[217, 218]
		OsDIR1	Significantly enhance drought and salinity stress	[219]
2.	Common bean	CA7, NCED	ABA-dependent expression	[220]
		bZIP	Transcription factor	[211]
		PvLEA3, PvLEA18	Protein stabilization	[221, 222]
		Aquaporin	Water allocation	[223]
		Asr1, Asr2	ABA signaling pathway	[224]
		Dreb2A, Dreb2B	Non-ABA dependent responses	[225]
		LTP	Lipid transfer protein (LTP) is ABA-induced responsive to environmental factors like water	[226]
3.	Soybean	P5CS	Proline biosynthesis	[227]
		GmPAP3	Reduces ROS accumulation and thereby alleviates osmotic stress	[228]
		GsAPK	ABA-mediated signal transduction	[229]
		GmCAM4	Activates a R2R3 type MYB transcription factor which in turn up-regulates several drought-responsive genes	[230]
		CDPK	Phosphorylates aquaporins thereby enhancing water permeability across the membranes	[231, 232]
		G93	Lipid signaling	[233, 234]

TABLE 11.1 *(Continued)*

S.no.	Crop species	Gene	Function	References
		GmUBC2	Up-regulates the expression of genes encoding ion transporters (AtNHX1 and AtCLCa), a proline biosynthetic enzyme (AtP5CS), and a copper chaperone (AtCCS)	[235]
		Histidine kinases (GmHK)		[236–238]
		Protein kinases (GmCLV1A, GmCLV1B)		
		GmRLK1,	Candidates of osmo sensors	
		GmRLK2,		
		GmRLK3,		
		GmRLK4		
		HSPs	Prevent the aggregation of denatured proteins	[239]
		GmDREB,		[240]
		GmDREB,		[241]
		GmDREB3		[242]
		GmERF3 b,		[243]
		GmERF4,		[244]
		GmERF089	Transcription factors	[245]
		GmbZIP1		[246]
		GmGT-2A,		[247]
		GmGT-2B		
		GsZFP1		[248]
		GmWRKY54,		[249]

TABLE 11.1 *(Continued)*

S.no.	Crop species	Gene	Function	References
		GmWRKY57B		[250]
4.	Wheat	TaSTRG	Tolerance to drought and salinity stress	[251]
5.	Arabidopsis	kin1, rd298, rab18	Cold and ABA-induced stress genes	[252–254]
		Rd29a	Stress induced genes through ABA-independent pathway	[255]
		LEA14	Encodes late embryogenesis abundant proteins provide tolerance to abiotic stresses	[256]
		SOS 1	Elevates Na$^+$/H$^+$ antiporter	[257]
		AtNCED3	Controls the level of endogenous ABA	[258]
		Gols (galactinol synthase) genes	Encodes galactinol synthase responsible for biosynthesis of raffinose	[259]
		ABF3/ABF4	Reduced transpiration and better survival under drought condition	[260]
		DREB1A (AP2, a TF)	Activated the expression of genes involved in stress tolerance (rd29A)	[261, 262]
		AtMYC2/AtMYB2	Less electrolyte leakage in transgenic plants	[263]
		ERECTA	Improves water use eficiency/transpiration efficiency	[264]
		DREB2C	Non-ABA dependent responses	[265]
		ERD11, GST	Drought tolerance	[266]
		ERA1	Mutant of ERA1 stimulate stomatal closure and enhance drought tolerance (negative effect)	[267]
		NFYA5	ABA-dependent expression	[268]
		NF-YB1, HDG11, SDIR1	Improve drought tolerance	[269–271]

TABLE 11.1 *(Continued)*

S.no.	Crop species	Gene	Function	References
		CYP707A3	Involved in ABA catabolism pathway (negative effect on drought tolerance)	[272]
		Sal1	Negative regulator of drought tolerance	[273]
6.	Tomato	CBF1 (DREB1B)	Activated gene expression, catalase1 coupled with decreased accumulation of H2O2	[274]
		NCED genes	ABA biosynthesis	[275]
		LeNCED1	Elevates endogenous ABA	[276]
		TSRF1	Improve osmotic and drought stress	[277]
		JERF1, JERF3	Enhance drought tolerance	[278]
7.	Barley	HVA1	Tolerance to soil water deficit	[279]
		*Hv*CBF4	Improved drought and salinity stress	[280]
8.	Cowpea	VuNCED1, VuABA1 (ZEP gene)	ABA biosynthesis	[281]
9.	Oat	ADC (arginine decarboxylase)	Polyamine biosynthesis	[282]
10.	Transgenic tobacco	TPS1 (from yeast)	Encodes trehalose-6-phosphate	[283, 284]
		SacB (from *Bacillus subtilis*)	Encodes for levan sucrase, which takes part in fructan synthesis	[285]
		bet B (from *E. coli*)	Encodes for betaine aldehyde dehydrogenase, involved in the biosynthesis of glycine betaine	[286]
		ODC (from yeast and mouse)	Encodes ornithine decarboxylase	[287, 288]
		SAMDC (from human)	Encodes S-adenosyl methionine decarboxylase	[289]

TABLE 11.1 *(Continued)*

S.no.	Crop species	Gene	Function	References
		SOD (superoxide dismutase)	Provide tolerance to oxidative stress	[216, 290, 291]
		Cu/Zn SOD	Improves photosynthetic performance under chilling stress	[292]
		Mn SOD	Tolerance to oxidative stress in presence of other antioxidant enzymes	[293]
		PMA 1959	Tolerance to drought and salinity	[294]
11.	Transgenic rice	PMA80	Tolerance to drought and salinity	[294]
		pdc1(pyruvate decarboxylase)	Higher PDC activity with survival after submergence	[295]
12.	Transgenic melon	HAL 1	Helps in retaining more K^+ ions under salt stress	[296]
13.	Transgenic potato	Cu/Zn SOD	Improves photosynthetic performance under chilling stress	[297]
14.	Transgenic tomato	HAL 1	Helps in retaining more K^+ ions under salt stress	[298]

11.5.2.2   REGULATING SIGNALING

Plants cope up with the stress by activating numerous responses which are initiated by the signal transduction pathways and involve complex networks that are interconnected at different levels.[104,299–302] However, the mechanism by which plant responds to the adverse conditions is not well understood.[104,303–305] Signal transduction pathways for different abiotic stresses (drought, cold, and salinity) are categorized into major three types:

    i)   Osmotic/oxidative stress signaling
    ii) Ca^{2+} dependent signaling
    iii) Ca^{2+} dependent SOS signaling.

Osmotic/oxidative stress signaling includes MAPK modules, which are involved in the generation of ROS scavenging enzymes and antioxidant compounds as well as osmolytes. While Ca^{2+} dependent signaling lead to the activation of LEA-type genes (such as the DRE/CRT class of genes) which are involved in the production of stress-responsive proteins and Ca^{2+} dependent SOS signaling helps in regulating ion homeostasis.

ABA is considered as the key endogenous messenger known for its regulatory role in integration of environmental adverse effect and plant development programs like embryo and seed development, acquisition of desiccation tolerance and dormancy, flowering, and organogenesis. It also regulates plant growth under non-stressful condition and is essential for vegetative growth in several organs.[306,307] Stomatal behavior is regulated by the transport of ABA from the root to shoot.[308,309] It is also reported that the accumulation of ABA in leaves would directly regulate the stomatal movement.[10,309,310] ABA regulates signals by two ways (a) ABA-dependent signal transduction pathways and (b) ABA- independent signal transduction pathways.[311,312] In ABA-dependent signaling systems two types of TFs mediate stress adaptation (a) the ABA-responsive element-binding protein/ABA-binding factor (AREB/ABF); and (b) the myelocytomatosis oncogene (MYC)/myeloblastosis oncogene (MYB).[155,313,314] Whereas in ABA-independent signal transduction pathway: (a) the CBF/DREB; and (b) the NAC and zincfinger homeodomain (ZF-HD) regulates stress response.[206,314] However, AP2/EREBP (ERF) TFs have been identified which exist in both ABA-dependent and independent pathways.[315,316] ATHK, Histidine kinase receptors belonging to the two-component family are induced in early drought stress in Arabidopsis.[317] Other protein kinases like receptor-like protein kinase 1 (RPK1) (Arabidopsis leucine rich protein)[318] and SPK1 from *P. vulgaris*[223] are the most abundant transcript under drought. At the time of drought, salinity, and cold stress SPK1 is induced by ABA in Arabidopsis.[319] Further the identification of RCARs/PYR1/PYLs which inactivate type 2C protein phosphatases (ABI 1 and ABI 2) illustrates the mechanism of ABA-dependent gene expression and ion channels[320] Identification of other kinases such as MAPKs which are also induced by abiotic stress, suggest that reversible phosphorylation plays an important role in drought signaling.[321,322] Abiotic stresses also induced CDPKs and inositol triphosphate (ITP, a phospholipid) which regulates ABA mediated stomatal closure in plants.[323-325]

In soybean (*Glycine max* L. Merr.) a decrease in relative leaf expansion rate, stomatal conductance, and leaf turgor whereas an increase in leaf and xylem (ABA) and xylem pH was observed under drought stress. It was also found that decrease in stomatal conductance coincided with an increase in xylem (ABA) before any significant change of leaf turgor could be detected, which indicates that chemical signals (seemingly root-originated ABA) regulate stomatal behavior at moderate drought.[311] Yabba and Foster studied the effect of ABA on the root length in eight common bean genotype (BAT 477 (nodulating (nod)), DOR 364 (nod), DOR364 (non-nodulating (nn)), BAT 477 (nn), SEA5, 8-42-M-2, PR9603-22, and XAN176) and found that ABA induced significantly greater root length in the drought suscep-tible check (8-42-M-2) than in the drought resistant check (BAT 477).[326] It has been reported that over expression of *Rhizobium etli* cbb3 oxidase (CFNX713) can alleviate the negative effect of drought and can increase the tolerance of symbiotic nitrogen fixation to drought.[327] Drought reduces the impact of Al toxicity on the root tip by reducing the formation of Al-induced callose and multidrug and toxin extrusion (MATE) expression. Moreover, Al and drought stress together accelerate up regulation of 1-Aminocyclopro-pane-1-carboxylic acid (ACC) oxidase (ACCO) expression and synthesis of zeatin riboside, whereas reduce the concentration of drought-enhanced ABA and expression of nine-cis-epoxycarotenoid dioxygenase (NCED) involved in ABA biosynthesis and the TFs bZIP and MYB, which affect the regulation of ABA-dependent genes (SUS, PvLEA18, KS-DHN, and LTP) in root tips. Thus it may be concluded that drought alleviates Al injury, but Al renders the root apex more drought-sensitive, particularly by impacting the gene regula-tory network involved in ABA signal transduction and cross-talk with other phytohormones necessary for maintaining root growth under drought.[226] Drought stress induces oxidative stress related to membrane damage. The tolerance of a genotype may be evaluated on the basis of its lower lipid peroxidation (LPO) and higher membrane stability, with the levels of its anti-oxidant enzymatic activity. In 2008, three contrasting cultivars of common bean (Dobrudjanski ran, Plovdiv 10, and Prelom) were evaluated for oxidative stress and plant antioxidant system in the response to drought and it was found that cvs. Plovdiv 10 and Prelom can be considered as drought tolerant and cv. Dobrudjanski ran as sensitive to water stress on the basis of the various parameters for stress tolerance evaluation of plants such as electrolyte leakage, LPO, and activity of antioxidant enzymes. It was also suggested that antioxidant enzymes ascorbate peroxidase (APOX), *superoxide dismutase* (SOD), and cata-lase (CAT) might be involved in overcoming of oxidative stress. The increased activities of antioxidant enzymes can act as a damage control system and provide protection from oxidative stress which results in lower LPO and higher membrane stability in tolerant genotypes.[328] The biochemical activities including the water channel and the response of plants to water changing may be adversely affected due to the damaged tissue under drought.[329] Since the last few years, several studies have been made to understand the complex gene expression cascades in stress conditions which can be further used in improving breeding strategies in crops under stress adverse conditions.

11.5.2.3   TRANSGENIC

Genetic transformation, one of the biotechnological approaches which is being widely used throughout the world for abiotic and biotic stress tolerance in various crops from a decade. In this approach, a desire gene is transferred from one species to other which results in the development of new varieties with improved traits.[157] The process of development of transgenic plant is shown in Figure 11.5. Genes that encoded several TFs, heat shock, LEA proteins, and compatible organic osmolytes have been incorporated in various crops by using this approach. The main objective of genetic transformation is to understand the functional role of stress-induced genes in the tolerance response and to improve the tolerance trait in the target genotype.[330,331] This approach works on the principle of introduction of a gene of interest or foreign gene in a desired individual and studying the expression (over expression or suppression) of that gene only under the stress conditions.[169] Transgenic approach overcomes the drawback of all other breeding approaches as in this approach one can introgress the genes that do not exist in gene pool of a species, that are tightly linked with some undesirable trait, or that have a desirable trait with an unknown underlying physiology. Many countries do not favor the use of transgenic products because of some ethical and religious issues in spite of this, there is a 100-fold increase in the biotech crop hectares from 1.7 million ha in 1996 to over 170 million ha in 2012.[332] To develop stress tolerant variety, researchers are transferring stress-inducible genes from one variety to another. This is possible only when the functions of stress-inducible genes are well understood. With this approach, drought tolerant common bean plants can be developed. But this approach is difficult to handle and the rate of recovery of transgenic lines is still limited, due to the lack of availability of efficient protocols for regeneration of transformed undifferentiated tissue. Under drought conditions, plants lose their cellular turbidity causing protein aggregation and misfolding.[333] The stress-tolerant genes encode LEA proteins, a class of heat shock proteins. Through an ultrasonic plus vacuum infiltration-assisted Agrobacterium-mediated transformation technique, the *Brassica napus* group 3 LEA gene was introduced into a local cultivar of Japan common bean cv. Green Light.[334] The transgenic lines showed enhanced growth ability under salt and water deficit stress conditions. The stress tolerance power was more highly correlated at expression level than gene integration level because transgenic lines with high levels of LEA gene expression showed higher tolerance than lines with lower expression level. Kwapata et al. transferred the *Hordeum vulgare* LEA protein type III encoding gene (hva1) into different varieties of *P. vulgaris* and reported development of drought-tolerant transgenic plants.[335] Hence, transgenics and transgenic breeding are an option for improvement of the common bean.

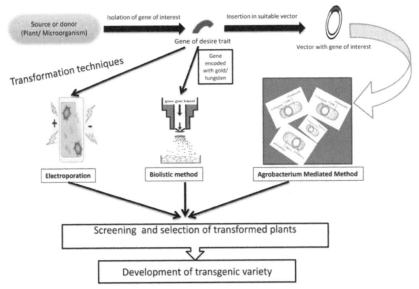

FIGURE 11.5 Various transformation methods used for generation of transgenic plants to tackle different stresses.

11.5.2.4   PROTEOMICS APPROACH

Water deficit stress significantly reduces the yield of economically important crops by altering the plant growth and metabolism which ultimately lead to death of plant.[336–341] One of the recent biotechnological approaches to tackle various biotic and abiotic stresses is to study the structural and functional determinants of cells which help us in interpreting the stress responses occurring in plants.[342,343] The term "proteome" was coined by Marc Wilkins in 1994 which stands for the PROTEins expressed by genOME.[333,344] Over the last decade, applications of proteomics at the technical level in biological fields have greatly increased.[345–353] It has been reported that research carried out in proteomics results in the identification of various new proteins and their function whereas it also unfolds the regulatory networks controlling protein expression.[354] Since the introduction of proteomics it has been used in understanding various mechanisms both in plants and animals. In plant sciences, this approach can be used in studying numerous aspects—cellular and sub cellular, structural and developmental, physiological and genetic studies—for protein profiling of an organism, for comparative expression analysis of two or more protein samples, for understanding of posttranslational modifications (PTMs), for studying protein–protein interactions, and for identification of new biomarkers to detect and monitor specific stress manifestations.[333,344,355] Recently some studies in the field of proteomics (organelle proteome) revealed abiotic stress tolerance mechanism in plants[356] and it has also been reported that proteomics has become a major tool for understanding protein variation of physiological changes taking place in different plant organs.[305,357–362] A comparative proteomic approach which can be used to overcome the stress is explained in Figure 11.6.

Biotic and abiotic stress conditions change the various physiological processes that are directly controlled by genes and activated by different proteins in plants. Proteomic studies in legumes have increased significantly in the last years but few studies have been performed till to date in *P. vulgaris*. Some of the studies conducted in *Phaseolus* sp. are discussed here under this section. Nitric oxide signaling pathway was studied in *Phaseolus aureus* (mung bean) using proteomic approach. In this study, comparative two-dimensional gel electrophoresis (2-DGE) method was used to identify nine proteins including seven downregulated and two upregulated proteins. Six out of these nine proteins found were involved in either photosynthesis or cellular metabolism. Moreover, this study suggested that chloroplasts might be one of the main subcellular targets of NO as it decreases the level of glucose in mung bean[363] In another study a proteomics-based approach

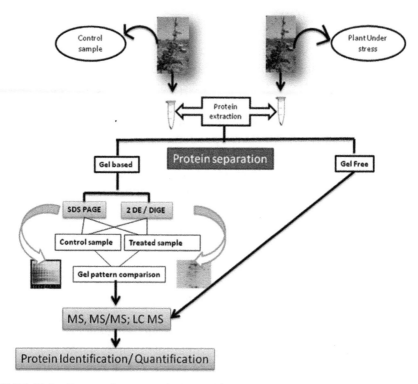

FIGURE 11.6   Comparative proteomics approach.

was used on cultivated bean (*P. vulgaris* L. cv. IDIAP R-3) and maize (*Zea mays* L. cv. Guarare 8128) plants to investigate responses of ozone (O_3) stress. Gel-based proteomics showed a clear modulation of oxidative stress, heat shock, and secondary metabolism-related proteins by O_3. Moreover, potential novel protein markers of ozone stress in leaves of cultivated bean and maize species were identified.[364] The effect of drought stress on early pod fill stage on seed sink strength of a drought-resistant inbred line (SEA 15) and a drought susceptible cultivar (BrSp) was studied. 2-DGE approach was used to assess the quantitative and qualitative changes in leaf protein patterns and a total of 550 proteins out of which 230 differentially expressed leaf proteins were identified. These differentially expressed proteins include 23.5, 15.1, 4.3, and 3.5% were downregulated, upregulated, newly appeared, and disappeared, respectively.[365] Further in 2010 proteomic approach was used to study the nutritional quality of the common bean. Marsolais et al. conducted a study on common bean seed with storage protein deficiency. The results revealed upregulation of sulfur-rich proteins, starch, and raffinose metabolic enzymes whereas secretory pathway was downregulated. This provided information on the pleiotropic phenotype associated with storage protein deficiency in a dicotyledonous seed.[366] In 2010, Mensack et al. used three different omics approaches, transcriptomics, proteomics, and metabolomics, to qualitatively evaluate the diversity of the common bean from two centers of domestication (COD). They found that all three approaches were able to classify the common bean according to their COD.[367] 2D-based proteomics approach was used to analyze common bean seeds. They used three different protein extraction methods (trichloroacetic acid (TCA)–acetone, phenol, and the commercial clean-up kit) to compare impact on the level of quality of downstream protein separation and identification. A total of 50 spots were used in protein identification by mass spectrometry (MS) (MALDI-TOF MS and MALDI-TOF/TOF). Seventy percent of the spots were identified in spite of incomplete genome/protein databases for bean and other legume species. Most of the identified proteins include storage protein, carbohydrate metabolism, defense and stress response, phaseolin, phytohemagglutinin, and lectin-related α-amylase inhibitor.[368] Zadražnik et al. in 2013 used 2-D DIGE technique to compare differences in protein abundance between control and stressed bean plants and identified drought responsive proteins in leaves of drought tolerance and sensitive common bean cultivars differing in drought response. The majority of proteins they identified were classified into functional categories that include energy metabolism, photosynthesis, adenosine triphosphate (ATP) interconversion, protein synthesis and proteolysis, and stress and defense-related proteins.[369] Natarajan et al.[370] identified 237 protein spots from a wild bean, Mexico. TCA-acetone precipitation method, two-dimensional polyacrylamide gel electrophoresis (2D-PAGE) for protein separation, and subsequent MS were used for identification of protein. The identified proteins were digested with trypsin, and analyzed by MALDI/TOF/TOF MS. A total of 141 protein spots were identified by searching the National Center for Biotechnology Information (NCBI) non-redundant databases using the Mascot search engine and 43 unique proteins were found. The majority of proteins are involved in binding, catalytic, nutrient reservoir, antioxidant, transporter, enzyme regulator, structural molecule, and electron carrier activity. This study however revealed that TCA/Acetone extraction and 2D-PAGE are efficient in separating common bean proteins. Pandurangan et al. studied differential response of sulfur nutrition in two genetically related lines of common bean SARC1 and SMARC1NPN1 differing in seed storage protein composition. SMARC1NPN1 lacks phaseolin and lectins but has more sulfur rich proteins resulting in increased concentration of cysteine and methionine. To study the potential effects associated with an increased concentration of sulfur amino acids in the protein pool, the response of the two genotypes to low and high sulfur nutrition was evaluated under controlled conditions. Increase in seed yield, total cysteine and extractible globulins were observed in SMARCNPN1 whereas seed concentrations of sulfur, sulfate, and S methyl cysteine were altered in both the genotypes by treatment with high sulfate. Moreover, proteomic analysis shows an increased level of identified arcelin-like protein4, lipoxygenase3, albumin2, and alpha amylase inhibitor beta chain under high sulfur conditions. It was also observed that Lipxygenase3 accumulation was sensitive to sulfur nutrition in SMARC1NPN1. Further, both the genotypes exhibited a slight increase in yield in response to sulfur treatment under field conditions. By combining proteomics with other omics (genomics, transcriptomics, and metabolomics) approaches we can study the genes, transcripts, proteins, and metabolites of an organism and investigate the various regulatory pathways that are involved in stress tolerance, nutritional fortification, and yield which lead to better understanding of biological processes and improvements in the common bean.[371]

11.6   OUR VISION AND PERSPECTIVE

As discussed in the above sections the impact of abiotic stress especially drought on common bean is evident. Various strategies have been used to tackle these stresses and to develop drought tolerant varieties. At present, we are in process of evaluation of diverse common bean germplasm under rain fed condition, in order to identify some desirable genotypes that can withstand drought stress. Such genotypes will act as a valuable genetic resource for development of drought tolerant common bean varieties. Further genetics/genomics studies are being conducted and we hope this study will help us in tracking novel genes/QTLs for improving drought tolerance of common bean through molecular breeding involving MAS. Moreover, proteomics research will be carried out to identify novel proteins.

11.7   CONCLUSION

The nutritional and medicinal value of common bean makes it food of choice for many individuals in developing countries. Keeping in view the impact of abiotic stress, climate change, and the decrease in common bean production, there is an urgent need to breed common bean drought tolerant varieties that can sustain under low water conditions by using various breeding and biotechnological approaches.

KEYWORDS

- **abiotic stress**
- **common bean**
- **drought stress**
- **conventional breeding**
- **proteomics**

REFERENCES

1. Reddy, G. L. Importance of Agriculture in Indian Economy: Some Issues. *PIJR.* **2015,** *4*, 47–49.
2. Zargar, S. M.; Nazir, M.; Cho, K.; Kim, D. W.; Jones, O. A. H.; Sarkar, A.; Agrawal, S. B.; Shibato, J.; Kubo, A.; Jwa, N. S.; Agrawal, G. K.; Rakwal R. Impact of Climatic Changes on Crop Agriculture: OMICS for Sustainability & Next Generation Crops. In *Sustainable Agriculture and New Bio-Technologies;* Noureddine, B., Ed.; Taylor & Francis (CRC Press): Boca Raton, FL, 2011.
3. Zargar, S. M.; Nazir, M.; Agrawal, G. K.; Rakwal R. OMICS Based Interventions for Climate Proof Crops. *Genomics Appl. Biol.* **2011,** *2* (14), 98–100.
4. Spaldon, S.; Samnotra, R. K.; Chopra, S. Climate Resilient Technologies to Meet the Challenges in Vegetable Production. *Int. J. Curr. Res. Aca. Rev.* **2015,** *3* (2), 28–47.
5. Bates, B. C.; Kundzewicz, Z. W.; Wu, S.; Palutikof, J. P. *Climate Change and Water*; IPCC Technical Paper VI: Geneva, 2008; p 210.
6. Tirado, M. C.; Clarke, R.; Jaykus, L. A.; McQuatters-Gollop, A.; Frank, J. M. Climate Change and Food Safety: A Review. *Food Res. Int.* **2010,** *43,* 1745–1765.
7. IPCC (Intergovernmental Panel on Climate Change); *The Physical Science Basis*; Contribution of Working Group I to the Fourth Assessment Report of the IPCC: Solomon, S., Qin, D., Manning, M., Chen, Z., Marquis, M., Averyt, K. B., Tignor, M., Miller, H. L. Eds.; Cambridge University Press: Cambridge, UK, 2007; p 996.
8. Fischlin, A.; Midgley, G. F.; Price, J. T.; Leemans, R.; Gopal, B.; Turley, C., et al. *Ecosystems, their Properties, Goods and Services. Climate Change 2007: Impacts, Adaptation and Vulnerability*; Contribution of Working Group II to the Fourth Assessment Report of the Intergovernmental Panel on Climate Change: Parry, M. L., Canziani, O. F., Palutikof, J. P., van der Linden, P. J., Hanson, C. E., Eds.; Cambridge University Press: Cambridge, 2007; pp 211–272.

9. Knapp, A. K.; Briggs, J. M.; Koelliker, J. K. Frequency and Extent of Water Limitation to Primary Production in a Mesic Temperate Grassland. *Ecosystems.* **2001,** *4,* 19–28.

10. Chaves, M. M.; Maroco, J. P.; Pereira, J. S. Understanding Plant Response to Drought-from Genes to the Whole Plant. *Funct. Plant Biol.* **2003,** *30,* 239–264.

11. McMichael, A. J.; Campbell-Lendrum, D. H.; Corvalan, C. F.; Ebi, K. L.; Githeko, A. K.; Scheraga, J. D.; Woodward, A. *Climate Change and Human Health: Risks and Responses;* World Health Organization: Geneva, 2003; Chapter 6, pp 103–127.

12. Rao, G. G. S. N.; Rao, A. V. M. S.; Rao, V. U. M. Trends in Rainfall and Temperature in Rainfed India in Previous Century. In *Global Climate Change and Indian Agriculture Case Studies from ICAR Network Project;* Aggarwal, P. K., Ed.; ICAR Publication: New Delhi, India, 2009; pp 71–73.

13. IMD; *Annual Climate Summary*; India Meteorological Department, Government of India, Ministry of Earth Sciences: Pune, India, 2010; p 27.

14. Afroza, B.; Wani, K. P.; Khan, S. H.; Jabeen, N.; Hussain, K.; Mufti, S.; Amit, A. Various Technological Interventions to Meet Vegetable Production Challenges in View of Climate Change. *Asian J. Hort.* **2010,** *5* (2), 523–529.

15. Harvest Plus. Iron-Bean 2009. Available online: http://www.harvestplus.org/sites/default/files/ HarvstPlus_Bean_Strategy.pdf (accessed on 26 January 2015).

16. Ibarra-Perez, F.; Ehdaie, B.; Waines, G. Estimation of Out Crossing Rate in Common Bean. *Crop. Sci.* **1997,** *37,* 60–65.

17. Burle, M. L.; Fonseca, J. R.; Kami, J. A.; Gepts, P. Microsatellite Diversity and Genetic Structure Among Common Bean (*Phaseolus vulgaris* L.) Landraces in Brazil, a Secondary Center of Diversity. *Theor. Appl. Genet.* **2010,** *121,* 801–813.

18. Broughton, W. J.; Hernandez, G.; Blair, M.; Beebe, S.; Gepts, P.; Vanderleyden, J. Beans (*Phaseolus* spp.)-Model Food Legumes. *Plant Soil* **2003,** *252,* 55–128.

19. Welch, R. M.; House, W. A.; Beebe, S.; Cheng, Z. Genetic Selection for Enhanced Bioavailable Levels of Iron in Bean (*Phaseolus vulgaris* L.) Seeds. *J. Agric. Food Chem.* **2000,** *48,* 3576–3580.

20. Blair, M.; Gonzales, L. F.; Kimani, P. M.; Butare, L. Genetic Diversity, Inter-Gene Pool Introgression and Nutritional Quality of Common Beans (*Phaseolus vulgaris* L.) from Central Africa. *Theor. Appl. Genet.* **2010,** *121,* 237–248.

21. Duke J., A. Y. Ayensa E. S., *Medicinal Plants of China;* Reference Publications: Algonac, MI, 1985.

22. Pennington, J. A. T.; Young, B. Iron Zinc Copper Manganese Selenium and Iodine in Foods from the United States Total Diet Study. *Food Compost. Anal.* **1990,** *3,* 166–184.

23. Souci, S. W.; Fachmann, W.; Kraut, H. *Food Composition and Nutrition Tables;* Medpharm: Stuttgart, Germany, 1994; Vol. 5.

24. Guzman N. E.; Infante J. A. G. Antioxidant Activity in Cotyledon of Black and Yellow Common beans (*Phaseolus vulgaris* L), *Res. J. Biol. Sci.* **2007,** *2* (1), 112–117.

25. WMO; *The Global Climate System*; Climate System Monitoring Dec 1988–May 1991. WMO World Climate Data and Monitoring Programme, United Nations Environment Programme (UNEP): Nairobi, Kenya, 1992; pp 73–74.

26. Beebe, S.; Gonzalez, A. V.; Rengifo, J. Research on Trace Minerals in the Common Bean. *Food Nutr. Bull.* **2000,** *21* (4), 387–391.

27. Tryphone, G. M.; Nchimbi-Msolla, S. Diversity of Common Bean (*Phaseolus vulgaris* L.) Genotypes in Iron and Zinc Contents Under Screen House Conditions. *Afr. J. Agr. Res.* **2010,** *5* (8), 738–747.

28. Nouet, C.; Motte, P.; Hanikenne, M. Chloroplastic and Mitochondrial Metal Homeostasis. *Trends Plant Sci.* **2001,** *16,* 395–404.

29. Yruela, I. Transition Metals in Plant Photosynthesis. *Metallomics.* **2013,** *5,* 1090– 1109.

30. Marschner, H. *Mineral Nutrition of Higher Plants;* Academic Press: Salt Lake City, UT, 1995.

31. Tewari, R. K.; Kumar, P.; Neetu.; Sharma, P. N. Signs of Oxidative Stress in the Chlorotic Leaves of Iron Starved Plants. *Plant Sci.* **2005,** *169,* 1037–1045.

32. Bashir, K.; Nagasaka, S.; Itai, R.; Kobayashi, T.; Takahashi, M.; Nakanishi, H.; Mori, S.; Nishizawa, N. K. Expression and Enzyme Activity of Glutathione Reductase is Upregulated by Fe Deficiency in Graminaceous Plants. *Plant Mol. Biol.* **2007,** *65* (3), 277–284.

33. Ishimaru, Y.; Bashir, K.; Nishizawa, N. Zn Uptake and Translocation in Rice Plants. *Rice.* **2011,** *4,* 21–27.

34. Rhodes, D.; Klug, A. Zinc Fingers. *Sci. Am.* **1993,** *268,* 56–65.

35. Vallee, B. L.; Falchuk, K. H. The Biochemical Basis of Zinc Physiology. *Physiol. Rev.* **1993,** *73,* 79–118.

36. Beard, J. L. Iron Biology in Immune Function, Muscle Metabolism and Neuronal Functioning. (Suppl.). *Am. J. Clin. Nutr.* **2001,** *131,* 568S–580S.

37. Rasmussen, K. M. Is there a Causal Relationship between Iron Deficiency or Iron Deficiency Anemia and Weight at Birth, Length of Gestation and Perinatal Mortality? (Suppl.). *Am. J. Clin. Nutr.* **2001,** *131,* 590S–603S.

38. King, J. C.; Keen, C. L. Zinc. In *Modern Nutrition in Health and Disease;* Shils, M. E., Olson, J. A., Shike, M., Ross, A. C., Eds.; Lippincott Willimans & Wilkins: Baltimore, MD, 1999; pp 223–249.

39. Bouis, H. E. Micronutrient Fortification of Plants Through Plant Breeding: Can it Improve Nutrition in Man at Low Cost? *Proc. Nutr. Soc.* **2003,** *62,* 403–411.

40. Solomons, N. W.; Ruz, M. Zinc and Iron Interaction: Concepts and Perspectives in the Developing World. *Nutr. Res.* **1997,** *17,* 177–185.

41. Beninger, C. W.; Hosfield, G. L. Antioxidant Activity of Extracts, Condensed Tannin Fractions, and Pure Flavonoids from *Phaseolus vulgaris* L. Seed Coat Color Genotypes. *J. Agric. Food Chem.* **2003,** *51,* 7879–7883.

42. Akond, M.; Golam, A. S. M.; Khandaker, L.; Berthold, J.; Gates, L.; Peters, K.; Delong, H.; Hossain, K. Anthocyanin, Total Polyphenols and Antioxidant Activity of Common Bean. *Am. J. Food Technol.* **2011,** *6* (5), 385–394.

43. González de Mejía, E.; Hanzkins, C. N.; Paredes, L. O; y Shannon, A. M. The Lectins and Lectins-Like Proteins of Tepary beans (*Phaseolus acutifolius*) and Tepary-Common Bean (*Phaseolus vulgaris*) Hybrids. *J. Food Biochem.* **1990,** *14,* 117–126.

44. Blumenthal, M.; Busse, W. R.; Goldberg, A.; Grenwald, J.; Hall, T.; Riggins, C. W.; Rister, R. S. *The Complete German Commission E Monographs-Therapeutic Guide to Herbal Medicines;* American Botanical Council; Austin, TX; Integrative Medicine Communications: Boston, MA, 1998; p 157.

45. Rafi, M. M.; Vastano, B. C. Novel Polyphenol Molecule Isolated from Licorice Root (*Glycrrhizaglabra*) Induces Apoptosis, G2/M Cell Cycle Arrest, and Bcl-2 Phosphorylation in Tumor Cell Lines. *J. Agric. Food Chem.* **2002,** *50,* 677–684.

46. Abass, S. M.; Mohamed, H. I. Alleviation of Adverse Effects of Drought Stress on Common Bean (*Phaseolus vulgaris* L.) by Exogenous Application of Hydrogen Peroxide. *Bangladesh J. Bot.* **2011,** *41*(1), 75–83.

47. Foyer, C. H.; Noctor, G. Oxygen Processing in Photosynthesis: Regulation and Signaling. *New Phytol.* **2002,** *146,* 359–388.

48. Graham, P. H.; Ranalli, P. Common Bean (*Phaseolus vulgaris* L.). *Field Crops Res.* **1997,** *53,* 131–146.

49. Torres, F. M. L.; Contour, A. D.; Zuily, F. Y.; Pham, T. A. T. Molecular Cloning of Glutathione Reductase cDNAs and Analysis of GR Gene Expression in Cowpea and Common Bean Leaves during Recovery from a Moderate Drought Stress. *J. Plant Physiol.* **2008,** *165,* 514–521.

50. Xoconostle-Cazares, B.; Ramires-Ortega, F. A.; Leonardo, F. E.; Ruiz-Medrano, R. Drought Tolerance in Crop Plants. *Am. J. Plant Physiol.* **2011,** *5,* 1–14.

51. Anonymous. *FAO Production Year Book;* FAO: Rome, Italy, 2003; Vol. 56, pp 109–110.

52. Grant, O. M. Understanding and Exploiting the Impact of Drought Stress on Plant Physiology. In *Abiotic Stress Responses in Plants: Metabolism, Productivity and Sustainability;* Ahmad, P.; Prasad, M. N. V.; Eds.; Springer Science + Business Media: Berlin, Germany, 2012; pp 89–104.

53. Naeem, M.; Khan, M. N.; Masroor, M.; Khan, M. M. A.; Moinuddin. Adverse Effects of Abiotic Stresses on Medicinal and Aromatic Plants and their Alleviation by Calcium. In *Plant Acclimation to Environmental Stress;* Tuteja, N., Gill, S. S., Eds.; Springer Science + Business Media: New York, **2013,** pp 101–146.

54. Acosta-Gallegos, J. A.; Kohashi-Shibata, V. Effect of Water Stress on Growth and Yield of Indeterminate Dry Bean (*Phaseolus vulgaris*) Cultivars. *Field Crops Res.* **1989,** *20,* 81–90.

55. Acosta-Gallegos, J. A.; Adams, M. W. Plant Traits and Yield Stability of Dry Bean (*Phaseolus vulgaris*) Cultivars under Drought Stress. *J. Agr. Sci.* **1991,** *117,* 213–219.

56. Muñoz-Perea, C. G.; Terán, H.; Allen, R. G.; Wright, J. L.; Westermann, D. T.; Singh, S. P. Selection for Drought Resistance in Dry Bean Landraces and Cultivars. *Crop Sci.* **2006,** *46,* 2111–2120.

57. Asfaw, A.; Almekinders, C. J. M.; Blair, M. W.; Struik, P. C. Participatory Approach in Common Bean (*Phaseolus vulgaris* L.) Breeding for Drought Tolerance for Southern Ethiopia. *Plant Breed.* **2012,** *131,* 125–134.

58. Beebe, S. E.; Rao, I. M.; Blair, M. W.; Acosta-Gallegos, J. A. Phenotyping Common Beans for Adaptation to Drought. *Front. Physiol.* **2013,** *4,* 35.

59. Valentine, A. J.; Benedito, V. A.; Kang, Y. Legume Nitrogen Fixation and Soil Abiotic Stress: From Physiology to Genomic and Beyond. *Annu. Plant Rev.* **2011,** *42,* 207–248.

60. Toker, C.; Canci, H.; Yildirim, T.; Evalution of Perennial Wild Cicer Species for Drought Resistance. *Gen. Res. Crop Evol.* **2007,** *54,* 1781–1786.

61. Charlson, D. V.; Bhatnagar, S.; King, C. A., et al. Polygenic Inheritance of Canopy Wilting in Soybean [*Glycine max* (L.) Merr.]. *Theor. Appl. Gen.* **2009,** *119,* 587–594.

62. Khan, H. R.; Paull, J. G.; Siddique, K. H. M.; Stoddard, F. L. Faba Bean Breeding for Drought-Affected Environments: A Physiological and Agronomic Perspective. *Field Crops Res.* **2010,** *115,* 279–286.

63. Toker, C.; Mutlu, N. Breeding for Abiotic Stress. In *Biology and Breeding of Food Legumes;* Pratap, A., Kumar, J., Eds.; CAB International: Wallingford, Oxfordshire, 2011; pp 241–260.

64. Impa, S. M.; Nadaradjan, S.; Jagadish, S. V. K. Drought Stress Induced Reactive Oxygen Species and Anti-Oxidants in Plants. In *Abiotic Stress Responses in Plants: Metabolism, Productivity and Sustainability;* Ahmad, P., Prasad, M. N. V., Eds.; Springer Science + Business Media: Berlin, Germany, 2012; pp 131–147.

65. Hasanuzzaman, M.; Gill, S. S.; Fujita, M. Physiological Role of Nitric Oxide in Plants Grown Under Adverse Environmental Conditions. In *Plant Acclimation to Environmental Stress;* Tuteja, N., Gill, S. S., Eds.; Springer Science + Business Media: New York, 2013; pp 269–322.

66. Pagano, M. C. Drought Stress and Mycorrhizal Plants. In *Use of Microbes for the Alleviation of Soil Stresses;* Miransari, M., Ed.; Springer Science + Business Media: New York, 2014; pp 97–110.

67. Asfaw, A.; Blair, M. W.; Struik, P. C. Multi Environment Quantitative Trait Loci Analysis for Photosynthate Acquisition, Accumulation and Remobilization Traits in Common Bean Under Drought Stress. *G3(Bethesda).* **2012,** *2,* 579–595.

68. Nielsen, D. C.; Nelson, N. Black Bean Sensitivity to Water Stress at Various Growth Stages. *Crop Sci.* **1998,** *38,* 422–427.

69. Vallejo, P. R.; Kelly, J. D. Traits Related to Drought Resistance in Common Bean. *Euphytica.* **1998,** *99,* 127–136.

70. Asfaw, A.; Blair, M. W. Quantitative Trait Loci for Rooting Pattern Traits of Common Beans Grown Under Drought Stress Versus Nonstress. *Mol. Breed.* **2012,** *30,* 681–695.

71. Singh, S. P. Broadening the Genetic Base of Common Bean Cultivars. *Crop Sci.* **2001,** *41*(6), 1659–1675.

72. Masaya, P.; White, J. W. Adaptation to Photoperiod and Temperature. In *Common Beans: Research for Crop Improvement;* Schoonhoven, A., Voysest, O., Eds.; CAB. International: Wallingford; CIAT: Cali, Colombia, 1991; pp 445–500.

73. Serraj, R.; Sinclair, T. R. N2 Fixation Response to Drought in Common Bean (*Phaseolus vulgaris* L.). *Ann. Bot.* **1998,** *82,* 229–234.

74. Guida dos Santos, M.; Vasconcelos, R.; Ferraz, R.; Pimentel, C. Gas Exchange Aad Yield Response to Foliar Phosphorus Application in *Phaseolus vulgaris* L. under Drought. *Braz. J. Plant Physiol.* **2004,** *16,* 171–179.

75. Cornic, G. Drought Stress Inhibits Photosynthesis by Decreasing Stomatal Aperture: Not by Affecting ATP Synthesis. *Trend Plant Sci.* **2000,** *5,* 187–188.

76. Parry, M. A. J.; Androlojc, P. J.; Khan, S.; Lea, P. J.; Keys, A. J. Rubisco Activity: Effects of Drought Stress. *Ann. Bot.* **2002,** *89,* 833–839.

77. Bota, J.; Medrano, H.; Flexas, J. Is Photosynthesis Limited by Decreased Rubisco Acivity and RuBP Content under Progressive Water Stress? *New Phytol.* **2004,** *162,* 671–681.

78. Kim, J. Y.; Mahé, A.; Brangeon, J.; Prioul, J. L. A Maize Vacuolar Invertase, IVR2, is Induced by Water Stress. Organ/ Tissue Specificity and Diurnal Modulation of Expression. *Plant Physiol.* **2000,** *124,* 71–84.

79. Jaleel, C. A.; Sankar, B.; Murali, P. V.; Gomathinayagam, M.; Lakshmanan, G. M. A.; Panneerselvam, R. Water Deficit Stress Effects on Reactive Oxygen Metabolism in *Catharanthus roseus*; Impacts on Ajmalicine Accumulation. *Colloids Surf. B.* **2008,** *62,* 105–111.

80. Farooq, M.; Basra, S. M. A.; Wahid, A.; Cheema, Z. A.; Cheema, M. A.; Khaliq, A. Physiological Role Of Exogenously Applied Glycinebetaine in Improving Drought Tolerance of Fine Grain Aromatic Rice (*Oryza. sativa* L.). *J. Agron. Crop Sci.* **2008,** *194,* 325–333.

81. Liu, F.; Jensen, C. R.; Andersen, M. N. Drought Stress Effect on Carbohydrate Concentration in Soybean Leaves and Pods During Early Reproductive Development: Its Implication in Altering Pod Set. *Field Crops Res.* **2004,** *86* (1), 1–13.

82. Vadez, V.; Sinclair, T. R.; Serraj, R. Asparagine and Ureide Accumulation in Nodules and Shoot as Feedback Inhibitors of N2 Fixation in Soybean. *Physiol Plant.* **2000,** *110,* 215–223.

83. Ladrera, R.; Marino, D.; Larrainzar, E.; Gonzalez, E. M.; Arrese-Igor, C. Reduced Carbon Availability to Bacteroids and Elevated Ureides in Nodules, but not in Shoots, are Involved in Nitrogen Fixation Response to Early Drought in Soybean. *Plant Physiol.* **2007,** *145,* 539–546.

84. Gonzalez, E. M.; Gordon, A. J.; Hames, C. L.; Arrese-Igor, C. The Role of Sucrose Synthase in the Response of Soybean Nodules to Drought. *J. Exp. Bot.* **1995,** *46,* 1515–1523.

85. Gonzalez, E. M.; Aparicio-Tejo, P. M.; Gordon, A. J.; Minchin, F. R.; Royuela, M.; Arrese-Igor, C. Water-Deficit Effects on Carbon and Nitrogen Metabolism of Pea Nodules. *J. Exp. Bot.* **1998,** *49,* 1705–1714.

86. Ramos, M. L. G.; Gordon, A. J.; Michin, F. R.; Sprent, J. I.; Parsons, R. Effect of Water Stress on Nodule Physiology and Biochemistry of a Drought Tolerant Cultivar of Common Bean (*Phaseolus vulgaris* L.). *Ann. Bot.* **1999,** *83,* 57–63.

87. Streeter, J. G. Effects of Drought on Nitrogen Fixation in Soybean Root Nodules. *Plant Cell Environ.* **2003,** *26,* 1199–1204.

88. Galvez, L.; Gonzalez, E. M.; Arrese-Igor, C. Evidence for Carbon Flux Shortage and Strong Carbon/Nitrogen Interactions in Pea Nodules at Early Stages of Water Stress. *J. Exp. Bot.* **2005,** *56,* 2551–2561.

89. De Carvalho, M. H. C. Drought Stress and Reactive Oxygen Species: Production, Scavenging and Signaling. *Plant Signal. Behav.* **2008,** *3,* 156–165.

90. Arrese-Igor, C.; Gordon, C.; Gonzalez, E. M.; Marino, D.; Ladrera, R.; Larrainzer, E.; Gil-Quintana, E. Physiological Response of Legume Nodules to Drought. *Plant Stress.* **2011,** *5* (1), 24–31.

91. Neslihan-Ozturk, Z.; Talam'el, V.; Deyholos, M.; Michalowski, C. B.; Galbraith, D. W.; Gozukirmizi, N.; Tuberosa, R.; Bohnert, H. J. Monitoring Large-Scale Changes in Transcript Abundance in Drought- and Salt Stressed Barley. *Plant Mol. Biol.* **2002,** *48,* 551–573.

92. Ishitani, M.; Nakamura, T.; Han, S. Y.; Takabe, T. Expression of the Betaine Aldehyde Dehydrogenase Gene in Barley in Response to Osmotic Stress and Abscisic Acid. *Plant Mol. Biol.* **1995,** *27,* 307–315.

93. Lisse, T.; Bartels, D.; Kalbitzer, H. R.; Jaenicke, R. The Recombinant Dehydrin-Like Desiccation Stress Protein from the Resurrection Plant *Craterostigma plantagineum* Displays no Defined Three Dimensional Structure in its Native State. *Biol. Chem.* **1996,** *377,* 555–561.

94. Ingram, J.; Bartels, D. The Molecular Basis of Dehydration Tolerance in Plants. *Ann. Rev. Plant Physiol. Plant Mol. Biol.* **1996,** *47,* 377–403.

95. Izanloo, A.; Condon, A. G.; Langridge, P.; Tester, M.; Schnurbusch, T. Different Mechanisms of Adaptation to Cyclic Water Stress in Two South Australian Bread Wheat Cultivars. *J. Exp. Bot.* **2008,** *59,* 3327–3346.

96. Xu, Z. Z.; Zhou, G. S.; Shimizu, H. Effects of Soil Drought with Nocturnal Warming on Leaf Stomatal Traits and Mesophyll Cell Ultrastructure of a Perennial Grass. *Crop Sci.* **2009,** *49,* 1843–1851.

97. Geber, M. A.; Dawson, T. E. Genetic Variation in and Covariation Between Leaf Gas Exchange, Morphology and Development in *Polygonum arenastrum*, an Annual Plant. *Oecologia.* **1990,** *85,* 153–158.

98. Schulze, E. D. Carbon Dioxide and Water Vapor Exchange in Response to Drought in the Atmosphere and the Soil. *Annu. Rev. Plant Physiol.* **1986,** *37,* 247–274.

99. Jackson, R. B.; Sperry, J. S.; Dawson, T. E. Root Water Uptake and Transport: Using Physiological Processes in Global Predictions. *Trends Plant Sci.* **2000,** *5,* 482–488.

100. Morgan, J. M. Osmoregulation and Water Stress in Higher Plants. *Ann. Rev. Plant Physiol.* **1984,** *35,* 299–319.

101. Bartoli, C. G.; Simontacchi, M.; Tambussi, E.; Beltrano, J.; Montaldi, E.; Puntarulo, S. Drought and Watering-Dependent Oxidative Stress: Effect on Antioxidant Content in *Triticum aestivum* L. Leaves. *J. Exp. Bot.* **1999,** *50,* 375–85.

102. Peñuelas, J.; Munné-Bosch, S.; Llusià, J.; Filella, I. Leaf Reflectance and Photo- and Antioxidant Protection in Field-Grown Summer-Stressed *Phillyrea angustifolia.* Optical Signals of Oxidative Stress? *New Phytol.* **2004,** *162,*115–24.

103. Shen, B.; Jensen, R. G.; Bohnert, H. J. Increased Resistance to Oxidative Stress in Transgenic Plants by Targeting Mannitol Biosynthesis to Chloroplasts. *Plant Physiol.* **1997,** *113,*1177–1183.

104. Shinozaki, K.; Yamaguchi-Shinozaki, K. Gene Networks Involved in Drought Stress Tolerance and Response. *J. Exp. Bot.* **2007,** *58,* 221–227.

105. Siddiqui, M. H.; Khan, M. N.; Mohammad, F.; Khan, M. M. A. Role of Nitrogen and Gibberellin (GA3) in the Regulation of Enzyme Activities and in Osmoprotectant Accumulation in *Brassica juncea* L. Under Salt Stress. *J. Agron. Crop Sci.* **2008,** *194,* 214–224.

106. Hoffmann, A. A.; Merilä, J. Heritable Variation and Evolution Under Favourable and Unfavourable Conditions. *Trends Ecol. Evol.* **1999,** *14,* 96–101.

107. Sherrard, M. E.; Maherali, H.; Latta, R. G. Water Stress Alters the Genetic Architecture of Functional Traits Associated with Drought Adaptation in *Avena barbata. Evolution.* **2009,** *63,* 702–715.

108. Maherali, H.; Caruso, C. M.; Sherrard, M. E.; Latta, R. G. Adaptive Value and Costs of Physiological Plasticity to Soil Moisture Limitation in Recombinant Inbred Lines of *Avena barbata. Am. Nat.* **2010,** *175,* 211–224.

109. Bray, E. A. Plant Responses to Water Deficit. *Trends Plant Sci.* **1997,** *2,* 48–54.

110. Barnabás, B.; Jäger, K.; Fehér, A. The Effect of Drought and Heat Stress on Reproductive Processes in Cereals. *Plant Cell Environ.* **2008,** *31,* 11–38.

111. Acosta-Gallegos, J. A.; White, J. W. Phenological Plasticity as an Adaptation by Common Bean to Rainfed Environments. *Crop Sci.* **1995,** *35,*199–204.

112. Rao, I. M. Role of Physiology in Improving Crop Adaptation to Abiotic Stresses in the Tropics: The Case of Common Bean and Tropical Forages. In *Handbook of Plant and Crop Physiology;* Pessarakli, M., Ed.; Marcel Dekker: New York, 2001; pp 583–613.

113. Beebe, S. E.; Rao, I. M.; Blair, M. W.; Acosta-Gallegos, J. A. Phenotyping Common Beans for Adaptation to Drought. Drought Phenotyping in Crops: from Theory to Practice. In *Generation Challenge Program Special Issue on Phenotyping;* Ribaut, J. M., Monneveux, P., Eds.; CGIAR: Texcoco, Mexico. 2010; pp 311–334.

114. Beebe, S. E.; Rao, I. M.; Cajiao, C., Grajales, M. Selection for Drought Resistance in Common Bean Also Improves Yield in Phosphorus Limited and Favorable Environments. *Crop. Sci.* **2008,** *48,* 582–592.

115. McCouch, S. Diversifying Selection in Plant Breeding. *PLoS ONE.* **2004,** *2* (10), 1507–1512.

116. MariotE. J. Ecofisiologia do Feijoeiro. *in O feijão no Paraná;* IAPAR: Londrina, 1989; pp 27–38.

117. White, J. W. In *Preliminary Results of the Bean International Drought Yield Trial (BIDYT)*, Proceedings of the International Bean Drought Workshop, Cali, Colombia. 1987; CIAT: Cali, Colombia, 1987; pp 126–145.

118. White, J. W.; Ochoa, R.; Ibarra, F.; Singh, S. P. Inheritance of Seed Yield, Maturity and Seed Weight of Common Bean (*Phaseolus vulgaris* L.) Under Semi-Arid Rainfed Conditions. *J Agric Sci.* **1994,** *122,* 265–273.

119. Ter'an, H.; Singh, S. P. Comparison of Sources and Lines Selected for Drought Resistance in Common Bean. *Crop Sci.* **2002,** *42,* 64–70.

120. Frahm, M. A.; Rosas, J. C.; Mayek-P'erez, N.; L'opez-Salinas, E.; Acosta-Gallegos, J. A.; Kelly, J. D. Breeding Beans for Resistance to Terminal Drought in the Lowland Tropics. *Euphytica.* **2004,** *136,* 223–232.

121. Ishitani, M.; Rao, I.; Wenz, P.; Beebe, S.; Tohme, J. Integration of Genomics Approach with Traditional Breeding Towards Improving Abiotic Stress Adaptation: Drought and Aluminum Toxicity as Case Studies. *Field Crop Res.* **2004,** *90,* 35–45.

122. Acosta-Gallegos, J. A.; Ochoa-Marquez, R.; Arrieta-Montiel, M. P.; Ibarra-Perez, F.; Pajarito-Ravelero, A.; Sanchez-Valdez, I. Registration of 'Pinto Villa' Common Bean. *Crop Sci.* **1995,** *35,* 1211.

123. Blair M. W., Galeano C. H., Tovar E., Muñoz-Torres M. C., Castrillón A. V., et al. "Development of a Mesoamerican Intra-Genepool Genetic Map for Quantitative Trait Detection in a Drought Tolerant · Susceptible Common Bean (*Phaseolus vulgaris* L.) Cross," *Mol. Breed.* **2012,** *29* (1), 71–88.

124. Fleury, D.; Jefferies, S.; Kuchel, H.; Langridge, P. Genetic and Genomic Tools to Improve Drought Tolerance in Wheat. *J. Exp. Bot..* **2010,** *61* (12), 3211–3222.

125. Zargar, S. M., et al. Molecular Marker Assisted Approaches (MMAA) for Enhancing Low Water Stress Tolerance in Common Bean: An Update. *Mol. Plant Breed.* **2014,** *5* (14), 1–12.

126. Sinclair, T. R.; Purcell, L. C.; Vadez, V.; Serraj, R.; King, C. A.; Nelson, R. Identification of Soybean Genotypes with N_2 Fixation Tolerance to Water Deficits. *Crop Sci.* **2000,** *40,* 1803–1809.

127. Sinclair, T. R. Is Transpiration Efficiency a Viable Plant Trait in Breeding for Crop Improvement? *Funct. Plant Biol.* **2012,** *39,* 359–365.

128. White, J. W.; Castillo, J. A. Evaluation of Diverse Shoot Genotypes on Selected Root Genotypes of Common Bean Under Soil Water Deficits. *Crop Sci.* **1992,** *32,* 762–765.

129. Devi, J. M.; Sinclair, T. R.; Beebe, S. E.; Rao, I. M. Comparison of Common Bean (*Phaseolus vulgaris* L.) Genotypes for Nitrogen Fixation Tolerance to Soil Drying. *Plant Soil.* **2013,** *364,* 29–37.

130. White, J. W.; Singh S. P. Sources and Inheritance of Earliness in Tropically Adapted Indeterminate Common Bean. *Euphytica.* **1991,** *55,* 15–19.

131. Sponchiado, B. N.; White, J. W.; Castillo, J. A.; Jones, P. G. Root growth of Four Common Bean Cultivars in Relation to Drought Tolerance in Environments with Contrasting Soil Types. *Exp. Agr.* **1989,** *25,* 249–257.

132. Beebe, S.; Velasco, A.; Pedraza, F. In *Marcaje de Genes Para Rendimiento en Condiciones de alto y Bajo fosforo en las Accesiones de Frijol G21212 y BAT 881,* Poster Presented in the VI Reuni~ao Nacional de Pesquisa de Feij~ao, Salvador, Brazil, Nov 21–26, 1999.

133. Blair, M. W.; Mu~noz, M. C.; Beebe, S. E. *QTL Analysis of Drought and Abiotic Stress Tolerance in Common Bean RIL Populations*; Annual Report, Biotechnology Research Project: CIAT: Cali, Colombia. 2002; pp 68–72.

134. Blum, A. *Plant Breeding for Water-Limited Environments;* Springer: New York, 2010; p 272.

135. Blum, A. Effective Use of Water (EUW) and not Water-Use Efficiency (WUE) is the Target of Crop Yield Improvement Under Drought Stress. *Field Crops Res.* **2009,** *112,* 119–123.

136. Munns, R.; James, R. A.; Sirault, X. R. R.; Furbank, R. T.; Jones, H. G. New Phenotyping Methods for Screening Wheat and Barley for Beneficial Responses to Water Deficit. *J. Exp. Bot.* **2010,** *61,* 3499–3507.

137. Klaedke, S. M.; Cajiao, C.; Grajales, M.; Polania, J.; Borrero, G.; Guerrero, A.; Rivera, M.; Rao, I.; Beebe, S. E.; L'eon, J. Photosynthate Remobilization Capacity from Drought-Adapted Common Bean (*Phaseolus vulgaris* L.) Lines can Improve Yield Potential of Interspecific Populations within the Secondary Gene Pool. *J. Plant Breed. Crop Sci.* **2012,** *4,* 49–61.

138. Habibi, G. Influence of Drought on Yield and Yield Components in White Bean. *World Acad. Sci. Engin. Technol.* **2011,** *55,* 244–253.

139. Rao, I. M.; Beebe, S. E.; Polan'ia, J.; Grajales, M.; Cajiao, C.; Garc'ia, R.; Ricaurte, J.; Rivera, M. In *Physiological Basis of Improved Drought Resistance in Common Bean: The Contribution of Photosynthate Mobilization to Grain,* Inter Drought III: The 3rd International Conferemce on Integrated Approaches to Improve Crop Production under Drought-Prone Environments, Shanghai, China, Oct 11–16, 2009; FAO: Rome, Italy, 2009.

140. Rao, I. M.; Beebe, S. E.; Polania, J.; Ricaurte, J.; Cajiao, C.; Garc'ia, R.; Rivera, M. Can Tepary Bean be a Model for Improvement of Drought Resistance in Common Bean? *Afr. Crop Sci. J.* **2013,** *21,* 265–281.

141. Assefa, T.; Beebe, S. E.; Rao, I. M.; Cuasquer, J. B.; Duque, M. C.; Rivera, M.; Battisti, A.; Lucchin, M. Pod Harvest Index as a Selection Criterion to Improve Drought Resistance in White Pea Bean. *Field Crops Res.* **2013,** *148,* 24–33.

142. Araújo, S. S.; Beebe, S.; Crespi, M.; Delbreil, B.; González, E. M.; Gruber, V.; Lejeune-Henaut, I.; Maria, W. L.; Monteros, J.; Prats, E.; Rao, I.; Vadez, V.; Vaz Patto, M. C. Abiotic Stress Responses in Legumes: Strategies Used to Cope with Environmental Challenges, *Crit. Rev. Plant Sci.* **2015,** *34* (1–3), 237–280.

143. Mohamed, M. F.; Schmitz-Eiberger, N.; Keutgen, N.; Noga, G. Comparative Drought Postponing and Tolerance Potentials of Two Tepary Bean Lines in Relation to Seed Yield. *Afr. Crop Sci. J.* **2005,** *13,* 49–60.

144. Mejía-Jiménez, A.; Muñoz, C.; Jacobsen, H. J.; Roca, W. M.; Singh, S. P. Interspecific Hybridization Between Common and Tepary Beans: Increased Hybrid Embryo Growth, Fertility, and Efficiency of Hybridization Through Recurrent and Congruity Backcrossing. *Theor. Appl. Genet.* **1994,** *88,* 324–331.

145. Muñoz, L. C.; Blair, M. W.; Duque, M. C.; Tohme, J.; Roca, W. Introgression in Common Bean x Tepary Bean Interspecific Congruity-Backcross Lines as Measured by AFLP Markers. *Crop Sci.* **2004,** *44,* 637–645.

146. Singh, S. P. Selection for Water Stress Tolerance Interracial Populations of Common Bean. *Crop Sci.* **1995,** *35,* 118–124.

147. Beebe, S.; Lynch, J.; Galwey, N.; Tohme, J.; Ochoa, I. A Geographical Approach to Identify Phosphorus-Efficient Genotypes among Landraces and Wild Ancestors of Common Bean. *Euphytica.* **1997,** *95,* 325–336.

148. Schneider, K. A.; Brothers, M. E.; Kelly, J. D. Marker-Assisted Selection to Improve Drought Tolerance in Common Bean. *Crop Sci.* **1997,** *37,* 51–60.

149. Miklas, P. N.; Kelly, J. D.; Beebe, S. E.; Blair, M. W. Common Bean Breeding for Resistance against Biotic and Abiotic Stresses: From Classical to MAS Breeding. *Euphytica.* **2006,** *147,* 105–131.

150. Schneider, K. A.; Rosales-Serna, R.; Ibarra-P′erez, F.; Cazares- Enriquez, B.; Acosta-Gallegos, J. A.; Ramirez-Vallejo, P.; Wassimi, N.; Kelly, J. D. Improving Common Bean Performance Under Drought Stress. *Crop Sci.* **1997,** *37,* 43–50.

151. Shinozaki, K.; Yamaguchi-Shinozaki, K. Molecular Responses to Dehydration and Low Temperature: Differences and Cross-Talk Between two Stress Signalling Pathways. *Curr. Opin. Plant. Biol.* **2000,** *3,* 217–223.

152. Seki, M.; Kamei, A.; Yamaguchi-Shinozaki, K.; Shinozaki, K. Molecular Responses to Drought, Salinity and Frost: Common and Different Paths for Plant Protection. *Curr. Opin. Biotech.* **2003,** *14,*194–199.

153. Bartels, D.; Sunkar, R. Drought and Salt Tolerance in Plants. *Crit. Rev. Plant. Sci.* **2005,** *24,* 23–58.

154. Marcotte, W. R.; Russell, S. H.; Quatrano, R. S. Abscisic Acid-Responsive Sequences from the Em Gene of Wheat. *Plant Cell.* **1989,** *1,* 969–976.

155. Abe, H.; Yamaguchi-Shinozaki, K.; Urao, T.; Iwasaki, T.; Hosokawa, D.; Shinozaki, K. Role of Arabidopsis MYC and MYB Homologs in Drought- and Abscisic Acid-Regulated Gene Expression. *Plant Cell.* **1997,** *9,* 1859–1868.

156. Ashraf, M. J.; Athar, H. R.; Harris, P. J. C.; Kwon, T. R. Some Prospective Strategies for Improving Crop Salt Tolerance. *Adv. Agron.* **2008,** *97,* 45–110.

157. Ashraf, M. Inducing Drought Tolerance in Plants: Recent Advances. *Biotech. Adv.* **2010,** *28,* 169–183.

158. Bohnert, H. J.; Ayoubi, P.; Borchert, C.; Bressan, R. A.; Burnap, R. L.; Cushman, J. C.; Cushman, M. A.; Deyholos, M.; Fischer, R.; Galbraith, D. W. A Genomics Approach Towards Salt Stress Tolerance. *Plant Physiol. Biochem.* **2001,** *39,* 295–311.

159. Seki, M.; Narusaka, M.; Abe, H.; Kasuga, M.; Yamaguchi-Shinozaki, K.; Carninci, P., et al. Monitoring the Expression Pattern of 1300 Arabidopsis Genes Under Drought and Cold Stresses Using a Full-Length cDNA Microarray. *Plant Cell.* **2001,** *13,* 61–72.

160. Knight, H.; Knight, M. R. Abiotic Stress Signalling Pathways: Specificity and Cross-talk. *Trends Plant Sci.* **2001,** *6,* 262–267.

161. Chen, W.; Provart, N. J.; Glazebrook, J.; Katagiri, F.; Chang, H. S.; Eulgem, T., et al. Expression Profile Matrix of Arabidopsis Transcription Factor Genes Suggests their Putative Functions in Response to Environmental Stresses. *Plant Cell.* **2002,** *14,* 559–574.

162. Fowler, S.; Thomashow, M. F. Arabidopsis Transcriptome Pro-Filing Indicates that Multiple Regulatory Pathways are Activated during Cold Acclimation in Addition to the CBF Cold Response Pathway. *Plant Cell.* **2002,** *14,* 1675–1690.

163. Rudrabhatla, P.; Rajasekharan, R. Developmentally Regulated Dual-Specificity Kinase from Peanut that is Induced by Abiotic Stresses. *Plant Physiol.* **2002,** *130,* 380–390.

164. Yamaguchi-Shinozaki, K.; Kasuga, M.; Liu, Q.; Nakashima, K.; Sakuma, Y.; Abe, H., et al. Biological Mechanisms of Drought Stress Response. *JIRCAS Working Report.* **2002,** *23,* 1–8.

165. Shinozaki, K.; Yamaguchi-Shinozaki, K.; Seki, M. Regulatory Network of Gene Expression in the Drought and Cold Stress Responses. *Curr. Opin. Plant Biol.* **2003,** *6,* 410–417.

166. Bray, E. A.; Bailey-Serres, J.; Weretilnyk, E. Responses to Abiotic Stresses. In *Biochemistry and Molecular Biology of Plants;* Gruissem, W., Buchannan, B., Jones, R., Eds; American Society of Plant Physiologists: Rockville, MD, 2000; pp 1158–1249.

167. Wang, W. X.; Vinocur, B.; Shoseyov, O.; Altman, A. Biotechnology of Plant Osmotic Stress Tolerance: Physiological and Molecular Considerations. *Acta Hort.* **2001**, *560*, 285–292.

168. Zhu, J. K. Cell Signaling Under Salt, Water and Cold Stresses. *Curr. Opin. Plant Biol.* **2001**, *4*, 401–406.

169. Ludwig, A.; Romeis, T.; Jones, J. D. CDPK Mediated Signalling Pathways: Specificity and Cross-Talk. *J. Exp. Bot.* **2004**, *55*,181–188.

170. Frank, W.; Munnik, T.; Kerkmann, K.; Salamini, F.; Bartels, D. Water Deficit Triggers Phospholipase D Activity in the Resurrection Plant *Craterostigma plantagineum*. *Plant Cell.* **2000**, *12*, 111–124.

171. Choi, H. I.; Hong, J. H.; Ha, J.; Kang, J. Y.; Kim, S. Y. ABFs, a Family of ABA-Responsive Element Binding Factors. *J. Biol. Chem.* **2000**, *275*, 1723–1730.

172. Blumwald, E. Sodium Transport and Salt Tolerance in Plants. *Curr. Opin. Cell Biol.* **2000**, *12*, 431–434.

173. Riechmann, J. L.; Heard, J.; Martin, G.; Reuber, L.; Jiang, C.; Keddie, J., et al. Arabidopsis Transcription Factors: Genome-Wide Comparative Analysis among Eukaryotes. *Science.* **2000**, *290*, 2105–2110.

174. Guo, A.; He, K.; Liu, D.; Bai, S.; Gu, X.; Wei, L.; Luo, J. DATF: A Database of Arabidopsis Transcription Factors. *Bioinformatics.* **2005**, *21*, 2568–2569.

175. Iida, K.; Seki, M.; Sakurai, T.; Satou, M.; Akiyama, K.; Toyoda, T.; Konagaya, A.; Shinozaki, K. RARTF: Database and Tools for Complete Sets of Arabidopsis Transcription Factors. *DNA Res.* **2005**, *12*, 247–256.

176. Su, J.; Shen, Q.; Ho, T-H.; Wu, R. Dehydration- Stress-Regulated Transgene Expression in Stably Transformed Rice Plants. *Plant Physiol.* **1998**, *117*, 913–922.

177. Xiong, L.; Zhu, J. K. Abiotic Stress Signal Transduction in Plants: Molecular and Genetic Perspectives. *Physiol. Plant.* **2001**, *112*, 152–166.

178. Kirch, H. H.; Philips, J.; Bartels, D.; Scheel, D.; Wasternack, C. Eds.; *Plant Signal Transduction: Frontiers in Molecular Biology;* Oxford University Press: Oxford, 2001.

179. Bartels, D.; Salamini, F. Desiccation Tolerance in the Resurrection Plant *Craterostigma plantagineum*. A Contribution to the Study of Drought Tolerance at the Molecular Level. *Plant Physiol.* **2001**, *127*,1346–1353.

180. Xiong, L.; Schumaker, K. S.; Zhu, J. K. Cell Signaling during Cold, Drought, and Salt Stress. *Plant Cell.* **2002**, *14*, S165–S183.

181. Mahalingam, R.; Gomez-Buitrago, A. M.; Eckardt, N.; Shah, N.; Guevara-Garcia, A.; Day, P.; Raina, R.; Fedoroff, N. V. Characterizing the Stress/Defense Transcriptome of Arabidopsis. *Genome Biol.* **2003**, *4*, R20.

182. Seki, M.; Narusaka, M.; Ishida, J.; Nanjo, T.; Fujita, M.; Oono, Y., et al. Monitoring the Expression Profiles of 7000 Arabidopsis Genes Under Drought, Cold, and High-Salinity Stresses Using a Full-Length cDNA Microarray. *Plant J.* **2002**, *31*,2 79–92.

183. Close, T. J.; Kortt, A. A.; Chandler, P. M. A cDNA-Based Comparison of Dehydration-Induced Proteins (Dehydrins) in Barley and Corn. *Plant Mol. Biol.* **1989**, *13*, 95–108.

184. Hanin, M.; Brini, F.; Ebel, Ch.; Toda, Y.; Takeda, S.; Masmoudi, K. Plant Dehydrins and Stress Tolerance. Versatile Proteins for Complex Mechanisms. *Plant Signal. Behav.* **2011**, *6*,1503–1509.

185. Sun, X.; Xi, D. H.; Feng, H.; Du, J.B.; Lei, T.; Liang, H. G.; Lin, H. H. The Dual Effects of Salicylic Acid on Dehydrin Accumulation in Water-Stressed Barley Seedlings. *Russ. J. Plant Physiol.* **2009**, *56*, 3488–3594.

186. Mouillon, J. M.; Eriksson, S. K.; Harryson, P. Mimicking the Plant Cell Interior Under Water Stress by Macromolecular Crowding: Disordered Dehydrin Proteins are Highly Resistant to Structural Collapse. *Plant Physiol.* **2008,** *148,*1925–1937.

187. Souer, E.; vanHouwelingen, A.; Kloos, D.; Mol, J.; Koes, R. The No Apical Meristem Gene of Petunia is Required for Pattern Formation in Embryos and Flowers and is Expressed at Meristem and Primordia Boundaries. *Cell.* **1996,** *85,*159–170.

188. Aida, M.; Ishida, T.; Fukaki, H.; Fujisawa, H.; Tasaka, M. Genes Involved in Organ Separation in Arabidopsis: An Analysis of the Cup-Shaped Cotyledon Mutant. *Plant Cell.* **1997,** *9,* 841–857.

189. John, I.; Hackett, R.; Cooper, W.; Drake, R.; Farrell, A.; Grierson, D. Cloning and Characterization of Tomato Leaf Senescence-Related cDNAs. *Plant Mol. Biol.* **1997,** *33,* 641–651.

190. Sablowski, R. W. M.; Meyerowitz, E. M. A Homolog of NO APICAL MERISTEM is an Immediate Target of the Floral Homeotic Genes APETALA3/PISTILLATA. *Cell.* **1998,** *92,* 93–103.

191. Xie, Q.; Frugis, G.; Colgan, D.; Chua, N. H. Arabidopsis NAC1 Transduces Auxin Signal Downstream of TIR1 to Promote Lateral Root Development. *Genes. Dev.* **2000,** *14,* 3024–3036.

192. Collinge, M.; Boller, T. Differential Induction of Two Potato Genes, *Stprx2* and *StNAC,* in Response to Infection by *Phytophthora infestans* and to Wounding. *Plant Mol. Biol.* **2001,** *46,* 521–529.

193. Duval, M.; Hsieh, T. F.; Kim, S. Y.; Thomas, T. L. Molecular Characterization of AtNAM: A Member of the Arabidopsis NAC Domain Superfamily. *Plant Mol. Biol.* **2002,** *50,* 237–248.

194. Hegedus, D., Yu, M.; Baldwin, D.; Gruber, M.; Sharpe, A.; Parkin, I., et al. Molecular Characterization of *Brassica napus* NAC Domain Transcriptional Activators Induced in Response to Biotic and Abiotic Stress. *Plant Mol. Biol.* **2003,** *53,* 383–397.

195. Vroemen, C. W.; Mordhorst, A. P.; Albrecht, C.; Kwaaitaal, M.; de Vries, S. C. The CUP-SHAPED COTYLEDON3 Gene is Required for Boundary and Shoot Meristem Formation in Arabidopsis. *Plant Cell.* **2003,** *15,* 1563–1577.

196. Fujita, M.; Fujita, Y.; Maruyama, K.; Seki, M.; Hiratsu, K.; Ohme-Takagi, M., et al. A Dehydration-Induced NAC Protein, RD26, is Involved in a Novel ABA-Dependent Stress-Signaling Pathway. *Plant J.* **2004,** *39,* 863–876.

197. Tran, L. S. P.; Nakashima, K.; Sakuma, Y.; Simpson, S. D.; Fujita, Y.; Maruyama, K., et al. Isolation and Functional Analysis of Arabidopsis Stress-Inducible NAC Transcription Factors that Bind to a Drought-Responsive Cis-Element in the Early Responsive to Dehydration Stress 1 Promoter. *Plant Cell.* **2004,** *16,* 2481–2498.

198. Mitsuda, N.; Seki, M.; Shinozaki, K.; Ohme-Takagi, M. The NAC Transcription Factors NST1 and NST2 of Arabidopsis Regulate Secondary Wall Thickenings and Are Required for Anther Dehiscence. *Plant Cell.* **2005,** *17,* 2993–3006.

199. Oh, S. K.; Lee, S.; Yu, S. H.; Choi, D. Expression of a Novel NAC Domain-Containing Transcription Factor (CaNAC1) is Preferentially Associated with Incompatible Interactions between Chili Pepper and Pathogens. *Planta.* **2005,** *222,* 876–887.

200. Uauy, C.; Distelfeld, A.; Fahima, T.; Blechl, A.; Dubcovsky, J. A NAC Gene Regulating Senescence Improves Grain Protein, Zinc and Iron Content in Wheat. *Science.* **2006,** *314,* 1298–1301.

201. Mitsuda, N.; Iwase, A.; Yamamoto, H.; Yoshida, M.; Seki, M.; Shinozaki, K.; et al. NAC Transcription Factors, NST1 and NST3, are Key Regulators of the Formation of Secondary Walls in Woody Tissues of Arabidopsis. *Plant Cell.* **2007,** *19,* 270–280.

202. Yoo, Y. S.; Kim, Y.; Kim, Y. S.; Lee, J. S.; Ahn, H. J. Control of Flowering Time and Cold Response by a NAC Domain Protein in Arabidopsis. *PLoS ONE.* **2007,** *2,* 642.

203. Jensen, M. K.; Hagedorn, P. H.; de Torres-Zabala, M.; Grant, M. R.; Rung, J. H.; Collinge, D. B., et al. Transcriptional Regulation by an NAC (NAM-ATAF1,2-CUC2) Transcription Factor Attenuates ABA Signalling for Efficient Basal Defence Towards *Blumeria graminis F* Sp Hordei in Arabidopsis. *Plant J.* **2008,** *56,* 867–880.

204. Yamaguchi-Shinozaki, K.; Shinozaki, K. Transcriptional Regulatory Networks in Cellular Responses and Tolerance to Dehydration and Cold Stresses. *Annu. Rev. Plant Biol.* **2006,** *57,* 781–803.

205. Udvardi, M. K.; Kakar, K.; Wandrey, M.; Montanari, O.; Murray, J.; Andriankaja, A., et al. Legume Transcription Factors: Global Regulators of Plant Development and Response to the Environment. *Plant Physiol.* **2007,** *144,* 538–549.

206. Nakashima, K.; Ito, Y.; Yamaguchi-Shinozaki, K. Transcriptional Regulatory Networks in Response to Abiotic Stresses in Arabidopsis and Grasses. *Plant Physiol.* **2009,** *149,* 88–95.

207. Tran, L-S. P.; Quach, T.; Guttikonda, S. K.; Aldrich, D. L.; Rajesh, K.; Neelakandan, A., et al. Molecular Characterization of Stress-Inducible *GmNAC* Genes in Soybean. *Mol. Gen. Genomics.* **2009,** *281,* 647–664.

208. Mochida, K.; Yoshida, T.; Sakurai, T.; Yamaguchi- Shinozaki, K.; Shinozaki, K.; Tran, L-S. P. In Silico Analysis of Transcription Factor Repertoire and Prediction of Stress Responsive Transcription Factors in Soybean. *DNA Res.* **2009,** *16,* 353–369.

209. Pinheiro, G. L.; Marques, C. S.; Costa, M. D.; Reis, P. A.; Alves, M. S.; Carvalho, C. M., et al. Complete Inventory of Soybean NAC Transcription Factors: Sequence Conservation and Expression Analysis Uncover their Distinct Roles in Stress Response. *Gene.* **2009,** *444,* 10–23.

210. Shinozaki, K.; Dennis, E. S. Cell Signalling and Gene Regulation: Global Analyses of Signal Transduction and Gene Expression Profiles. *Curr. Opin. Plant Biol.* **2003,** *6,* 405–409.

211. Rodriguez-Uribe, L.; O'Connell, M. A. A Root-Specific bZIP Transcription Factor is Responsive to Water Deficit Stress in Tepary Bean (*Phaseolus acutifolius*) and Common Bean (*P. vulgaris*). *J. Exp. Bot.* **2006,** *57,* 1391–1398.

212. Guo, Y.; Gan, S. AtNAP, a NAC Family Transcription Factor, has an Important Role in Leaf Senescence. *Plant J.* **2006,** *46,* 601–612.

213. Kavar, T.; Maras, M.; Kidrič, M.; Šuštar-Vozlič, J.; Meglič, V. Identification of Genes Involved in the Response of Leaves of *Phaseolus vulgaris* to Drought Stress. *Mol. Breed.* **2008,** *21,* 159–172.

214. Hu, H. H.; Dai, M. Q.; Yao, J. Y.; Xiao, B. Z.; Li, X. H.; Zhang, Q. F.; Xiong, L. Z. Over-expressing a NAM, ATAF, and CUC (NAC) Transcription Factor Enhances Drought Resistance and Salt Tolerance in Rice. *PNAS.* **2006,** *103* (35), 12987–12992.

215. Xiao, B. Z.; Huang, Y. M.; Tang, N.; Xiong, L. Z. Over-Expression of *LEA* Gene in Rice Improves Drought Resistance Under Field Conditions. *TAG.* **2007,** *115* (1), 35–46.

216. Oh, S. J.; Kim, Y. S.; Kwon, C. W.; Park, H. K.; Jeong, J. S.; Kim, J. K. Over Expression of the Transcription Factor AP37 in Rice Improves Grain Yield Under Drought Conditions. *Plant Physiol.* **2009,** *150* (3), 1368–1379.

217. Cui, M.; Zhang, W. J.; Zhang, Q.; Xu, Z.Q.; Zhu, Z. G.; Duan, F. P.; Wu, R. Induced Over-Expression of the Transcription Factor OsDREB2A Improves Drought Tolerance in Rice, *Plant Physiol. Biochem.* **2011,** *49* (12), 1384–1391.

218. Mallikarjuna, G.; Mallikarjuna, K.; Reddy, M. K.; Kaul, T. Expression of OsDREB2A Transcription Factor Confers Enhanced Dehydration and Salt Stress Tolerance in Rice (*Oryza sativa* L.). *Biotechnol. Lett.* **2011,** *33* (8), 1689–1697.

219. Gao, T.; Wu, Y. Y.; Zhang, Y. Y.; Liu, L. J.; Ning, Y. S.; Wang, D. J.; Tong, H. N.; Chen, S. Y.; Chu, C. C.; Xie Q. OsSDIR1 Overexpression Greatly Improves Drought Tolerance in Transgenic Rice. *Plant Mol. Biol.* **2011,** *76 (1–2),* 145–156.

220. Khodambashi, M.; Shiran, B.; Gharaghanipour, N. Differential Expression of *CA7* and *NCED* Genes in Common Bean Genotypes under Drought Stress. *J. Agr. Sci. Technol.* **2013,** *15* (1), 1491–1499.

221. Colmenero-Flores, J. M.; Campos, F.; Garciarrubio, A.; Covarrubias, A. A. Characterization of *Phaseolus vulgaris* cDNA Clones Responsive to Water Deficit: Identification of a Novel Late Embryogenesis Abundant-Like Protein. *Plant Mol. Biol.* **1997,** *35* (4), 393–405.

222. Barrera-Figueroa, C. E.; Peña-Castro, J. M.; Acosta-Gallegos, J. A.; Ruiz-Medrano, R.; Xoconostle-Cázares, B. Isolation of Dehydration-Responsive Genes in a Drought Tolerant Common Bean Cultivar and Expression of a Group 3 Late Embryogenesis Abundant mRNA in Tolerant and Susceptible Bean Cultivars. *Fun. Plant Biol.* **2007,** *34* (4), 368–381.

223. Montalvo-Hernández, L.; Piedra-Ibarra, E.; Gómez-Silva, L.; Lira-Carmona, R.; Acosta-Gallegos, J. A.; Vazquez Medrano, J.; Xoconostle Cázares, B.; Ruíz Medrano, R. Differential Accumulation of mRNAs in Drought-Tolerant and Susceptible Common Bean Cultivars in Response to Water Deficit. *New Phytol.* **2008,** *177,* 102–113.

224. Cortés, A. J.; Chavarro, M. C.; Madriñán, S.; This, D.; Blair, M. W. Molecular Ecology and Selection in Tte Drought-Related Asr Gene Polymorphisms in Wild and Cultivated Common Bean (*Phaseolus vulgaris*L.), *BMC Genet.* **2012,** *13 (1),* 58.

225. Cortés, A. J.; This, D.; Chavarro, C.; Madriñán, S.; Blair, M. W. Nucleotide Diversity Patterns at the Drought-Related DREB2 Encoding Genes in Wild and Cultivated Common Bean (*Phaseolus vulgaris*L.). *Theor. Appl. Genetic.* **2012,** *125* (5), 1069–1085.

226. Yang, Z.; Eticha, D.; Albacete, A.; Rao, I. M.; Roitsch, T.; Horst, W. J. Physiological and Molecular Analysis of the Interaction Between Aluminium Toxicity and Drought Stress in Common Bean (*Phaseolus vulgaris*). *J. Exp. Bot.* **2012,** *63* (8), 3109–3125.

227. Porcel, R.; Azcón, R.; Ruiz-Lozano, J. M. Evaluation of the role of Genes Encoding for Δ-Pyrroline-5-Carboxylate Synthetase (P5CS) during Drought Stress in Arbuscular Mycorrhizal and Plants. *Physiol. Mol. Plant Pathol.* **2004,** *65* (4), 211–221.

228. Li, W. Y. F.; Shao, G.; Lam, H. M. Ectopic Expression of GmPAP3 Alleviates Oxidative Damage Caused by Salinity and Osmotic Stresses. *New Phytol.* **2008,** *178* (1), 80–91.

229. Yang, L.; Ji, W.; Gao, P.; Li, Y.; Cai, H.; Bai, X.; Chen, Q.; Zhu Y. GsAPK, an ABA-Activated and Calcium-Independent SnRK2-Type Kinase from G. Soja, Mediates the Regulation of Plant Tolerance to Salinity and ABA Stress. *PLoS ONE.* **2012,** *7* (3), 338.

230. Yoo, J. H.; Park, C. Y.; Kim, J. C.; Do Heo, W.; Cheong, M. S.; Park, H. C.; Kim, M. C.; Moon, B. C.; Choi, M. S.; Kang, Y. H. Direct Interaction of a Divergent Cam Isoform and the Transcription Factor, MYB2, Enhances Salt Tolerance in Arabidopsis. *J. Biol. Chem.* **2005,** *280* (5), 3697–3706.

231. Guenther, J. F.; Chanmanivone, N.; Galetovic, M. P.; Wallace, I. S.; Cobb, J. A.; Roberts, D. M. Phosphorylation of Soybean Nodulin 26 on Serine 262 Enhances Water

Permeability and is Regulated Developmentally and by Osmotic Signals. *Plant Cell.* **2003,** *15* (4), 981–991.

232. Rivers, R. L.; Dean, R. M.; Chandy, G.; Hall, J. E.; Roberts, D. M.; Zeidel, M. L. Functional Analysis of Nodulin 26, an Aquaporin in Soybean Root Nodule Symbiosomes. *J. Biol. Chem.* **1997,** *272* (26), 16256–16261.

233. Drobak, B. K. PARF-1, an *Arabidopsis thaliana* FYVE-Domain Protein Displaying a Novel Eukaryotic Domain Structure and Phosphoinositide Affinity. *J. Exp. Bot.* **2002,** *53* (368), 565–567.

234. Clement, M.; Boncompagni, E.; de Almeida-Engler, J.; Herouart, D. Isolation of a Novel Nodulin: A Molecular Marker of Osmotic Stress in *Glycine max/Bradyrhizobium japonicum* Nodule. *Plant Cell Environ.* **2006,** *29* (9), 1841–1852.

235. Zhou, G. A.; Chang, R. Z.; Qiu, L. J. Overexpression of Soybean Ubiquitin-Conjugating Enzyme Gene GmUBC2 Confers Enhanced Drought and Salt Tolerance Through Modulating Abiotic Stress-Responsive Gene Expression in Arabidopsis. *Plant Mol. Biol.* **2010,** *72* (4–5), 357–367.

236. Yamamoto, E.; Karakaya, H. C.; Knap, H. T. Molecular Characterization of Two Soybean Homologs of *Arabidopsis thaliana* CLAVATA1 from the Wild Type and Fasciation Mutant. *Biochim. Biophys. Acta. (BBA).* **2000,** *1491* (1), 333–340

237. Yamamoto, E.; Knap, H. T. Soybean Receptor-Like Protein Kinase Genes: Paralogous Divergence of a Gene Family. *Mol. Biol. Evol.* **2001,** *18* (8), 1522–1531.

238. Le, D. T.; Nishiyama, R.; Watanabe, Y.; Mochida, K.; Yamaguchi-Shinozaki, K.; Shinozaki, K.; Tran, L. S. P. Genome-Wide Expression Profiling of Soybean Two-Component System Genes in Soybean Root and Shoot Tissues Under Dehydration Stress. *DNA Res.* **2011,** *18* (1), 17–29.

239. Horwitz, J. Alpha-crystallin can Function as a Molecular Chaperone. *Proc. Natl. Acad. Sci. USA.* **1992,** *89,* 10449–10453.

240. Gao, F.; Xiong, A.; Peng, R.; Jin, X.; Xu, J.; Zhu, B.; Chen, J.; Yao, Q. OsNAC52, a Rice NAC Transcription Factor, Potentially Responds to ABA and Confers Drought Tolerance in Transgenic Plants. *Plant Cell. Tiss. Org. Culture.* **2010,** *100* (3), 255–262.

241. Chen, M.; Wang, Q. Y.; Cheng, X. G.; Xu, Z. S.; Li, L. C.; Ye, X. G.; Xia, L. Q.; Ma, Y. Z. GmDREB2, a Soybean DRE-Binding Transcription Factor, Conferred Drought and High-Salt Tolerance in Transgenic Plants. *Biochem. Biophys. Res.Commun.* **2007,** *353* (2), 299–305.

242. Chen, M.; Xu, Z. S.; Xia, L. Q.; Li, L. C.; Cheng, X. G.; Dong, J. H.; Wang, Q. Y.; Ma Y. Z. Cold-Induced Modulation and Functional Analyses of the DRE-Binding Transcription Factor Gene, GmDREB3, in Soybean (*Glycine max* L.). *J. Exp. Bot.* **2009,** *60* (1), 121–135.

243. Zhang, G.; Chen, M.; Li, L.; Xu, Z.; Chen, X.; Guo, J.; Ma, Y. Overexpression of the Soybean GmERF3 Gene, an AP2/ERF Type Transcription Factor for Increased Tolerances to Salt, Drought, and Diseases in Transgenic Tobacco. *J Exp. Bot.* **2009,** *60* (13), 3781–3796.

244. Zhang, G.; Chen, M.; Chen, X.; Xu, Z.; Li, L.; Guo, J.; Ma, Y. Isolation and Characterization of a Novel EAR-Motif-Containing Gene GmERF4 from Soybean (*Glycine max* L.), *Mol. Biol. Rep.* **2010,** *37* (2), 809–818.

245. Liao, Y.; Zhang, J. S.; Chen, S. Y.; Zhang, W. K. Role of Soybean GmbZIP132 Under Abscisic Acid and Salt Stresses. *J. Integr. Plant Biol.* **2008,** *50* (2), 221–230.

246. Gao, S. Q.; Chen, M.; Xu, Z. S.; Zhao, C. P.; Li, L.; Xu, H.; Tang, Y.; Zhao, X.; Ma, Y. Z. The Soybean GmbZIP1 Transcription Factor Enhances Multiple Abiotic Stress Tolerances in Transgenic Plants. *Plant Mol. Biol.* **2011,** *75* (6), 537–553.

247. Xie, Z. M.; Zou, H. F.; Lei, G.; Wei, W.; Zhou, Q. Y.; Niu, C. F.; Liao, Y.; Tian, A. G.; Ma, B; Zhang, W. K. Soybean Trihelix Transcription Factors GmGT-2A and GmGT-2B Improve Plant Tolerance to Abiotic Stresses in Transgenic Arabidopsis. *PLoS ONE.* **2009,** *4* (9), 6898.

248. Luo, X.; Bai, X.; Zhu, D.; Li, Y.; Ji, W.; Cai, H.; Wu, J.; Liu, B.; Zhu, Y. GsZFP1, a New Cys2/His2-Type Zinc-Finger Protein, is a Positive Regulator of Plant Tolerance to Cold and Drought Stress. *Planta.* **2011,** *235* (6), 1–15.

249. Zhou, Q. Y.; Tian, A. G.; Zou, H. F.; Xie, Z. M.; Lei, G.; Huang, J.; Wang, C. M.; Wang, H. W.; Zhang, J. S.; Chen S. Y. Soybean WRKY-Type Transcription Factor Genes, GmWRKY13, Drought Stress and Tolerance in Soybean GmWRKY21, and GmWRKY54, Confer Differential Tolerance to Abiotic Stresses in Transgenic Arabidopsis Plants. *Plant Biot. J.* **2008,** *6* (5), 486–503.

250. Zhang, L.; Wang, X. P.; Bi, Y. D.; Zhang, C. Y.; Fan, Y. L.; Wang, L. Isolation and Functional Analysis of Transcription Factor GmWRKY57B from Soybean, *Chin. Sci. Bull.* **2008,** *53* (22), 3538–3545.

251. Zhou, W.; Li, Y.; Zhao, B. C.; Ge, R. C.; Shen, Y. Z.; Wang, G.; Huang, Z. J. Overexpression of *TaSTRG* Gene Improves Salt and Drought Tolerance in Rice. *J. Plant Physiol.* **2009,** *166* (15), 1660–1671.

252. Kurkela, S.; Franck, M. Cloning and Characterization of a Cold-and ABA-Inducible Arabidopsis Gene. *Plant Mol. Biol.* **1990,** *15,* 137–144.

253. Lang, V.; Palva, E. T. The Expression of a Rab-Related Gene, Rab18, is Induced by Abscisic Acid during the Cold Acclimation Process of *Arabidopsis thaliana* (L.) Heynh. *Plant Mol. Biol.* **1992,** *20,* 951–962.

254. Yamaguchi-Shinozaki, K.; Shinozaki, K.; Arabidopsis DNA Encoding Two Desiccation-Responsive rd29 Genes. *Plant Physiol.* **1993,** *101,* 1119–1120.

255. Maruyama, K.; Sakuma, Y.; Kasuga, M.; Ito, Y.; Seki, M.; Goda, H.; Shimada, Y.; Yoshida, S.; Shinozaki, K.; Yamaguchi-Shinozaki, K. Identification of Cold-Inducible Downstream Genes of the Arabidopsis DREB1A/CBF3 Transcriptional Factor Using two Microarray Systems. *Plant J.* **2004,** *38,* 982–993.

256. Singh, S.; Cornilescu, C. C.; Tyler, R. C.; Cornilescu, G.; Tonelli, M.; Lee, M. S.; Markley, J. L. Solution Structure of a Late Embryogenesis Abundant Protein (LEA14) from *Arabidopsis thaliana*, a Cellular Stress-Related Protein. *Protein Sci.* **2005,** *14* (10), 2601–2609.

257. Shi, H.; Ishitani, M.; Kim, C.; Zhu, J. K. The *Arabidopsis thaliana* Salt Tolerance Gene SOS1 Encodes A Putative Na+/H+ Antiporter. *Proc. Nat. Acad. Sci. USA.* **2000,** *97,* 6896–6901.

258. Iuchi, S.; Kobayashi, M.; Taji, T.; Naramoto, M.; Seki, M.; Kato, T.; Tabata, S.; Kakubari, Y.; Yamaguchi-Shinozaki, K.; Shinozaki, K. Regulation of Drought Tolerance by Gene Manipulation of 9-Cis-Epoxycarotenoid Dioxygenase, a Key Enzyme in Abscisic Acid Biosynthesis in Arabidopsis. *Plant J.* **2001,** *27* (4), 325–333.

259. Taji, T.; Seki, M.; Satou, M.; Sakurai, T. Comparative Genomics in Salt Tolerance between Arabidopsis and Arabidopsis-Related Halophyte Salt Cress using Arabidopsis Microarray. *Plant Physiol.* **2004,** *135,* 1697–1709.

260. Kang, J. Y.; Choi, H. I.; Im, M. Y.; Kim, S. Y. Arabidopsis Basic Leucine Zipper Proteins that Mediate Stress-Responsive Abscisic Acid Signaling. *Plant Cell.* **2002,** *14* (2), 343–357.

261. Liu, Q.; Kasuga, M.; Sakuma; Abe; Miura, S.; Yamaguchi-Shinozaki K.; Shinozaki K. Two Transcription Factors, DREB1 and DREB2, with an EREBP/AP2 DNA Binding Domain Separate Two Cellular Signal Transduction Pathwaysin Drought-and Low-Temperature-Responsive Gene Expression, Respectively, in Arabidopsis. *Plant Cell Online.* **1998,** *10* (8), 1391–1406.

262. Kasuga, M.; Liu, Q.; Miura, S.; Yamaguchi-Shinozaki, K.; Shinozaki, K. Improving Plant Drought, Salt, and Freezing Tolerance by Gene Transfer of a Single Stress-Inducible Transcription Factor. *Nat. Biotech.* **1999,** *17* (3), 287–291.

263. Abe, H.; Urao, T.; Ito, T.; Seki, M.; Shinozaki, K.; Yamaguchi-Shinozaki, K. Arabidopsis AtMYC2 (bHLH) and AtMYB2 (MYB) Function as Transcriptional Activators in Abscisic Acid Signaling. *Plant Cell Online.* **2003,** *15* (1), 63–78.

264. Masle, J.; Gilmore, S. R.; Farquhar, G. D. The ERECTA Gene Regulates Plant Transpiration Efficiency in Arabidopsis. *Nature.* **2005,** *436* (7052), 866–870.

265. Lee, S.; Kang, J.; Park, H.; Kim, M. D.; Bae, M. S.; Choi, H.; Kim, S. Y. DREB2C Interacts with ABF2, a bZIP Protein Regulating Abscisic Acid-Responsive Gene Expression, and its Overexpression Affects Abscisic Acid Sensitivity. *Plant Physiol.* **2010,** *153* (2), 716–727.

266. Chini, A.; Grant, J. J.; Seki, M.; Shinozaki, K.; Loake, G. J. Drought Tolerance Established by Enhanced Expression of the CC–NBS–LRR Gene, ADR1, Requires Salicylic Acid, EDS1 and ABI1. *Plant J.* **2004,** *38* (5), 810–822.

267. Cutler, S.; Ghassemian, M.; Bonetta, D.; Cooney, S.; McCourt, P. A Protein Farnesyl Transferase Involved in Abscisic Acid Signal Transduction in Arabidopsis. *Science.* **1996,** *273,* 1239–1241.

268. Li, W.; Oono, Y.; Zhu, J.; He, X.; Wu, J.; Iida, K.; Lu, X.; Cui, X.; Jin, H.; Zhu, J. The Arabidopsis NFYA5 Transcription Factor is Regulated Transcriptionally and Post Transcriptionally to Promote Drought Resistance. *Plant Cell Online.* **2008,** *20* (8), 2238–2251.

269. Nelson, A. E.; Repetti, P. P.; Adams, T. R.; Creelman, R. A.; Wu, J.; Warner, D. C.; Anstrom, D. C.; Bensen, R. J.; Castiglioni, P. P.; Donnarummo, M. G. Plant Nuclear Factor Y (NF-Y) B Subunits Confer Drought Tolerance and Lead to Improved Corn Yields on Water-Limited Acres. *Proc. Nat. Acad. Sci.* **2007,** *104* (42), 16450–16455.

270. Yu, H.; Chen, X.; Hong, Y.; Wang, Y.; Xu, P.; Ke, S.; Liu, H.; Zhu, J.; Oliver, D. J.; Xiang, C. Activated Expression of an Arabidopsis HD-START Protein Confers Drought Tolerance with Improved Root System and Reduced Stomatal Density. *Plant Cell Online.* **2008,** *20* (4), 1134–1151.

271. Zhang, Y.; Li, Y.; Gao, T.; Zhu, H.; Wang, D.; Zhang, H.; Ning, Y. Arabidopsis SDIR1 Enhances Drought Tolerance in Crop Plants. *Biosci. Biotechnol. Biochem.* **2008,** *72* (8), 2251.

272. Umezawa, T.; Okamoto, M.; Kushiro, T.; Nambara, E.; Oono, Y.; Seki, M.; Kobayashi, M.; Koshiba, T.; Kamiya, Y.; Shinozaki, K. CYP707A3, a Major ABA 8′-Hydroxylase Involved in Dehydration and Rehydration Response in *Arabidopsis thaliana. Plant J.* **2006,** *46* (2), 171–182.

273. Manmathan, H.; Shaner, D.; Snelling, J.; Tisserat, N.; Lapitan, N. Virus-Induced Gene Silencing of *Arabidopsis thaliana* Gene Homologues In Wheat Identifies Genes Conferring Improved Drought Tolerance. *J. Exp. Bot.* **2013,** *64 (5),* 1381–1392.

274. Hsieh, T.; Lee, J.; Charng, Y.; Chan, M. Tomato Plants Ectopically Expressing Arabidopsis CBF1 Show Enhanced Resistance to Water Deficit Stress. *Plant Physiol.* **2002,** *130* (2), 618–626.

275. Burbidge, A.; Grieve, T. M.; Jackson, A.; Thompson, A.; Taylor, I. B. Structure and Expression of a cDNA Encoding a Putative Neo Xanthin Cleavage Enzyme (NCE) Isolated from a Wilt-Related Tomato (*Lycopersicon esculentum Mill.*) Library. *J. Exp. Bot.* **1997,** *47,* 2111–2112.

276. Thompson, A. J.; Jackson, A. C.; Symonds, R. C.; Mulholland, B. J.; Dadswell, A. R.; Blake, P. S.; Burbidge, A.; Taylor, I. B. Ectopic Expression of a Tomato 9-Cis-Epoxy-carotenoid Dioxygenase Gene Causes Over-Production of Abscisic Acid. *Plant J.* **2000,** *23,* 363–374.

277. Quan, R.D.; Hu, S. J.; Zhang, Z. L.; Zhang, H. W.; Zhang, Z. J.; Huang, R. F. Over-expression of an ERF Transcription Factor TSRF1 Improves Rice Drought Tolerance. *Plant Biotechnol. J.* **2010,** *8* (4), 476–488.

278. Zhang, H. W.; Liu, W.; Wan, L. Y.; Li, F.; Dai, L. Y.; Li, D. J.; Zhang, Z. J.; Huang R. F. Functional Analyses of Ethylene Response Factor JERF3 with the Aim of Improving Tolerance to Drought and Osmotic Stress in Transgenic Rice. *Transgenic Res.* **2010,** *19* (5), 809–818.

279. Xu, D.; Duan, X.; Wang, B.; Hong, B.; Ho, T. H. D.; Wu, R. Expression of Late Embryo-genesis Abundant Protein Gene HVA1, from Barley Confers Tolerance to Water Deficit and Salt Stress in Transgenic Rice. *Plant Physiol.* **1996,** *110* (1), 249–257.

280. Oh, S. J.; Kwon, C. W.; Choi, D. W.; Song, S. I.; Kim, J. K. Expression of Barley HvCBF4 Enhances Tolerance to Abiotic Stress in Transgenic Rice. *Plant Biotech. J.* **2007,** *5* (5), 646–656.

281. Iuchi, S.; Kobayashi, M.; Yamaguchi-Shinozaki, K.; Shinozaki, K. A Stress-Inducible Gene for 9-Cis-Epoxycarotenoid Dioxygenase Involved in Abscisic Acid Biosynthesis under Water Stress in Drought-Tolerant Cowpea. *Plant Physiol.* **2000,** *123,* 553–562.

282. Masgran, C.; Altabella, T.; Fascas, R.; Flores, D.; Thompson, A. J.; Besford, R. T.; Tiburcio, A. F. Inducible Over Expression of Oat Arginine Decarboxylase in Transgenic Tobacco Plants. *Plant J.* **1997,** *11,* 463–473.

283. Holmstrom, K. O.; Mantyia, E.; Welin, B.; Mandal, A.; Palva, E. T.; Tunnela, O. E.; Londesborough, J. Drought Tolerance in Tobacco. *Nature.* **1996,** *379,* 683–684.

284. Romero, C.; Belles, J. M.; Vaya, J. L.; Serrrano, R.; Culianez-Macia, F. A. Expression of the Yeast Trehalose-6-Phosphate Synthase Gene in Transgenic Tobacco Plants: Pleio-tropic Phenotypes Include Drought Tolerance. *Planta.* **1997,** *201,* 293–297.

285. Pilon-Smith, E. A. H.; Ebskamp, M. J. M.; Paul, M. J.; Jeuken, M. J. W.; Weisbeek, P. J.; Smeekens, S. C. M. Improved Performance of Transgenic Fructan-Accumulating Tobacco Under Drought Stress. *Plant Physiol.* **1995,** *107,* 125–130.

286. Holmstrom, K. O.; Weilin, B.; Mandal, A.; Kristiamdottir, I.; Teeri, T. H.; Lamark, T.; Strom, A. R.; Palva, E. T. Production of Escherichia Coli Betaine-Aldehyde Dehydro-genase an Enzyme Required for the Synthesis of Osmoprotectant Glycine Betaine in Transgenic Plants. *Plant J.* **1994,** *6,* 749–758.

287. Hamill, J. D.; Robins, R. J.; Parr, A. J.; Evans, D. M.; Furze, J. M.; Rhodes, M. J. C. The Use of Polymerase Chain Reaction in Plant Transformation Studies. *Plant Mol. Biol.* **1990,** *15,* 27–38.

288. Descenzo, R. A.; Minocha, S. C. Modulation of Cellular Polyamines in Tobacco by Transfer and Expression of Mouse Ornithine Decarboxylase cDNA. *Plant Mol. Biol.* **1993,** *22,* 113–127.

289. Noh, E.; Minocha, S. C. Expression of Human S-Adenosyl Methionine Decarboxylase in Transgenic Tobacco and its Effect on Polyamine Biosynthesis. *Transgenic Res.* **1994,** *3,* 26–35.

290. Bowler, C.; Slooten, L.; Vandenbranden, S.; Rycke, R. D.; Botterman, J.; Sybesma, C.; van Montagu, M.; Inze, D. Manganese Superoxide Dismutase can Reduce Cellular Damage Mediated by Oxygen Radicals in Transgenic Plants. *EMBO. J.* **1991**, *10,* 1723–1732.

291. Van Camp, W.; Capiau, K; Van Montagu, M.; Inzé, D.; Slooten, L. Enhancement of Oxidative Stress Tolerance in Transgenic Tobacco Plants Overproducing Fe-Superoxide Dismutase in Chloroplasts. *Plant Physiol.* **1996**, *112,*1703–1714.

292. Sen Gupta, A.; Heinen, J. L.; Holady, A. S.; Burke, J. J.; Allen, R. D. Increased Resistance to Oxidative Stress in Transgenic Plants that Over-Express Chloroplastic Cu/Zn Superoxide Dismutase. *Proc. Nat. Acad. Sci. USA.* **1993**, *90,* 1629–1633.

293. Slooten, L.; Capiau, K.; Van Camp, W.; Montagu, M. V.; Sybesma, C.; Inzé, D. Factors Affecting the Enhancement of Oxidative Stress Tolerance in Transgenic Tobacco Over-expressing Manganese Superoxide Dismutase in the Chloroplasts. *Plant Physiol.* **1995**, *107, 737–775.*

294. Cheng, W. H.; Endo, A.; Zhou, L.; Penney, J.; Chen, H. C.; Arroyo, A.; Leon, P.; Nambara, E.; Asami, T.; Seo, M. A Unique Short-Chain Dehydrogenase/Reductase in Arabidopsis Glucose Signaling and Abscisic Acid Biosynthesis and Functions. *Plant Cell.* **2002**, *14,* 2723–2743.

295. Quimlo, C. A.; Torrizo, L. B.; Setter, T. L,; Ellis, M.; Grover, A.; Abrigo, E. M.; Oliva, N. P.; Ella, E. S.; Carpena A. L.; Ito, O.; Peacock, W. J.; Dennis, E.; Datta, S. K. Enhancement of Submergence Tolerance in Transgenic Rice Plants Overproducing Pyruvate Decarboxylase. *J. Plant Physiol.* **2000**, *156,* 516–521.

296. Bordás, M.; Montesinos, C.; Dabauza, M.; Salvador, A.; Roig, L. A.; Serrano, R.; Moreno, V. Transfer of the Yeast Salt Tolerance Gene HAL1 to *Cucumis melo* L. Cultivars and *In Vitro* Evaluation if Salt Tolerance. *Transgenic Res.* **1997**, *5,* 1–10.

297. Perl, A.; Perl-Treves, R.; Galili, S.; Aviv, D.; Shalgi, E.; Malkin, S.; Galun, E. Enhanced Oxidative Stress Defense in Transgenic Potato Expressing Tomato Cu, Zn Superoxide Dismutases. *Theor. Appl. Genet.* **1993**, *85,* 568–576.

298. Gisbert, C.; Rus, A. M.; Bolarin, M. C.; Lopez-Coronado, M.; Arrillaga, I.; Montesinos, C.; Caro, M.; Serrano, R.; Moreno, V. The Yeast HAL1 Gene Improves Salt Tolerance of Transgenic Tomato. *Plant Physiol.* **2000**, *123,* 393–402.

299. Sarwat, M.; Ahmad, P.; Nabi, G.; Hu, X. Ca2 + Signals: The Versatile Decoders of Environmental Cues. *Crit. Rev. Biotechnol.* **2012**, *33,* 97–109.

300. Tran, L. S. P.; Nakashima, K.; Shinozaki, K.; Yamaguchi- Shinozaki, K. Plant Gene Networks in Osmotic Stress Response: From Genes to Regulatory Networks. *Methods Enzymol.* **2007**, *428,* 109–128.

301. Tran, L. S. P.; Urao, T.; Qin, F., et al. Functional Analysis of AHK1/ATHK1 and Cytokinin Receptor Histidine Kinases in Response to Abscisic Acid, Drought and Salt Stress in Arabidopsis. *Proc. Natl. Acad. Sci. USA.* **2007**, *104,* 20623–20628.

302. Tran, L. S. P.; Nishiyama, R.; Yamaguchi-Shinozaki, K.; Shinozaki, K. Potential Utilization of NAC Transcription Factors to Enhance Abiotic Stress Tolerance in Plants by Biotechnological Approach. *GM Crops.* **2010**, *1,* 32–39.

303. Hadiarto, T.; Tran, L. S. Progress Studies of Drought-Responsive Genes in Rice. *Plant Cell Rep.* **2011**, *30,* 297–310.

304. Manavalan, L. P; Guttikonda, S. K.; Tran, L. S.; Nguyen, H. T. Physiological and Molecular Approaches to Improve Drought Resistance in Soybean. *Plant Cell Physiol.* **2009**, *50,* 1260– 1276.

305. Tran, L. S. P.; Mochida, K. Functional Genomics of Soybean for Improvement of Productivity in Adverse Conditions. *Funct. Integ. Genomics.* **2010,** *10,* 447–462.

306. Chow, B.; McCourt, P. Hormone Signalling from a Developmental Context. *J. Expt. Bot.* **2004,** *55,* 247–51.

307. Raghavendra, A. S.; Gonugunta, V. K.; Christmann, A.; Grill, E. ABA Pereception and Signalling. *Trends Plant Sci.* **2010,** *15,* 395–401.

308. Steudle, E. Water Uptake by Roots: Effects of Water Deficit. *J. Exp. Bot.* **2000,** *51,* 1531–1542.

309. Holbrook, N. M.; Shashidhar, V. R.; James, R. A.; Munns, R. Stomatal Control in Tomato with ABA-Deficit Roots: Response of Grafted Plants to Soil Drying. *J. Exp. Bot.* **2002,** *53,* 1503–1514.

310. Liu. F.; Jensen, C. R.; Andersen, M. N. Hydraulic and Chemical Signals in the Control of Leaf Expansion and Stomatal Conductance in Soybean Exposed to Drought Stress. *Funct. Plant Biol.* **2004,** *30,* 65–73.

311. Cutler, S. R.; Rodriguez, P. L.; Finkelstein, R. R.; Abrams, S. R. Abscisic Acid: Emergence of a Core Signaling Network. *Annu. Rev. Plant Biol.* **2010,** *61,* 651–679.

312. Kim. Y. H.; Kim, M. D.; Choi, Y. I.; Park, S. C.; Yun, D. J.; Noh, E. W.; Lee, H. S.; Kwak, S. S. Transgenic Poplar Expressing *Arabidopsis NDPK2* Enhances Growth as well as Oxidative Stress Tolerance. *Plant Biotechnol. J.* **2010,** *9,* 334–347.

313. Busk, P. K.; Page`s, M. Regulation of Abscisic Acid-Induced Transcription. *Plant Mol. Biol.* **1998,** *37,* 425–435.

314. Saibo, N. J. M.; Lourenco, T.; Oliveira, M. M. Transcription Factors and Regulation of Photosynthetic and Related Metabolism Under Environmental Stresses. *Ann. Bot.* **2009,** *103,* 609–623.

315. Yamaguchi-Shinozaki, K.; Shinozaki, K. A Novel Cis-Acting Element in an Arabidopsis Gene is Involved in Responsiveness to Drought, Low-Temperature, or High-Salt Stress. *Plant Cell.* **1994,** *6,* 251–264.

316. Kizis, D.; Page`s, M. Maize DRE-Binding Proteins Dbf1 and Dbf2 are Involved in Rab17 Regulation Through the Drought-Responsive Element in an Aba-Dependent Pathway. *Plant J.* **2002,** *30,* 679–689.

317. Urao, T.; Yakubov, B.; Satoh, R., et al. A Transmembrane Hybrid-Type Histidine Kinase in Arabidopsis Functions as an Osmosensor. *Plant Cell.* **1999,** *11,* 1743–1754.

318. Hong, S. W.; Jon, J. H.; Kwak, J. M.; Nam, H. G. Identification of a Receptor-Like Protein Kinase Gene Rapidly Induced by Abscisic Acid, Dehydration, High Salt and Cold Treatments in *Arabidopsis thaliana*. *Plant Physiol.* **1997,** *113* (4), 1203–1212.

319. Osakabe, Y.; Maruyama, K.; Seki, M.; Satou, M.; Shinozaki, K.; Yamaguchi-Shinozaki, K. Leucine-Rich Repeat Receptor-Like Kinase1 is a Key Membrane-Bound Regulator of Abscisic Acid Early Signaling in Arabidopsis. *Plant Cell.* **2005,** *17,* 1105–1119.

320. Kang, J.; Hwang, J. U.; Lee, M.; Kim, Y. Y.; Assmann, S. M.; Martinoia, E.; Lee, Y. PDR-type ABC Transporter Mediates Cellular Uptake of the Phytohormone Abscisic Acid. *Proc. Natl. Acad. Sci.* **2010,** *107,* 2355–2360.

321. Jonak, C.; Kiegerl, S.; Ligterink, W.; Barker, P. J.; Huskisson, N. S.; Hirt. H. Stress Signaling in Plants: A Mitogen-Activated Protein Kinase Pathway is Activated by Cold and Drought. *Proc. Natl. Acad. Sci. USA.* **1996,** *93,* 11274–11279.

322. Mizoguchi, T.; Irie, K.; Hirayama, T.; Hayashida, N.; Yamaguchi-Shinozaki, et al. A Gene Encoding a Mitogen- Activated Protein Kinase Kinase Kinase is Induced Simultaneously with Genes for a Mitogen- Activated Protein Kinase and an S6 Ribosomal

Protein Kinase by Touch, Cold and Water Stress in *Arabidopsis thaliana*. *Proc. Natl. Acad. Sci. USA.* **1996,** *93,* 765–769.

323. Schroeder, J. I.; Kwak, J. W.; Allen, G. J. Guard Cell Abscisic Acid Signalling and Engineering Drought Hardiness in Plants. *Nature.* **2001,** *410,* 327–330.

324. Ramanjulu, S.; Bartels, D. Drought and Desiccation-Induced Modulation of Gene Expression in Plants. *Plant Cell Environ.* **2002,** *25,* 141–151.

325. Hirayama, T.; Ohto, C.; Mizoguchi, T; Shinozaki, K. A Gene Encoding a Phosphatidylinositol-Specific Phospholipase C is Induced by Dehydration and Salt Stress in *Arabidopsis thaliana. Proc. Natl. Acad. Sci.* **1995,** *92,* 3903–3907.

326. Yabba, M. D.; Foster, E. F. Common Bean Root Response to Abscisic Acid Treatment. *Annu. Rep. Bean Improv. Coop.* **2003,** *46,* 85–86.

327. Talbi, C.; Sa´nchez, C.; Hidalgo-Garcia, A.; Gonza´ez, E. M.; Arrese-Igor, C.; Girard, L.; Bedmar, E. J.; Delgado, M. J. Enhanced Expression of *Rhizobium etli* cbb3 Oxidase Improves Drought Tolerance of Common Bean Symbiotic Nitrogen Fixation. *J. Expt. Bot.* **2012,** *63,* 1–9.

328. Zlatev, Z. S.; Lidon, F. C.; Ramalho, J. C.; Yordanov I. T. Comparison of Resistance to Drought of Three Bean Cultivars. *Biol. Plant.* **2006,** *50* (3), 389–394.

329. Xu, Z. S.; Ni, Z. Y.; Li, Z. Y.; Li, L. C.; Chen, M.; Gao, D. Y.; Yu, X. D.; Liu, P.; Ma, Y. Z. Isolation and Functional Characterization of HvDREB1, a Gene Encoding a Dehydration-Responsive Element Binding Protein in *Hordeum vulgare. J. Plant Res.* **2009,** *122,* 121–130.

330. Zhang, J. Z.; Creelman, R. A.; Zhu, J. K. From Laboratory to Field. Using Information from Arabidopsis to Engineer Salt, Cold, and Drought Tolerance in Crops. *Plant Physiol.* **2004,** *135,* 615–621.

331. Cuartero, J.; Bolarin, M. C.; Moreno, V.; Pineda, B. Molecular Tools for Enhancing Salinity Tolerance in Plants. In *Molecular Techniques in Crop Improvement;* Jain, S. M., Brar, D. S., Eds.; Springer: New York, 2010; pp 373–405.

332. Zargar, S. M.; Bhattacharjee, C.; Rai, R.; Fukao, Y.; Agrawal, G. K.; Rakwal, R. Omics-Based Approaches for Improvement of the Common Bean. In *Omics Technologies and Crops Improvement;* Noureddine, B., Ed.; Taylor & Francis (CRC Press): Boca Raton, FL, 2014; Chapter 12, pp. 271–301.

333. Zhu, J. K. Salt and Drought Stress Signal Transduction in Plants. *Ann. Rev. Plant Biol.* **2002,** *53,* 247–273.

334. Liu, Z. C., Park, B. J.; Kanno, A., et al. The Novel Use of a Combination of Sonication and Vacuum Infiltration in Agrobacterium-Mediated Transformation of Kidney Bean (*Phaseolus vulgaris* L.) with Lea Gene. *Mol Breed.* **2005,** *16,* 189–197.

335. Kwapata, K., Nguyen, T.; Sticklen, M. Genetic Transformation of Common Bean (*Phaseolus vulgaris* L.) with the Gus Colormarker, the Bar Herbicide Resistance, and the Barley (*Hordeum vulgare*) HVA1 Drought Tolerance Genes. *Int. J. Agron.* **2012,** *2012,* 8.

336. Ahmad, P.; Prasad, M. N. V. *Environmental Adaptations and Stress Tolerance in Plants in the Era of Climate Change;* Springer Science + Business Media: New York, 2012.

337. Ahmad, P.; Prasad, M. N. V. *Abiotic Stress Responses in Plants: Metabolism, Productivity and Sustainability;* Springer Science + Business Media: New York, 2012.

338. Ahmad, P.; Umar, S. *Oxidative Stress: Role of Antioxidants in Plants;* Studium Press Pvt. Ltd.: New Delhi, India, 2011.

339. Ahmad, P., Ashraf, M., Younis, M., et al. Role of Transgenic Plants in Agriculture and Biopharming. *Biotechnol. Adv.* **2012,** *30,* 524–540.

340. Ahmad, P.; Jaleel, C. A.; Salem, M. A.; Nabi, G.; Sharma, S. Roles of Enzymatic and Non-Enzymatic Antioxidants in Plants During Abiotic Stress. *Crit. Rev. Biotechnol.* **2010,** *30,* 161–175.

341. Tester, M.; Langridge, P. Breeding Technologies to Increase Crop Production in a Changing World. *Science.* **2010,** *327,* 818–822.

342. Bindschedler, L. V.; Cramer, R. Quantitative Plant Proteomics. *Proteomics.* **2011,** *11,* 756–775.

343. Boisvert, F. M.; Lam, Y. W.; Lamont, D.; Lamond, A. I. A Quantitative Proteomics Analysis of Subcellular Proteome Localization and Changes Induced by DNA Damage. *Mol. Cell Prot.* **2010,** *9,* 457–470.

344. Hakeem, K. R.; Chandna, R.; Ahmad, P.; Iqbal, M.; Ozturk, M. Relevance of Proteomic Investigations in Plant Abiotic Stress Physiology. *OMICS.* **2012,** *16,* 621–635.

345. Thelen, J. J.; Peck, S. C. Quantitative Proteomics in Plants: Choices in Abundance. *Plant Cell.* **2007,** *19,* 3339–33346.

346. Agrawal, G. K.; Rakwal, R. *Plant Proteomics: Technologies, Strategies, and Applications;* Wiley: Hoboken, NJ, **2008.**

347. Bindschedler, L. V.; Palmblad, M.; Cramer, R. Hydroponic Isotope Labelling of Entire Plants (HILEP) for Quantitative Plant Proteomics; an Oxidative Stress Case Study. *Phytochemistry.* **2008,** *69,* 1962–1972.

348. Agrawal, G. K.; Bourguignon, J.; Rolland, N., et al. Plant Organelle Proteomics: Collaborating for Optimal Cell Function. *Mass Spec. Revs.* **2011,** *30,* 772–853.

349. Chen, X.; Ronald, P. C. Innate Immunity in Rice. *Trends Plant Sci.* **2011,** *16,* 451–459.

350. Kaufmann, K.; Smaczniak, C.; de Vries, S.; Angenent, G. C.; Karlova, R. Proteomics Insights into Plant Signaling and Development. *Proteomics.* **2011,** *11,* 744–755.

351. Nanjo, Y.; Nouri, M. Z.; Komatsu, S. Quantitative Proteomics Analyses of Crop Seedlings Subjected to Stress Conditions; a Commentary. *Phytochemistry.* **2011,** *72,* 1263–1272.

352. Yang, L.; Bai, X.; Yang, Y.; Ahmad, P.; Yang, Y.; Hu, X. Deciphering the Protective Role of Nitric Oxide Against Salt Stress at the Physiological and Proteomic Levels in Maize. *J. Proteom. Res.* **2011,** *10,* 4349–4364.

353. Yokthongwattana, C.; Mahong, B.; Roytrakul, S.; Phaonaklop, N.; Narangajavana, J.; Yokthongwattana, K. Proteomic Analysis of Salinity-Stressed *Chlamydomonas reinhardtii* Revealed Differential Suppression and Induction of a Large Number of Important Housekeeping Proteins. *Planta.* **2012,** *235,* 649– 659.

354. Zhang, H.; Han, B.; Wang, T., et al. Mechanisms of Plant Salt Response: Insights from Proteomics. *J. Proteome. Res.* **2012,** *11,* 49–67.

355. Zheng, M.; Wang, Y.; Liu, K.; Shu, H.; Zhou, Z. Protein Expression Changes During Cotton Fiber Elongation in Response to Low Temperature Stress. *J. Plant Physiol.* **2012,** *169,* 399– 409.

356. Evers, D.; Legay, S.; Lamoureux, D.; Hausman, J. F.; Hoffmann, L.; Renaut, J. Towards a Synthetic View of Potato Cold and Salt Stress Response by Transcriptomic and Proteomic Analyses. *Plant Mol. Biol.* **2012,** *78,* 503–514.

357. Acero, F. J. F.; Carbu. M.; El-Akhal, M.; Garrido, C.; Gonza´lez- Rodrı´guez, V. E.; Cantoral, J. M. Development of Proteomics-Based Fungicides: New Strategies for Environmentally Friendly Control of Fungal Plant Diseases. *Int. J. Mol. Sci.* **2011,** *12,* 795–816.

358. Chandramouli, K.; Qian, P. Y. Proteomics: Challenges, Techniques and Possibilities to Overcome Biological Sample Complexity. *Hum. Genomics Proteomics.* **2009,** *8,* 23920.

359. Hossain, Z.; Nouri, M. Z.; Komatsu, S. Plant Cell Organelle Proteomics in Response to Abiotic Stress. *J. Proteome. Res.* **2012,** *11,* 37–48.

360. Yan, S. P.; Zhang, Q. Y.; Tang, Z. C.; Su, W. A.; Sun, W. N. Comparative Proteomic Analysis Provides New Insights into Stress Responses in Rice. *Mol. Cell Prot.* **2006,** *5,* 3–9.

361. Timperio, A. M.; Egidi, M. G.; Zolla, L. Proteomics Applied on Plant Abiotic Stresses: Role of Heat Shock Proteins (HSP). *J. Proteomics.* **2008,** *71,* 391–411.

362. Kosová, K.; Vitámvás, P.; Prášil, I. T., et al. Plant Proteome Changes Under Abiotic Stress Contribution of Proteomics Studies to Understanding Plant Stress Response. *J. Proteomics.* **2011,** *12,* 1301–1322.

363. Lum, H. K.; Lee, C. H.; Butt, Y. K., et al. Sodium Nitroprusside Affects the Level of Photosynthetic Enzymes and Glucose Metabolism in *Phaseolus aureus* (mung bean). *Nitric Oxide.* **2005,** *12,* 220–230.

364. Torres, N. L.; Cho, K.; Shibato, J., et al. Gel-Based Proteomics Reveals Potential Novel Protein Markers of Ozone Stress in Leaves of Cultivated Bean and Maize Species of Panama. *Electrophoresis.* **2007,** *28,* 4369–4381.

365. Gebeyehu, S.; Wiese, H.; Schubert, S. Effects of Drought Stress on Seed Sink Strength and Leaf Protein Patterns of Common Bean Genotypes. *Afr. Crop Sci. J.* **2010,** *18* (2), 75–88.

366. Marsolais, F.; Pajak, A.; Yin, F., et al. Proteomic Analysis of Common Bean Seed with Storage Protein Deficiency Reveals Up-Regulation of Sulfur-Rich Proteins and Starch and Raffinose Metabolic Enzymes, Aad Down-Regulation of the Secretory Pathway. *J. Proteomics.* **2010,** *73,* 1587–1600.

367. Mensack, M. M.; Fitzgerald, V. K.; Ryan, E. P., et al. Evaluation of Diversity among Common Beans (*Phaseolus vulgaris* L.) from Two Centres of Domestication Using "Omics" Technologies. *BMC Genomics.* **2010,** *11,* 686.

368. De Fuente, M. L.; Borrajo, A.; Bermúdez, J.; Lores, M.; Alonso, J.; López, M.; Santalla, M.; De Ron, A. M.; Zapata, C.; Alvarez, G. 2-DE-Based Proteomic Analysis of Common Bean (*Phaseolus vulgaris* L.) Seeds. *J. Proteomics.* **2011,** *7,* 4262– 4 267.

369. Zadražnik, T.; Hollung, K.; Egge-Jacobsen, W., et al. Differential Proteomic Analysis of Drought Stress Response in Leaves of Common Bean (*Phaseolus vulgaris* L.). *J. Proteomics.* **2012,** *78,* 254–272.

370. Natarajana, S. S.; Pastor-Corralesa, M. A.; Khan, F. H.; Garrett, W. M. Proteomic Analysis of Common Bean (*Phaseolus vulgaris* L.) by Two-Dimensional Gel Electrophoresis and Mass Spectrometry. *J. Basic Appl. Sci.* **2013,** *9,* 424–437.

371. Pandurangan, S.; Sandercock, M.; Beyaert, R.; Conn, K. L., Hou, A.; Marsolais, F. Differential Response to Sulfur Nutrition of Two Common Bean Genotypes Differing in Storage Protein Composition. *Plant Physiol.* **2015,** *6* (92), 1–11.

CHAPTER 12

PROTEOMICS OF SEED DEVELOPMENT: A CASE STUDY OF RICE AND SOYBEAN

RAVI GUPTA[1*], CHUL WOO MIN[1], YONG CHUL KIM[1],
SOON WOOK KWON[1], GANESH KUMAR AGRAWAL[2,3],
RANDEEP RAKWAL[2,3,4], and SUN TAE KIM[1**]

[1]*Department of Plant Bioscience, Life and Industry Convergence Research Institute, Pusan National University, Miryang 627707, South Korea*

[2]*Research Laboratory for Biotechnology and Biochemistry (RLABB), GPO Box 13265, Kathmandu, Nepal*

[3]*GRADE (Global Research Arch for Developing Education) Academy Private Limited, Adarsh Nagar 13, Main Road, Birgunj, Nepal*

[4]*Faculty of Health and Sport Sciences & Tsukuba International Academy for Sport Studies (TIAS), University of Tsukuba, 1-1-1 Tennoudai, Tsukuba 3058574, Ibaraki, Japan*

*Corresponding author. E-mail: ravigupta@pusan.ac.kr
**Co-corresponding author. E-mail: stkim71@pusan.ac.kr

CONTENTS

ABSTRACT

Seeds are the most imperative parts of a plant and are the next generation as they contain information critical to plant growth and development. Besides, the seeds of several species including soybean and rice, important food crops for human survival, have socio-economic value. Therefore, the study of seed itself is essential. As the seeds are generally rich in proteins, seed proteomics constitutes an important aspect of studying the seed biology. Proteomics of the seed is not an easy task, as there are difficulties in isolating and resolving its proteins due to the presence of high amounts of carbohydrates, oils, secondary metabolites, and so forth in the seeds. Seed proteomics involves the analysis of total protein component during not only development and germination but also under a variety of abiotic and biotic stress conditions. The analyses involve both gel-based and gel-free approaches, standard to any proteomic study. Looking at the differences between dicot (soybean) and monocot (rice) seeds, one can observe a high degree of variability among their proteomes. Nevertheless, they show a common trend of accumulation in the seed storage proteins (SSPs) that accounts for ~50–80% of the total seed proteins. β-conglycinin and glycinin-A in soybean, and glutelin in rice are some of the common examples of SSPs. It has been shown that these SSPs perform anarray of functions, besides acting as storage proteins. SSPs have been shown to be tightly associated with the nutritional quality of the seeds and hence have direct impact on humans and animals. In this chapter, we present an overview of the seed proteomewhile discussing the future perspective of seed proteomics in soybean and rice as examples.

12.1   INTRODUCTION

Seed development is one of the most crucial aspects of the plant's life cycle as it gives rise to a new generation. The process of seed development is initiated with the fertilization, which involves the fusion of male and female gametes.[1] Male gametes are enclosed in pollen grains which are transferred to the ovule, located inside the ovary, with the help of pollen tubes. In most of the angiosperms, there is always a double fertilization involving fusion of two male gametes with the egg and central cell of the ovule, giving rise to embryo and endosperm, respectively.[1] After fertilization, while ovary develops into the fruit, ovules mature as seeds. During the development of seeds, endosperm provides nutrition to the developing embryos. In some seeds, endosperm is completely consumed by the developing embryos with mature seeds containing no or negligible endosperm. These seeds are called as non-albuminous seeds while seeds which contain endosperm even after maturity, are known as albuminous seeds. Seeds of all the agricultural important legumes are non-albuminous while most of the monocot seeds are albuminous in nature.[2]

From the very beginning, humanity has been dependent on the seeds for their food and feeding the livestock because of their unique reserve composition. During the development, copious amounts of reserves including proteins, oils, carbohydrates (majorly starch), and fibers are accumulated in the seeds, which make these seeds the perfect staple food.[2,3] Nevertheless, the composition of these reserves varies drastically among seeds of different plants. Some seeds are rich in oil, some in proteins, and others in starch. As an example, rice seeds are rich in starch (~85%), rapeseeds in oil (~40%), and soybean in proteins (~50%).[4] However, the biochemical pathways that lead to the accumulation of these reserves in different seeds are not clearly understood.

Proteomics is the large-scale analysis of the whole set of proteins present in a tissue at a particular time point. Proteome of protein composition of a tissue is determined by the specific temporal and spatial expression of a particular set of genes. However, there is absence of 1:1 relationship between the genes and proteins due to various post-transcriptional and post-translational regulations.[5] Moreover, proteins are the final reflectors of gene expression and are real players of the metabolic reactions; therefore, analysis of the proteins can provide a better picture of the metabolic regulation. In the past decade, proteomics studies have significantly increased our understanding on how these reserves are accumulated in different seeds, the details of which are discussed in the following sections.[6]

12.2   SEED STORAGE PROTEINS

Seed development is a complex process and can be subdivided into different phases including fertilization, cell differentiation, reserve accumulation, endoduplication, and dehydration (Fig.12.1). Both monocots and dicots seeds accumulate many high-abundant proteins, commonly referred as seed storage proteins (SSPs), during the period of reserve accumulation, also known as the seed filling stage. Based on the amounts, these SSPs can constitute upto 50–80% of the total protein content of the seeds.[7] Initially it was thought that the only function of these SSPs is to provide nutrition during seed germination by acting as a source of carbon and nitrogen. However, a growing body of evidence suggests that these SSPs have diverse functions other than just acting as storage proteins. Based on their solubility, SSPs are classified as albumins, globulins, prolamins, and glutelins. Albumins are soluble in water, globulins in dilute saline, prolamins in alcohol-water mixtures, and glutelins in dilute alkali or acids.[7] Of these, globulins are further classified as 7S and 11S, based on their sedimentation coefficients. In general, albumins are found in all the seeds, prolamins and glutelins in monocots and globulins in the dicots.

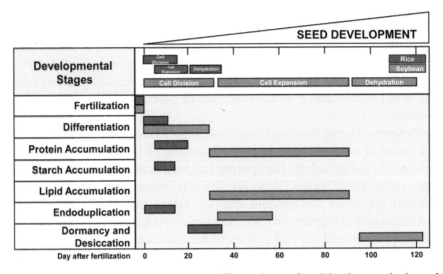

FIGURE 12.1 Graphical picture showing different phases of seed development in rice and soybean.

12.2.1   SEED STORAGE PROTEINS OF SOYBEAN AND RICE

Soybean seeds are highly rich in SSPs. Major SSPs include different subunits of β-conglycinin and glycinin-A. Both of these SSPs belong to globulin class of SSPs of which β-conglycinin is 7S globulin while glycinin is 11S globulins (Fig.12.2). Together these proteins account for ~70–80% of the total seed proteins. Glycinin is composed of five subunits (named G1–G5) which are encoded by five non-allelic genes while β-conglycinin consists of three subunits, α-subunit, α'-subunit, and β-subunit. Of these three subunits of β-conglycinin, α- and α'-subunits are encoded by same gene group while the β-subunit is encoded by another gene group.[8,9]

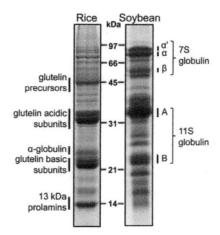

FIGURE 12.2 SDS-PAGE of rice and soybean seed proteins showing different isoforms/subunits of SSPs.

Similar to soybean, rice seeds are also rich in SSPs including glutelins, albumins, globulins, and prolamines. To be more specific, glutelins are the most abundant SSP in the rice seeds constituting upto 80% of the total protein content while the percentage of other SSPs is very less, and is 1–5% for albumins, 4–15% for globulins, and 2–8% for prolamines.[10] Based on the amino acid sequence similarity, glutelins are further classified into four groups (GluA, GluB, GluC, and GluD).[11] Prolamins have been classified into three groups based on their electromobility on sodium dodecyl sulfate-poly-acrylamide *gel* electrophoresis (SDS-PAGE) gels and molecular weights (10, 13, and 16 kDa). Among these 13 kDa prolamins are further classified into three subgroups (class I, II, and III) according to the cysteine residue content.[12] Based on the amino acid sequence similarity, 16 and 13 kDa prolamins correspond to maize γ_2-zein, and 10 kDa prolamins correspond to δ-zeins (Fig.12.2).[13]

12.2.2 THE LOW ABUNDANCE PROTEINS

Although considerable progress has been made in the field of seed physiology and seed proteomics in last decade, the major bottleneck in the analysis of seed proteome is the presence of these SSPs. As these SSPs accounts for a major portion of the total proteome, presence of these proteins hinders the detection and identification of low abundance proteins (LAPs), most of which are the critical components of signaling. In recent years, few techniques have been reported that can enrich the seed proteome with LAPs. These techniques are based on the specific depletion of SSPs using isopropanol,[14] calcium phytate,[15] and protamine sulfate.[16] However, none of these studies have been used for addressing any biological question related to the seed biology except for protamine sulfate based method, which was used in a recent study to address the biological question that why some soybean seeds are rich in proteins and others in oil.[17]

12.3 PROTEOMICS OF SOYBEAN SEEDS

Due to the high economic importance of soybean seeds, numerous studies have been conducted to understand the physiology of developing soybean seeds. These studies are focused on the whole seeds,[18,19] embryos, and seed coats.[8,20]

12.3.1 SOYBEAN SEED PROTEOME DURING DIFFERENT STAGES OF DEVELOPMENT

Hajduch et al. analyzed the seed filling proteome of soybean seeds using high-resolution 2D gels. Proteins were isolated from the developing seeds, harvested at 2, 3, 4, 5, and 6 weeks after flowering (WAF) and resolved on 2D gels. A total of 679 spots were observed on the gels of which 422 were successfully identified using matrix assisted laser desorption ionization-time of flight mass spectrometry (MALDI-TOF MS). All the identified proteins were grouped into 14 functional categories of which metabolism, protein destination and storage, metabolite transport, and disease/defense were the most abundant categories. An overall decrease of metabolism-related proteins was observed while proteins related to the destination and storage were found to be increased during seed development.[19] In another similar study published by the same group, a comparison was made between the rapeseed and soybean seed filling proteome to address one of the major biological questions that why some seeds are rich in proteins while others in oil.[18] Two-dimensional polyacrylamide gel electrophoresis (2-DGE) and multidimensional protein identification technology (MudPIT) approaches were used to analyze the soybean and rapeseed proteins during five different stages of seed development. Using both these approaches, a total of 478 non-redundant proteins were identified. In line with the previous study, proteins related to the metabolism, protein destination and storage, and energy were most abundant accounting for 42–61% of the total identified proteins. Interestingly, proteins related to the glycolytic and fatty acid biosynthetic pathways were highly enriched in the rapeseed as compared to the soybean, pointing toward a possible explanation of why rapeseeds accumulate higher percentage of oils in comparison with the soybean.[18]

In addition to total seed proteome, seed coat or testa proteome was also investigated during different stages of seed development. Testa is the outer covering of the seeds, which is developed by the integuments of the ovule, after fertilization. In case of soybean, seed coats develop prior to the development of the embryo.[8] At first, a shotgun proteomics approach was used to identify the seed coat proteins during five stages of seed development. Using this approach, a total of 306 (S2), 328 (S4), 273 (S6), 193 (S8), and 272 (S9) proteins were identified which were classified into 11 functional groups including primary metabolism, secondary metabolism, cellular structure, stress responses, nucleic acid metabolism, protein synthesis, protein folding, protein targeting, hormones and signaling, SSPs, and proteins of unknown functions.[21] In another study, seed coat proteins of two black colored soybean cultivars (Seonheuk and Geomjeong 2) were analyzed during seed development.[20] Using a 2-DGE approach, 36 differentially modulated proteins were identified according to the color changing stages of seed coats. Trypsin inhibitors and manganese superoxide dismutase were downregulated while isoflavone reductase homolog-1 was upregulated in Geomjeong 2 as compared to Seonheuk, indicating that antioxidant activity of Geomjeong 2 may be higher than Seouheuk.[20] In another recent study, a comparative proteome profiling of seed coat proteins of brown and yellow colored soybean seeds were carried out to investigate the pathways that leads to development of different seed coat color.[8] The study was focused on the soybean cultivars "Mallikong (M)," having yellow seed coat color and its naturally mutated cultivar "Mallikong mutant (MM)" that develops a brown seed coat color upon maturation. Resolution of seed coat proteins from three different stages of seed development led to the detection of 178 differentially modulated proteins between M and MM of which 172 were successfully identified. Out of these identified proteins, two isoforms of isoflavone reductase showed down-regulation in MM, suggesting lesser isoflavone production in the MM seed coats. High-performance liquid chromatography (HPLC)-based quantitative estimation of total and individual isoflavones confirmed low levels of isoflavones in the seed coat of MM as compared to M. Taken together, these results suggested that low levels of isoflavones in MM can be one of the reasons for the development of its brown seed coat color. As the pathway for production of isoflavones and proanthocyanidins are same, low levels of isoflavones may offer high substrate for the proanthocyanidin production which could result in the development of brown seed coat in MM.[8]

12.3.2   SOYBEAN SEED PROTEOME DURING DIFFERENT STAGES OF GERMINATION

For analyzing the soybean seed proteome during germination, a gel-based approach was initially used to reveal the mobilization of storage proteins in soybean seed (*Glycine max* L.) during germination and seedling growth. Out of the total identified proteins, more than 80% were identified as different subunits of glycinin and β-conglycinin, which were found to be degraded during germination. However, it was observed that degradation patterns of different subunits and isoforms of glycinin and β-conglycinin was different and was temporarily regulated during seed germination and seedling growth.[22] In addition to this, a MudPIT approach was also used to identify the complete repertoire of the proteins involved in the germination process.[4] Isolated proteins, from four different stages of seed germination, were resolved on the SDS-PAGE, subjected to in-gel digestion and subsequently identified by the LC-MS/MS. Using this approach, a total of 764 proteins were identified, and which were classified into 14 functional categories. Proteins identified in this study suggested that germination induced degradation of lipids is mediated by the lipoxygenase dependent pathway while degradation of the proteins is mediated by both protease and 26S proteosome system. Moreover, lipoxygenase was also found to quench the reactive oxygen species (ROS) during the mobilization of reserves of soybean germinating seeds.[4]

12.4   PROTEOMICS OF RICE SEED

12.4.1   RICE SEED PROTEOME DURING DIFFERENT STAGES OF DEVELOPMENT

Rice seeds are albuminous as it contains a starchy endosperm, surrounded by aleurone layer. The protein content of rice seeds is comparatively lesser than that of soybean and ranges from 5 to 12% only.[23,24] The first study utilized a gel-based proteomics approach to identify the developmental stage specific proteins of the rice seeds, isolated at eight different time points.[25] This study led to the identification of 345 differentially modulated proteins majorly involved in metabolism, protein synthesis/destination, defense response, and cell growth/division.[10,25] To get a detailed view of metabolic pathways, associated with the seed development, metabolism category proteins were further divided into 11 subcategories including glycolysis, alcoholic fermentation, tricarboxylic acid (TCA) cycle, starch synthesis, amino acid metabolism, and so forth. During the initial stages of seed development, proteins related to carbon metabolism; amino acid and protein synthesis; proteolysis; and cell division/growth were highly abundant, indicating the cell growth and protein turnover as primary tasks during the early stages of seed development. In the middle stages, proteins related to starch biosynthesis were highly enriched, suggesting middle stages as seed filling stages in rice. At the final stages, proteins related with the protein folding and modifications were highly accumulated in addition with the starch biosynthetic enzymes suggesting the pivotal roles of these processes during seed maturation.[25] To extend this observation, a 2D-DIGE approach was used which led to identification of 317 differentially modulated proteins from 12 to 18 days after fertilization (DAF) of which 22 were associated with the ROS scavenging. Moreover, the ROS burst during seed development was further confirmed using biochemical assays which showed hydrogen peroxide (H_2O_2) accumulation in the innermost part of endosperm at 12 DAF which were shifted to the outermost part of endosperm at 18 DAF. These results suggested the activation of ROS-induced redox mediated pathways during seed development in rice.[26] In addition to total seed and endosperm proteome, embryo proteome was also analyzed during different developmental stages of rice seeds.[27,28] Analysis of embryo proteome during 5, 7, 14, 21, and 30 DAF led to the identification of 275 differentially modulated proteins.[27] In another similar study, 53 differentially modulated proteins were identified among three developmental stages of rice seed development.[28] Identified embryo proteins in these studies were mainly involved in metabolism and proteins synthesis as observed in the endosperm proteome, suggesting that high metabolic activities and thus energy is required to sustain the growth of embryo, endosperm, and whole seed development.[10,27,28] Recently, a 2-DGE based approach was used to investigate the stress related proteins during the desiccation phase of rice seed development.[29] Accumulation of late embryogenesis abundant proteins, small heat shock proteins, and antioxidative proteins were observed during initial phases of seed desiccation, which maintained at higher level till the later stages, suggesting the important roles of these proteins in dehydration of rice seeds.[29]

12.4.2 RICE PROTEOME DURING DIFFERENT STAGES OF SEED GERMINATION

Similar to other seeds, germination of rice seeds follows a tri-phasic pattern. For analyzing the changes in rice seed proteome during different developmental stages, whole seed, embryo, and endosperm proteins were analyzed mainly using a gel-based approach. To understand the physiology of rice seed germination, initially a 2-DE based approach was used.[30,31] Using *Oryza sativa* cv. Indica 9311 and *O. sativa* cv. Nipponbare, 148 and 39 differentially modulated proteins were identified, respectively. Proteins with increased abundance were mainly related with the defense and catabolism, suggesting the mobilization of reserves and pathogen defense to be the primary functions of germinating seeds.[3,30,32] It was observed that seed maturation proteins and desiccation associated proteins degrade prior to the degradation of SSPs during germination of rice seeds, suggesting that even the catabolism of specific groups of proteins are tightly regulated and is not a general effect.[3] For analyzing the overall functional groups associated with the germination, a MudPIT approach was used which led to the identification of 673 proteins associated with 14 functional categories.[31] This study highlights that energy required for germination is not only acquired from the glycolysis and TCA cycles but also from the fermentation pathways.[31] In order to check the role of transcription and translation during germination of rice seeds, these were treated with both transcriptional (actinomycin D) and translational (*cyclo-heximide*) and their germination was observed.[33] Results obtained from this study suggested that de-novo transcription is not required for the germination as more than 80% of the seeds germinated even after actinomycin D treatment. In contrast, treatment with the *cycloheximide* resulted in the complete inhibition of the germination, suggesting that translation plays a crucial role in seed germination process in rice. Twenty proteins, involved in the carbohydrate metabolism and cytoskeleton formation showing increased abundance after actinomycin treatment, were identified. These results indicated that the germination-specific proteins are synthesized from long-lived mRNAs. Moreover, it was observed that the translation of eight up-regulated proteins was later than that of the other up-regulated proteins under conditions of actinomycin D mediated transcription inhibition, suggesting that translation of long-lived mRNAs in rice seeds is regulated according to germination phase.[33]

In case of embryo, 60 differentially expressed proteins were identified which were associated with the multiple functional groups including metabolism, oxygen-detoxifying, protein processing/degradation, stress/ defense, energy, and others suggesting key roles of embryo proteins during rice seed germination.[34] Moreover, to investigate the role of phytohormones on rice seed germination, embryonic tissue proteins were isolated after application of gibberellic acid (GA) and abscisic acid (ABA). A total of 16 differentially modulated proteins were identified of which two; rice isoflavone reductase (OsIFR) and rice PR10 (OsPR10), were selected for further validation.[35] Western blot and immuno-localization analysis OsIFR and OsPR10 showed that both proteins were specifically expressed in the embryo and were down-regulated by ABA, suggesting that embryo proteins are more sensitive to phytohormones than endosperm proteins.[3,35] Recently, a combination of both gel-based and gel-free proteomics approaches were used for an in-depth proteomics analysis of rice embryos during germination.[36] A total of 343 differentially modulated proteins were identified which were clustered into 11 functional categories. Interestingly, various starch biosynthesis-related enzymes including starch branching enzyme, granule-bound starch synthase 1 and starch synthase showed increased abundance in the initial stages and then gradually decreased during the later stages of germination. The TCA cycle enzymes showed an increased abundance during the later stages of germination, suggesting that sucrose might act as an intermediate for biosynthesis of starch in the embryos.[36]

In addition with the total proteome analysis, attempts have also been made to analyze the phosphoproteome during the germination of rice seeds.[37,38] Using Pro-Q diamond staining, a total of 168 differentially expressed proteins were observed which were identified by MALDI-TOF/TOF.[37] Phosphorylation of the signaling-related proteins were enhanced during the initial stages of germination while the phosphorylation of stress and storage associated proteins were enhanced during the later stages. Moreover, it was also observed that the phosphorylation of fructokinase, pyruvate kinase, malate dehydrogenase, GDP-mannose 3,5-epimerase1, ascorbate peroxidase, and glutathione S-transferase consistently enhanced their enzymatic activities.[37] Using a gel-free approach, it was shown that phosphoprotein abundance of three core components involved in brassinosteroid signaling significantly increases during the early stages of germination. Moreover, treatment of brassinolide resulted in increased seed germination rate; however, no difference in the rate of embryonic axis elongation was observed. Taken together, these findings suggested the key involvement of brassinosteroid signal transduction for seed germination in rice.[38]

12.5   CONCLUSION

Despite the difficulties in the isolation of seed proteins due to presence of high amounts of carbohydrates, lipids, and phenolic compounds; several studies have been conducted to understand the seed physiology during germination and different biotic and abiotic stress conditions. Analysis of seed proteome during seed development showed an overall decrease of metabolism-related proteins and up-regulation of the destination and storage related proteins. Germination studies clearly showed degradation of SSPs during germination and their degraded products are supposed to act as carbon and nitrogen source during seedling growth. Moreover, germination studies conducted in rice highlight the involvement of *de-novo* proteins synthesis, phosphorylation, and brassinosteroid signaling during seed germination in rice. In future, a system biology approach would be helpful to understand the seed physiology in more depth.

12.6   ACKNOWLEDGMENT

This study was financially supported by a grant from SSAC (grant no PJ011070032016) provided to STK.

KEYWORDS

- seed storage proteins
- rice
- soybean
- endosperm
- embryo
- seed coat
- seed germination
- seed development

REFERENCES

1. Qu, L. J; Li, L.; Lan, Z.; Dresselhaus, T. Peptide Signalling during the Pollen Tube Journey and Double Fertilization. *J. Exp. Bot.* **2015,** *66,* 5139–5150.
2. Wang, W. Q.; Liu, S. J.; Song, S. Q.; Møller, I. M. Proteomics of Seed Development, Desiccation Tolerance, Germination and Vigor. *Plant Physiol. Biochem.* **2015,** *86,* 1–15.
3. He, D.; Yang, P. Proteomics of Rice Seed Germination. *Front. Plant Sci.* **2013,** *9,* 4:246.
4. Han, C.; Yin, X.; He, D.; Yang, P. Analysis of Proteome Profile in Germinating Soybean Seed, and its Comparison with Rice Showing the Styles of Reserves Mobilization in Different Crops. *PLoS One.* **2013,** *8,* e56947.
5. Gupta, R.; Wang, Y.; Agrawal, G. K.; Rakwal, R.; Jo, I. H.; Bang, K. H.; Kim, S. T. Time to dig deep into the plant proteome: a hunt for low-abundance proteins. *Front. Plant Sci.* **2015,** 6–22.
6. Agrawal, G. K.; Rakwal, R. *Seed Development: OMICS Technologies Toward Improvement of Seed Quality and Crop Yield. Omics in Seeds Biology;* Springer: the Netherlands, 2012.
7. Shewry, P. R.; Napier, J. A.; Tatham, A. S. Seed Storage Proteins: Structures and Biosynthesis. *Plant Cell.* **1995,** *7,* 945–956.
8. Gupta, R.; Min, C. W.; Kim, S. W.; Wang, Y.; Agrawal, G. K.; Rakwal, R.; Kim, S. G.; Lee, B. W.; Ko, J. M.; Back, I. Y.; Bae, D. W.; Kim, S. T. Comparative Investigation of Seed Coats of Brown-versus Yellow-Colored Soybean Seeds Using an Integrated Proteomics and Metabolomics Approach. *Proteomics.* **2015,** *15,* 1706–1716.
9. Schuler, M. A.; Schmitt, E. S.; Beachy, R. N. Closely Related Families of Genes Code for the Alpha and Alpha' Subunits of the Soybean 7S Storage Protein Complex. *Nucleic Acids Res.* **1982,** *10,* 8225–8244.
10. Deng, Z. Y; Gong, C. Y; Wang, T. Use of Proteomics to Understand Seed Development in Rice. *Proteomics.* **2013,** *13,* 1784–1800.
11. Kawakatsu, T.; Yamamoto, M. P.; Hirose, S.; Yano, M.; Takaiwa, F. Characterization of a New Rice Glutelin Gene GluD-1 Expressed in the Starchy Endosperm. *J. Exp. Bot.* **2008,** *59,* 4233–4245.
12. Muench, D. G.; Ogawa, M., Okita, T. W. The Prolamins of Rice. In *Seed Proteins*; Shewry, P., Casey, R., Eds.; Kluwer Academic Publishers: Dordrecht, The Netherlands, 1999; pp 93–108.
13. Xu, J. H.; Messing, J. Amplification of Prolamin Storage Protein Genes in Different Subfamilies of the Poaceae. *Theor. Appl. Genet.* **2009,** *119,* 1397–1412.
14. Natarajan, S. S.; Krishnan, H. B.; Lakshman, S.; Garrett, W. M. An Efficient Extraction Method to Enhance Analysis of Low Abundant Proteins from Soybean Seed. *Anal. Biochem.* **2009,** *394,* 259–268.
15. Krishnan, H. B.; Oehrle, N. W.; Natarajan, S. S. A Rapid and Simple Procedure for the Depletion of Abundant Storage Proteins from Legume Seeds to Advance Proteome Analysis: A Case Study Using Glycine Max. *Proteomics.* **2009,** *9,* 3174–3188.
16. Kim, Y. J.; Wang, Y.; Gupta, R.; Kim, S. W.; Min, C. W.; Kim, Y. C.; Park, K. H.; Agrawal, G. K.; Rakwal, R.; Choung, M. G.; Kang, K. Y.; Kim. S. T. Protamine Sulfate Precipitation Method Depletes Abundant Plant Seed-Storage Proteins: A Case Study on Legume Plants. *Proteomics.* **2015,** *15,* 1760–1764.
17. Min, C. W.; Gupta, R.; Kim, S. W.; Lee, S. E.; Kim, Y. C.; Bae, D. W.; Han, W. Y.; Lee, B. W.; Ko, J. M.; Agrawal, G. K.; Rakwal, R.; Kim, S. T. Comparative Biochemical and

Proteomic Analyses of Soybean Seed Cultivars Differing in Protein and Oil Content. *J. Agric. Food Chem.* **2015**, *63*, 7134–7142. DOI: 10.1021/acs.jafc.5b03196

18. Agrawal, G. K.; Hajduch, M.; Graham, K.; Thelen, J. J. In-depth Investigation of the Soybean Seed-Filling Proteome and Comparison with a Parallel Study. *Plant Physiol.* **2008**, *148*, 504–518.

19. Hajduch, M.; Ganapathy, A.; Stein, J. W.; Thelen, J. J. A systematic Proteomic Study of Seed Filling in Soybean. Establishment of High-Resolution Two-Dimensional Reference Maps, Expression Profiles, and an Interactive Proteome Database. *Plant Physiol.* **2005**, *137*, 1397–1419.

20. Kim, Y. J.; Lee, S. J.; Lee, H. M.; Lee, B. W.; Ha, T. J.; Bae, D. W.; Son, B. Y.; Kim, Y. H.; Baek, S. B.; Kim, Y. C.; Kim, S. G.; Kim, S. T. Comparative Proteomics Analysis of Seed Coat from Two Black Colored Soybean Cultivars during Seed Development. *Plant Omics J.* **2013**, *6*, 456–463.

21. Miernyk, J. A.; Johnston, M. L. Proteomic Analysis of the Testa from Developing Soybean Seeds. *J. Proteomics.* **2013**, *89*, 265–272.

22. Krishnan, H. B.; White, J. A. Morphometric Analysis of Rice Seed Protein Bodies: Implication for a Significant Contribution of Prolamine to the Total Protein Content of Rice Endosperm. *Plant Physiol.* **1995**, *109*, 1491–1495.

23. Villareal, R. M.; Juliano, B. O. Properties of Glutelin from Mature and Developing Rice Grain. *Phytochemistry.* **1978**, *17*, 177–182.

24. Kim, H. T.; Choi, U. K.; Ryu, H. S.; Lee, S. J.; Kwon, O. S. Mobilization of Storage Proteins in Soybean Seed (*Glycine max* L.) During Germination and Seedling Growth. *Biochim. Biophys. Acta.* **2011**, *1814*, 1178–1187.

25. Xu, S. B.; Li, T.; Deng, Z. Y.; Chong, K. Xue, Y.; Wang, T. Dynamic Proteomic Analysis Reveals a Switch between Central Carbon Metabolism and Alcoholic Fermentation in Rice Filling Grains. *Plant Physiol.* **2008**, *148*, 908–925.

26. Xu, S. B.; Yu, H. T.; Yan, L. F.; Wang, T. Integrated Proteomic and Cytological Study of Rice Endosperms at the Storage Phase. *J. Proteome Res.* **2010**, *9*, 4906–4918.

27. Xu, H.; Zhang, W.; Gao, Y.; Zhao, Y.; Guo, L.; Wang, J. Proteomic Analysis of Embryo Development in Rice (*Oryza sativa*). *Planta.* **2012**, *235*, 687–701.

28. Zi, J.; Zhang, J.; Wang, Q.; Lin, L.; Tong, W.; Bai, X.; Zhao, J.; Chen, Z.; Fu, X.; Liu, S. Proteomics Study of Rice Embryogenesis: Discovery of the Embryogenesis Dependent Globulins. *Electrophoresis.* **2012**, *33*, 1129–1138.

29. Sano, N.; Masaki, S.; Tanabata, T.; Yamada, T.; Hirasawa, T.; Kanekatsu, M. Proteomic Analysis of Stress-Related Proteins in Rice Seeds during the Desiccation Phase of Grain Filling. *Plant Biotechnol.* **2013**, *30*, 147–156.

30. Yang, P.; Li, X.; Wang, X.; Chen, H.; Chen, F.; Shen, S. Proteomic Analysis of Rice (*Oryza sativa*) Seeds during Germination. *Proteomics.* **2007**, *7*, 3358–3368.

31. He, D.; Han, C.; Yang, P. Gene Expression Profile Changes in Germinating Rice. *J. Integr. Plant Biol.* **2011**, *53*, 835–844.

32. He, D.; Han, C.; Yao, J.; Shen, S.; Yang, P. Constructing the Metabolic and Regulatory Pathways in Germinating Rice Seeds through Proteomic Approach. *Proteomics.* **2011**, *11*, 2693–2713.

33. Sano, N.; Permana, H.; Kumada, R.; Shinozaki, Y.; Tanabata, T.; Yamada, T.; Hirasawa, T.; Kanekatsu, M. Proteomic Analysis of Embryonic Proteins Synthesized from Long-Lived mRNAs during Germination of Rice Seeds. *Plant Cell Physiol.* **2012**, *53*, 687–698.

34. Kim, S. T.; Wang, Y.; Kang, S. Y.; Kim, S. G.; Rakwal, R.; Kim, Y. C.; Kang, K. Y. Developing Rice Embryo Proteomics Reveals Essential Role for Embryonic Proteins in Regulation of Seed Germination. *J. Proteome Res.* **2009,** *8,* 3598–3605.

35. Kim, S. T.; Kang, S. Y.; Wang, Y.; Kim, S. G.; Hwang Du, H.; Kang, K. Y. Analysis of Embryonic Proteome Modulation by GA and ABA from Germinating Rice Seeds. *Proteomics.* **2008,** *8,* 3577–3587.

36. Han, C.; He, D.; Li, M.; Yang, P. In-depth Proteomic Analysis of Rice Embryo Reveals its Important Roles in Seed Germination. *Plant Cell Physiol.* **2014,** *55,* 1826–1847.

37. Han, C.; Wang, K.; Yang, P. Gel-Based Comparative Phosphoproteomic Analysis on Rice Embryo during Germination. *Plant Cell Physiol.* **2014,** *55,* 1376–1394.

38. Han, C.; Yang, P; Sakata, K; Komatsu, S. Quantitative Proteomics Reveals the Role of Protein Phosphorylation in Rice Embryos during Early Stages of Germination. *J. Proteome Res.* **2014,** *13,* 1766–1782.

CHAPTER 13

UNDERSTANDING ABIOTIC STRESS TOLERANCE IN PLANTS BY PROTEOMICS APPROACH

VASEEM RAJA, MUBASHIR AHMAD WANI, UMER MAJEED WANI, NELOFER JAN, and RIFFAT JOHN*

Plant Molecular Biology Laboratory, Department of Botany, University of Kashmir, Hazratbal, Srinagar190006, Jammu and Kashmir, India

Corresponding author. E-mail: riffat_iit@yahoo.com

CONTENTS

ABSTRACT

Increase in the crop yield is priority due to rise in population and global demand for food. There are plethora of factors that limit crop yield; however, the major factors limiting crop productivity are abiotic factors such as salinity, extremes of temperature, drought, intense light and oxidative stress caused by reactive oxygen species (ROS). Plants respond to these environmental assaults by differential expression of genes which can bring about changes at genomic, transcriptomic as well as at proteomic level. Changes in the protein levels are the result of the altered gene expression that may lead to adaptation or regulation of responses against abiotic stresses. Proteomics can help in identifying novel candidate genes that can be used for engineering crops for improved stress tolerance. It can also assist in selection of appropriate phenotypes for proteomic-based functional analyses which can play a key role in future crop breeding. This review highlights the proteomic approach elucidating various processes under abiotic stress.

13.1   INTRODUCTION

Abiotic stresses such as salinity, drought, cold, and so forth, are among the valuable factors that limit plant growth and development as well as negatively reduce the crop productivity.[1-3] Plants can cope up with these adverse factors through stress avoidance for example, surviving in the stress conditions by remaining in physiologically non-active stage such as mature seeds or stress tolerance, that is, adjusting themselves to the existing conditions through an active reversible adjustment popularly termed as acclimation.[4,5] Activation of multiple responsive cascades involving comprehensive networks interconnected to each other at many levels is among the notable features adopted by the plants in response to abiotic stress.[6] The process is arbitrated through changes up to the level of gene expression which in turn may result in altered composition of transcriptome, metabolome, and proteome. Changes occurring during gene expression at the level of transcription are not often accompanied with the changes at protein level.[7,8] Meanwhile proteins, unlike transcripts, are direct effectors of stress response consequently it is imperative to investigate the changes occurring in plant proteome. Proteins are involved in many vital processes such as in enzyme catalysis that triggers changes at metabolite level not only that proteins can also regulate stress response at transcript and protein level, because of being an essential component of transcription and translation machinery. Proteins are among the key players that directly play a role in stress-acclimation leading to alteration in composition and properties of plasma membrane, cytoplasm, and cytoskeleton as well as in intracellular compartmentation. From the above discussion it is evident that the changes in protein accumulation under stress conditions are meticulously interrelated to plant phenotypic response to stress determining plant stress tolerance. Thus proteomic studies are among the cutting edge tools that can ominously contribute to our understanding of several physiological mechanisms underlying plant stress tolerance. Proteomic studies can also pave the way for the identification of potential protein markers associated with quantitative changes occurring in some vital physiological processes used as a description of genotype's level of stress tolerance. During last few years a dramatic expansion is being observed in the proteomic analysis of plants.[9-16] Proteomics is now gaining momentum in three major aspects of plant science: in cellular and subcellular, structural and developmental, and physiological and genetic studies. Identification of proteins has become easy with the help of developing technology. Proteome analysis is touching new horizons possibly due to advancement in mass spectrometry technology, stains, software, and progress in bioinformatics. Sub-cellular localization of proteins and assignment of proteins to a particular organelle by the development of algorithms are among the examples that make a person understand how proteomic studies contribute to the plant science. In comparison to the transcriptomic analysis protein compliment of plant tissue in response to cold or other abiotic stress is still limited, but a breakthrough has been achieved during recent past with the help of mass spectrometry that has led to the separation as well as sensitive identification of proteins involved in cold stress.[17] A vast range of research studies have clearly demonstrated that the levels of transcripts and proteins are correlated.[7,18] In addition to that post-translational modifications such as ubiquitylation, phosphorylation, sumoylation, glycosylation, and many other modifications may occur in proteins which are very crucial in regulation of protein function.[17,19,20] In order to understand the mechanism underlying the cellular processes operating in response to environmental stresses a thorough analysis of cellular proteome compliment is required. Despite of limitations in separation capacity two-dimensional gel electrophoresis (2-DE) is the most frequently used method to explore differential protein profusion in large-scale proteomic experiments on crude proteins.[21-23] In order to improve sensitivity, linearity, and reproducibility vast number of technical developments have been carried out, including narrower pH ranges, improved dyes for staining purposes mostly based on covalent or non-covalent labeling of proteins,[24,25] not only this much but above all sophisticated softwares for gel evaluation patterns.

13.2 ABIOTIC STRESSES AND THE IMPACT ON AGRICULTURE

Today, worldwide 7 billion people, agriculture is facing great challenges to ensure sufficient food supply while maintaining high productivity and quality standards. In addition to an ever-increasing demographic demand, alterations in weather patterns due to changes in climate are impacting crop productivity globally. Warming and shifts in rainfall patterns caused an historically high $10.3 billion in crop insurance payments to cover agriculture losses in 2011 in US unfavorable climate (resulting in abiotic stresses) not only causes changes in agro-ecological conditions, but indirectly affects growth and distribution of incomes, and thus increasing the demand for agricultural production.[26] Adverse climatic factors, such as water scarcity (drought), extreme temperatures (heat and freezing), photon irradiance, and contamination of soils by high ion concentration (salt and metals), are the major growth stressors that significantly limit productivity and quality of crop species worldwide. As has been pointed out, current achievements in crop production have been associated with management practices that have degraded the land and water systems, soil and water salinity problems exist in crop lands in China, India, the United States, Argentina, Sudan, and many other countries in Western and Central Asia. Globally, an estimated 34 million irrigated hectares are salinized and the global cost of irrigation-induced salinity is equivalent to an estimated US$11 billion per year.

A promising strategy to cope with adverse scenario is to take advantage of the flexibility that biodiversity (genes, species, and ecosystems) offers and increase the ability of crop plants to adapt to abiotic stresses. The Food and Agricultural Organization (FAO) of the United Nations promotes the use of adapted plants and the selection and propagation of crop varieties adapted or resistant to adverse conditions. Global programs, such as the Global Partnership Initiative for Plant Breeding Capacity Building (GIPB), aim to select and distribute crops and cultivars with tolerance to abiotic stresses for sustainable use of plant genetic resources for food and agriculture.

13.3 PLANT RESPONSES TO ABIOTIC STRESS

Through the history of evolution, plants have developed a wide variety of highly sophisticated and efficient mechanisms to sense, respond, and adapt to a wide range of environmental changes. When in adverse or limiting growth conditions, plants respond by activating tolerance mechanisms at multiple levels of organization (molecular, tissue, anatomical, and morphological), by adjusting the membrane system and the cell wall architecture, by altering the cell cycle and rate of cell division, and by metabolic tuning.[27] At a molecular level, many genes are induced or repressed by abiotic stress, involving a precise regulation of extensive stress-gene networks.[28-30] Products of those genes may function in stress response and tolerance at the cellular level. Proteins involved in biosynthesis of osmoprotectant compounds, detoxification enzyme systems, proteases, transporters, and chaperones are among the multiple protein functions triggered as a first line of direct protection from stress. In addition, activation of regulatory proteins (e.g., transcription factors, protein phosphatases, and kinases) and signaling molecules are essential in the concomitant regulation of signal transduction and stress-responsive gene expression.[31,32] Early plant response mechanisms prevent or alleviate cellular damage caused by the stress and re-establish homeostatic conditions and allow continuation of growth.[33] Equilibrium recovery of the energetic, osmotic, and redox imbalances imposed by the stressor are the first targets of plant immediate responses observed, tolerance responses toward abiotic stress in plants are generally composed of stress specific response mechanisms and also more general adaptive responses that confer strategic advantages in adverse conditions. General response mechanisms related to central pathways are involved in energy maintenance and include calcium signal cascades[34,35] ROS scavenging/signaling elements[36,37] and energy deprivation (energy sensor protein kinase, SnRK1) signaling.[38] Induction of these central pathways is observed during plant acclimation toward different types of stress. For example, protein kinase SnRK1is a central metabolic regulator of the expression of genes related to energy-depleting conditions, but this kinase also becomes active when plants face different types of abiotic stress such as drought, salt, flooding, or nutrient depravation.[39-43] SnRK1 kinases modify the expression of over 1000 stress-responsive genes allowing the re-establishment of homeostasis by repressing energy consuming processes, thus promoting stress tolerance.[44,45] The optimization of cellular energy resources during stress is essential for plant acclimation; energetically expensive processes are partially arrested, such as reproductive activities, translation, and some biosynthetic pathways. For example, nitrogen and carbon assimilation are impaired in maize during salt stress and potassium-deficiency stress; the synthesis of free amino acids, chlorophyll, and protein are also affected.[46-48] Once energy-expensive processes are curtailed, energy resources can be redirected to activate protective mechanisms. This is exemplified by the decrease in *de novo* protein synthesis in *Brassica napus* seedlings, *Glycine max, Lotus japonicus*, and *Medicago truncatula* during heat stress accompanied by an increased translation of heat shock proteins (HSPs).[49]

13.4   ABIOTIC STRESS AND PROTEOME

Abiotic stress factors severely limit plant growth and development as well as crop yield. Two major plant strategies on how to react on stress lie either in plant stress avoidance, for example, a survival of stress in a physiologically non-active stage such as mature seeds, or in plant stress tolerance, that is, an active reversible adjustment to stress conditions generally termed acclimation.[4,5] Acclimation to stress is mediated via profound changes in gene expression which result in changes in composition of plant transcriptome, proteome, and metabolome. Several studies[7,8] have already proven that the changes in gene expression at transcript level do not often correspond with the changes at protein level. Therefore, investigation of changes in plant proteome is highly important since proteins, unlike transcripts, are direct effectors of plant stress response. Proteins not only include enzymes catalyzing changes in metabolite levels, but they also include components of transcription and translation machinery, that is, they also regulate plant stress response at transcript and protein levels. Proteins have also direct stress-acclimation functions leading to changes in plasma membrane, cell cytoplasm, cytoskeleton as well as intracellular compartment composition which involve changes in their properties, for example, affinity of cell cytoplasm to water. As suggested above, changes in protein accumulation under stress are closely interrelated to plant phenotypic response to stress determining plant tolerance to stress. Therefore, studies of plant reaction upon stress conditions at protein level can significantly contribute to our understanding of physiological mechanisms underlying plant stress tolerance. Proteomics studies could thus lead to identification of potential protein markers whose changes in abundance can be associated with quantitative changes in some physiological parameters used for a description of genotype's level of stress tolerance. Plant stress response is a dynamic process which is dependent on stress intensity and stress duration. Several stages of plant stress response could be distinguished[5] an initial alarm phase when stress causes a shock to a non-acclimated plant and the level of plant stress tolerance decreases, an acclimation phase which lasts several days and which leads to an establishment of a new homeostasis in plant metabolism under stress (the level of plant stress tolerance increases during acclimation phase), a maintenance phase when a newly established homeostasis is maintained under stress conditions (the level of plant stress tolerance remains stable upon maintenance phase), and an exhaustion phase if a stress treatment lasts too long and a plant fails to maintain a stress-induced homeostasis (the level of plant stress tolerance declines during exhaustion phase). After the cessation of plant stress treatment, a recovery phase could be observed when a re-establishment of a cellular homeostasis under non-stressed conditions occurs. Each stage of plant stress response could be characterized by its unique proteome composition. The process of plant stress acclimation which leads to an establishment of a new homeostasis in plant metabolism facing stress conditions is an active process which requires an extra energy. Therefore, plant metabolism upon stress is redirected from an active growth and development to stress acclimation. These profound changes in plant metabolism upon stress are mirrored also at proteome level.

13.4.1 SALINITY

High salinity is the most serious threat for crop production. Salt stress causes a change in plant metabolic processes, including ion imbalances, reduction in water potential, reduction in CO_2 assimilation, toxicity, oxidative stress, and susceptibility to injury. Adaptation to salt stress requires modification in gene expression, which can lead to the accumulation of certain metabolites responsible for salt tolerance.

In tolerant potato cultivar, HSP, osmotin-like protein, and calreticulin (CRT) were up-regulated under salt stress.[50,51] These are defense-related proteins. Osmotine-like proteins are also up-regulated in potato by abscisic acid, fungal stress, NaCl treatment, and low temperatures. TSI-1 proteins (stress-induced proteins) were first characterized in tomato. These are suggested as the major stress-responsive proteins overexpressed in tomato in response to a number of stress conditions such as fungal infection, wounding, salicylic acid, and other stresses. This protein has been shown to have ribonuclease (RNase) activity in some plants which acts against various biotic stresses. However, the precise function of TSI-1 response to various abiotic stresses is unknown. It can be speculated that its up-regulation in potato under salt stress might be involved in increasing salt tolerance. Up-regulation of defense proteins such as TSI-1 proteins, osmotine-like proteins, HSP proteins, N-succynilarginine, CRT, and cysteine protease inhibitor can produce a relative tolerance in the tolerant cultivar of potato to salt stress. The salt stress-induced alterations in the root proteome of barley genotypes and reported that the proteins involved in the glutathione (GSH)-based detoxification of ROS were abundant in the tolerant barley genotype emphasizing the role of these proteins stress tolerance during salt stress.[52] In the chloroplast proteome of salt-stressed mangrove, up-regulated proteins were the members of photosynthetic complexes of thylakoid membrane:

One from cytochrome complex, five from photosystem I (PSI), 14 subunits from photosystem II (PSII), and one in the adenosine triphosphate (ATP) synthase family, indicating that proteins involved in photosynthesis and energy metabolism play an important role in salt tolerance.[53]

The 14-3-3 proteins (a highly conserved family), are known to be involved in responses to various stresses including salinity.[53-55] The biological role of 14-3-3 complexes are in the signal transduction, regulation of primary metabolism, and subcellular and defense reactions.[56] These proteins also have been recognized as positive regulators of plasma membrane H⁺-ATPase in the regulation of cytoplasmic pH and ion transport.[57,58] In plants, the basic metabolism changes in response to stress. Glyceraldehyde 3-phosphate dehydrogenase (GPD) involved in glycolysis and gluconeogenesis is up-regulated under salt stress. GPD is a target gene regulating the cellular response to salt stress and can be used in the development of new salt-tolerant soybean cultivars. Salt tolerant transgenic potato plant was developed by transferring GPD gene in potato. Moreover, the overexpression of GSH peroxidases and glutathione S-transferases (GSTs) enhanced the growth of transgenic tobacco seedlings during salt and chilling stress. A seed maturation protein PM36 was up-regulated by salt stress in N2899 (variety of soybean). PM36 has a protective role in the plant cells under salt stress. In rice roots under salt stress, three membrane skeleton proteins including a putative remorin (U5), a putative alpha-soluble N-ethylmaleimide-sensitive factor (NSF) attachment protein (U9) and unknown protein similar to the myosin heavy-chain-related protein (U8) were identified as salt responsive proteins. Remorin belonging to a plant specific lipid raft/ PM-associated filamentous protein act as scaffolds for signaling in development or defense and may play an important role in stabilization of damaged plasma membrane under salt stress. Alpha-soluble NSF attachment protein is involved in membrane trafficking. Moreover, 14-3-3 proteins are involved in various signal transduction pathways, controlling the activities of phosphatases and kinases,[59,60] and hence suggest that 14-3-3 proteins regulate multiple pathways involved in salt stress response in higher plants.[55] In *Kandelia candel*, CRT is up-regulated in response to salt stress.[53] In soybeans under salt stress, ferritin (a class of iron-storage proteins) increased during seed germination. Ferritin may be involved in defense mechanism against iron-mediated oxidative stress caused by salt stress.[61] In soybean 20S proteasome α subunit A was up-regulated under osmotic stress caused by salt stress.[62] This 20S proteasome is a catalytic core of 26S proteasome (proteolytic complex) and is involved in recognizing and catabolizing the ubiquitin-protein to remove abnormal proteins.[63,64] It could be associated with the degradation of oxidatively-damaged proteins caused by salt stress. In wheat under salt stress, 52 salt-responsive proteins, classified into six functional categories that included transport-associated proteins, ATP synthase, detoxifying enzymes, protein folding, carbon metabolism, and proteins with unknown biological functions were identified. Among these 52 differentially expressed proteins, 21 were down-regulated, 26 were up-regulated, and five proteins showed multi-expression patterns.

13.4.2 DROUGHT

Among various abiotic stresses, drought stress is one of the major stresses in agriculture worldwide. Plants exhibit a wide range of responses at the cellular and molecular levels when exposed to drought stress.[65] Under drought stress, the expression pattern of many genes change, including the proteins involved in stress signaling pathway. 2-DE in combination with mass spectrometry is a powerful approach for identifying proteins playing an important role in the drought response. However, there are only a few published proteomic studies of drought-stressed plants. The response to drought was the accumulation of unusual metabolites such as polyamines and proline, increased level of abscisic acid, the induction of a specific set of genes and alteration in the activity of certain enzymes.[66] In sunflower under drought stress, proteins contributing to basic carbon metabolism were significantly increased.[67] Eleven proteins were identified in sugarbeet leaves under drought stress including eight up-regulated proteins (nucleoside-diphosphate kinase (NDPK), 2-cysteine peroxiredoxin (2-Cys Prx), hypothetical protein, superoxide dismutase (SOD), and four ribulose-1,5-bisphosphate carboxylase/oxygenase (*RuBisCO)* fragments), two proteins detected under water deficit conditions (putative oxidoreductase and HSP), and one down-regulated protein (a putative nascent polypeptide-associated complex a-chain (a-NAC)). HSPs (chaperones) play an important role in protecting plants against stress by re-establishing normal protein conformation and also cellular homeostasis.[68] They are also responsible for proper protein folding, assembly, degradation, and translocation in normal cellular processes, stabilizing membranes and proteins, and assist in protein refolding under various stress conditions. They are often involved in the translocation/import of precursor proteins and assist in the folding of *de novo* synthesized polypeptides. Two sunflower genotypes were analyzed (showing different tolerance level to drought), for proteomic analysis of their responses to drought stress. In susceptible genotype, there was decrease in the photosynthetic enzymes/protein: carbonic anhydrase, PSII oxygen-evolving complex protein 1 precursor, ferredoxin-NADP+ reductase, *RuBisCO*large and small subunits, those of the gluconeogenesis/glycolysis: GDP, phosphoglycerate kinase (PGK), aldolase (ALDO), and proteins involved in various functions including SOD and glyoxalase. While tolerant genotype showed increase in photosynthetic proteins/carbohydrate metabolism enzymes: *RuBisCO* large subunit and phosphoglycerate hydroxylase, and decrease in L-ascorbate peroxidase.

In sensitive barley genotype, photosynthesis related proteins:PSI reaction center II, *RuBisCO* binding protein, and *RuBisCO* activase decreased under drought stress; however, in the tolerant barley genotype, oxygen evolving complex precursor and PSI reaction center II were increased.[69] Drought stress leads to a decline in photosynthetic rate and the extent of this reduction depends on the genotypic differences and the osmotic adjustment.[70] In alfalfa, drought stress leads to a decrease in *RuBisCO* binding protein content at the leaf level.[71] In rice leaf sheath, SOD, SALT (a salt induced protein), oxygen evolving enhancer protein 2, light harvesting complex chain II, chloroplast ATPase, PSII oxygen evolving complex protein, 2-Cys peroxiredoxin, serine hydroxymethyltransferase 1, phosphoglucomutase cytoplasmic 2, and actin depolymerizing factor increased, and RuBisCO large and small subunits decreased under drought stress . SALT is a mannose-binding lectin localized to xylem parenchyma cells in the vascular bundles of the minor and major leaf veins, suggesting its accumulation under drought stress also apart from salt stress. Phosphoglucomutase catalyzes the interconversion of glucose1-phosphate and glucose6-phosphate leading to the synthesis and consumption of sucrose during drought stress.

Activity of malate dehydrogenase increases under drought stress to cope with the high-energy demand of plant under stress.[72] Malate dehydrogenase is an important enzyme of cellular metabolism catalyzing the conversion of malate and oxaloacetate.[73] The increased concentration of malate dehydroge-nase in the tolerant barley genotype indicates its role in drought stress accli-mation.[69] Similarly, in the tolerant barley genotype, GSH transferase also increased to enhance the stress tolerance in plant. Attenuated level of GSH transferase under drought stress conditions was related to the suppression of ROS.[74] In the tolerant genotype, higher concentration of GSH transferase along with malate dehydrogenase may activate ROS scavenging cascade for better survival of plant under drought stress. In rice peduncles under drought stress, newly induced annotated proteins include group 6 LEA protein, puta-tive ABA-responsive protein, LEA type 1 protein, putative phosphatidyl-ethanolamine-binding protein, LEA protein, ABA/WDS-induced protein, putative HSP, and three hypothetical/unknown proteins. Five up-regulated proteins were putative cold shock protein-1, actin-binding protein, LEA type 1 protein, putative Cu–Zn SOD, and dehydroascorbate reductase. Down-regulated proteins were putative nuclear transport factor, translationally controlled tumor like protein, S-adenosylmethionine synthetases (SAMSs), xyloglucan transglycosylase, and Bet v allergen-like protein. These differentially expressed proteins were involved in various cellular processes affecting cell division, cell elongation, lignin biosynthesis, ABA responsiveness, signaling, and so forth, in response to drought stress. Drought stress leads to an increase in the abundance of xyloglucan endotransglycosylase homolog and all isoforms of the aldose-1-epimerase-like protein (classified in the cell wall metabolism category) in the xylem sap of maize. During cell elongation, these enzymes are involved in cell wall loosening. Several peroxidases (classified in the cell wall metabolism category), but also participate in plant defense mechanisms, were also increased in abundance under drought stress.

The protein α-SNAP and methionine synthase increased twice in tolerant barley genotype. α-SNAP is involved in intracellular transport[75] which is important for the survival of plant under drought stress. Methionine synthase catalyzes the transfer of a methyl group from 5-methyltetrahydrofolate to homocysteine for methionine formation. This reaction occurs in an activated methyl cycle, which is important for the production of methylated polyols under water stress.[76] These methylated polyols have many functions, such as chromatin protection, maintaining ion balance, and decreasing the generation of ROS, which can otherwise lead to cellular damage resulting in cell death during drought stress. In pea leaf mitochondria, glycine decarboxylase, tetrahydrofolate polyglutamates, and serine hydroxymethyltransferase form a complex. The binding affinity of tetrahydrofolate polyglutamates for these proteins increased continuously with increase in the number of glutamates up to six residues. When bound to these proteins, tetrahydrofolate, a very O_2-sensitive molecule, was protected from oxidative degradation, suggesting that serine hydroxymethyltransferase is induced by drought stress for protection from oxidative degradation.

13.4.3   COLD

Energy metabolism is powerfully changed due to cold stress.[77] Cold stress leads to reduced water uptake procreating cellular dehydration.[78,79] The countenance of cold stress significantly perturbs plant's reaction at proteome level. The effect of cold on protein ampleness has been exhaustively studied in *Arabidopsis thaliana*.[80–82] *A. thaliana* cold and salt tolerant relative *Thellungiella halophila*,[83] rice,[84–88] chicory,[89] meadow fescue,[90] soyabean,[91] pea,[92,93] as well as in woody poplar and peach species. The proteomic study of leaf tissue[82,83] as well as root tissue[87,88] of the plants or simply trinucleate pollen in anthers,[84] plant embryos from germinated seeds[91] or plant seedlings[85] were employed. The proteomic analysis has been accomplished at cellular as well as at organellar level, for instance *A. thaliana* nuclear proteome[80] or pea mitochondrial proteome upon cold stress. The RNA binding protein cp29 has been found to increase significantly due to cold stress,[82,83] cp29 is localized in chloroplast stroma, its activeness could be regulated by phosphorylation and it is ramified in plastid RNA processing. Cold stress also significantly affects photosynthesis. Kosmala et al.[90] have studied the proteomic response of two genotypes of meadow fescue (*Festuca pratensis*) in cold at 2 °C in frost tolerance. When protein ampleness in the two genotypes of meadow fescue exposed to cold stress was juxtaposed, great differences (about 1.5 times) were observed for several components of thylakoid associated photosynthetic apparatus including light harvesting complexes (LHC), oxygen evolving complexes (OEC) (oxygen evolving enhancer protein1OEEP), cytochrome b6f complex, iron sulfur center, and Rieske Fe-s protein. The ampleness of Rieske Fe-S protein was raised as observed in cold treated plants of *A. thaliana* by Amme et al.[82] and *T. halophila* by Gao et al.[83] In *Thellungeila* multiplied level of plastocyanin was also observed. The components of C3 cycle which are present in stroma are over expressed by cold stress, mainly *RuBisCO* subunit and *RuBisCO* activase. Big chunk of studies have shown the increased levels of different enzymes involved in carbohydrate metabolism under cold stress. Generally it has been observed that under cold stress condition there is over expression of catabolic pathways and repression of anabolic pathways. The expression of enzymes involved in the catalysis of terminal steps of glycolysis such as triosephosphateisomerase (TPI), glyceraldehyde3-phosphatedehydrogenase (GAPDH), 3-PGK, phosphoglyceratemutase (PGM), and especially enolase (ENO) were up-regulated. It would be great to mention that ENO is not only involved in glycolysis but also it has its own glycolytic pathway in chloroplast.[94] Studies have shown that in *A. thaliana* ENO is also found in nucleus where it is involved as a transcriptional repressor of STZ/ZAT10, a repressor of CBF/DREB1 pathway.[95] The repression of C-repeat binding factor (CBF) regulon, ENO thereby functions indirectly as an activator of CBF regulon, that is, an activator of COR gene expression. ENO thus have a great role not only in carbohydrate metabolism but also in the expression of cold inducible genes. It is, however, important to mention that nucleus located ENO plays role in regulation of cold inducible genes and not cytoplasm located ENO which has role only in glycolytic pathway. It was reported that short-term cold treatment also induces the many anabolic enzymes as uridine diphosphate (UDP) glucose pyrophosphorylase or sucrose synthase 1 which results in the formation of UDP-glucose, a starting unit of many glycoproteins and polysaccharides, mainly cellulose, the main polysaccharide component of cell walls. It is, however, important to mention that the plants reaction to cold stress bank on many factors such as the length of cold treatment, plant growth stage, water supply, and so forth. It has been reported that the activity of enzymes involved in the biosynthesis of S-adenosylmethionine was enhanced, which serves as the source of methyl groups in methylation reaction. The levels of ethylene and polyamines increased eloquently upon cold stress due to increase in the synthesis of S-adenosyl methionine (SAM) which serves as precursor of these two hormones. The increased expression of SAMS has been observed by Amme et al.,[82] Cui et al.,[85] Yan et al.,[86] and others. In addition, the cold stress also harbingers the enhanced aggregate of cold regulated/late embryogenesis-abundant (COR/LEA) proteins which are highly hydrophobic and act as high molecular osmoprotectants. Cold also results in the increased accumulation of specific dehydration inducible LEAII proteins called as dehydrins has been reported by several authors.[81,82,89,91,96,97] The elevated levels of cold activated transcription activators has been observed, which in turn harbingers the expression of COR/LEA proteins. The transcriptional activators belonging to families MYB (AtMYB2), bZIP (OBF4), and bHLH similar to MYC groups were over expressed in nuclear proteome of *A. thaliana* plants exposed to 4 °C for 6 h[80] and in poplar leaves exposed to 4 °C for 7 or14 days CRT/DRE binding factor 1 was over expressed. It was reported that over expression of many transcription factors involved in stress-response gene expression, up-regulation of RNA binding proteins involved in RNA splicing (U2RNP-A) and assembly of ribosomal subunits (60S ribosomal protein P2_A).[80] In presence of cold stress, increased accumulation of several ROS scavenging enzymes was found which included dehydroascorbate reductase (DHAR), GST, thioredoxin h-type (Trx h), and chloroplast precursors of SOD involved in metabolism of ascorbate and GSH have been detected by many workers.[81,90] Degand et al. reported the elevated levels of Cu/Zn SOD[89] and increased levels of ascorbateperoxidase has been reported by Imin et al.[84] in trinucleate pollen under cold stress. Low temperature stress also results in potential risk of protein misfolding which results in non-functional protein. The increased accumulation of proteins with chaperone functions has been reported.[80,81] It was found that cold adapted winter wheat led to increased amounts of some HSP proteins with chaperone functions, namely HSP70[80,92] and down-regulation of HSP90. It was found that HSP90 could act as a capacitor of phenotypic variations.[80] HSP90 helps to maintain normal allelomorphs of different proteins in genome of a given population due to regulation of proper protein folding.

Temperature stress significantly affects the metabolism of plants.[98] High and low temperature stresses significantly reduce the production of many important crops such as rice, wheat, and maize. Temperate plants experienced low temperature (0–12 °C) during their growing season which reduced their crop production, as chilling temperatures eloquently alter plant metabolism and physiology.[99] To understand the changes which occur inside plants because of cold stress, proteomics approach has been wonderfully used to observe these changes.[100–103] Yan et al.[104] showed through their experiments that the protein expression pattern in rice leaf changes in response to cold stress. In their study, they showed that over 1000 protein spots were resolved when leaf proteins were separated by 2D-PAGE, and among them 31 proteins showed down-regulation and 65 proteins show up-regulation. Yan et al.[104] observed that good number of spots were consistently expressed either up- or down-regulated till 24 h of recovery from temperature stress. Kosmala et al.[105] reported that most affected proteins were of photosynthesis with 35.3% of cold responsive proteins belong to this category. Lee et al.[106] reported some new proteins involved in energy production and metabolism, vesicular trafficking, and detoxification (e.g., acetyl transferase, phosphogluconate dehydrogenase, fructokinase, NADP-specific isocitrate dehydrogenase, putative alpha-soluble NSF attachment protein, and glyoxalase 1) and found that these proteins showed higher sensitivity to cold stress in rice. Hashimoto and Komatsu[107] reported down-regulation of some proteins at the organ level under cold stress. For example, 5-methyletetrahydropteroyltriglutamate-homocysteine S-methyl transferase was down-regulated not only in the roots, but also in leaf sheaths, signifying that common cold responsive pathways exist for the regulation of some proteins in roots and leaf sheaths. Kamal et al.[108] reported that wheat under cold stress induced some cold related proteins, such as cold acclimation proteins (9000–22,000), cold shock proteins (16,000–38,000), ABA-inducible proteins (10,000–41,000), cyclophilin (13,000–18,000), low-temperature-regulated proteins (7000–14,000), kinase-like protein (6000–74,000), nitrogen-activated proteins (40,000–80,000), transcriptional adaptors (7000–29,000), and translation initiation proteins (12,000–17,000). Rinalducci et al.[109] revealed that out of 1000 protein spots that were consistently observed on each gel in

Cheyenne, a winter wheat variety, 31 protein spots were down-regulated and 65 were over-expressed. Mass spectroscopy (MS) analysis of these protein spots showed85 differentially expressed proteins, including RNA-binding proteins, lectin protein, and peptide methionine sulfoxide reductase (PMSR). PMSR is important in catalyzing the reduction of protein bound methionine sulfoxide groups to methionine, which inhibits the inactivation and degradation of proteins by ROS. Cold stress results in the degradation of proteins related to photosynthetic activity especially *RuBisCO* large subunit. Heat stress is observed in plants due to increase in the emission of greenhouse gases. A significant part of crop loss is due to increase in temperature worldwide, especially when combined with other stresses such as drought. Heat stress results in improper protein folding leading to denaturation of intracellular proteins and membrane components. High temperature stress results in over expression of proteins with chaperon functions, such as HSPs, and so called small HSPs (sHSPs). It was reported by Peng et al.[110] that rice grain yields decreased by 10% for each 1 °C increase in minimum temperature during the growing season. Lee et al.[111] identified some new high temperature stress responsive genes and determined their expression patterns. Heat stress results in the oxidative damage to biomolecules. In addition to it, several enzymatic and non-enzymatic antioxidants, which aid in redox homeostasis are over expressed. These enzymes include DHAR, GST, Trx h, and chloroplast precursors of SOD. Heat stress induced significant changes in cytoskeleton composition, indicating its reorganization. An increased accumulation of some eukaryotic translation initiation factors (eIF4F and eIF5A-3) was observed, indicating significant cellular organization leading to programmed cell death (PCD) under heat stress.

13.4.4   HEAT STRESS

Heat stress is associated with an enhanced risk of improper protein folding and denaturation of several intracellular protein and membrane complexes. It has long been well known that heat leads to increased expression of several proteins with chaperone functions, especially several members of large family of HSPs which are classified into five distinct sub-families according to their molecular weight (HSP110, HSP90, HSP70, HSP60, so-called small HSPs or sHSPs).[112] However, proteomics analyses of heat stress response have uncovered several other proteins crucial for acclimation to heat besides HSPs. Heat-stress response at proteome level has been studied predominantly in rice,[113] wheat grain during grain filling period[114,115] a heat- and drought-tolerant poplar (*Populus euphratica*)[116] but also in wild plant *Carissa spinarum* inhabiting hot and dry valleys in Central China.[117] In all cases, a heat-induced increase in several HSPs including proteins from HSP100, HSP70, and sHSP families has been observed. Small HSPs belonging to cytoplasmic-located sHSPs as well as mitochondrial-targeted and chloroplast-targeted sHSPs were detected. In heat-treated grains of two genotypes of common wheat with contrasting tolerance to high temperatures, Skylas et al.[114] detected seven sHSPs unique to a tolerant genotype which have been proposed biomarkers of heat tolerance and dough strength. Moreover, enhanced accumulation of several other proteins with chaperone functions (chaperonin 60 β subunit CPN60-β, HSP90, chaperonin CPN10, chloroplast chaperonin) was regularly observed.[116,117] Another characteristic feature of heat stress is oxidative damage. Up-regulation of several enzymes involved in redox homeostasis such as GST, DHAR, Trx h, and chloroplast precursors of SOD was reported.[113] Concerning energy metabolism, an increased accumulation of enzymes involved in biosynthesis of UDP-glucose pyrophosphorylase (UGPase), thiamine and dehydrogenation of pyruvate (pyruvate dehydrogenase), and transketolase was detected upon heat.[113] These results indicated an up-regulation of the activity of pyruvate dehydrogenase complex involved in transferring carbon from glycolysis to Krebs cycle and an increase in the activity of *RuBisCO* activases and pentose phosphate pathway (PPP) involved in regeneration of ribulose-1,5-bisphosphate (RuBP), the substrate for *RuBisCO*.[113,117] Increased activity of enzymes involved in thiamine biosynthesis (biotin carboxylase) and other enzymes of sulfur metabolism (ATP sulfurylase ATPS) is in accordance with thiamine role as a cofactor of transketolase and pyruvate dehydrogenase. Although transketolase was up-regulated by heat, two other major enzymes of pentose phosphate pathway, phosphoribulokinase, and *RuBisCO* small subunit (*RuBisCOSSU*), were down-regulated. In wheat grain, increased accumulation of β-amylase indicating starch degradation was found.[115] Heat stress also induces profound changes in cytoskeleton composition indicating its reorganization.[116] In addition, an increased accumulation of some eukaryotic translation initiation factors (eIF4F and eIF5A-3) indicates significant cellular reorganization leading to PCD under a long-term heat treatment.[115,117]

13.4.5 HEAVY METALS

From past 100 years there has been release of heavy metals in environment due to industrial activity. The main sources of heavy metal pollution are power stations, zinc smelting, paint factories and continuous use of phosphate fertilizers. Some of the heavy metals like nickel, chromium, and mercury are health threatening. Heavy metals enter to human and animal foods through crop plants.[118–120] Due to heavy metal toxicity plants respond to different ways such as immobilization, chelation, and compartmentalization of metal ions.[121] Excess free metal ions can damage at molecular level by stimulating the ROS.[122] It has already been confirmed that when there is excess generation of ROS it induces the expression of HSP and chaperonins and it protects the cell from oxidative damage, if we consider these HSPs as stress markers they mark the morphological changes in the aerial parts of plants. Nearly 20% of stressed plants have got necrotic spots and this was mostly observed in younger leaves. It has already been confirmed that large number of stress proteins that are induced by heavy metal stress have the molecular mass of 10,000 –70,000 Da in plants,[123] so it has been analyzed that proteomic approach can deliver us relevant information for better understanding of plants to heavy metal stress. In *Zea maize* the use of heavy metal like potassium dichromate led to up-regulation of different cellular and metabolic pathways.[124] They suggested that activation of oxidative stress mechanisms affects sugar metabolism and ATP synthesis. Requejo and Tena,[125] analyzed that about 10% of detected maize root proteins were up-regulated or down-regulated by arsenic. Twenty proteins that showed pronounced effect with metal were identified by matrix assisted laser desorption ionization-time of flight mass spectrometry (MALDI-TOF/MS). It was reported from this study that differentially expressed proteins that were found in maize included major homogenous group of seven enzymes. Twenty proteins showed reproducible effects of the metal, and were selected for further analysis by MALDI-TOF/MS. Out of these 20 proteins, 11 were identified by comparing their peptide mass fingerprints to a protein and expressed sequence tag database. Their study found that maize root proteins highly responsive to arsenic exposure included a major homogenous group of seven enzymes that are involved in cellular homeostasis for redox perturbation (e.g., three SODs, two GSH peroxidases, one peroxiredoxin, and one p-benzoquinone reductase), in addition to four additional functionally heterogeneous proteins (e.g., ATP synthase, succinyl-CoA synthetase, cytochrome P450, and guanine nucleotide-binding protein b subunit). From these findings it was clear that oxidative stress is main problem resulted by heavy metal. With increase in cadmium (Cd) concentration up-regulation of enzymes with reduced GSH was observed.[126] When the concentration of Cd increased there occurs accumulation of proteins like copper chaperone (CCH), ROS-scavenging enzymes and those that are involved in biosynthesis of GSH, while as the activity of *RuBisCO* such as *RuBisCO*LSU-binding proteins and ribulose phosphate 3-epimerase decreases significantly in response to Cd stress. In different heavy metals a number of studies has been conducted on elevated concentration of arsenic, copper,[127] nickel,[128] mercury,[129] and zinc. From these studies it was confirmed that formation of necrotic spots associated with degradation of *RuBisCO* especially *RuBisCO*LSU. With the proteomic approach some molecules were revealed that play an important role in plants during exposure to heavy metals. This information helps in selection of plants that resist deleterious metal accumulation in food chain and simultaneously it may help in overcome the essential metal deficiencies in humans. Heavy metals such as Cd, lead, mercury, or metalloids such as arsenic revealed inhibitory effects on many vital processes in plant cells since they act as an inhibitor of many enzymes with metal cofactors. With the help of proteomic study it was found that higher concentration of heavy metals have focused on the impact of arsenic in rice roots,[129] Cd in poplar,[130-132] *A. thaliana*,[133-135] leaves and roots in rice[136,137] and comparison of the impacts of Cd, cobalt, copper, lithium, mercury, strontium, and zinc on rice leaves.[138] There are some tolerant plants that grow on soils enriched with heavy metals and can serve as hyperaccumulators, for example, *A. thaliana* relatives *Alyssum lesbiacum*[139] and *Thlaspi caerulescens*.[140] In rice roots increased concentration of arsenic induced enhance lipid peroxidation, GSH concentration, and H202 content.[129] Due to high concentration of arsenic (As), in leaf proteome of *Agrostis tenitis*, especially As (III), led to degradation several proteins involved in photosynthetic reactions (*RuBisCO*LSU and SSU, components of OEC complex, and components of ATP synthase) which corresponded to leaf chlorosis observed upon As treatments.[141-142]

Leeet al. demonstrated that[137] approximately half of the up-regulated proteins in rice roots belonged to the proteins involved in oxidative stress response and GSH metabolism (GST, GS, GSH reductase (GR), peroxidase, putative ferredoxin: NADP (H) oxidoreductase, APX1). In poplar, however, a decrease in some ROS scavenging enzymes (peroxidase, putative Trx peroxidase, and Cu/Zn-SOD) was also reported under Cd stress.[131] Cd induced elevated accumulation of protective proteins such as CCH, enzymes involved in ROS scavenging, but also enzymes involved in biosynthesis of GSH such as GS like 1 since GSH is a precursor of phytochelatins (PCs), important proteins with chelating function.[133-135]In rice roots, induction of several proteins involved in protein synthesis (chloroplast translational elongation factor Tu, and elongation factor P), protein folding (putative chaperonin 60 beta), or energy metabolism (ATP synthase F0 subunit 1 and putative vacuolar proton-ATPase) was reported.[137] In poplar, a strong increase in various PR proteins, namely β-1,3-glucanases, class I chitinases, and putative thaumatin-like proteins in response to Cd treatment was observed by Kieffer et al.[131]In contrast, Cd treatments revealed a repressive effects on the expression of proteins involved in photosynthesis and carbon fixation. Proteins of thylakoid-membrane-located photosynthetic electron transport chain (OEE1, OEE2, cytochrome b6-f complex Fe-S subunit, chloroplast precursor, chloroplast ferredoxin NADP+ oxidoreductase, and precursor) were down-regulated under stress.[130–132,135] Similarly, enzymes involved in carbon fixation and the activity of *RuBisCO* such as *RuBisCO*LSU binding protein, *RuBisCO* activase 2, a chloroplast precursor of *RuBisCO* activase, ribulose-phosphate 3-epimerase, or carbonic anhydrase decreased significantly in response to Cd stress. The adverse effects of Cd on the expression of proteins involved in photosynthesis, carbon fixation, and synthesis of photoassimilates have been confirmed by later studies.[133,134]In Cd-treated flax cell culture, enhanced accumulation of proteins including heavy-metal binding proteins (lipocalin-1 and ferritin-2), fiber annexin, GS, isoflavone reductase-like protein, HSP70, formate dehydrogenase, chitinase, and enzymes involved in SAM biosynthesis (SAMS and methionine synthetase), and enzymes involved in glycolysis ALDO was observed. The down-regulated proteins included HSP83 and α-tubulin 3.

13.4.6   OZONE

Plants are aerobic organisms with oxygenic type of photosynthesis (i.e., photosynthesis associated with evolvement of molecular oxygen), thus they have to face oxidative stress every day. Organelles where aerobic respiration (mitochondria) or oxygenic photosynthesis (chloroplasts) take place had to evolve efficient ROS scavenging machinery. However, in recent years, not only ROS of plant cell origin (superoxide anion radical, hydrogen peroxide, hydroxyl radical, and singlet oxygen), but also exogenous ROS as pollutants produced by humans adversely affect plant tissues, especially leaves where photosynthesis takes place. Stratospheric ozone has important protective functions due to absorption of harmful UV-B radiation; in contrast, tropospheric ozone is a secondary pollutant arising from light-catalyzed decomposition of nitric oxide (NO) produced by cars. Ozone reveals severe damaging effects on plant tissues, especially leaf tissue where it leads to chlorophyll and xanthophyll degradation resulting in chlorosis and formation of necrotic lesions.[143]At proteome level, ozone causes a severe oxidative stress which results in accumulation of several ROS scavenging enzymes (APX, SOD,

GSH reductase GR, and isoflavone reductase) and also proteins involved in regulation of protein redox status and folding (HSP70 and protein disulfide isomerase).[144,145] As indicated by leaf chlorosis, a drastic decrease of *RuBisCO* and other proteins involved in photosynthetic reactions (proteins of LHC I and II, OEC components OEE1 and 2, carbonic anhydrase, chloroplast precursor of *RuBisCO* activase, several enzymes of Calvin cycle, namely TPI, aldolase (ALDO), sedoheptulose-1,7-bisphosphatase, ribulose-5-phosphate 3-epimerase, and PRK) has repeatedly been observed.[144–146] In contrast to Calvin cycle, enzymes involved in glucose catabolism (phosphoglycerate mutase, a glycolytic enzyme, and malate dehydrogenase MDH, and a component of Krebs cycle) have increased under ozone stress. However, some acclimation to stress could be detected in later stages since after 35 days of treatment, the level of some isoforms of *RuBisCO* activase increased after the initial drop, and after the recovery period, most spots belonging to *RuBisCO* activase isoforms returned to their initial levels before the treatment.[145] Recovery process in rice after ozone treatment (24 h post-treatment) was associated with enhanced accumulation of a class 5 pathogenesis-related (PR) protein, three PR10 proteins, APX, SOD, calcium-binding protein, CRT, and a novel ATP-dependent CLP protease.[144] However, in poplar, PCA analysis of protein abundance at the individual sampling dates has revealed that a 10-day recovery period was not sufficient to compensate the ozone treatment.[145]

13.4.7   *LIGHT*

Plants are photosynthetic organisms and therefore, light plays a crucial role in their lives. Quantity of photosynthetically active radiation (PAR), irradiance, has crucial effects on primary photosynthetic processes which are localized in thylakoid membranes.[147–148]The effect of excessive irradiance on thylakoid proteome in *A. thaliana* wild-type plants as well as vtc2-2 mutants revealing ascorbate deficiency was studied by Giacomelli et al.[149] After five days of plant exposure to an increased irradiance, both wild-type plants and vtc2-2 mutants revealed enhanced accumulation of anthocyanins and ascorbate although mutants exhibited significantly lower levels than wild type. Proteome analysis detected 45 differentially expressed proteins which included four fibrillins (FIB1a, 1b, 2, and 7), fructose bisphosphate ALDO1, flavinreductase-related protein, some chloroplast isoforms of HSP70 (cpHSP70-1 and 2), PsbS protein and YCF37 protein. PsbS protein is known as a component of PSII and it reveals protective functions. Chloroplast isoforms of HSP70 are predominantly found in chloroplast stroma, but they can also associate with thylakoid membranes. YCF37 protein belongs to so-called tetratricopeptide repeat proteins and participates in assembly of PSI. In ascorbate-deficient mutant vtc2, excessive irradiance led to increased expression of some isoforms of SOD (Fe-SOD and Cu/Zn-SOD), possibly as a compensation of ascorbate deficiency. High irradiance also led to a general increase in accumulation of several fibrillins which function as coat proteins in plastoglobules. Inadequate light (darkness) or excessive irradiance on plastoglobule proteome of *A. thaliana* was studied by Ytter-berg et al.[150] An increased accumulation of fibrillins in response to both the conditions (darkness and excessive irradiance) was found, although there were some differences between the individual fibrillin proteins. There was a decrease in accumulation of FIB1b and an increase in FIB7b in response to darkness while a decrease in FIB2 and FIB4 was observed in response to excessive irradiance. Excessive irradiance also led to an increase in enzymes catalyzing biosynthesis of carotenoids (NCED4—9-cisepoxy- carotenoid dioxygenase, ZDS—ζ-caroten desaturase), linoleic acid and jasmonic acid (AOS—allene oxide synthase, lipoxygenase), tocopherol (VTE1—tocoph-erol cyclase), quinones (UbiE—ubiquinone methyltransferase), and four isoforms of ABC1 kinase. It is thus becoming evident that plastoglobules play an important role in metabolism of hydrophobic compounds (carot-enoids, tocopherols, quinones, linoleic acid, and jasmonic acid) and mediate a transport of these compounds between the chloroplast inner membrane and the thylakoid membranes.[150]

13.5   POST-TRANSLATIONAL MODIFICATIONS (PTMS)

Post-translational modifications (PTMs) significantly modulate protein function. Currently, the most studied PTM in the field of plant proteomics upon abiotic stress is phosphorylation, since phosphorylation plays an important role in signal transduction upon abiotic stress and modulates function of a wide array of enzymes. There are several approaches to study protein phosphorylation status at plant proteome level. Previously, an *in vitro* phosphoprotein radioactive labeling by 32P (in the form of γ-32P ATP) in crude protein extracts was used. An alternative to this approach is *in vivo* 32P labeling, though suitable only for plant cell suspension cultures, but not for whole plants. Recently, several studies dealing with changes in plant phosphoproteome under stress have become published due to an easy detec-tion of phosphoproteins in 2-DE gels by Pro-Q Diamond phosphoprotein dye staining.[151,152] Advances in plant phosphoproteomics are summarized in *Ref*.[153–156] Khan et al.[157] have studied changes in protein phosphorylation in roots, leaf sheaths, and leaf blades of young rice seedlings exposed to various stress treatments—1.0 µM ABA, 50 mM NaCl, or 5 °C for 24 h. All three-stress treatments led to phosphorylation of CRT, a Ca^{2+} binding protein involved in stress signaling. Stress-induced phosphorylation was also reported for cytoplasmic MDH (ABA and NaCl), a zinc finger family protein (ABA), glyoxylate-I, 5-aminolevulinate synthase, GAPDH, and a calmodulin-related protein (cold). Differential phosphorylation pattern in different rice organs indicated differential roles of various plant organs under stress. Taken together, glycolytic metabolism and Ca^{2+} signaling processes have been found to represent important targets of phosphorylation cascades in rice under stress conditions. Salt-induced changes in rice phosphoproteome during an early phase (10 and 24 h) of salt treatment have been studied by Chitteti and Peng.[158] Pro-Q Diamond staining revealed differential phosphorylation pattern for putative ribosomal protein S29, dnaK-type chaperone HSP70, GAPDH, endo-β-1, 3-glucosidase, and others. Zhou et al.[152] have studied the impacts of salt stress (300 mM NaCl for 6 h) on root phosphoproteome of Arabidopsis-related halophyte *T. halophila* (ecotype Shandong). Under salt stress, an increased phosphorylation has been reported for a Ran-binding protein Atranbp1 implicated in nucleocytoplasmic transport, thioredoxin-dependent peroxidase 1 (TPX1), and other enzymes (Cu/Zn-SOD and GST) involved in ROS scavenging and several enzymes involved in glycolysis and energy metabolism (ENO and NADH dehydrogenase subunit 9). These results indicated an enhanced activity of enzymes involved in ROS scavenging and energy metabolism under salt stress. Another important area of plant abiotic-stress proteomics is redox proteomics. Since oxidative stress is a common feature associated with abiotic stress factors, proteins involved in maintenance of redox homeostasis in plant cells play a crucial role in plant stress response. Hajheidari et al.[159] analyzed changes in redox proteome in wheat grain during drought and found a crucial role of Trx h in grain stress response.

13.6 PROTEOMICS; PRESENT STATUS AND CHALLENGES

Proteomics is gaining much importance in crop plants to facilitate breeding or to understand the basis of increased yield in hybrids. With few exceptions, however, protein identification in most crop species at present cannot take advantage of fully sequenced genomes with high quality gene annotation.

Proteome research in unsequenced crops therefore relies on genomes of phylogenetically related species, on EST databases, or databases compiling protein sequences from all plant species.[160,161] Nevertheless, several advances have been made in understanding crop physiology and function using proteomics approaches, including mineral deficiency,[162] abiotic stress response[163] plant disease, heavy metal hyperaccumulation,[164] flood tolerance,[165] and fruit development and ripening.[166] Single crop-focused proteomics studies have also been reviewed recently[167,168] and they will not be considered here. Large scale identification of crop proteins even with the most advanced MS technologies and state of the-art proteome databases[169–172] a full proteome coverage has not been achieved even in model plants. Nearly 300,000 non-redundant peptides matching to about 25,000 unique proteins are currently available via MASCP Gator, representing 70% of the expected Arabidopsis proteome.[173] Although the analysis of crop proteomes has benefited from the technical advances and novel methods developed for model plant species such as Arabidopsis and rice, cropproteomics is similarly constrained by the limitations inherent to currently established plant proteomics workflows. Large datasets of transcript sequences from crop species using next-generation sequencing (NGS) will facilitate the annotation of crop genomes.[174] For example, nearly 3500 plantain proteins were identified based on plantain RNA-Seq data. NGS will also provide information on sequence diversity in crop accessions. Arabidopsis and rice protein databases are currently being updated with proteome information from new accessions.[169,172] The Arabidopsis genome, and to some extent the rice genome, has also benefited from gene-by-gene curation. This type of curation is necessary for reliable annotation and the removal of errors generated by annotation software, as best illustrated by the first estimate of genes encoded in the rice genome.[175] Alternative approaches have been developed for proteomics studies of unsequenced plants to increase protein identification. For example, unassigned high quality MS spectra can be submitted to *de novo* sequencing and the newly identified peptides blasted against general plant protein databases, followed by filtering and removal of redundant and incorrectly assigned peptides.[176,177] Such analysis pipelines help extracting additional information from MS spectra that would otherwise not be exploited. Recent analysis pipelines also include translated transcriptomes from related species to increase protein identification. Most plants have experienced polyploidization of their genome during evolution,[178] and important crops such as wheat have allopolyploid genomes. Genomes of polyploidy species, especially those in which polyploidization have occurred recently, are more difficult to sequence and assemble because of gene redundancies.[179] Protein sequence redundancy in polyploidy genomes makes protein inference challenging, especially when using shotgun proteomics.[180] Nevertheless, recent studies in genotype-specific variability of potato[181] as well as protein accumulation in allopolyploid cotton,[182] Arabidopsis[183] and *Tragopogon mirus*[184] illustrated that advances in mass accuracy precursor alignment improves the detection of nearly identical proteins. It may also be possible, especially for protein quantification in polyploidy species, to apply multi hit and normalization strategies similar to those developed for NGS data, such as RPKM,[185] TMM[186] DEseq,[187] or RSEM.[188] Targeting of low-abundance proteins for useful crop traits, regulatory proteins that have key roles in plant responses to biotic and abiotic stress are often difficult to detect even with the most advanced MS instruments because of their low abundance. Pre-fractionation of protein extracts combined with RuBisCO depletion can facilitate the detection of LAPs. Protein fractionation and enrichment generate additional variability in biological samples; however, which is not compatible with high-throughput quantitative proteomics methods. Recently, combinatorial peptide ligand libraries (CPLL) have proven to be effective for removing abundant proteins from plant extracts to increase proteome coverage.[189,190] In some cases proteomics using purified organelles can facilitate the detection of LAPs such as nuclear transcription factors, which would otherwise escape detection because of their relative low abundance in complex cell protein extracts. Targeted approaches such as SRM combine the higher sensitivity of MS with reduced variation to detect LAPs.[191] However, to date there are only few studies reporting on the use of SRM in plant proteomics. SRM remains a challenging approach because it requires peptides with adequate transitions for identification and monitoring. Nevertheless, SRM strategies in crop proteomics will be particularly relevant in the future for quantitation of protein candidates previously identified in model species as key regulators for useful traits in crop breeding programs.

13.7 CROP IMPROVEMENT AND PROTEOMICS

According to an estimate, there are ~925 million people on the globe who live in a state of hunger.[192] Moreover, an additional 2 billion people are expected to be added by the year 2050. In an effort to eradicate that ugly spot of hunger from the beautiful face of the humanity, we need to significantly increase the production and supply of food by integrating different elements and strengthening the plant breeding tools[193] for crop improvement. A major hurdle for crop improvement programs faced by the plant breeders is a limited gene pool of domesticated crop species. The identification of potential useful genes across the animal and plant kingdom that could play key roles toward the improvement of important crop traits, generally derived from research in molecular biology including genomics and proteomics, is a crucial step. Such newly discovered genes, when placed into a desired crop species and then utilized for breeding programs, could be a boon to human society. Comparatively, it is now a fully mature science and is proud to be on the list of most quickly adopted crop technologies in world. Biotechnology provides the capabilities to breeders to achieve certain goals that would otherwise be impossible through conventional plant breeding approaches. Globally, today genetically modified crops are grown in fields at a commercial scale. Thus, the biotech crop area has increased from 1.7 million ha in 1996 to 160 million ha in 2011.[194] This trend was well expected by Dixon[195] when he stated that "Genomics (originally DNA- and transcript-based, but recently extended to integrate the proteome and metabolome) would play a major role in driving plant biotechnology." This review corroborates his long vision and focuses on the use of proteomics for genetic improvements in food and biofuel crops including food quality, safety, and nutritional values, tolerance to abiotic and biotic stresses, manufacturing plant-based vaccines, and proteomics-based fungicides. Apart from these, proteomics is being used for several other crop improvement programs such as, pre- and post-harvest losses, and crop quality characteristics but that is not a part of this review because of space constraints.

The knowledge of key proteins that play crucial roles in the proper growth and development of a plant are critical to propel the biotechnological improvement of crop plants. These proteins maintain cellular homeostasis under a given environment by controlling physiological and biochemical pathways. A search of the published research literature revealed that genomics and proteomics are the two major wheels that keep the discovery of novel genes rolling, which can eventually be placed into the pipeline for crop improvement programs. 2-DE and mass spectroscopy (MS), two of the most widely used proteomics methods, are used to catalog and identify proteins in different proteome states or environments. Advances in 2-DE have been extremely helpful in bringing proteomics close to biotechnological programs; however, due to some drawbacks and disadvantages associated with gel-based proteomics, for example, labor intensiveness, insensitiveness to low-copy number proteins, low reproducibility and the inability to characterize complete proteomes and many gel-free proteomic techniques have also become a valuable tool for scientists.[196-198]

13.8 CONCLUDING REMARKS

Proteomic studies in response to various abiotic stresses could provide vital information in understanding of physiological mechanisms underlying stress responses such as stress perception gene expression involving various signaling events, changes occurring at proteomic, transcriptional and metabolic levels underlying plant acclimation to particular stress and procurement of an elevated plant stress tolerance. More attention should also paid to the dynamics of plant stress response, that is, an initial shock phase, an acclimation phase, a maintenance phase, an exhaustion phase and/or a recovery phase after the cessation of stress treatment.

Currently, the information on proteome changes under stress is often fragmental, comparing only non-stressed (control) plants and stressed plants. Quite a lot of information on changes in cellular metabolism as well as stress-protective proteins in plant proteome under stress is available while much less is known about less abundant regulatory proteins involved in stress signaling and regulation of gene expression. Proteomics studies could also significantly contribute to identification and further characterization of key proteins underlying plant tolerance to a given stress which can be used as protein biomarkers of a given stress. Recently, proteomics studies dealing with plant response to abiotic stress have been focused mainly on a description of quantitative changes in plant proteome or specific subcellular proteome (e.g., nuclear, mitochondrial, and plastid proteomes) and they have been based on comparative proteomics approach using 2-DE followed by MS analysis. However, along with a boost of new methodological approaches, it could be expected that study of PTMs, especially phosphoproteomics and redox proteomics, as well as protein interactions (interactomics) will become applied in the field of plant stress proteomics. These approaches will contribute to a detailed protein functional characterization which will surely help us to better understand the processes of plant stress acclimation and stress tolerance acquisition. The proteomic strategy involved in the study of protein localization in cells is a necessary first step toward understanding protein functions in complex cellular networks. These fascinating events taking place during the metabolism that governs various physiological processes have been revealed by the dynamic nature of the proteome, and we believe that proteomics represents an important advance in the study of plant physiological processes.

KEYWORDS

- **abiotic stress**
- **proteomics**
- **crop productivity**
- **osmo-protactants**
- **phosphorylation**

REFERENCES

1. Ahmad, P.; Prasad, M. N. V. *Environmental Adaptations and Stress Tolerance in Plants in the Era of Climate Change;* Springer Science + Business Media: New York, 2012.

2. Ahmad, P.; Prasad, M. N. V. *Abiotic Stress Responses in Plants: Metabolism, Productivity and Sustainability;* Springer Science + Business Media: New York, 2012.

3. Ahmad, P.; Umar, S. *Oxidative Stress: Role of Antioxidants in Plants;* Studium Press Pvt. Ltd.: New Delhi, India, 2011.

4. Levitt, J. *Responses of Plants to Environmental Stress. Chilling, Freezing and High Temperature Stresses;* 2nd ed.; Academic Press: New York, 1980.

5. Larcher, W. *Physiological Plant Ecology;* 4th ed.; Springer Verlag: Berlin, Heidelberg, 2003.

6. Sarwat, M.; Ahmad, P.; Nabi, G.; Hu, X. Ca2 + Signals: The Versatile Decoders of Environmental Cues. *Crit. Rev. Biotechnol.* **2013,** *33*(1), 97–109.

7. Gygi, S. P.; Rochon, Y.; Franza, B. R.; Aebersold, R. Correlation between Protein and mRNA Abundance in Yeast. *Mol. Cell. Biol.***1999,** *19,* 1720–1730.

8. Bogeat-Triboulot, M. B.; Brosché, M.; Renaut, J.; Jouve, L.; Le Thiec, D.; Fayyaz, P. Gradual Soil Water Depletion Results in Reversible Changes of Gene Expression, Protein Profiles, Ecophysiology, and Growth Performance in *Populus euphratica,* a Poplar Growing in Arid Regions. *Plant Physiol.* **2007,** *143,* 876–892.

9. Chen, Q.; Yang, L.; Ahmad, P.; Wan, X.; Hu, X. Proteomic Profiling and Redox Status Alteration of Recalcitrant Tea (*Camellia sinensis*) Seed in Response to Desiccation. *Planta.* **2011,** *233,* 593–609.

10. Evers, D.; Legay, S.; Lamoureux, D.; Hausman, J. F.; Hoffmann, L.; Renaut, J. Towards a Synthetic View of Potato Cold and Salt Stress Response by Transcriptomic and Proteomic Analyses. *Plant Mol. Biol.* **2012,** *78,* 503–514.

11. Kaufmann, K.; Smaczniak, C.; de Vries, S.; Angenent, G. C.; Karlova, R. Proteomics Insights into Plant Signaling and Development. *Proteomics.* **2011,** *11,* 744–755.

12. Nanjo, Y.; Nouri, M. Z.; Komatsu, S. Quantitative Proteomics Analyses of Crop Seedlings Subjected to Stress Conditions; a Commentary. *Phytochemistry.* **2011,** *72,* 1263–1272.

13. Yang, L.; Bai, X.; Yang, Y.; Ahmad, P.; Yang, Y.; Hu, X. Deciphering the Protective Role of Nitric Oxide Against Salt Stress at the Physiological and Proteomic Levels in Maize. *J. Proteom. Res.* **2011,** *10,* 4349–4364.

14. Yokthongwattana, C.; Mahong, B.; Roytrakul, S.; Phaonaklop, N.; Narangajavana, J.; Yokthongwattana, K. Proteomic Analysis of Salinity-Stressed *Chlamydomonas reinhardtii* Revealed Differential Suppression and Induction of a Large Number of Important Housekeeping Proteins. *Planta* .**2012**, *235,* 649– 659.

15. Zhang, H.; Han, B.; Wang, T., et al. Mechanisms of Plant Salt Response: Insights from Proteomics. *J. Proteome Res.* **2012,** *11,* 49–67.

16. Zheng, M.; Wang, Y.; Liu, K.; Shu, H.; Zhou, Z. Protein Expression Changes during Cotton Fiber Elongation in Response to Low Temperature Stress. *J. Plant Physiol.* **2012,** *169,* 399–409.

17. Canovas, F. M.; Dumas-Gaudot, E.; Recorbet, G.; Jorrin, J.; Mock, H. P.; Rossignol, M. Plant Proteome Analysis. *Proteomics.* **2004,** *4,* 285–298.

18. Ideker, T.; Thorsson, V.; Ranish, J. A.; Christmas, R.; Buhler, J.; Eng, J. K.; Bumgarner, R.; Goodlett, D. R.; Aebersold, R.; Hood, L. Integrated Genomic and Proteomic Analysis of a Systematically Perturbed Metabolic Network. *Science.* **2001,** *292,* 929–934.

19. Mann, M.; Jensen, O. N. Proteomic Analysis of Post-Translational Modifications. *Nat. Biotechnol.* **2001,** *21,* 255–261.

20. Schweppe, R. E.; Haydon, C. E.; Lewis, T. S.; Resing, K. A.; Ahn, N. G. The Characterisation of Protein Post-Translational Modifications by Mass Spectrometry. *Acc. Chem. Res.* **2003,** *36,* 453–461.

21. O'Farrell. P. High Resolution Two-Dimensional Electrophoresis of Proteins. *J. Biol. Chem.* **1975,** *250,* 4007–4021.

22. Görg, A.; Obermaier, C.; Boguth, G.; Harder, A.; Scheibe, B.; Wildgruber, R.; Weiss, W. The Current State of Two Dimensional Electrophoresis with Immobilized pH Gradients. *Electrophoresis.* **2000,** *21,* 1037–1053.

23. Rabilloud, T. Two-Dimensional Gel Electrophoresis in Proteomics: Old, Old Fashioned, but it Still Climbs up the Mountains. *Proteomics.* **2002,** *2,* 3–10.

24. Patton, W. F. Detection Technologies in Proteome Analysis. *J. Chromatogr. B Analyt. Technol. Biomed. Life Sci.* **2002,** *771,* 3–31.

25. Patton, W. F. A Thousand Points of Light: The Application of Fluorescence Detection Technologies to Two-Dimensional Gel Electrophoresis and Proteomics. *Electrophoresis.* **2000,** *21,* 1123–1144.

26. Schmidhuber, J.; Tubiello, F. N. Global Food Security under Climate Change. *Proc. Natl. Acad. Sci. USA.* **2007,** *104*(50), 19703–19708.

27. Atkinson, N. J.; Urwin, P. E. The Interaction of Plant Biotic and Abiotic Stresses: From Genes to the Field. *J. Exp. Bot.* **2012,** *63*(10), 3523–3543.

28. Delano-Frier, J. P.; Aviles-Arnaut, H.; Casarrubias-Castillo, K.; Casique-Arroyo, G.; Castrillon-Arbelaez, P. A.; Herrera-Estrella, L., et al. Transcriptomic Analysis of Grain Amaranth (*Amaranthus hypochondriacus*) Using 454 Pyrosequencing: Comparison with *A. tuberculatus*, Expression Profiling in Stems and in Response to Biotic and Abiotic Stress. *BMC Genomics.* **2011,** *12,* 363.

29. Grativol, C.; Hemerly, A. S.; Ferreira, P. C. Genetic and Epigenetic Regulation of Stress Responses in Natural Plant Populations. *Biochim. Biophys. Acta.* **2012,** *1819*(2), 176–185.

30. Shinozaki, K.; Yamaguchi-Shinozaki, K. Gene Networks Involved in Drought Stress Response and Tolerance. *J. Exp. Bot.* **2007,** *58*(2), 221–227.

31. Krasensky, J.; Jonak, C. Drought, Salt, and Temperature Stress-Induced Metabolic Rearrangements and Regulatory Networks. *J. Exp. Bot.* **2012,** *63*(4), 1593–1608.

32. Wang, X. Q.; Yang, P. F.; Liu, Z.; Liu, W. Z.; Hu, Y.; Chen, H., et al. Exploring the Mechanism of *Physcomitrella* Patens Desiccation Tolerance Through a Proteomic Strategy. *Plant Physiol.* **2009,** *149*(4), 1739–1750.

33. Peleg, Z.; Apse, M. P.; Blumwald, E. Engineering Salinity and Water Stress Tolerance in Crop Plants: Getting Closer to the Field. *Adv. Bot. Res.* **2011,** *57,* 405–443.

34. Pan, Z.; Zhao, Y.; Zheng, Y.; Liu, J.; Jiang, X.; Guo, Y. A High-Throughput Method for Screening *Arabidopsis* Mutants with Disordered Abiotic Stress Induced Calcium Signal. *J. Genet. Genomics.* **2012,** *39*(5), 225–235.

35. Reddy, A. S.; Ali, G. S.; Celesnik, H.; Day, I. S. Coping with Stresses: Roles of Calcium- and Calcium/Calmodulin-Regulated Gene Expression. *Plant Cell.* **2011,** *23*(6), 2010–2032.

36. Ahmad, P.; Jaleel, C. A.; Salem, M. A.; Nabi, G.; Sharma, S. Roles of Enzymatic and Nonenzymatic Antioxidants in Plants during Abiotic Stress. *Crit. Rev. Biotechnol.* **2010,** *30*(3), 161–175.

37. Loiacono, F. V.; De Tullio, M. C. Why We Should Stop Inferring Simple Correlations between Antioxidants and Plant Stress Resistance: Towards the Antioxidomic Era. *OMICS.* **2012,** *16*(4), 160–167.

38. Baena-Gonzalez, E.; Sheen, J. Convergent Energy and Stress Signaling. *Trends Plant Sci.* **2008,** *3*(9), 474–482.

39. Umezawa, T.; Yoshida, R.; Maruyama, K.; Yamaguchi-Shinozaki, K.; Shinozaki, K. SRK2C, a SNF1-Related Protein Kinase 2, Improves Drought Tolerance by Controlling Stress-Responsive Gene Expression in *Arabidopsis thaliana. Proc. Natl. Acad. Sci. USA.* **2004,** *101*(49), 17306–17311.

40. Hey, S.; Mayerhofer, H.; Halford, N. G.; Dickinson, J. R. DNA Sequences from *Arabidopsis,* which Encode Protein Kinases and Function as Upstream Regulators of Snf1 in Yeast. *J. Biol. Chem.* **2007,** *282*(14), 10472–10479.

41. Ghillebert, R.; Swinnen, E.; Wen, J.; Vandesteene, L.; Ramon, M.; Norga, K., et al. The AMPK/SNF1/SnRK1 Fuel Gauge and Energy Regulator: Structure, Function and Regulation. *FEBS J.* **2011,** *278*(21), 3978–3990.

42. Lovas, A.; Bimbo, A.; Szabo, L.; Banfalvi, Z. Antisense Repression of Stub-GAL83 Affects Root and Tuber Development in Potato. *Plant J.* **2003,** *33*(1), 139–147.

43. Cho, Y. H.; Hong, J. W.; Kim, E. C.; Yoo, S. D. Regulatory Functions of SnRK1 in Stress-Responsive Gene Expression and in Plant Growth and Development. *Plant Physiol.* **2012,** *158*(4), 1955–1964.

44. Baena-Gonzalez, E. Energy Signaling in the Regulation of Gene Expression during Stress. *Mol. Plant.* **2010,** *3*(2), 300–313.

45. Good, A. G.; Zaplachinski, S. T. The Effects of Drought Stress on Free Amino Acid Accumulation and Protein Synthesis in *Brassica napus. Physiologia Plantarum.* **1994,** *90*(1), 9–14.

46. Qu, C.; Liu, C.; Ze, Y.; Gong, X.; Hong, M.; Wang, L., et al. Inhibition of Nitrogen and Photosynthetic Carbon Assimilation of Maize Seedlings by Exposure to a Combination of Salt Stress and Potassium-Deficient Stress. *Biol. Trace. Elem. Res.* **2011,** *144*(1–3), 1159–1174.

47. Holcik, M.; Sonenberg, N. Translational Control in Stress and Apoptosis. *Nat. Rev. Mol. Cell. Biol.* **2005,** *6*(4), 318–327.

48. Dhaubhadel, S.; Browning, K. S.; Gallie, D. R.; Krishna, P. Brassinosteroid Functions to Protect the Translational Machinery and Heat-Shock Protein Synthesis Following Thermal Stress. *Plant J.* **2002,** *29*(6), 681–691.

49. Soares-Cavalcanti, N. M.; Belarmino, L. C.; Kido, E. A.; Pandolfi, V.; Marcelino-Guimaraes, F. C.; Rodrigues, F. A., et al. Overall Picture of Expressed Heat Shock Factors in *Glycine max*, *Lotus japonicus* and *Medicago truncatula*. *Genet. Mol. Biol.* **2012,** *35*(1), 247–259.

50. Aghaei, K.; Akbar, A. E.; Komatsu, S. Proteome Analysis of Potato under Salt Stress. *J. Proteome Res.* **2008,** *7,* 4858–4868.

51. Everse, D.; Overney, S.; Simon, P.; Greppin, H.; Hausman, J. F. Salt Tolerance of *Solanum tuberosum* L. Overexpressing an Heterologous Osmotin-Like Protein. *Biol. Plant.* **1999,** *42,* 105–112.

52. Witzel, K.; Annette, W.; Giridara-Kumar, S.; Andreas, B.; Hans-Peter, M. Salt Stress-Induced Alterations in the Root Proteome of Barley Genotypes with Contrasting Response Towards Salinity. *J. Exp. Bot.* **2009,** *60*(12), 3545–3557.

53. Wang, L.; Liu, X.; Liang, M.; Tan, F.; Liang, W., et al. Proteomic Analysis of Salt-Responsive Proteins in the Leaves of Mangrove *Kandelia candel* during Short-Term Stress. *PLoS ONE.* **2014,** *9*(1), e83141. doi:10.1371/journal.pone.0083141.

54. Zhang, H.; Han, B.; Wang, T.; Chen, S. X.; Li, H. Y., et al. Mechanisms of Plant Salt Response: Insights from Proteomics. *J. Proteome Res.* **2011,** *11,* 49–67.

55. Wang, W. F.; Shakes, D. C. Molecular Evolution of the 14–3–3 Protein Family. *J. Mol. Evol.* **1996,** *43,* 384–398.

56. You, X. R.; Wang, L. X.; Liang, W. Y.; Gai, Y. H.; Wang, X. Y., et al. Floral Reversion Mechanism in Longan (*Dimocarpus longan* Lour.) Revealed by Proteomic and Anatomic Analyses. *J. Proteomics.* **2012,** *75,* 1099–1118.

57. Finnie, C.; Borch, J.; Collinge, D. B.14–3–3 Proteins: Eukaryotic Regulatory Proteins with Many Functions. *Plant Mol. Biol.* **1999,** *40,* 545–554.

58. Palmgren, M. G. Proton Gradients and Plant Growth: Role of the Plasmamembrane H+-ATPase. *Adv. Bot. Res.* **1998,** *28,* 1–70.

59. Camoni, L.; Harper, J. F.; Palmgren, M. G 14–3–3 Proteins Activate a Plant Calcium-Dependent Protein Kinase (CDPK). *FEBS Lett.* **1998,** *430,* 381–384.

60. Aitken, A. 14–3–3 Proteins on the MAP. *Trends Biochem. Sci.* **1995,** *20,* 95.

61. Briat, J. F.; Lobréaux, S.; Grignon, N.; Regulation of Plant Ferritin Synthesis: How and Why. *Cell. Mol. Life Sci.* **1999,** *56*(1–2), 155–166.

62. Toorchi, M.; Yukawa, K.; Nouri, M. Z.; Komatsu, S. Proteomics Approach for Identifying Osmotic-Stressrelated Proteins in Soybean Roots. *Peptides.* **2009,** *30*(12), 2108–2117.

63. Smalle, J.; Vierstra, R. D. The Ubiquitin 26sProteasome Proteolytic Pathway. *Annu. Rev. Plant Biol.* **2004,** *55*(1), 555–590.

64. Sassa, H.; Oguchi, S.; Inoue, T.; Primary Structural Features of the 20S Proteasome Subunits of Rice (*Oryza sativa*). *Gene.* **2000,** *250*(1–2), 61–66.

65. Hasegawa, P. M.; Bressan, R. A.; Zhu, J. K.; Bohnert, H. J. Plant Cellular and Molecular Responses to High Salinity. *Ann. Rev. Plant Physiol. Plant Mol. Biol.* **2000,** *51,* 463–499.

66. Skriver, K.; Mundy, J. Gene Expression in Response to Abscisic Acid and Osmotic Stress. *Plant Cell.* **1990,** *2,* 503–512.

67. Fulda, S.; Mikkat, S.; Stegmann, H.; Horn, R. Physiology and Proteomics of Drought Stress Acclimation in Sunflower (*Helianthus annuus* L.). *Plant Biol.* **2011,** *13*(4), 632–642.

68. Wang, W.; Vinocur, B.; Shoseyov, O.; Altman, A. Role of Plant Heat-Shock Proteins and Molecular Chaperones in the Abiotic Stress Response. *Trends Plant Sci.* **2004,** *9,* 244–252.

69. Kausar, R.; Muhammad, A.; Armghan, S.; Setsuko, K. Proteomics analysis of Sensitive and Tolerant Barley Genotypes under Drought Stress. *Amino Acids.* **2013**, *44,* 345–359.

70. Arnau, G.; Monneveux, P.; This, D.; Alegre, L. Photosynthesis of Six Barley Genotypes as Affected by Water Stress. *Photosynthetica.* **1997, 34,** 67–76.

71. Aranjuelo, I.; Molero, G.; Erice, G.; Avice, J. C.; Nogues, S. Plant Physiology and Proteomics Reveals the Leaf Response of Drought in Alfalfa (*Medicago sativa* L.). *J. Exp. Bot.* **2010,** *62,* 111–123.

72. Guicherd, P.; Peltier, J. P.; Gout, E.; Bligny, R.; Marigo, G. Osmotic Adjustment in *Fraxinus excelsior* L. Malate and Mannitol Accumulation in Leaves Under Drought Conditions. *Trees.* **1997,** *11,* 155–161.

73. Musrati, R. A.; Kollarova, M.; Mernik, N.; Mikulasova, D. Malate Dehydrogenase: Distribution, Function and Properties. *Gen. Physiol. Biophys.* **1998,** *17,* 193–210.

74. Dhindsa, R. S. Drought Stress, Enzymes of Glutathione Metabolism, Oxidation Injury, and Protein Synthesis in *Tortula ruralis. Plant Physiol.* **1991,** *95,* 648–651.

75. Peter, F.; Wong, S. H.; Subramaniam, V. N.; Tang, B. L.; Hong, W. A- SNAP but not C-SNAP is Required for ER-Golgi Transport after Vesicle Budding and the Rab1-Requiring Step but before the EGTA-Sensitive Step. *J. Cell. Sci.* **1998,** *111,* 2625–2633.

76. Bohnert, H. J.; Jensen, R. G. Strategies for Engineering Water-Stress Tolerance in Plants. *Trends Biotechnol.* **1996,** *14,* 89–97.

77. Apel, K.; Hirt, H. Reactive Oxygen Species: Metabolism, Oxidative Stress, and Signal Transduction. *Annu. Rev. Plant Biol.* **2004,** *55,* 373–399.

78. Thomashow, M. F. Plant Cold Acclimation: Freezing Tolerance Genes and Regulatory Mechanisms. *Annu. Rev. Plant Physiol. Plant Mol. Biol.* **1999,** *50,* 571–599.

79. Ruelland, E.; Vaultier, M. N.; Zachowski, A.; Hurry, V. Cold Signalling and Cold Acclimation in Plants. In *Advances in Botanical Research;* Kader, J. C., Delseny, M., Eds.; Elsevier Ltd: Amsterdam, The Netherlands, 2009; pp 35–150.

80. Bae, M. S.; Cho, E. J.; Choi, E. Y.; Park, O. K. Analysis of the *Arabidopsis* Nuclear Proteome and its Response to Cold Stress. *Plant J.* **2003,** *36,* 652–663.

81. Kawamura, Y.; Uemura, M. Mass Spectrometric Approach for Identifying Putative Plasma Membrane Proteins of *Arabidopsis* Leaves Associated with Cold Acclimation. *Plant J.* **2003,** *36,* 141–154.

82. Amme, S.; Matros, A.; Schlesier, B.; Mock, H. P. Proteome Analysis of Cold Stress Response in *Arabidopsis thaliana* Using DIGE-Technology. *J. Exp. Bot.* **2006,** *57,* 1537–1546.

83. Gao, F.; Zhou, Y.; Zhu, W.; Li, X.; Fan, L.; Zhang, G. Proteomic Analysis of Cold Stress-Responsive Proteins in *Thellungiella rosette* Leaves. *Planta.* **2009,** *230,* 1033–1046.

84. Imin, N.; Kerim, T.; Rolfe, B. G.; Weinman, J. J. Effect of Early Cold Stress on the Maturation of Rice Anthers. *Proteomics.* **2004,** *4,* 1873–1882.

85. Cui, S.; Huang, F.; Wang, J.; Ma, X.; Cheng, Y.; Liu, J. A Proteomic Analysis of Cold Stress Responses in Rice Seedlings. *Proteomics.* **2005,** *5,* 3162–3172.

86. Yan, S. P.; Zhang, Q. Y.; Tang, Z. C.; Su, W. A.; Sun, W. N. Comparative Proteomic Analysis Provides New Insights into Chilling Stress Responses in Rice. *Mol. Cell. Proteomics.* **2006,** *5,* 484–496.

87. Hashimoto, M.; Komatsu, S. Proteomic Analysis of Rice Seedlings during Cold Stress. *Proteomics.* **2007,** *7,* 1293–1302.

88. Lee, D. G.; Ahsan, N.; Lee, S. H.; Lee, J. J.; Bahk, J. D.; Kang, K. Y., et al. Chilling Stress-Induced Proteomic Changes in Rice Roots. *J. Plant Physiol.* **2009,** *166,* 1–11.

89. Degand, H., Faber, A. M.; Dauchot, N.; Mingeot, D.; Watillon, B.; Van Cutsem, P., et al. Proteomic Analysis of Chicory Root Identifies Proteins Typically Involved in Cold Acclimation. *Proteomics.* **2009,** *9,* 2903–2907.

90. Kosmala, A.; Bocian, A.; Rapacz, M.; Jurczyk, B.; Zwierzykowski, Z. Identification of Leaf Proteins Differentially Accumulated during Cold Acclimation between *Festuca pratensis* Plants with Distinct Levels of Frost Tolerance. *J. Exp. Bot.* **2009,** *60,* 3595–3609.

91. Cheng, L.; Gao, X.; Li, S.; Shi, M.; Javeed, H.; Jing, X., et al. Proteomic Analysis of Soybean *Glycine max* (L.) *Meer.* Seeds during Imbibition at Chilling Temperature. *Mol. Breed* **2010,** *26,* 1–17.

92. Taylor, N. L.; Heazlewood, J. L.; Day, D. A.; Millar, A. H. Differential Impact of Environmental Stresses on the Pea Mitochondrial Proteome. *Mol. Cell. Proteomics.* **2005,** *4,* 1122–1133.

93. Dumont, E.; Bahrman, N.; Goulas, E.; Valot, B.; Sellier, H.; Hilbert, J. L., et al. A Proteomic Approach to Decipher Chilling Response from Cold Acclimation in Pea (*Pisum sativum* L.). *Plant Sci.* **2011,** *180,* 86–98.

94. Andriotis, V. M. E.; Kruger, N. J.; Pike, M. J.; Smith, A. M. Plastidial Glycolysis in Developing *Arabidopsis embryos. New Phytol.* **2010,** *185,* 649–662.

95. Lee, H.; Guo, Y.; Ohta, M.; Xiong, L.; Stevenson, B.; Zhu, J. K. LOS2, a Genetic Locus Required for Cold-Responsive Gene Transcription Encodes a Bi-Functional Enolase. *EMBO J.* **2002,** *21,* 2692–2702.

96. Vítámvás, P.; Saalbach, G.; Prášil, I. T.; Čapková, V.; Opatrná, J.; Jahoor, A. WCS120 Protein Family and Proteins Soluble upon Boiling in Cold-Acclimated Winter Wheat. *J. Plant Physiol.* **2007,** *164,* 1197–1207.

97. Vítámvás, P.; Prášil, I. T. WCS120 Protein Family and Frost Tolerance during Cold Acclimation, Deacclimation and Reacclimation of Winter Wheat. *Plant Physiol. Biochem.* **2008,** *46,* 970–976.

98. Suzuki, N.; Mittler, R. Reactive Oxygen Species and Temperature Stresses: A Delicate Balance between Signaling and Destruction. *Physiol. Plant.* **2006,** *126,* 45–51.

99. Foyer, C. H.; Noctor, G. Oxidant and Antioxidant Signaling in Plants: A Re-Evaluation of the Concept of Oxidative Stress in a Physiological Context. *Plant Cell. Environ.* **2005,** *28,* 1056–1071.

100. Afroz, A.; Ali, G. M.; Mir, A.; Komatsu, S. Application of Proteomics to Investigate Stress-Induced Proteins for Improvement in Crop Protection. *Plant Cell. Rep.* **2011,** *30,* 745–763.

101. Evers, D.; Legay, S.; Lamoureux, D.; Hausman, J. F.; Hoffmann, L.; and Renaut, J. Towards a Synthetic View of Potato Cold and Salt Stress Response by Transcriptomic and Proteomic Analyses. *Plant Mol. Biol.* **2012,** *78,* 503–514.

102. Kosová, K.; Vítámvás, P.; Prášil, I. T.; Renaut, J. Plant Proteome Changes Under Abiotic Stress—Contribution of Proteomics Studies to Understanding Plant Stress Response. *J. Proteomics.* **2011,** *74,* 1301–1322.

103. Zheng, M.; Wang, Y.; Liu, K.; Shu, H.; Zhou, Z. Protein Expression Changes during Cotton Fiber Elongation in Response to Low Temperature Stress. *J. Plant Physiol.* **2012,** *169,* 399–409.

104. Yan, S. P.; Zhang, Q. Y.; Tang, Z. C.; Su, W. A.; Sun, W. N. Comparative Proteomic Analysis Provides New Insights into Stress Responses in Rice. *Mol. Cell. Prot.* **2006,** *5,* 3–9.

105. Kosmala, A.; Bocian, A.; Rapacz, M.; Jurczyk, B.; Zwierzykowski, Z. Identification of Leaf Proteins Differentially Accumulated during Cold Acclimation between *Festuca pratensis* Plants with Distinct Levels of Frost Tolerance. *J. Exp. Bot.* **2009**, *60*, 3595–3609.

106. Lee, D. G.; Ahsan, N.; Lee, S. H., et al. Chilling Stress Induced Proteomic Changes in Rice Roots. *J. Plant Physiol.* **2009**, *166*, 1–11.

107. Hashimoto, M.; Komatsu, S. Proteomic Analysis of Rice Seedlings during Cold Stress. *Proteomics.* **2007**, *7*, 1293–1302.

108. Kamal, A. H. M.; Kim, K. H.; Shin, K. H.; Shin, D. H.; Seo, H. S. In *Functional-Proteome Analysis of Wheat: Systematic Classification of Abiotic Stress Responsive Proteins,* 19th World Congress of Soil Science, Soil Solutions for a Changing World, Brisbane, Australia, Aug 1–6, 2010.

109. Rinalducci, S.; Egidi, M. G.; Mahfoozi, S.; Godehkahriz, S. J.; Zolla, L. The Influence of Temperature on Plant Development in a Vernalization-Requiring Winter Wheat: A 2-DE Based Proteomic Investigation. *J. Proteomics.* **2011**, *74*, 643–659.

110. Peng, S.; Huang, J.; Sheehy, J. E.; Laza, R. C.; Visperas, R. M.; Zhong, X. Rice Yields Decline with Higher Night Temperature from Global Warming. *Proc. Natl. Acad. Sci. USA.* **2004**, *101*, 9971–9975.

111. Lee, D. G.; Ahsan, N.; Lee, S. H., et al. A Proteomic Approach in Analyzing Heat-Responsive Proteins in Rice Leaves. *Proteomics.* **2007**, *7*, 3369–3383.

112. Baniwal, S. K.; Bharti, K.; Chan, K. Y.; Fauth, M.; Ganguli, A.; Kotak, S., et al. Heat Stress Response in Plants: A Complex Game with Chaperones and More than Twenty Heat Stress Transcription Factors. *J. Biosci.* **2004**, *29*, 471–487.

113. Lee, D. G.; Ahsan, N.; Lee, S. H.; Kang, K. Y.; Bahk, J. D.; Lee, I. J.; et al. A Proteomic Approach in Analyzing Heat-Responsive Proteins in Rice Leaves. *Proteomics.* **2007**, *7*, 3369–3383.

114. Skylas, D. J.; Cordwell, S. J.; Hains, P. G.; Larsen, M. R.; Basseal, D. J.; Walsh, B. J., et al. Heat Shock of Wheat during Grain Filling: Proteins Associated with Heat-Tolerance. *J. Cereal Sci.* **2002**, *35*, 175–188.

115. Majoul, T.; Bancel, E.; Triboi, E.; Ben Hamida, J.; Branlard, G. Proteomic Analysis of the Effect of Heat Stress on Hexaploid Wheat Grain: Characterization of Heat-Responsive Proteins from Non-Prolamins Fraction. *Proteomics.* **2004**, *4*, 505–513.

116. Ferreira, S.; Hjernø, K.; Larsen, M.; Wingsle, G.; Larsen, P.; Fey, S., et al. Proteome Profiling of *Populus euphratica* Oliv. Upon Heat Stress. *Ann. Bot.* **2006**, *98*, 361–377.

117. Zhang, M. H.; Li, G. W.; Huang, W.; Bi, T.; Chen, G. Y.; Tang, Z. C. Proteomic Study of *Carissa spinarum* in Response to Combined Heat and Drought Stress. *Proteomics.* **2010**, *10*, 3117–3129.

118. Ahmad, P.; Nabi, G.; Ashraf, M. Cadmium-Induced Oxidative Damage in Mustard *Brassica juncea* (L.) (Czern. & Coss.) Plants can be Alleviated by Salicylic Acid. *South Afr. J. Bot.* **2011**, *77*, 36–44.

119. John, R.; Ahmad, P.; Gadgil, K.; Sharma, S. Heavy Metal Toxicity: Effect on Plant Growth, Biochemical Parameters and Metal uptake by *Brassica juncea* L. *Int. J. Plant Produc.* **2009**, *3*, 65–76.

120. John, R.; Ahmad, P.; Gadgil, K.; Sharma, S. Cadmium and Lead Induced Changes in Lipid Peroxidation, Antioxidative Enzymes and Metal Accumulation in *Brassica juncea* L. at Three Different Growth Stages. *Arch. Agro. Soil Sci.* **2009**, *55*, 395–405.

121. Ahsan, N.; Lee, D. G.; Alam, I., et al. Comparative Proteomic Study of Arsenic-Induced Differentially Expressed Proteins in Rice Roots Reveals Glutathione Plays a Central Role during as Stress. *Proteomics.* **2008,** *8,* 3561–3576.

122. Kieffer, P.; Schro¨ der, P.; Dommes, J.; Hoffmann, L.; Renaut, J.; Hausman, J. F. Proteomic and Enzymatic Response of Poplar to Cadmium Stress. *J. Proteomics.* **2009,** *72,* 379–396.

123. Cheng, Y.; Qi, Y.; Zhu, Q., et al. New Changes in the Plasma-Membrane-Associated Proteome of Rice Roots Under Salt Stress. *Proteomics.* **2009,** *9,* 3100–3114.

124. Labra, M.; Gianazza, E.; Waitt, R., et al. *Zea mays* L. Protein Changes in Response to Potassium Dichromate Treatments. *Chemosphere.* **2006,** *60,* 1234–1244.

125. Requejo, R.; Tena, M. Proteome Analysis of Maize Roots Reveals that Oxidative Stress is a Main Contributory Factor to Plant Arsenic Toxicity. *Phytochemistry.* **2005,** *66,* 1519–1528.

126. Lee, K.; Bae, D. W.; Kim, S. H., et al. Comparative Proteomic Analysis of the Short-Term Responses of Rice Roots and Leaves to Cadmium. *J. Plant Physiol.* **2010,** *167,* 161–168.

127. Ahsan, N.; Lee, D. G.; Lee, S. H.; Kang, K. Y.;Joo, J. Excess Copper Induced Physiological and Proteomic Changes in Germinating Rice Seeds. *Chemosphere.* **2007,** *67,* 1182–1193.

128. Ingle, R. A.; Smith, J. A. C.; Sweetlove, L. J. Responses to Nickel in the Proteome of the Hyperaccumulator Plant *Alyssum lesbiacum. Bio. Metals.* **2005,** *18,* 627–641.

129. Isarankura-Na-Ayudhya, P.; Isarankura-Na-Ayudhya, C.; Treeratanapaiboon, L.; Kasikun, K.; Thipkeaw, K.; Prachayasittikul, V. Proteomic Profiling of Escherichia Coli in Response to Heavy Metals Stress. *Europ. J. Sci. Res.* **2009,** *25,* 679–688.

130. Kieffer, P.; Dommes, J.; Hoffmann, L.; Hausman, J. F.; Renaut, J. Quantitative Changes in Protein Expression of Cadmium-Exposed Poplar Plants. *Proteomics.* **2008,** *8,* 2514–2530.

131. Kieffer, P.; Planchon, S.; Oufir, M.; Ziebel, J.; Dommes, J.; Hoffmann, L., et al. Combining Proteomics and Metabolite Analyses to Unravel Cadmium Stress-Response in Poplar Leaves. *J Proteome Res.* **2009,** *8,* 400–417.

132. Durand, T. C.; Sergeant, K.; Planchon, S.; Carpin, S.; Label, P.; Morabito, D., et al. Acute Metal Stress in *Populus tremula*×P. alba (717–1B4 genotype): Leaf and Cambial Proteome Changes Induced by Cadmium. *Proteomics.* **2010,** *10,* 349–368.

133. Roth, U.; von Roepenack-Lahaye, E.; Clemens, S. Proteome Changes in *Arabidopsis thaliana* Roots upon Exposure to Cd2+. *J. Exp. Bot.* **2006,** *57,* 4003–4013.

134. Sarry, J. E.; Kuhn, L.; Ducruix, C.; Lafaye, A.; Junot, C.; Hugouvieux, V., et al. The Early Responses of *Arabidopsis thaliana* Cells to Cadmium Exposure Explored by Protein and Metabolite Profiling Analyses. *Proteomics.* **2006,** *6,* 2180–2198.

135. Semane, B.; Dupae, J.; Cuypers, A.; Noben, J. P.; Tuomainen, M.; Tervahauta, A., et al. Leaf Proteome Responses of *Arabidopsis thaliana*Exposed to Mild Cadmium Stress. *J. Plant Physiol.* **2010,** *167,* 247–254.

136. Aina, R.; Labra, M.; Fumagalli, P.; Vannini, C.; Marsoni, M.; Cucchi, U., et al. Thiol-Peptide Level and Proteomic Changes in Response to Cadmium Toxicity in *Oryza sativa* L. Roots. *Env. Exp. Bot.* **2007,** *59,* 381–392.

137. Lee, K.; Bae, D. W.; Kim, S. H.; Han, H. J.; Liu, X.; Park, H. C., et al. Comparative Proteomic Analysis of the Short-Term Responses of Rice Roots and Leaves to Cadmium. *J. Plant Physiol.* **2010,** *167,* 161–168.

138. Hajduch, M.; Rakwal, R.; Agrawal, G. K.; Yonekura, M.; Pretova, A. High-Resolution Two-Dimensional Electrophoresis Separation of Proteins Frommetal-Stressed Rice (*Oryza sativa* L.) Leaves: Drastic Reductions/Fragmentation of Ribulose-1,5-Bisphosphate Carboxylase/Oxygenase Andinduction of Stress-Relatedproteins. *Electrophoresis.* **2001,** *22,* 2824–2831.

139. Ingle. R. A.; Smith, J. A. C.; Sweetlove, L. J. Responses to Nickel in the Proteome of the Hyperaccumulator Plant *Alyssum lesbiacum. Biometals.* **2005,** *18,* 627–641.

140. Tuomainen, M. H.; Nunan, N.; Lehesranta, S. J.; Tervahauta, A. I.; Hassinen, V. H.; Schat H., et al. Multivariate Analysis of Protein Profiles of Metal Hyperaccumulator *Thlaspi caerulescens* Accessions. *Proteomics.* **2006,** *6,* 3696–3706.

141. Duquesnoy, I.; Goupil, P.; Nadaud, I.; Branlard, G.; Piquet-Pissaloux, A.; Ledoigt, G. Identification of *Agrostis tenuis* Leaf Proteins in Response to As(V) and As(III) Induced Stress Using a Proteomics approach. *Plant Sci.* **2009,** *176,* 206–213.

142. Hradilová, J.; Řehulka, P.; Řehulková, H.; Vrbová, M.; Griga, M.; Brzobohatý, B. Comparative Analysis of Proteomic Changes in Contrasting Flax Cultivars upon Cadmium Exposure. *Electrophoresis.* **2010,** *31,* 421–431.

143. Renaut, J.; Bohler, S.; Hausman, J. F.; Hoffmann, L.; Sergeant, K.; Ahsan, N., et al. The Impact of Atmospheric Composition on Plants: A Case Study of Ozone and Poplar. *Mass Spectrom. Rev.* **2009,** *28,* 495–516.

144. Agrawal, G. K.; Rakwal, R.; Yonekura, M.; Kubo, A.; Saji, H. Proteome Analysis Of Differentially Displayed Proteins as a Tool for Investigating Ozone Stress in Rice (*Oryza sativa* L.) Seedlings. *Proteomics.* **2002,** *2,* 947–959.

145. Bohler, S.; Bagard, M.; Oufir, M.; Planchon, S.; Hoffmann, L.; Jolivet, Y., et al. A DIGE Analysis of Developing Poplar Leaves Subjected to Ozone Reveals Major Changes in Carbon Metabolism. *Proteomics.* **2007,** *7,* 1584–1599.

146. Bohler, S.; Sergeant, K.; Lefèvre, I.; Jolivet, Y.; Hoffmann, L.; Renaut, J., et al. Differential Impact of Chronic Ozone Exposure on Expanding and Fully Expanded Poplar Leaves. *Tree Physiol.* **2010,** *30,* 1415–1432.

147. Chang, W. W. P.; Huang, L.; Shen, M.; Webster, C.; Burlingame, A. L.; Roberts, J. K. M. Patterns of Protein Synthesis and Tolerance to Anoxia in Root Tips of Maize Seedlings Acclimated to a Low-Oxygen Environment, and Identification of Proteins by Mass Spectrometry. *Plant Physiol.* **2000,** *122,* 295–317.

148. Murchie, E. H.; Horton, P. Acclimation of Photosynthesis to Irradiance and Spectral Quality in British Plant Species: Chlorophyll Content, Photosynthetic Capacity and Habitat Preference. *Plant Cell. Environ.* **1997,** *20,* 438–448.

149. Giacomelli, L.; Rudella, A.; van Wijk, K. J. High Light Response of the Thylakoid Proteome in *Arabidopsis* Wild-Type and the Ascorbate-Deficient Mutant Vtc2–2. A Comparative Proteomics Study. *Plant Physiol.* **2006,** *141,* 685–701.

150. Ytterberg, A. J.; Peltier, J. B.; van Wijk, K. J. Protein Profiling of Plastoglobules in Chloroplasts and Chromoplasts. A Surprising Site for Differential Accumulation of Metabolic Enzymes. *Plant Physiol.* **2006,** *140,* 984–997.

151. El-Khatib, R. T.; Good, A. G.; Muench, D. G. Analysis of the *Arabidopsis* Cell Suspension Phosphoproteome in Response to Short-Term Low Temperature and Abscisic Acid Treatment. *Physiol. Plant.* **2007,** *129,* 687–697.

152. Zhou, Y. J.; Gao, F.; Zhan, J.; Zhang, G. F. Alterations in Phosphoproteome Under Salt Stress in *Thellungiella* Roots. *Chin. Sci. Bull.* **2010,** *55,* 3673–3679.

153. de la Fuente van Bentem, S.; Roitinger, E.; Anrather, D.; Chazar, E.; Hirt, H. Phospho-proteomics as a Tool to Unravel Plant Regulatory Mechanisms. *Physiol. Plant.* **2006,** *126,* 110–119.

154. Kerst, N. B.; Agrawal, G. K.; Iwahashi, H.; Rakwal, R. Plant Phosphoproteomics: A Long Road Ahead. *Proteomics.* **2006,** *6,* 5517–5528.

155. Kersten, B.; Agrawal, G. K.; Duek, P.; Neigenfind, J.; Schulze, W.; Walther, D., et al. Plant Phosphoproteomics: An Update. *Proteomics.* **2009,** *9,* 964–988.

156. Ytterberg, A. J, Jensen, O. N. Modification-Specific Proteomics in Plant Biology. *J. Proteomics.* **2010,** *73,* 2249–2266.

157. Khan, M.; Takasaki, H.; Komatsu, S. Comprehensive Phosphoproteome Analysis in Rice and Identification of Phosphoproteins Responsive to Different Hormones/Stresses. *J. Proteome Res.* **2005,** *4,* 1592–1599.

158. Chitteti, B. R.; Peng, Z. Proteome and Phosphoproteome Differential Expression under Salinity Stress in Rice (*Oryza sativa*) Roots. *J. Proteome. Res.* **2007,** *6,* 1718–1727.

159. Hajheidari, M.; Eivazi, A.; Buchanan, B. B.; Wong, J. H.; Majidi, I.; Salekdeh, G. H. Proteomics Uncovers a Role for Redox in Drought Tolerance in Wheat. *J. Proteome. Res.* **2007,** *6,* 1451–1460.

160. Carpentier, S. C.; Panis, B.; Vertommen, A.; Swennen, R.; Sergeant, K.; Renaut, J., et al. Proteome Analysis of Non-Model Plants: A Challenging but Powerful Approach. *Mass. Spectrom. Rev.* **2008,** *27,* 354–377.

161. Varshney, R. K.; Close, T. J.; Singh, N. K.; Hoisington, D. A.; Cook, D. R. *Orphan legume* Crops Enter the Genomics Era! *Curr. Opin. Plant Biol.* **2009,** *12,* 202–210.

162. Liang, C.; Tian, J.; Liao, H. Proteomics Dissection of Plant Responses to Mineral Nutrient Deficiency. *Proteomics.* **2012,** *13*(3–4), 624–636.

163. Zhang, H.; Han, B.; Wang, T.; Che, S.; Li, H.; Zhang, Y., et al. Mechanisms of Plant Salt Response: Insights from Proteomics. *J. Proteome. Res.* **2012,** *11,* 49–67.

164. Visioli, G.; Marmiroli, N. The Proteomics of Heavy Metal Hyperaccumulation by Plants. *J. Proteomics.* **2012,** *79C,* 133–145.

165. Komatsu, S.; Hira, A. S.; Yanagawa, Y. Proteomics Techniques for the Development of Flood Tolerant Crops. *J. Proteome. Res.* **2012,** *11,* 68–78.

166. Palma, J. M.; Corpas, F. J.; del Rio, L. A. Proteomics as an Approach to the Understanding of the Molecular Physiology of Fruit Development and Ripening. *J. Proteomics.* **2011,** *74,* 1230–1243.

167. Pechanova, O.; Takac, T.; Samaj, J.; Pechan, T. Maize Proteomics: An Insight to Biology of Important Cereal Crop. *Proteomics.* **2013,** *13*(3–4), 637–662.

168. Agrawal, G. K.; Rakwal, R. Rice Proteomics: A Move Toward Expanded Proteome Coverage to Comparative and Functional Proteomics Uncovers the Mysteries of Rice and Plant Biology. *Proteomics.* **2011,** *11,* 1630–1649.

169. Joshi, H. J.; Christiansen, K. M.; Fitz, J.; Cao, J.; Lipzen, A.; Martin, J., et al. 1001 Proteomes: A Functional Proteomics Portal for the Analysis of *Arabidopsis thaliana* Accessions. *Bioinformatics.* **2012,** *28,* 1303–1306.

170. Hirsch-Hoffmann, M.; Gruissem, W.; Baerenfaller, K. Pep2pro: The High-Throughput Proteomics Data Processing, Analysis, and Visualization Tool. *Front. Plant Sci.* **2012,** *3,* 123.

171. Joshi, H. J.; Hirsch-Hoffmann, M.; Baerenfaller, K.; Gruissem, W.; Baginsky, S.; Schmidt R., et al. MASCP Gator: An Aggregation Portal for the Visualization of *Arabidopsis* Proteomics Data. *Plant Physiol.* **2011,** *155,* 259–270.

172. Helmy, M.; Tomita, M.; Ishihama, Y. OryzaPG-DB: Rice Proteome Database Based on Shotgun Proteogenomics. *BMC Plant Biol.* **2011,** *11,* 63.

173. Mann, G. W.; Joshi, H. J.; Petzold, C. J.; Heazlewood, J. L. Proteome Coverage of the Model Plant *Arabidopsis thaliana*: Implications for Shotgun Proteomic Studies. *J. Proteomics.* **2012,** *79C,* 195–199.

174. Yandell, M.; Ence, D. A Beginner's Guide to Eukaryotic Genome Annotation. *Nat. Rev. Genet.* **2012,** *13,* 329–342.

175. Bennetzen, J. L.; Coleman, C.; Liu, R.; Ma, J.; Ramakrishna, W. Consistent Over-Estimation of Gene Number in Complex Plant Genomes. *Curr. Opin. Plant. Biol.* **2004,** *7,* 732–736.

176. Grossmann, J.; Fischer, B.; Baerenfaller, K.; Owiti, J.; Buhmann, J. M.; Gruissem, W., et al. A Workflow to Increase the Detection Rate of Proteins from Unsequenced Organisms in High-Throughput Proteomics Experiments. *Proteomics.* **2007,** *7,* 4245–4254.

177. Vertommen, A.; Moller, A. L.; Cordewener, J. H.; Swennen, R.; Panis, B.; Finnie, C., et al. A Workflow for Peptide-Based Proteomics in a Poorly Sequenced Plant: A Case Study on the Plasma Membrane Proteome of Banana. *J. Proteomics.* **2011,** *74*(8), 1218–1229.

178. Adams, K. L.; Wendel, J. F. Polyploidy and Genome Evolution in Plants. *Curr. Opin. Plant Biol.* **2005,** *8,* 135–141.

179. Jackson, S. A.; Iwata, A.; Lee, S. H.; Schmutz, J.; Shoemaker, R. Sequencing Crop Genomes: Approaches and Applications. *New Phytol.* **2011,** *191,* 915–925.

180. Claassen, M. Inference and Validation of Protein Identifications. *Mol. Cell. Proteomics.* **2012,** *11,* 1097–1104.

181. Hoehenwarter, W.; Larhlimi, A.; Hummel, J.; Egelhofer, V.; Selbig, J.; van Dongen, J. T., et al. MAPA Distinguishes Genotype-Specific Variability of Highly Similar Regulatory Protein Isoforms in Potato Tuber. *J. Proteome. Res.* **2011,** *10,* 2979–2991.

182. Hu, G.; Houston, N. L.; Pathak, D.; Schmidt, L.; Thelen, J. J.; Wendel, J. F. Genomically Biased Accumulation of Seed Storage Proteins in Allopolyploid Cotton. *Genetics.* **2011,** *189,* 1103–1115.

183. Ng, D. W.; Zhang, C.; Miller, M.; Shen, Z.; Briggs, S. P.; Chen, Z. J. Proteomic Divergence in *Arabidopsis* Autopolyploids and Allopolyploids and their Progenitors. *Heredity.* **2012,** *108,* 419–430.

184. Koh, J.; Chen, S.; Zhu, N.; Yu, F.; Soltis, P. S.; Soltis, D. E. Comparative Proteomics of the Recently and Recurrently Formed Natural Allopolyploid *Tragopogon mirus* (Asteraceae) and its Parents. *New Phytol.* **2012,** *196,* 292–305.

185. Mortazavi, A.; Williams, B. A.; McCue, K.; Schaeffer, L.; Wold, B. Mapping and Quantifying Mammalian Transcriptomes by RNA-Seq. *Nat. Methods.* **2008,** *5,* 621–628.

186. Robinson, M. D.; Oshlack A. A Scaling Normalization Method for Differential Expression Analysis of RNA-seq Data. *Genome Biol.* **2010,** *11,* R25.

187. Anders, S.; Huber, W. Differential Expression Analysis for Sequence Count Data. *Genome Biol.* **2010,** *11,* R106.

188. Li, B.; Dewey, C. N. RSEM: Accurate Transcript Quantification from RNA-Seq Data with or without a Reference Genome. *BMC Bioinformatics.* **2011,** *12,* 323.

189. Boschetti, E.; Bindschedler, L. V.; Tang, C.; Fasoli, E.; Righetti, P. G. Combinatorial Peptide Ligand Libraries and Plant Proteomics: A Winning Strategy at a Price. *J. Chromatogr. A.* **2009,** *1216,* 1215–1222.

190. Frohlich, A.; Gaupels, F.; Sarioglu, H.; Holzmeister, C.; Spannagl, M.; Durner, J., et al. Looking Deep Inside: Detection of Low-Abundance Proteins in Leaf Extracts of *Arabidopsis* and Phloem Exudates of Pumpkin. *Plant Physiol.* **2012,** *159,* 902–914.

191. Lange, V.; Picotti, P.; Domon, B.; Aebersold, R. Selected Reaction Monitoring for Quantitative Proteomics: A Tutorial. *Mol. Syst. Biol.* **2008,** *4,* 222.

192. Karimizadeh, R.; Mohammadi, M.; Ghaffaripour, S.; Karimpour, F.; Shefazadeh, M. K. Evaluation of Physiological Screening Techniques for Drought-Resistant Breeding of Durum Wheat Genotypes in Iran. *Afr. J. Biotechnol.* **2011,** *10,* 12107–12117.

193. Beddington, J.; Asaduzzaman, M.; Clark, M.; Bremauntz, A.; Guillou, M.; Jahn, M., et al. The Role for Scientists in Tackling Food Insecurity and Climate Change. *Agric. Food Sec.* **2012,** *1,* 10.

194. Khush, G. Genetically Modified Crops: The Fastest Adopted Crop Technology in the History of Modern Agriculture. *Agric. Food Sec.***2012,** *1,* 14.

195. Dixon, R. A. Plant Biotechnology Kicks off into the 21stcentury. *Trends Plant Sci.* **2005,** *10,* 560–561.

196. Baggerman, G.; Vierstraete, E.; Loof, A. D.; Schoofs, L. Gel Based Versus Gel-Free Proteomics: A Review. *Comb. Chem. High Throughput Screen.* **2005,** *8,* 669–677.

197. Lambert, J. P.; Ethier, M.; Smith, J. C.; Figeys, D. Proteomics: From Gel Based to Gel-Free. *Anal. Chem.* **2005,** *77,* 3771–3788.

198. Scherp, P.; Ku, G.; Coleman, L.; Kheterpal, I. Gel-based and Gel-Free Proteomic Technologies. *Methods Mol. Biol.* **2011,** *702,* 163–190.

CHAPTER 14

POTENTIALITIES OF PROTEOMICS FOR GENERATING ABIOTIC STRESS TOLERANT CROP SPECIES

RAM KUMAR[1], NEHA JAIN[1], VAGISH MISHRA[1], NISHA SINGH[1], PRAGYA MISHRA[1], SAJAD MAJEED ZARGAR[2], AJAY JAIN[1], NAGENDRA KUMAR SINGH[1], and VANDNA RAI[1*]

[1]*National Research Centre on Plant Biotechnology, Indian Agriculture Research Institute, New Delhi 110012, India*

[2]*Centre for Plant Biotechnology, Division of Biotechnology, SKUAST-K, Shalimar, Srinagar 190025, Jammu and Kashmir, India*

Corresponding author. E-mail: vandnarai2006@gmail.com

CONTENTS

ABSTRACT

Enhancing crop resistance toward various abiotic stresses and improved nutritional quality of staple food crops is an urgent need to meet an ever-increasing global demand of feeding population. Quantitative gel-free holistic proteomic approach, facilitating both quantification and identification of relatively less abundant proteins, has expedited in an identification of post-transcriptionally regulated array of functionally diverse genes playing a pivotal role in conferring resistance toward different abiotic stresses and nutritional quality of agronomically important crop species. These genes potentially serve as a rich repository that is amenable for engineering smart plants that show enhanced tolerance toward abiotic stresses and improved nutritional quality for sustainable food production. A holistic proteomic approach is an option to attain sustainability in crop yield vis-a-vis nutritional security. In this review we will focus on the proteomic approaches currently being used for improvement of staple food crops for their abiotic stress tolerance and nutritive values.

14.1 INTRODUCTION

Cereals (rice and wheat) and legumes (soybean and pigeon pea) are major sources of carbohydrate and/or proteins in dietary food particularly in Indian subcontinent. Among these crop species, 90% of rice is consumed largely in tropical and subtropical Asian countries contributing two thirds of the total caloric intake.[1,2] In India cultivation of rice is practiced in varied agroclimatic regions ranging from rain fed to irrigated. On the contrary, cultivation of wheat is rather confined to north-western states.

The average rice and wheat growing area is 5 and 6.24 Mha, respectively, with an average yield of 43.96 and 13.70 q/ha for rice and wheat.[3] In India total wheat producing area is 30 Mha. In recent year, 2013–2014, wheat is the second most-produced cereal; world production in this year is 714 million tons whereas world total rice production is on third rank, approx 478 million tons (GMR, 2015). Demand of wheat from 1.2 to 2.5 billion is projected to increase by 60% by the year 2050. Wheat (*Triticum aestivum* L.) constitutes approximately 30% of the total cereals consumed in the world, making it a major source of minerals (FAO, 2003). Wheat provides nearly 20% of the calorific energy and 55% of the carbohydrates worldwide.[4] Its cultivation in India started 5000 years ago.[5]

Among pulses pigeonpea is the major legume consumed by Indian population and is cultivated on 4.64 Mha with an annual production of 3.43 million tons and a mean productivity of 780 kg/ha. India accounts for total 3.53 Mha area and 2.51 million ton of pigeonpea production. Whether consumption of rice, wheat, and pulses would have a similar influence on Indian population is a matter of conjecture and indeed warrants detailed studies. Whatever the outcome of this survey could be, the matter of fact is that both these cereal crops play vital roles in food habits and socio-economic profiles of Indian population. Therefore, it assumes greater importance for making efforts toward sustainable production.

14.1.1   IMPACT OF ABIOTIC STRESS ON CROP PRODUCTION

Often plants are subjected to various abiotic stresses comprising salinity, drought, submergence, and deficiency of various micro and macro nutrients which adversely affect production and grain yield. Around 6% of the world's total land area and 20% of irrigated land are affected by high salinity (FAO, 2008). Cereal agricultural production is limited by a wide array of abiotic and biotic stress factors including drought,[6] cold,[7,8] heat, and salinity.[9,10] Crop shows its maximum yield potential in stress free environment while environmental stress causes reduction in this potential. Abiotic factors are the major threat to wheat productivity and causing up to 71% reduction in yield. Among abiotic stresses, drought is one of the most challenging threats with worldwide distribution that directly reduces the growth and development of crop plants. Bread wheat is one of the most important crop plants worldwide, occupying 17% (one sixth) of crop acreage of the world,[11] a staple food for 35% of the world's population, providing more calories and protein in the global diet than that of any other crop.[12] Wheat is probably the most important crop in the world, daily average temperature for optimal growth conditions is 22–25 °C, high temperature reduces the vegetative growth and seed setting. Changes in ambient temperature occur within hours, unlike drought and salinity stresses. Therefore, plants need to suppress and respond to the adverse effects of heat in a very short time. Gradual temperature increase in a day could cause some alterations in antioxidant metabolism or in other physiological responses and leads to increase in the expression of heat shock proteins (HSPs). Heat stress influences photosynthesis, cellular and subcellular membrane components, protein content in cell, and antioxidant enzyme activity; thereby significantly limits crop production. Heat stress also induces oxidative stress in plants caused by the generation and the accumulation of super-oxides (O_2^-), hydrogen peroxide (H_2O_2), and hydroxyl radicals (OH^-), which are commonly known as reactive oxygen species (ROS).[13]

Rice has mainly grown under rain-fed conditions[14] and subjected to various types of abiotic stresses such as salinity,[15,16] drought,[17,18] submergence,[19] and nutrition. In India, 9.04 Mha of rice growing area is affected by salinity leading to considerable loss of grain yield. Not only salinity stress drastically affects rice production but frequently occurred severe droughts also drastically affect rice production[20] In India rice production could not meet its demand from 1965 to 2009 due to drought (DES, 2009). Not only drought and salinity have been identified as the most important constraints for rice production but submergence has also been identified as the third most important constraint for rice production in Eastern India. Among 29% of total area under rice production about 13 Mha is rain-fed lowland. Approximately 3 Mha is deep waterlogged and remaining 3 Mha is submergence or flood-prone which leads to severe crop damage[21] Another major crop wheat can be grown in irrigated and rain-fed environments which are threatened by salinity stress.[22,23] In India approximately 8–10% of the area for wheat cropping is affected by salinity.[23] Wheat production is greatly affected (20–30%) by climate-change-induced rise in temperature.

Among legumes soybean is the most widely cultivated seed and an economical source of protein diet and important vegetable oil for human consumption. But soybean production has been affected by various abiotic stresses as flooding, drought, and salinity[24] and seedling stage is more prone to flooding and drought stress. In India, pigeonpea (*Cajanus cajan*) is a major source of dietary protein. However, under different agroclimatic conditions, pigeonpea is often subjected to various abiotic stresses such as waterlogging and salinity, thereby severely limiting its productivity.

14.1.2 POTENTIAL CROP IMPROVEMENT BY PROTEOMIC APPROACH

It is important to fathom the nature of stress responsive mechanism for improvement of food crops in marginal area. Conventionally attempts have been made to understand morpho-physiological and molecular responses to various abiotic stresses using molecular breeding techniques. In rice major quantitative trait locus (QTL) Saltol[25] for salinity, Sub1[26] for flood tolerance, drought grian yield (DTY)[27] for yield under drought, and Pstol[28] for low phosphorus tolerance were identified. Although conventional breeding could be a viable alternative to circumvent this problem, it is often a limitation due to either non-availability of suitable genetic material or being labor-intensive and time consuming. Therefore, it is pertinent to use a molecular strategy for dissecting various molecular entities that may play a pivotal role in regulation of sensing and signaling cascades that are triggered during abiotic stresses. Arabidopsis has been a favored model species for dissecting pivotal components of the salt stress sensing and signaling network in root cells. It is encouraging that genomic DNA of major crops rice,[29] wheat[12] soybean,[30] and pigeonpea[31,32] has been sequenced. Information generated form genome sequencing can be further utilized for functional annotation of genes through proteomics. This development has now been made it feasible to take initiative toward unraveling some of the molecular intricacies and/or enigmas associated with abiotic stress. Further, transcriptomics provides useful information on differentially regulated genes that subsequently encode the protein. However, mRNA species generated from transcription are often modified post-transcriptionally and post-translationally (phosphorylation, acylation, myrisotylation, etc.), and thus transcriptomic analysis does not always provide the realistic assessment of the gene function. Therefore, functional translated fraction approach is more of a viable alternative for global and functional annotation of the genes and identifying master regulators of that play a crucial role in deciphering pathways altered during abiotic stress and cross talk if any operated during salinity, drought and submergence stress tolerance. Capturing those master regulator switches is one of very useful components for breeding of improved crop varieties as well as developing genetically engineered crops. Therefore, an integrated endeavor across the agricultural, biotechnology, and molecular biology sectors has required to translate proteomic data into enhanced food productivity.[33]

14.2 TECHNIQUES DEVELOPED FOR STUDY OF PROTEOMICS

Plant proteomics most often investigates total protein populations. Extracting proteins from plant samples is challenging, because plant cells generally contain low amounts of protein protected by cell walls that require extreme measures to disrupt. There are several protein extraction methods based on either trichloroacetic acid (TCA)/acetone precipitation or phenol extraction, many of which are widely used in the plant proteomics. For extraction of proteins from different tissues of plants; phenol,[34] TCA-acetone,[35] and phosphate TCA-acetone[36] methods have been employed. There are two different approaches have been used for proteomics, one is gel based (two-dimensional (2D) gel electrophoresis) and another one is gel free.

14.2.1   GEL BASED TECHNIQUE

Two-dimensional gel electrophoresis (2DE) is one of the most powerful tools for large-scale protein separation and quantification.[37] In the first dimension, proteins are separated according to isoelectric point and in second according to molecular weight.[38] Currently, 2DE combined with mass spectrometry (MS) is the major platform for proteomics.[39] The aim of most 2DE analyses is to maximize the number of polypeptides that can be resolved, especially for comparative proteomics, which generally involves looking for differences between experimental and control samples.[40] The 2D gels can be scanned using a transmissive scanner with a 32 bit pixel depth and 300 dpi resolution. Protein spots on 2D gels are detected and processed using Image Master 2D Platinum software 6.0 (GE Healthcare, Waukesha, WI, USA). Intensity of each spot was determined from each biological replicates. The significant spots are only selected for further identification process. For data analysis Delta 2D software, Image Master Platinum, and PDQuest can be employed. For analysis, three gel images arising from three different cultures and nuclear preparations warp for each group onto a master image. The master gel image should then warp onto another master gel image and a union fusion image of all the gel images was then made. The detection will be carried out on this fusion image, and the detection will be then propagated to each individual image. The resulting quantification table should be analyzed using the student's t-test function of the software, and the spots having both a p-value less than 0.05 and an induction/repression ratio of two or greater will be selected for further analysis by MS after all have been manually verified. The spots of interest are excised from a stained gel by a scalpel blade. Destaining of the spots should carry out by the ferricyanide–thiosulfate method on the same day as silver staining to improve sequence coverage in the MS analysis. The 2DE is a mature and well-established technique; on the other hand it suffers from some ongoing concerns regarding quantitative reproducibility and limitations on the ability to study certain classes of proteins. Therefore, in recent years, most developments have focused on alternative approaches, such as promising gel-free proteomics.

14.2.2   GEL-FREE TECHNIQUE

With the appearance of MS-based proteomics, an entirely new toolbox has become available for quantitative analysis. The protein spots are excised from the gels and further process for in-gel tryptic digestion and quadrupole time-of-flight (Q-TOF) tandem mass spectrometry (MS/MS) analysis.[41] Excess trypsin solution was removed and replaced with the same volume of ammonium bicarbonate without trypsin. Peptides are extracted by repeated cycles of sonication in 0.1% trifluoroacetic acid (TFA)/50% acetonitrile (ACN) aqueous solution, dried completely in a vacuum centrifuge, and reconstituted in 0.1% (v/v) formic acid in water. The reconstituted peptide samples can be separated by high performance liquid chromatography (HPLC) using C18 reverse phase column and further analyzed in a matrix assisted laser desorption ionization-time of flight (MALDI-TOF)/TOF. The obtained peak lists are further searched using the available online Mascot (http://www. matrixscience.com) algorithm against NCBInr database. Proteins are finally identified on the basis of their two or more peptide matches whose ions scores exceeded the threshold, $p < 0.05$, which is indicated by the 95% confidence level for peptides. Proteins will accept as identified by their matches if the threshold will exceed and the protein spot possess the correct molecular mass and isoelectric point value in the corresponding gel. The protein spots can be excised and analyzed from more than one gel to ensure the accuracy of protein identification. In shotgun proteomics complex peptide fractions, generated after protein proteolytic digestion, can be resolved using different fractionation strategies, which offer high-throughput analyses of the proteome of an organelle or a cell type and provide a snapshot of the major protein constituents. The isobaric tag for relative and absolute quantitation (iTRAQ) technology offers several advantages, which include the ability to multiplex several samples, quantification, simplified analysis, and increased analytical precision and accuracy.[42] In this technique four or eight samples[43] can be analyzed simultaneously. Introduction of stable isotopes using iTRAQ reagents occurs on the level of proteolytic peptides. Due to the isobaric mass design of the iTRAQ reagents, differentially labeled peptides appear as a single peak in MS, reducing the possibility of overlapping peak. When iTRAQ-tagged peptides are subjected to MS/MS analysis, the mass balancing carbonyl moiety is released as a neutral fragment, liberating the isotope-encoded reporter ions which provide relative quantitative information on proteins. An inherent drawback of the reported iTRAQ technology is due to the enzymatic digestion of proteins prior to labeling, which artificially increases sample complexity, and therefore this requires a multidimensional fractionation method of peptides before MS identification.

Although these novel approaches were initially pitched as replacements for gel-based methods, they should probably be regarded as complements of 2DE. There are many points of comparison and contrast between the standard 2DE and shotgun analyses, such as sample consumption, depth of proteome coverage, analyses of isoforms, and quantitative statistical power. Both platforms have the ability to resolve hundreds to thousands of features, so the choice between the different platforms is often determined by the biological question addressed. Currently there is no single method, which can provide qualitative and quantitative information of all protein components of a complex mixture. Ultimately, these approaches are both of great value to a proteomic study and often provide complementary information for an overall richer analysis.

Targeted proteomics has become progressively more popular because of its ability to precisely quantify selected proteins. Linear trap quadrupole (LTQ)-Orbitrap mass spectrometer can be used in targeted parallel reaction monitoring despite its unconventional dual ion trap configuration. The instrument duty cycle is a critical parameter limiting sensitivity, necessitating peptide retention time scheduling. We evaluate synthetic peptide and recombinant peptide standards to predict or experimentally determine target peptide retention times. We can optimize parallel reaction monitoring to protein degradation in signaling regulation, an area that is receiving increased attention in plant physiology and molecular biology.

14.3   USE OF PROTEOMIC APPROACH FOR UNDERSTANDING ABIOTIC STRESS TOLERANCE IN RICE, WHEAT, AND LEGUMES

Though most of the work elucidating mechanistic details has been targeted in model species, now global consensus is to target some of the important crop species. Among them major crops are rice, wheat, and soybean. Since last one decade significant progress has been achieved in proteomics of rice using high-throughput techniques.[44] A comparative study on different abiotic stresses in rice, wheat, and soybean is presented (Fig. 14.1, Table 14.1). Salinity stress responses at protein level were analyzed by[45] in rice after short (24 h) and long (7 d) term exposure to salinity and phosphoglycerate kinase expression was induced after short-term exposure but there was no change for longer period while superoxide dismutase (SOD) and S-adenosyl-L-methionine synthetase induced after seven days of salinity stress. A total of 24 differentially expressed proteins were identified from root plasma membrane of salt tolerant rice IR6591 which was involved in salt stress adaptation responses.[46] In control condition IR64 (salt sensitive) and Pokkali (salt tolerant) rice varieties depicted much higher protein spots ~1088 and ~1076, respectively, that was decreased to ~482 and ~654 when 200 mM NaCl stress was applied. In halophytic rice *Porteresia coarctata*[47]

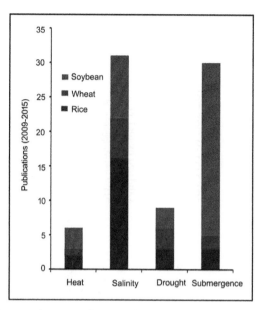

FIGURE 14.1   Progress in proteomic research toward abiotic stress tolerance in rice, wheat, and soybean in last six years.

~700 protein spots were detected under control condition, that were declined under salt stress, however, the reduction in *P. coarctata* was not as severe as in other varieties. When roots of rice were treated with abscisic acid (ABA) together with salinity, 40 protein spots were identified among which 78% were induced, however, only 16 spots were detected in salt stress condition representing that plant may save their energy by expressing fewer proteins.[48] With the advent of technology the CyDye™ DIGE fluorescent labeling method that is more sensitive than coomasi brilliant blue (CBB) had also been used for salt stress study in rice. A temporal study of salt stress on alterations in apoplastic proteins of rice stem was performed with CyDye™ DIGE,[49] where 117 protein spots were identified, among them 64 were apoplastic proteins. When roots were exposed to salt stress the proteins involved in following processes got activated: (a) signal transduction, (b) reactive oxygen scavenging, (c) exclusion and compartmentalization of ions, (d) carbohydrate metabolism, and (e) other salt stress mediated mechanism, and all those processes work in a network mode to maintain cellular homeostasis of plant roots and govern plant growth and development under salt stress condition.[50] The reproductive stage is considerable sensitive for salt stress. To know molecular mechanisms for salt tolerance at reproductive

TABLE 14.1 Proteins Identified for Salinity, Drought, and Submergence Stress in Rice, Wheat, and Soybean.

S. no.	Stress	Species	Protein/gene name	Function	Identification	Reference
1	Salt	Rice	Putative thioredoxin peroxidase, putative glutathione S-transferase, OsGSTF3, putative peroxiredoxin Q, thioredoxin M-like, thioredoxin x	Antioxidant activity	iTRAQ and validation by RT	[77]
2	Salt	Rice	SOD, ferredoxinthiore doxin reductase, fiber protein, inorganic pyrophosphatase	Stress	2DE MALDI-TOF/TOF validated by qRT PCR	[78]
3	Salt	Rice	2,3-bisphosphoglycerate independent phospho glycerate mutase, L-ascorbate peroxidase, chlorophyll A-B binding protein, and harpin binding protein	Energy and defence	2DE MALDI-TOF/TOF validated by qRT PCR	[79]
4	Salt	Rice	1,4-Benzoquinone reductase, putative remorin, hypersensitive induced response protein	Stress response	2DE MALDI-TOF/TOF validated by qRT PCR	[80]
5	Salt	Rice	31kDa protein	Stress	2DElectrphoresis and validated by western blot	[81]
6	Salt	Rice	Class III peroxidase 29 precursor, B1-3 glucanase precursor, putative ranscription factor, chaprone21 precursor, rubisco activase, ribonuclease	Antioxidant stress	2DE MALDI-TOF and ESI-MS/MS analysis	[82]
7	Drought	Wheat	Plastid glutamine synthetase, heat shock protein 70kDa, phospho ribulokinase, carbonic anhydrase, fructose-1,6-bisphosphate aldolase	Aminoacid metabolism Photosynthesis Stress response	MALDI-TOF-TOF validation by western blot	[83]
8	Drought	Wheat	Glyceraldehyde-3-phosphate dehydrogenase, protein disulfide isomerase, RUBISCO, ATP synthase alpha subunit, phosphoglyceratekinase, SOD(leaf)	Stress response	nano LC-ESI-MS / MS and validation by qRT and	[84]

TABLE 14.1 *(Continued)*

S. no.	Stress	Species	Protein/gene name	Function	Identification	Reference
9	Drought	Wheat	ATP synthase beta subunit, ascorbate peroxidase, Cu/Zn superoxide dismutase, 16.9 kDa class I heat shock protein 1, glutathione-S-transferase(root)	Stress response	nano LC-ESI-MS/MS and validation by qRT	[84]
10	Drought	Wheat	Pathogenesis-related proteins, peroxidases, and chitinases	Stress response	2DE MALDI-TOF/TOF validated by qRT PCR	[85]
11	Drought	Wheat	Heat stress associated proteins, (HSP70) (HSP90) and calcium dependent protein kinase	Stress response	2-DE MALDI TOF/MS, nLC/MS/MS validation by qRTand western blot	[86]
12	Drought	Wheat	Putative AdoMet synthase 2, glutathione transferase F6, glucan endo-1,3-beta-D-glucosidase	Amino acid and carbohydrate metabolism	2DE MALDI-TOF/TOF validated by qRT PCR	[87]
13	Flooding	Soybean	Plant stearoyl acyl-carrier protein, secretion-associated RAS superfamily 2, ascorbate peroxidase 1, ferretin 1	Lipid metabolism, Cell signaling Redox Metal handling	nano-LC-MS/MS validation by qRT	[88]
14	Flooding	Soybean	Urease and copper chaperone	Protein metabolism	2DE MALDI-TOF/TOF, nano-LC–MS/MS validated by qRT PCR	[89]
15	Flooding	Soybean	Protein disulfide isomerase, luminal binding protein 5, calreticulin, N-glycosylation related genes, glucosaminephosphotransferase, STT3 subunit, alpha-1,2glucosyltransferase, mannosyl-oligosaccharide glucosidase and oligosaccharyltransferase	Amino acid and lipid biosynthesis	LC-MS/MS and validated by qRT PCR	[90]

TABLE 14.1 *(Continued)*

S. no.	Stress	Species	Protein/gene name	Function	Identification	Reference
16	Flooding	Soybean	Sucrose phosphate synthase, eukaryotic translation initiation factor 4G, UDP-glucose 6-dehydrogenase	Sugar, protein and cell wall synthesis	LC-MS/MS and validated by qRT PCR	[91]
17	Flooding	Soybean	Peroxidase super family protein	Scavenging of toxic radicals	LC-MS/MS and validated by qRT PCR	[92]
18	Flooding	Soybean	Ascorbate peroxidase, antipyruvate dehydrogenase, RuBisCOlarge subunit	Salt stress	2DE MALDI-TOF/TOF validated by Western blot	[59]
19	Flooding	Soybean	C2H2 zinc finger protein SERRATE, CCCH type zinc finger family protein, and transducin	Bind to DNA, RNA and proteins	LC-MS/MS and validated by Western blot	[93]
20	Salt	Soybean	Aspartate aminotransferase, S-adenosylmethioninesynthetase, ACC synthetase	Amino acid metaboilsm	2DE MALDI-TOF/TOF validated by qRT PCR	[94]

stage, proteomic analysis of two contrasting rice genotypes IR64 (salt sensitive) and Cheriviruppu (salt tolerant) was performed under salt stress and identified 18 proteins which were involved in different processes to salt tolerance such as carbohydrate/energy metabolism, anther wall remodeling and metabolism, and protein synthesis and assembly. It was suggested that which were upregulated in salt tolerant to increased starch content in pollen, which would support pollen growth and development under salt stress.[51] In our lab[52] total 67 protein spots were detected which were differentially expressed in CSR27 (salt tolerant), MI48 (salt sensitive), and their tolerant and sensitive recombinant inbred lines (RILs). Those proteins belong to metabolic activities (45%), cell defense (16%), transcriptional factors (17%), miscellaneous proteins (16%), and unknown proteins (6%).

In wheat one of the major problems is temperature stress. Heat stress tolerant genes and its translation may be a strategy to tailor tolerant plant. To decipher the mechanism of heat stress tolerance wheat genotypes (WH 730-heat tolerant; Raj 4014-heat intolerant) along with their 10 extreme RILs were given heat treatment and stress proteins were identified using 2DE coupled with MALDI TOF/TOF.[53] Three proteins, Rubisco activase A, Con A, and PEP carboxylase 1, were differentially regulated only in WH 730 (heat tolerant) suggesting its significant role in heat tolerance.

Soybean is very much susceptible to flood condition. Other than morphological and physiological responses, several molecular studies revealing stress-induced changes in transcriptome and metabolites have been documented.[54] Ubiquitination of protein is a typical post-translational regulatory mechanism in flooding condition[55] which requires ubiquitin (Ub)-activating enzyme (E1), Ub-conjugating enzyme (E2 or UBC), and Ub ligase (E3)[56] Ub/proteasome-mediated proteolysis of glycolytic and fermentation pathway enzymes was inhibited in hypoxic condition caused by submergence stress in soybean.[57] Differential regulation of 20S proteasome subunits was reported through gel-base and gel-free proteomic techniques in soybean that may lead to its tolerance under flooding stress.[58] Mitochondrial genes with altered expression were reported for soybean under flooding stress, among them the significant increase in expression of peptidase and inhibited expression ATPase beta subunit was observed.[59] In roots and hypocotyls of soybean proteins from tricarboxylic acid cycle and gamma-aminobutyric acid (GABA) shunt were enhanced, while proteins of electron transport chain (ETC) were decreased under flooding stress. It means flooding stress directly damages the ETC despite of increase in nicotineamide adenine dinucleotide hydride (NADH) generation.[59] It was suggested that roots of soybean respond via signaling cascade resulted through reduction of hydrogen peroxide and ascorbate.

Alterations in the protein level are required for insight of stress signals and its transduction into the cell at plasma membrane. Using 2DE MS/sequencer-based proteomics and nano LC-MS/MS-based proteomic techniques 35 flood stress inducible proteins were identified in soybean containing SOD.[24,60] SODs are essential enzymes for ascorbate glutathione pathway and first level of defense against ROS (superoxide radicals and hydrogen peroxide). HSP (70 kDa) was also induced under flooding stress in soybean.[61] Some down-regulated proteins of lignin and isoflavonoids pathways were also reported which have an important role in mitigating oxidative injuries.[62]

Uridine diphosphate (UDP)-glucose pyro-phosphorylase, fructose-bisphosphate aldolase, and alcohol dehydrogenase (Adh) are expressed under flooding stress in soybean.[54,58,59] Adh was expressed only under flooding stress which is involved in fermentation pathway of ethanol to withstand hypoxic environment and its overexpression in soybean improved growth under flood stress.[63] S-adenosylmethionine synthesis and proteins of cell wall modification were implicated recovery mechanism.[64] Upregulation of proteins of glycolysis and fermentation was reported in roots of short-term submergence stressed soybean, while ROS scavenger proteins, peroxidases, cytosolic ascorbic peroxidase (cAPX), and SOD were decreased.[65,66] Conclusively, soybean proteomics under flood stress depicts that metabolic adjustment for energy conservation through flooded roots was a vital strategy for its resistance to the same.

14.4 EMPLOYING PROTEOMICS IN ENHANCING NUTRITIONAL QUALITY OF FOOD CROPS

The major concerns at global level are food and nutrition. In order to attain food security various strategies are being enforced, however, very limited efforts are being made to enhance nutritional quality of foods. Nutritional security is must. Among various nutrients micronutrients play a vital role. Micronutrients, also called as "magic wands," are needed in minute quantities, which are important in proper functioning of proteins, enzymes, hormones, and other substances. Vitamins and minerals are essential for the physical as well as mental development of human beings; their involvement in metabolic processes and functioning of immune system is well known.[67] Here we discuss the importance of iron (Fe) and zinc (Zn). They have a very vital role to play in both plant and animal metabolism. According to WHO, 30% of world's population is affected by Fe deficiency and more than 2 billion people around the globe are affected by deficiency of Zn (http://

www.who.int/nutrition/topics/ida/en/index.html). This is an alarm indicating the need to work for increasing the micronutrient content in staple foods. Several scientific groups around globe are working for improving the availability of Fe and Zn in rice through molecular breeding. There is a further need to undertake the research to improve transport of these micronutrients in food grains. Nutrigenomics and proteogenomics based approaches may help in enhancing the micronutrient contents in foods. Nutrigenomics based approach will help us in tracking gene(s) and genetic components that may be contributing for increasing the micronutrient contents whereas proteomics or proteogenomics based strategy may help us in identification of molecular switches regulating uptake and accumulation of micronutrients. Lot of work has been done in case of Arabidopsis for understanding impact of various Fe and Zn levels on genetic regulation using proteomics based strategies.[68,69,70] Proteomics based approach coupled with physiological and biochemical observations has helped in dissecting many metabolic pathways and elucidating the underlined mechanisms for various biological process. As such this approach can answer the problem of nutritional insecurity.

14.5   PROTEOGENOMICS

DNA sequencing of genome explores basics of the genetic structure but not enough alone for expounding biological functions. To circumvent this, proteogenomics emerged from the confluence of genomics and proteomics to identify protein-coding regions in the genome.[71] The state-of-the-art technologies using a combination of comparative (homologous sequence data) and non-comparative (ab initio gene prediction algorithms) methods can be utilized for identifying protein-coding genes in whole genome sequencing data. It complements the nucleotide-based annotation by unambiguously predicting the reading frame, translation start and stop sites, splice boundaries, and the validity of short open reading frames (ORFs) facilitating in providing a comprehensive protein-coding catalog.[72,73] To substantiate coding regions of genome sequence MS/MS-based proteomic techniques are worth to incorporate for improvement of the quality of genome annotations.[74] For validation *Arabidopsis thaliana* gene models were used and sequences of amino acid with 144,079 distinct peptides were analyzed by MS/MS. However, 18,024 novel peptides were not corresponds to annotated genes. Augustus, a gene finding program, was used to discover novel peptides with > 99% correct annotations in Arabidopsis.[75] A data repository for shotgun proteogenomics for genome re-annotation for rice (Oryza

PG-DB) was generated using 27 nano LC-MS/MS runs on ion trap-orbitrap mass spectrometer and 3200 genes were enclosed, among which 40 had novel genomic structures.[76] The OryzaPG database was constructed and is freely available at http://oryzapg.iab.keio.ac.jp/.

14.6   CONCLUSION

Increasing population and growing life standards have led to higher global demand of food crops necessitating high productivity from stress prone marginal lands. Soil salinity, drought, and flood are amongst the most detrimental stresses limiting productivity. Proteomics have largely been used to decipher mechanisms operating during abiotic stress conditions to find out the potential target for crop improvement.

KEYWORDS

- **abiotic stress**
- **proteomics**
- **rice**
- **wheat**
- **proteogenomics**

REFERENCES

1. Young, V. R.; Pellett, P. L. Wheat Proteins in Relation to Protein Requirements and Availability of Amino Acids. *Am. J. Clin. Nutr.* **1985**, *41*, 1077–1090.
2. Maclean, J. L.; Dawe, D. C.; Hardy, B.; Hettel, G. P. Eds.; *Rice Almanac, Source Book for the Most Important Economic Activity on Earth;* 3rd ed.; CABI Publishing: Wallingford, Oxon, 2002.
3. Basavaraja, N. Freshwater Fish Seed Resources in India. In *Assessment of Freshwater Fish Seed Resources for Sustainable Aquaculture;* Fisheries Technical Paper. No. 501. Bondad-Reantaso, M. G., Ed.; FAO: Rome, Italy, 2007; pp 267–327.
4. Breiman, A.; Graur, D. Wheat Evolution. *Isr. J. Plant Sci.* **1995**, *43*, 85–89.
5. Feldman, M. Origin of Cultivated Wheat. In *The World Wheat Book: A History of Wheat Breeding;* Bonjean, A, P., Angus, W. J., Eds.; Lavoisier Publishing: Paris, France, 2001; pp 3–56.

6. Cattivelli, L.; Rizza, F.; Badeck, F. W.; Mazzucotelli, E.; Mastrangelo; A. M.; Francia E., et al. Drought Tolerance Improvement in Crop Plants: An Integrated View from Breeding to Genomics. *Field Crops Res.* **2008,** *105,* 1–14.

7. Thomashow, M. F. Plant Cold Acclimation. Freezing Tolerance Genes and Regulatory Mechanisms. *Annu. Rev. Plant Physiol. Plant Mol. Biol.* **1999,** *50,* 571–599.

8. Kosova, K.; Prášil, I. T.; Vítámvás, P. The Relationship between Vernalization- And Photoperiodically-Regulated Genes and the Development of Frost Tolerance in Wheat and Barley. *Biol. Plant.* **2008,** *52,* 601–615.

9. Munns, R. Genes and Salt Tolerance: Bringing Them Together. *New Phytol.* **2005,** *167,* 645–663.

10. Kosova, K.; Prasil, I. T.; Vítámvás, P. Protein Contribution to Plant Salinity Response and Tolerance Acquisition. *Int. J. Mol. Sci.* **2013,** *14,* 6757–6789.

11. Gupta, P. K.; Mir, R. R.; Mohan, A.; Kumar, J. Wheat Genomics: Resent Status and Future prospects. *Int. J. Plant Genomics.* **2008,** *2008,* 1–36.

12. Kaitao, L.; Paul, J, B.; Michal, T. L.; Christopher, D.; Lars, S.; Sahana, M.; Jiri, S.; David, E. Wheat Genome.Info: An Integrated Database and Portal for Wheat Genome Information. *Plant Cell Physiol.* **2012,** *53* (2), e2.

13. Kumar, R. R.; Goswami, S.; Sharma, S. K.; Singh, K.; Gadpayle, A. K.; Kumar, N.; Rai, G. K.; Singh, M.; Rai, R. D. Protection Against Heat Stress in Wheat Involves Change in Cell Membrane Stability, Antioxidant Enzymes, Osmolyte, H_2O_2 and Transcript of Heat Shock Protein. *Int. J. Plant Physiol. Biochem.* **2012,** *4* (4), 83–91.

14. Bailey-Serres, J.; Colmer, T. D. Plant Tolerance of Flooding Stress-Recent Advances. *Plant Cell Environ.* **2014,** *37* (10), 2211–2215.

15. Deinlein, U.; Stephan, A. B.; Horie, T.; Luo, W.; Xu, G.; Schroeder, J. I. Plant Salt-Tolerance Mechanisms. *Trends Plant Sci.* **2014,** *19,* 371–379.

16. Roy, S. J.; Negrao, S.; Tester, M. Salt Resistant Crop Plants. *Curr. Opin. Biotechnol.* **2014,** *26,* 115–124.

17. Hadiarto, T.; Tran, L. S. Progress Studies of Drought-Responsive Genes in Rice. *Plant Cell Rep.* **2011,** *30* (3), 297–310.

18. Swamy, B. P.; Kumar, A. Genomics-Based Precision Breeding Approaches to Improve Drought Tolerance in Rice. *Biotechnol Adv.* **2013,** *31* (8), 1308–1318.

19. Fukao, T.; Yeung, E.; Bailey-Serres, J. The Submergence Tolerance Regulator SUB1A Mediates Crosstalk between Submergence and Drought Tolerance in Rice. *Plant Cell.* **2011,** *23* (1), 412–427.

20. Pandey, G.; Yoshikawa, K.; Hirasawa, T.; Nagahisa, K.; Katakura, Y.; Furusawa, C.; Shimizu, H.; Shioya, S. Extracting the Hidden Features in Saline Osmotic Tolerance in Saccharomyces Cerevisiae from DNA Microarray Data Using the Self-Organizing Map: Biosynthesis of Amino Acids. *Appl. Microbiol. Biotechnol.* **2007,** *75* (2), 415–426.

21. Sarkar, M.; Leventis, P. A.; Silvescu, C. I.; Reinhold, V. N.; Schachter, H.; Boulianne, G. L. Null Mutations in Drosophila N-acetylglucosaminyltransferase I Produce Defects in Locomotion and a Reduced Life Span. *J. Biol. Chem.* **2006,** *281* (18), 12776–12785.

22. Ghassemi, F.; Jakerman, A. J.; Nix, H. A. *Salinization of Land Water Resources;* CAB International: Wallingford, Oxon, 1995.

23. Mujeeb, K.; Diaz de, L. Conventional and Alien Genetic Diversity for Salt Tolerant Wheats: Focus on Current Status and New Germplasm Development. In *Prospects for Saline Agriculture;* Ahmad, R., Malik, K. A., Eds.; Kluwer Academic Publishers: Dordrecht, The Netherlands, 2002; Vol. 37, pp 69–82.

24. Hossain, Z.; Khatoon, A.; Komatsu, S. Soybean Proteomics for Unraveling Abiotic Stress Response Mechanism. *J. Proteome. Res.* **2013**, *12,* 4670–4684.
25. Lin, H. X.; Zhu, M. Z.; Yano, M.; Gao, J. P.; Liang, Z. W.; Su, W. A., et al. QTLs for Na+ and K+ Uptake of the Shoots and Roots Controlling Rice Salt Tolerance. *Theor. Appl. Genet.* **2004**, *108,* 253–260.
26. Xu, K.; Mackill, D. J. A Major Locus for Submergence Tolerance Mapped on Rice Chromosome 9. *Mol. Breeding.* **1996**, *2,* 219–224.
27. Kumar, A.; Dixit, S.; Ram, T.; Yadaw, R. B.; Mishra, K. K.; Mandal, N. P. Breeding High-Yielding Drought-Tolerant Rice: Genetic Variations and Conventional and Molecular Approaches. *J. Exp. Bot.* **2014**, *65* (21), 6265–6278.
28. Gamuyao, R.; Chin, J. H.; Pariasca-Tanaka, J.; Pesaresi, P.; Catausan, S.; Dalid, C.; Slamet-Loedin, I.; Tecson-Mendoza, E. M.; Wissuwa, M.; Heuer, S. The Protein Kinase Pstol1 from Traditional Rice Confers Tolerance of Phosphorus Deficiency. *Nature.* **2012**, *488* (7412), 535–539.
29. Kikuchi, S.; Satoh, K.; Nagata, T.; Kawagashira, N.; Doi, K.; Kishimoto, N.; Yazaki, J.; Ishikawa, M.; Yamada, H.; Ooka, H.; Hotta, I.; Kojima, K.; Namiki, T.; Ohneda, E.; Yahagi, W.; Suzuki, K.; Li, C. J.; Ohtsuki, K.; Shishiki, T. Foundation of Advancement of International Science Genome Sequencing and Analysis Group. Otomo, Y.; Murakami, K.; Iida, Y.; Sugano, S.; Fujimura, T.; Suzuki, Y.; Tsunoda, Y.; Kurosaki, T.; Kodama, T.; Masuda, H.; Kobayashi, M.; Xie, Q.; Lu, M.; Narikawa, R.; Sugiyama, A.; Mizuno, K.; Yokomizo, S.; Niikura, J.; Ikeda, R.; Ishibiki, J.; Kawamata, M.; Yoshimura, A.; Miura, J.; Kusumegi, T.; Oka, M.; Ryu, R.; Ueda, M.; Matsubara, K.; RIKEN; Kawai, J.; Carninci, P.; Adachi, J.; Aizawa, K.; Arakawa, T.; Fukuda, S.; Hara, A.; Hashizume, W.; Hayatsu, N.; Imotani, K.; Ishii, Y.; Itoh, M.; Kagawa, I.; Kondo, S.; Konno, H.; Miyazaki, A.; Osato, N.; Ota, Y.; Saito, R.; Sasaki, D.; Sato, K.; Shibata, K.; Shinagawa, A.; Shiraki, T.; Yoshino, M.; Hayashizaki, Y.; Yasunishi, A. Collection, Mapping, and Annotation of over 28,000 cDNA Clones from Japonica Rice. *Science.* **2003**, *301* (5631), 376–379.
30. Kim, M. Y.; Lee, S.; Van, K.; Kim, T. H.; Jeong, S. C.; Choi, I. Y.; Kim, D. S.; Lee, Y. S.; Park, D.; Ma, J.; Kim, W. Y.; Kim, B. C.; Park, S.; Lee, K. A.; Kim, D. H.; Kim, K. H.; Shin, J. H.; Jang, Y. E.; Kim, K. D.; Liu W. X.; Chaisan, T.; Kang, Y. J; Lee, Y. H; Kim, K. H.; Moon, J. K.; Schmutz, J.; Jackson, S. A.; Bhak, J.; Lee, S. H. Whole-Genome Sequencing and Intensive Analysis of the Undomesti-Cated Soybean (*Glycine soja Sieb.* and *Zucc.*) Genome. *Proc. Natl. Acad. Sci. USA.* **2010**, *107* (51), 22032–22037.
31. Singh, N. K.; Gupta, D. K.; Jayaswal, P. K.; Mahato, A. K.; Dutta, S.; Singh, S.; Bhutani, S.; Dogra, V.; Singh, B. P.; Kumawat, G.; Pal, J. K.; Pandit, A.; Singh, A.; Rawal, H.; Kumar, A.; Rama Prashat, G.; Khare, A.; Yadav, R.; Raje, R. S.; Singh. M. N.; Datta, S.; Fakrudin, B.; Wanjari, K. B.; Kansal, R.; Dash, P. K.; Jain, P. K.; Bhattacharya, R.; Gaikwad, K.; Mohapatra, T.; Srinivasan, R.; Sharma, T. R. *J. Plant Biochem. Biotechnol.* **2012**, *21,* 98–112.
32. Varshney, R. K, Chen, W.; Li, Y.; Bharti, A. K.; Saxena, R. K.; Schlueter, J. A.; Donoghue, M. T.; Azam, S.; Fan, G.; Whaley, A. M.; Farmer, A. D.; Sheridan, J.; Iwata, A.; Tuteja, R.; Penmetsa, R. V.; Wu, W.; Upadhyaya, H. D.; Yang, S. P.; Shah, T.; Saxena, K. B.; Michael, T.; McCombie, W. R.; Yang, B.; Zhang, G.; Yang, H.; Wang, J.; Spillane, C.; Cook, D. R.; May, G. D.; Xu, X.; Jackson, S. A. Draft Genome Sequence of Pigeonpea (*Cajanus cajan*), an Orphan Legume Crop of Resource-Poor Farmers. *Nat. Biotechnol.* **2011**, *30* (1), 83–89.

33. Ngara, R.; Ndimba, B. K. Understanding the Complex Nature of Salinity and Drought-Stress Response in Cereals Using Proteomics Technologies. *Proteomics.* **2014**, *14* (4–5), 611–621.
34. Hurkman, W. J.; Tanaka, C. K. Solubilization of Plant Membrane Proteins for Analysis by Two-Dimensional Gel Electrophoresis. *Plant Physiol.* **1986**, *81* (3), 802–806.
35. Deng, L.; Abdel-Hamid, O.; Yu, D. A In *Deep Convolutional Neural Network Using Heterogenous Pooling for Trading Acoustic Invariance with Phonetic Confusion,* Proceedings of the IEEE International Conference on Acoustic, Speech and Signal Process, Vancouver, Canada, 2013; IEEE: Vabcouver, Canada, 2013.
36. Singh, N.; Jain, N.; Kumar, R.; Jain, A.; Singh, N. K.; Rai, V. A Comparative Method for Protein Extraction and 2-D Gel Electrophoresis from Different Tissues of *Cajanus cajan. Front. Plant Sci.* **2015**, *6,* 606.
37. Weiss, W.; Gorg, A.; Sample Solublization Buffers for Two Dimensional Electrophoresis. *Methods Mol. Biol.* **2008**, *424,* 35–42.
38. Rabilloud, T.; Lelong, C.; Two-Dimensional Gel Electrophoresis in Proteomics: A Tutorial. *J. Proteomics.* **2011**, *74,* 1829–1841.
39. Cottrell, J. S. Protein Identification Using MS/MS Data. *J. Proteomics.* **2011**, *74,* 1842–1851.
40. Mechin, V.; Damerval, C.; Zivy, M.; Total Protein Extraction with TCA–Acetone. *Methods Mol. Biol.* **2007**, *355,* 1–8.
41. Asif, A. R.; Oellerich, M.; Armstrong, V.; Riemenschneider, B.; Monod, M.; Reichard, U. Proteome of Conidial Surface Associated Proteins of *Aspergillus fumigatus* Reflecting Potential Vaccine Candidates and Allergens. *J. Proteome Res.* **2006**, *5,* 954–962.
42. Agrawal, G. K.; Jwa, N. S.; Iwahashi, Y.; Yonekura, M., Iwahashi, H.; Rakwal, R. Rejuvenating Rice Proteomics: Facts, Challenges, and Visions. *Proteomics.* **2006**, *6,* 5549–5576.
43. Pierce, S. E.; Davis, R. W.; Nislow, C.; Giaever, G. Genome-Wide Analysis of Barcoded Saccharomyces Cerevisiae Gene-Deletion Mutants in Pooled Cultures. *Nat. Protoc.* **2007**, *2,* 2958–2974.
44. Agrawal, G. K.; Rakwal, R. Rice Proteomics: A Move Toward Expanded Proteome Coverage to Comparative and Functional Proteomics Uncovers the Mysteries of Rice and Plant Biology. *Proteomics.* **2011**, *11* (9), 1630–1649.
45. Parker, R.; Flowers, T. J.; Moore, A. L.; Harpham, N. V. J. An Accurate and Reproducible Method for Proteome Profiling of the Effects of Salt Stress in the Rice Leaf Lamina. *J. Exp. Bot.* **2006**, *57,* 1109–1118.
46. Malakshah, S. N.; Rezaei, M. H.; Heidai, M.; Salekdeh, G. H. Proteomics Reveals New Salt Responsive Proteins Associated with Rice Plasma Membrane. *Biosci. Biotechnol. Biochem.* **2007**, *71* (9), 2144–2154.
47. Sengupta, S.; Majumder, A. L. Insight into the Salt Tolerance Factors of a Wild Halophytic Rice, *Porteresia coarctata*: A Physiological and Proteomic Approach. *Planta.* **2009**, *229,* 911–929.
48. Li, G.; Vissers Johannes, P. C.; Silva, J. C.; Golick, D.; Gorenstein, M. V.; Geromanos, S. J. Database Searching and Accounting of Multiplexed Precursor and Product Ion Spectra from the Data Independent Analysis of Simple and Complex Peptide Mixtures. *Proteomics.* **2009**, *9,* 1696–1719.
49. Song, Y.; Zhang, C.; Gea, W.; Zhang, Y.; Burlingame, A. L.; Guo, Y. Identification of NaCl Stress-Responsive Apoplastic Proteins in Rice Shoot Stems by 2D-DIGE. *J. Proteomics.* **2011**, *74* (7), 1045–1067.

50. Zhao, Q.; Zhang, H.; Wang, T.; Chen, S.; Dai, S. Proteomics-Based Investigation of Salt-Responsive Mechanisms in Plant Roots. *J. Proteomics.* **2013,** *82,* 230–253.

51. Sarhadi, E.; Bazargani, M. M.; Sajise, A. G.; Abdolahi, S.; Vispo, N. A.; Arceta, M.; Nejad, G. M.; Singh, R. K.; Salekdeh, G. H. Proteomic Analysis of Rice Anthers Under Salt Stress. *Plant Physiol. Biochem.* **2012,** *58,* 280–287.

52. Mishra, V.; Mishra, P.; Takabe, T, Rai, V.; Singh, N. K. Elucidation of Salinity Stress-Responsive Metabolic Pathways in Contrasting Rice Genotypes. *Plant cell Reports.* **2015,** (in press).

53. Gupta, O. P; Mishra, V.; Singh, N. K.; Tiwari, R.; Sharma, P.; Gupta, R. K.; Sharma, I. Deciphering the Dynamics of Changing Proteins of Tolerant and Intolerant Wheat Seedlings Subjected to Heat Stress. *Mol. Biol. Rep.* **2014,** *33* (14), 3738–3739.

54. Komatsu, S.; Yamamoto, R.; Nanjo, Y.; Mikami, Y.; Yunokawa, H.; Sakata, K. A Comprehensive Analysis of the Soybean Genes and Proteins Expressed Under Flooding Stress Using Transcriptome and Proteome Techniques. *J. Proteome Res.* **2009,** *8,* 4766–4778.

55. Devoto, A.; Muskett, P. R.; Shirasu, K. Role of Ubiquitination in the Regulation of Plant Defense Against Pathogens. *Curr. Opin. Plant Biol.* **2003,** *6,* 307–311.

56. Pickart, C. M. Mechanisms Underlying Ubiquitination. *Annu. Rev. Biochem.* **2001,** *70,* 503–33.

57. Komatsu, S.; Hiraga, S.; Yanagawa, Y. Proteomics Techniques for the Development of Flood Tolerant Crops. *J. Proteome Res.* **2012,** *11,* 68–78.

58. Nanjo, Y.; Skultety, L.; Ashraf, Y.; Komatsu, S. Comparative Proteomic Analysis of Early-Stage Soybean Seedlings Responses to Flooding by Using Gel and Gel-Free Techniques. *J. Proteome Res.* **2010,** *9,* 3989–4002.

59. Komatsu, S.; Yamamoto, A.; Nakamura, T.; Nouri, M. Z.; Nanjo, Y.; Nishizawa, K.; Furukawa, K. Comprehensive Analysis of Mitochondria in Roots and Hypocotyls of Soybean under Flooding Stress using Proteomics and Metabolomics Techniques. *J. Proteome Res.* **2011,** *10,* 3993–4004.

60. Hossain, Z.; López-Climent, M. F.; Arbona, V.; Pérez-Clemente, R. M.; Gómez-Cadenas, A. Modulation of the Antioxidant System in Citrus under Waterlogging and Subsequent Drainage. *J. Plant Physiol.* **2009,** *166* (13), 1391–1404.

61. Yin, G.; Sun, H.; Xin, X.; Qin, G.; Liang, Z.; Jing, X. Mitochondrial Damage in the Soybean Seed Axis during Imbibitions at Chilling Temperatures. *Plant Cell Physiol.* **2009,** *50,* 1304–1318.

62. Kajikawa, M.; Hirai, N.; Hashimoto, T. A PIP-Family Protein is Required for Biosynthesis of Tobacco Alkaloids. *Plant Mol. Biol.* **2009,** *69,* 287–298.

63. Tougou, M.; Hashiguchi, A.; Yukawa, K.; Nanjo, Y.; Hiraga, S.; Nakamura, T.; Nishizawa, K.; Komatsu, S. Responses to Flooding Stress in Soybean Seedlings with the Alcohol Dehydrogenase Transgene. *Plant Biotech.* **2012,** *29* (3), 301–305.

64. Salavati, A.; Khatoon, A.; Nanjo, Y.; Komatsu, S. Analysis of Proteomic Changes in Roots of Soybean Seedlings during Recovery after Flooding. *J. Proteomics.* **2012,** *75,* 878–893.

65. Hossain, M. M; Li, X.; Evans, I. H.; Rahman, M. A. A Proteomic Analysis of Seed Proteins Expressed in a *Brassica* Somatic Hybrid and its Two Parental Species. *J. Plant Tissue Cult. Biotechnol.* **2014,** *24,* 11–26.

66. Shi, F.; Yamamoto, R.; Shimamura, S.; Hiraga, S.; Nakayama, N.; Nakamura, T.; Yukawa, K.; Hachinohe, M.; Matsumoto, H.; Komatsu, S. Cytosolic Ascorbate

Peroxidase 2 (cAPX 2) is Involved in the Soybean Response to Flooding. *Phytochemistry.* **2008,** *69* (6), 1295–1303.

67. Burchi, F.; Fanzo, J.; Frison, E. The Role of Food and Nutrition System Approaches in Tackling Hidden Hunger Review. *Int. J. Environ. Res. Public Health.* **2011,** *8,* 358–373.

68. Zargar, S. M.; Kurata, R.; Inaba, S.; Fukao, Y. Unraveling the Iron Deficiency Responsive Proteome in *Arabidopsis* Shoot by iTRAQ OFF GEL Approach. *Plant Signal. Behav.* **2013,** *8* (10), e26892.

69. Zargar, S. M.; Kurata, R.; Inaba, S.; Oikawa, A.; Fukui, R.; Ogata, Y.; Agrawal, G. K.; Rakwal, R.; Fukao, Y. Quantitative Proteomics of Arabidopsis Shoot Microsomal Proteins Reveals a Cross-Talk Between Excess Zinc and Iron Deficiency. *Proteomics.* **2015,** *15,* 1196–1201.

70. Fukao, Y.; Yoshida, M.; Kurata, R.; Kobayashi, M.; Nakanishi, M.; Fujiwara, M.; Nakajima, K.; Ferjani, A. Peptide Separation Methodologies for in-Depth Proteomics in Arabidopsis. *Plant Cell Physiol.* **2013,** *54* (5), 808–815.

71. Castellana, N. E.; Shen, Z.; He, Y.; Walley, J. W.; Cassidy, C. J.; Briggs, S. P.; Bafna, V. An Automated Proteogenomic Method Uses Mass Spectrometry to Reveal Novel Genes in Zea Mays. *Mol. Cell Proteomics.* 2014, *13* (1), 157–167.

72. Gupta, R.; Kus, B.; Fladd, C.; Wasmuth, J.; Tonikian, R.; Sidhu, S.; Krogan, N. J.; Parkinson, J.; Rotin, D. Ubiquitination Screen Using Proteinmicroarrays for Comprehensive Identification of Rsp5 Substrates in Yeast. *Mol. Syst. Biol.* **2007,** *3,* 116.

73. Tanner, S.; Shen, Z.; Ng, J.; Florea, L.; Guigó, R.; Briggs, S. P.; Bafna, V. Improving Gene Annotation Using Peptide Mass Spectrometry. *Genome Res.* **2007,** *17,* 231–239.

74. Ansong, C.; Purvine, S. O.; Adkins, J. N.; Lipton, M. S.; Smith, R. D. Proteogenomics: Needs and Roles to be Filled by Proteomics in Genome Annotation. *Brief Funct. Genomic Proteomics.* **2008,** *7* (1), 50–62.

75. Castellanaa, N. E.; Payneb, S. H.; Shenc, Z.; Stanked, M.; Bafnaa, V.; Briggsc, S. P. Discovery and Revision of Arabidopsis Genes by Proteogenomics. *PNAS.* **2008,** *105* (52), 21034–21038.

76. Helmy, M.; Tomita, M.; Ishihama, Y. OryzaPG-DB: Rice Proteome Database Based on Shotgun Proteogenomics. *BMC Plant Biol.* **2011,** *11,* 63.

77. Xu, J.; Lan, H.; Fang, H.; Huang, X.; Zhang, H.; Huang, J. Quantitative Proteomic Analysis of the Rice (*Oryzasativa* L.) Salt Response. *PLoS ONE.* **2015,** *10,* e0120978.

78. Hosseini, S. A.; Gharechahi, J.; Heidari, M.; Koobaz, P.; Abdollah, S.; Mirzaei, M.; Nakhoda, B.; Salekdeh, G. H. Comparative Proteiomic and Physiological Characterization of Two Closely Related Rice Genotype with Contrasting Responses to Salt Stress. *Funct. Plant Biol.* **2015,** *42* (6), 527. DOI:10.1071/FP14274

79. Yang, Y.; Zhu, K.; Xia, H.; Chen, L.; Chen, K. Comparative Proteomic Analysis of Indica and Japonica Rice Varieties. *Genet. Mol. Biol.* **2014,** *37* (4), 652–661.

80. Malakshah, S. N.; Rezaei, M. H.; Heidari, M.; Salekdeh, G. H. Proteomics Reveals New Salt Responsive Proteins Associated with Rice Plasma Membrane. *Biosci. Biotechnol. Biochem.* **2007,** *71,* 2144–2154. 10.1271/bbb.7002.

81. Kong-ngern, K.; Sakda, D.; Chaisiri, W.; Sumontip, B.; Manit, K.; Piyada, T. Protein Profiles in Response to Salt Stress in Leaf Sheaths of Rice Seedlings. *Sci. Asia.* **2005,** *31,* 403–408.

82. Lee, D. G.; Park, K. W.; An, J. Y.; Sohn, Y. G.; Ha, J. K.; Kim, H. Y.; Bae, D. W.; Lee, K. H.; Kang, N. J.; Lee, B. H.; Kang, K. Y.; Lee, J. J. Proteomics Analysis of Salt-Induced Leaf Proteins in Two Rice Germplasms with Different Salt Sensitivity. *Can. J. Plant Sci.* **2011,** *91* (2), 337–349. 10.4141/CJPS10022.

83. Liu, H.; Sultan, M. A. R. F.; Liu, X. I.; Zhang, J.; Yu, F.; Zhao, H. X. Physiological and Comparative Proteomic Analysis Reveals Different Drought Responses in Roots and Leaves of Drought- Tolerant Wild Wheat (*Triticumboeoticum*) *PLoS ONE*. **2015**, *10* (4), e0121852. DOI:10.1371/ journal.pone.0121852.

84. Faghani, H. E.; Gharechahi, J.; Komatsu, S.; Mirzaei, M.; Khavarinejad, R. A.; Najafi, F.; Farsad, L. K.; Salekdeh, G. H. Comparative Physiology and Proteomic Analysis of Two Wheat Genotypes Contrasting in Drought Tolerance. *J. Proteomics*. **2015**, *114*, 1–15.

85. Peremarti, A.; Marè, C.; Aprile, A.; Roncaglia, E.; Cattivelli, L.; Villegas, D.; Royo, C. Transcriptomic and Proteomic Analyses of a Pale-Green Durum Wheat Mutant Shows Variations in Photosystem Components and Metabolic Deficiencies Under Drought Stress. *BMC Genomics*. **2014**, *15*, 125. doi: 10.1186/1471-2164-15-125

86. Kumar, R, R.; Singh, G. P.; Goswami, S.; Pathak, H.; Rai, R. D. Proteome Analysis of Wheat (*Triticum aestivum*) for the Identification of Differentially Expressed Heat-Responsive Proteins. *AJCS*. **2014**, *8* (6), 973–986.

87. Peng, Z.; Wang, M.; Li, F.; Lv, H.; Li, C.; Xia, G. A Proteomic Study of the Response to Salinity and Drought Stress in an Introgression Strain of Bread Wheat. *Mol. Cell. Proteomics*. **2009**, *8*, 2676–2686. 10.1074/mcp.M900052-MCP200.

88. Kamal, A. H. M.; Rashid, H.; Sakata, K.; Komatsu, S. Gel-Free Quantitative Proteomic Approach to Identify Cotyledon Proteins in Soybean Under Flooding Stress. *J. Proteomics*. **2015**, *112*, 1–13.

89. Oh, M. W.; Nanjo, Y.; Komatsu, S. Gel-Free Proteomic Analysis of Soybean Root Proteins Affected by Calcium Under Flooding Stress. *Front. Plant Sci*. **2014**, *5*, 559. doi: 10.3389/fpls. 2014.00559.

90. Mustafa, G.; Komatsu, S. Quantitative Proteomics Reveals the Effect of Protein Glycosylation in Soybean Root Under Flooding Stress. *Front. Plant Sci*. **2014**, *5*, 627.

91. Yin, X.; Sakata, X.; Komatsu, S. Phosphoproteomics Reveals the Effect of Ethylene in Soybean Root under Flooding Stress. *J. Proteome. Res.* **2014**, *13* (12), 5618–5634. dx.doi.org/10.1021/pr500621c|

92. Khan, M. N.; Sakata, K.; Hiraga, S.; Komatsu, S. Quantitative Proteomics Reveals that Peroxidases Play Key Roles in Post-Flooding Recovery in Soybean Roots. *J. Proteome. Res.* **2014**, *13* (12), 5812–5828. doi: 10.1021/pr5007476.

93. Komatsu, S.; Han, C.; Nanjo, Y.; Altaf-Un-Nahar, M.; Wang, K.; He, D.; Yang, P. Label-Free Quantitative Proteomic Analysis of Abscisic Acid Effect in Early-Stage Soybean under Flooding. *Proteome. Res.* **2013**, *12* (11), 4769–4784. DOI: 10.1021/pr4001898

94. Ma, H.; Song, L.; Shu, Y.; Wang, S.; Niu, J.; Wang, Z.; Yu, T.; Gu, W.; Ma, H. Comparative Proteomic Analysis of Seedling Leaves of Different Salt Tolerant Soybean Genotypes. *J. Proteomics*. **2012**, *75*, 5. doi:10.1016/j.jprot.2011.11.026.

CHAPTER 15

STRATEGIES FOR BREEDING CEREAL CROPS TO ATTAIN SUSTAINABILITY WITH MAJOR EMPHASIS ON RICE

R. K. SALGOTRA[1*], B. B. GUPTA[2], and MEENAKSHI RAINA[1]

[1]*School of Biotechnology, Sher-e-Kashmir University of Agricultural Sciences & Technology of Jammu, Chatha, Jammu 180009, Jammu and Kashmir, India*

[2]*Division of Plant Breeding & Genetics, Sher-e-Kashmir University of Agricultural Sciences & Technology of Jammu, Chatha, Jammu 180009, Jammu and Kashmir, India*

*Corresponding author. E-mail: rks_2959@rediffmail.com

CONTENTS

ABSTRACT

Cereal crops are the main food staples for much of the world, with wheat, rice, and maize the most widely cultivated. Demand for cereal crops will rise because of global population growth, increased demand for cereal-based animal feed, and diversion of crops from foods to biofuels. The yields of cereal crops have already plateaued and most of the productivity gains in the future will have to be achieved through better management of natural resources and crop improvement. Among the various cereal crops rice is the most important food crop of the world. Global rice demand is estimated to rise from 676 million tons in 2013 to 852 million tons in 2035. To meet this challenge rice production on existing land must be increased. Rice has become a model plant for genetics and breeding research. Advances in molecular biology and genomics, proteomics and metabolomics have opened new avenues to apply innovative approaches to rice breeding. A large number of genes/quantitative trait loci (QTLs) for various traits have been tagged with molecular markers to apply molecular-assisted selection (MAS) for trait improvement. Genome sequence data have become an important source for detecting allelic variation. Applying these tools to crop improvement for cereal crops had potential to accelerate genetic progress. All these advances when integrated with conventional breeding will result in designer rice varieties to meet the challenges of rice food security on sustainable basis.

15.1   INTRODUCTION

The ever-increasing demand for food, feed, nutrition, and other basic needs has put a lot of pressure on the agriculture and the accompanied challenges associated with increasing crop productivity. The realistic utilization of genetic resources in agriculture has not only brought about profound changes in the crop productivity and quality, but also opened up newer and hitherto unforeseen potential vistas including improvement of novel traits. Due to overgrowing and burgeoning human population there is an enormous pressure on food security. The changes in climate cause leading to uneven rainfall patterns and other issues affect the plant growth. The most widely cultivated cereal crops are wheat, rice, and maize and are the main food staples for most of the world population. But in past few years or decades agriculture production has decreased. Due to this decrease in production the major goal of rice breeders is to increase rice productivity by maintaining the environment.

Rice is one of the major staple food crops for more than a third of the world's population. Rice is grown in well puddled and irrigated condition and requires two or three times more water than that of other food crops. World's most population is dependent on rice so there is an urgent need to increase the production of rice to meet the global demand. Various yield-attributing traits in rice include plant height, tiller number, grain weight, and panicle type. Among all these traits plant height is the most important trait related to plant architecture and it linearly correlates with biomass. Panicle number, which consists of planting density and effective tiller number, is a major influencing factor of the total grain production per unit area. Panicle characters are directly linked with the yield of rice. There are a number of factors that affect the rice production like abiotic stress, climate change, biotic stress, and so forth (Fig. 15.1). The major diseases of rice are; bacterial blight (*Xanthomonas oryzae pv.oryzae*), blast (*Magnaporthe griesa*), sheath blight (*Rhizoctonia solani*), tungro viruses (rice tungro spherical virus (RTSV) and rice tungro bacilliform virus (RTBV)), brown spot (*Bipolaris oryzae*), and number of other diseases.[1]

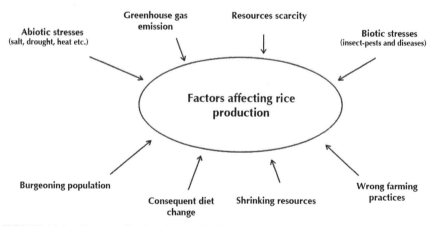

FIGURE 15.1 Factors affecting rice production.

15.1.1 SUSTAINABLE DEVELOPMENT AND ITS IMPORTANCE

In order to attain the goal of food security there is a need for sustainable development. Sustainable development will fulfill the human needs and protect environment so that food security is attained at present and in future also (Fig. 15.2). To make today's agriculture sustainable, it is necessary to adopt advanced technologies, exploit these technologies for improvement of agriculture, and increase the efficiency of method, for example selection process. More precision should be incorporated in the selection process. To meet the security of food supply, there is a need to increase the crop yield per unit area by 50% before 2030.[2] Before the whole genome sequencing of rice, researchers made lot of breakthroughs like development of semi-dwarf rice variety, development of hybrid rice variety, cytoplasmic male sterile (CMS) lines, and so forth and that gained the attention of the world. Breeders used conventional methods before the introduction of genomics to improve the rice yield and quality. These methods were based on qualitative traits only but not for quantitative traits. Conventional breeding selects genotype indirectly through phenotype. Later molecular techniques were developed, that allowed application of molecular breeding technology in rice and integrated the modern technology into conventional breeding.

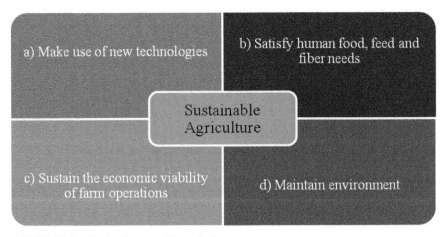

FIGURE 15.2   Goals of sustainable agriculture.

15.2   STRATEGIES FOR SUSTAINABLE AGRICULTURE IN RICE

The ultimate goal of crop breeding is to develop varieties having high yield and with other desirable agronomic characteristics. Breeders try to attain these goals but emphasis on one trait leads to instability in another, for example emphasis on high-quality tends to result in unstable yields, likewise introgressing the disease or insect resistant traits minimizes the quality.[3] So, to make the production sustainable all these points should be considered and should concentrate on varieties with maximum potential. The different breeding strategies used are enlisted in the Figure 15.3.

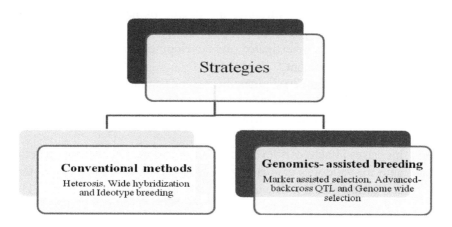

FIGURE 15.3   Different breeding strategies used for sustainable agriculture.

15.2.1   CONVENTIONAL METHODS

In this method the hybridization between diverse parents is done and in this way variability is created. Another way was selecting desirable individuals and then making cross between them. The individuals were selected on the basis of field observations and yield trials. This method is very time consuming and increase in production per year is very low. Recurrent selection and pedigree selection were traditional breeding methods for crop improvement. Population improvement through recurrent selection is not a widely used method. However, pedigree selection was widely used in rice. In general, every year the breeders used best breeding lines to make new elite varieties.

15.2.1.1   HETEROSIS

Heterosis, hybrid vigor, or outbreeding enhancement is the natural phenomenon whereby offspring exhibits enhanced trait quality as compared to both of its parents. George Harrison Shull coined the term "heterosis" in 1948. The concept of heterosis is also applied in the production of commercial plants and livestock and is used to increase yields, uniformity, and vigor. Heterosis has some unique characteristics that include:

(1) Heterosis is highly variable; the degree of heterosis varies with respect to the genetic distance of the parents, their reproductive mode, the traits investigated, the developmental stage of the plants, and the environment.[4]

(2) Heterosis is a universal phenomenon and can increase crop yields by 15–50% depending on crop type. Hybrid rice has a yield advantage of 20–30% over the best available inbred rice cultivars, facilitating a 44.1% increase in production.[5]

(3) Increases in heterosis level diminish over time. On average, genetic gain for yield was gradually increasing by 1.5–2.0% per year at the end of last century. The average yield gain in rice went from 3.1% per year in the 1980s to 1.4% per year in the 1990s and then 0.8% per year in the 2000s.

The three categories of heterosis that have been defined on the basis of genetic distance of parental lines include: (a) intra-specific heterosis, resulting from crosses between two accessions belonging to the same species, (b) inter-subspecific heterosis, resulting from crosses between two subspecies, and (c) wide-hybridization heterosis, resulting from crosses between two individuals of a different species or genus.

Hybrid breeding methods are used in many food crops such as maize, sorghum, rice, sugar beet, onion, spinach, sunflowers, and broccoli to create a more productive crop. Heterosis in rice was first reported by Jones (1926) who observed marked increase in culm number and grain yield in some F_1 hybrids in comparison to their parents. Since then, several rice researchers have reported the occurrence of this phenomenon for various agronomic traits such as yield, grain weight, grains per panicle, panicles per plant, plant height, days to flower, kernel size, and so forth.[6]

Hybrid breeding is faced with some problems. First, not every hybrid combination exhibits strong heterosis. Second, negative heterotic loci may occur simultaneously in the F_1 generation and must be removed in subsequent generations without compromising the degree of positive heterosis. Third, with time the heterosis level diminishes. Moreover, the degree of heterosis tends to increase with increasing genetic diversity of the parents; this also increases the likelihood of meiosis abnormalities, such as poor chromosome pairing. Hence, the divergence and stability of both parental and F_1 genomes influence seed yield and the stable inheritance of agronomic traits.

15.2.1.2 WIDE HYBRIDIZATION

Wide hybridization constitutes an important tool in cytogenetic and plant breeding to combine diverged desirable genomes into one nucleus. Wide hybridization includes the crossing or hybridization between plants from different species, belonging to the same or two different genera. It also provides the possibility of breaking the species barrier for the transfer of gene or whole genome from wild to cultivated species, thus helps in introducing the alien variation into the cultivated population. This transfer of genes from one species to another provides the raw material for species evolution and speciation, thus contributing toward the improvement of that crop species.[7] Wide hybridization includes both inter-specific (hybridization between two species of the same genus by sexual fusion) and inter-generic hybridization (hybridization between two different genera of the same family).

Inter-specific hybridization provides an important means of breeding many crops species, including many cereals, like wheat, barley, and oats. However, the situation is very different in rice, where the introduction of wild genes in cultivated varieties has been very rare, and the use of hybridization has been restricted to closely related species sharing the same genome[8]. Rice being the major food crop for more than half of the world's population attracts plant breeder around the world for its trait improvement. The cultivated rice (*Oryza sativa*) and its closely related wild species *Oryza perennis, Oryza nivara,* and *Oryza longistaminata* share the AA genome. These wild species can be easily crossed with *O. sativa* and genes from them can be transferred to cultivated rice by conventional crossing and backcrossing procedures. However, problem arises when the genomes other than AA are used for hybridization which is overcome by using embryo rescue method.

The first successful examples of transfer of a useful gene from wild species include the introgression of a gene for grassy stunt virus resistance from *O. nivara* to cultivated rice varieties[9] and transfer of a CMS source from wild rice to develop CMS lines for commercial hybrid rice production.[10] Moreover, genes for resistance to brown planthopper (BPH), bacterial blight, blast, tungro, acid sulfate soils, and iron toxicity have been introgressed from AA, BBCC, CC, CCDD, EE, and FF genomes into rice. Genes introgressed from wild species (*Bph10, Bph18, Xa21, Pi-9*) have been mapped and also used in marker-assisted selection (MAS).[11]

15.2.1.3   IDEOTYPE BREEDING

Donald (1968) proposed the ideotype approach to plant breeding in contrast to the empirical breeding approach of defect elimination and selection for yield per se.[12] He defined "crop ideotype" as an idealized plant type with a specific combination of characteristics favorable for photosynthesis, growth, and grain production based on knowledge of plant and crop physiology and morphology. In rice, Tsunoda (1962) compared yield potential and yield response to nitrogen (N) fertilizer in relation to the plant type of rice geno-types.[13] Varieties with high yield potential and greater responsiveness to applied N had short sturdy stems and leaves that were erect, short, narrow, thick, and dark green. The close association between certain morphological traits and yielding ability in response to N led to the "plant type concept" as a guide for breeding improved varieties.[14]

Simulation models predicted that a 25% increase in yield potential was possible by modification of the following traits of the current plant type.[15]

(1) Enhanced leaf growth combined with reduced tillering during early vegetative growth.
(2) Reduced leaf growth and greater foliar N concentration during late vegetative and reproductive growth.
(3) A steeper slope of the vertical N concentration gradient in the leaf canopy with a greater proportion of total leaf N in the upper leaves.
(4) Increased carbohydrate storage capacity in stems.
(5) A greater reproductive sink capacity and an extended grain-filling period. These traits are both physiological and morphological.

To break the yield potential barrier, International Rice Research Institute (IRRI) scientists proposed modifications to the high-yielding indica plant type in the late 1980s and early 1990s.[16] The newly designed plant type was mainly based on the results of simulation modeling and new traits were mostly morphological since they are relatively easy to select for compared with physiological traits in a breeding program. The proposed new plant type (NPT) has low tillering capacity (3–4 tillers when direct seeded); few unproductive tillers; 200–250 grains per panicle; a plant height of 90–100 cm; thick and sturdy stems; leaves that are thick, dark green, and erect; a vigorous root system; 100–130 days' growth duration; and increased harvest index.[17]

15.2.2 IDENTIFICATION OF GENETIC VARIATION CONTROLLING RICE AGRONOMIC TRAITS

To develop the rice with superior genotypes there is a need to understand the genetic basis of agronomically important traits and the allelic variation at those loci. Genomics assisted breeding has made the platform more easy and loci involved in agronomic traits have been mapped and their allelic variations have been assessed. Next generation sequencing (NGS) has been used to construct a genetic map for 150 rice recombinant inbred lines (RILs), which was 35 times more precise in recombination breakpoint determination.[18] Further, 49 QTLs with phenotypic effect ranging from 3.2 to 46.0% for 14 agronomic traits have been detected in these RILs, which indicated that NGS could provide a powerful solution to map QTL with high resolution.[19] On the basis of low-coverage sequences of RILs, a high-density linkage map was constructed using high-quality SNPs in RILs without genotype data of the parental lines.[20] The new SNP map detected more QTL with precise map locations, showing advantages in detecting power and resolution compared to the RFLP/SSR map.[21]

Another most preferred method for rice mapping is association mapping. It is easy to perform genome-wide association studies (GWAS) in natural population of rice. Bin Han and his colleagues performed a GWAS using 517 China landraces that had been selected from 50,000 rice accessions. Their analysis of 14 agronomic traits localized six previously cloned genes to regions less than 26 kb.[22] Furthermore, they extended the use of this methodology to 950 worldwide collected rice varieties and performed an additional GWAS that identified 32 new loci associated with flowering time and ten grain related traits.[23] Another study generating genome sequences from 446 geographically diverse accessions of the wild rice species *Oryza rufipogon* and 1083 cultivated *indica* and *Japonica* varieties constructed a comprehensive map of rice genome variation and identified 55 selective sweeps that have occurred during domestication.[24]

Gene cloning using T-DNA and EMS induced mutants is also used for mapping. In T-DNA mutants, tail polymerase chain reaction (PCR) is used for identification of insertion position. Mutmap, a new method developed for gene cloning is based on whole genome resequencing of pooled DNA. In this method mutant is crossed into its original wild type lines and the DNA of F_2 progenies with phenotypic differences is pooled and sequenced.[25] With the use of these methods and accumulation of gene information functional and geneic markers can be developed.

15.2.3 GENOMICS ASSISTED BREEDING

Plant genomics technologies have made a lot of breakthrough in today's agriculture. It has helped breeders in achieving targeted objectives to improve the quality and productivity of crops. The DNA based molecular markers have made the selection process easier by enabling early generation selection for key traits and thus, overcoming the drawbacks of conventional methods. By using this technology both genotype and phenotype of new varieties can be analyzed. The goal of these technologies is to integrate molecular tools in classical breeding and attain the sustainability. Some of the genomics based methods are:

15.2.3.1 MAS

MAS refers to the use of DNA markers that are tightly linked to target loci to assist phenotypic screening. By determining the allele of a DNA marker, plants that possess particular genes or quantitative trait loci (QTLs) may be identified based on their genotype rather than their phenotype. There are five characteristics for use of DNA markers in MAS. These are (a) reliability, (b) quantity, (c) quality, (d) technical procedures for marker assay, and (e) Level of polymorphism and cost.

The most widely used marker in rice is simple sequence repeat (SSR) marker or microsatellites. Some of the features of these markers are (a) highly reproducible, (b) co-dominant, (c) relatively simple, (d) cheap, and (e) highly polymorphic. MAS is based on marker instead of trait itself and which relies in the linkage of the marker and gene or QTL of interest.[26] The availability of markers spread throughout the genome facilitates introgression of individual or multiple genes or QTLs from one or several donors into the genetic background of a recurrent parent, as well as the gene pyramiding (i.e., the introduction of several genes or QTLs controlling traits of interest in a single genetic background). "Another powerful application of MAS is the breeding by design approach,[27] which allows predicting the outcome of a set of crosses on the basis of molecular markers information." There are three major steps involved in MAS: (a) identification of molecular marker(s) associated with trait(s) of interest, (b) validation of identified marker(s) in the genetic background of the targeted genotypes to be improved, and (c) marker-assisted backcrossing (MABC) to transfer the QTL/gene from the donor genotype into the targeted genotype.

The selection process can be independent from phenotype, which allows selection in off-season nurseries, making the technique more cost effective to grow for more generations per year. Another benefit of MAS is the sharp reduction of required population size because many lines can be discarded in earlier breeding generations after MAS. The efficiency and usefulness of MAS for traits of simple inheritance (i.e., qualitative traits controlled by one or a few genes) have been well proven in many crops, including rice.[26]

15.2.3.2   MARKER ASSISTED GENE PYRAMIDING

Gene pyramiding is a method aimed at assembling multiple desirable genes from multiple parents into a single genotype. The end product of a gene-pyramiding program is a genotype with all of the target genes. Using conventional phenotypic selection, individual plants must be phenotypically screened for all traits tested. Therefore, it may be very difficult to assess plants from certain population types (e.g., F_2) or for traits with destructive bioassays. DNA markers may facilitate selection because DNA marker assays are non-destructive and markers for multiple specific genes/QTLs can be tested using a single DNA sample without phenotyping (Fig. 15.4).

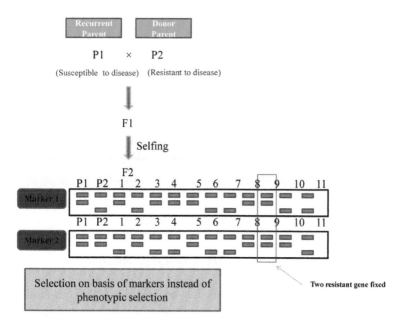

FIGURE 15.4   Marker assisted gene pyramiding.

15.2.3.3   MABC SELECTION

MABC is a method that is precise and effective to introgress a single locus controlling a trait of interest while retaining the genome of recurrent parent. This approach has been used with great success for "enhancing" rice varieties for traits such as bacterial blight resistance gene *XA21*,[28] the waxy locus for grain quality, and submergence tolerance SUB1.[29,30] Using markers located throughout the genome of the recurrent parent, the offspring with the highest percentage of recurrent parent genes can be identified quickly and accurately and within just a few generations, individuals with 95% or more of the recurrent parent genome can be identified, a level that would be nearly impossible to reach with conventional breeding (Fig. 15.5).

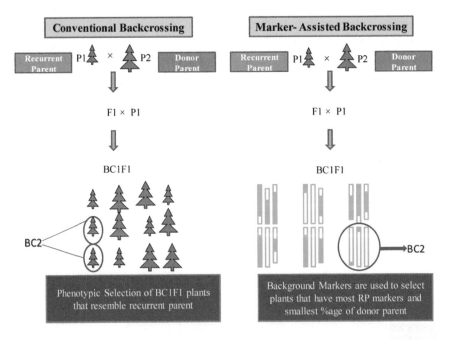

FIGURE 15.5   Marker assisted backcross selection.

15.2.3.3.1   *Advanced-Backcross QTL (AB-QTL) Analysis*

AB-QTL analysis was proposed by Tanksley and Nelson (1996) to simultaneously identify and introgress favorable alleles from unadapted donors into an elite background. The general steps that are involved in AB-QTL analysis consist of following experimental phases: (a) generating an elite parent through donor hybrid, (b) backcrossing to the elite parent to produce a BC_1 population that is subjected to marker or phenotypic selection against undesirable donor alleles, (c) genotyping BC_2 or BC_3 population with polymorphic molecular markers, (d) evaluating the segregated BC_2 or BC_2F_2 population for traits of interest and QTL analysis, (e) selecting target genomic regions bearing useful donor alleles for the production of near isogenic lines (NILs) in the elite genetic background, and (f) evaluation of the agronomic traits of the NILs and elite parent controls in replicated environments.[31–32]

The effectiveness of AB-QTL in rice was well demonstrated by the parallel AB-QTL studies for yield and yield components conducted by Cornell University. The same wild accession of *O. rufipogon* (IRGC 105491) was used as donor parent. The recurrent parents are elite varieties including the high-yielding Chinese hybrid V20/Ce64, the upland *O. sativa* subsp. japonica rice variety Caiapo from Brazil, the US long grain tropical japonica cultivar Jefferson, and the elite tropical cultivar IR64. In these studies, 30 to 50% of QTLs identified in each of the advanced backcross (BC_2F_2) populations showed improved performance for the target traits. Wild QTLs improve recurrent parent performance by 5 to 20% for most of the characters examined.[33] In several cases, specific *O. rufipogon* introgressions were found to be associated with superior performance across several genetic backgrounds and environments. By contrast, some yield and flowering time QTLs are associated with a positive effect in one genetic background and/or environment, but not in others.[33] In parallel experiments using recurrent parents from India, Korea, China, and the Philippines, essentially identical results were reported, with similar levels of transgressive variation and comparative percentages of "favorable wild QTLs" coming from other wild or weedy accessions of *O. rufipogon*. Subsequently, this method was used to develop populations with other *O. rufipogon* accessions.[34] Successful applications of the AB-QTL method have been reported by using another inter-specific population, which is derived from crosses between *O. sativa* and the wild species *Oryza glaberrima* (acc. IRGC#103544 from Mali),[35] *Oryza Glumaepatula*,[36] and *O. nivara*.[37] It is anticipated that the use of AB-QTL will be accelerated in a range of crops for improving important traits such as disease resistance as well as yield traits.

15.2.3.3.2   Genomic Selection (GS)

GS is a new breeding method in which genome-wide markers are used to predict the breeding value of individuals in a breeding population. GS can be used to pyramid favorable alleles for minor effect QTL at the whole genome level.[38] By analyzing the phenotypes and high-density marker scores GS predicts the breeding values of lines in a population. It calculates the marker effects across the entire genome that explains the entire phenotypic variation. If we explain in simple terms in GS we do not have to identify the marker associated with the trait. GS is a form of MAS that simultaneously estimates all locus, haplotype, or marker effects across the entire genome to calculate genomic estimated breeding values (GEBVs).[38] This approach contrasts greatly with traditional MAS because in this technology significant markers are not used for selection. Instead, GS analyzes jointly all markers on a population attempting to explain the total genetic variance with dense genome-wide marker coverage through summing marker effects to predict breeding value of individuals.[38] In this method, GS calculates GEBVs for every individuals of the progeny based on genotyping data using a model that was "trained" from the individuals of another training populations having both phenotyping and genotyping data. These GEBVs are then used to select the progeny lines for advancement in the breeding cycle. Thus, GS provides a strategy for selection of an individual without phenotypic data by using a model to predict the individual's breeding value.[39]

15.3 CONCLUSION

In summary, the future world food security requires continued and sustainable increase in the world rice production. With the development of genomics assisted breeding, it has made lot of progress in rice breeding. But still plant breeders have to face many challenges while adopting the new techniques. This may be due to shortage of trained personnel, inadequate access to genotyping, unaffordable bioinformatics system, inappropriate phenotyping infrastructure, and lack of integrating new technologies with traditional breeding. Increasing rice production to the level where we meet the demands of food security needs lot of efforts. Rice research should deal with the problems and find the keys to sustainability in intensive rice cultivation. A major problem that plant breeders have to face is to enhance food security in an environmental and sustainable manner. Genomics based breeding is likely to play a key role in the future practice of rice breeding. The development of genomics helps conventional breeding in exploiting the natural or artificially created diversity for developing new cultivars with improved characteristics. The future advances and development in rice will increase the production of rice and meet the need of food security.

KEYWORDS

- genomics
- hybrid breeding methods
- molecular breeding technology
- genome-wide markers
- next generation sequencing

REFERENCES

1. Mew, T.; Leung, H.; Savary, S.; Vera Cruz, C. M.; Leach, J. E. Looking Ahead in Rice Disease Research and Management. *Crit. Rev. Plant Sci.* **2004,** *23,* 103–127.
2. Cheng, S.; Hu, P. Development Strategy of Rice Science and Technology in China. *Chin. J. Rice Sci.* **2008,** *22,* 223–226.
3. Khan, M. H.; Dar, Z. A.; Sher, A. D. Breeding Strategies for Improving Rice Yield. *Agric. Sci.* **2015,** *6,* 467–478.
4. Groszmann, M.; Greaves, I. K.; Fujimoto, R.; James Peacock, W.; Dennis, E. S. The Role of Epigenetics in Hybrid Vigour. *Trends Genet.* **2013,** *29* (12), 684–690.
5. Cheng, S. H.; Zhuang, J. Y.; Fan, Y. Y.; Du, J. H.; Cao, L. Y. Progress in Research and Development on Hybrid Rice: A Super Domesticate in China. *Ann. Bot.* **2007,** *100* (5), 959–966.
6. Fu, D.; Xiao, M.; Hayward, A.; Fu, Y.; Liu, G.; Jiang, G.; Zhang, H. Utilization of Crop Heterosis: a Review. *Euphytica.* **2014,** *197,* 161–173.
7. Abbott, R.; Albach, D.; Ansell, S.; Arntzen, J. W. Hybridization and Speciation. *J. Evol. Biol.* **2013,** *26,* 229–246.
8. Bouharmont, J. Embryo Culture for Wide Hybridization in Rice. In *Rice;* Biotechnology in Agriculture and Forestry; Springer: Berlin, Germany, 1991, Vol. 14, pp 95–104.
9. Khush, G. S.; Ling, K. C.; Aquino, R. C.; Aquiero, V. M. In *Breeding for Resistance to Grassy Stunt in Rice,* Proceeding of 3rd International Congress, SABRAO, Plant Breeding Papers 1[4] Canberra, Australia, 1977, 3–9.
10. Lin, S. C.; Yuan, L. P. A Mass Screening Method for Testing Grassy Stunt Disease of Rice. Hybrid Rice Breeding in China. In *Innovative Approaches to Rice Improvement;* International Rice Research Institute: Manila, 1980; pp 35–51.
11. Zeliang, P. K.; Pattanayak, A. Wide Hybridization in the Genus *Oryza*: Aspcts and Prospects. *Indian J. Hill Fmg.* **2013,** *26* (2), 71–77.
12. Donald, C. M. The Breeding of Crop Ideotypes. *Euphytica.* **1968,** *17,* 385–403.
13. Tsunoda, S. A Developmental Analysis of Yielding Ability in Varieties of Field Crops. IV. Quantitative and Spatial Development of the Stem-System. *Jpn. J. Breed.* **1962,** *12,* 49–55.
14. Yoshida, S. Physiological Aspects of Grain Yield. *Annu. Rev. Plant Physiol.* **1972,** *23,* 437–464.

15. Dingkuhn, M.; Penning de Vries, F. W. T.; De Datta, S. K.; van Laar, H. H. Concepts for a New Plant Type for Direct Seeded Flooded Tropical Rice. In *Direct Seeded Flooded Rice in the Tropics;* International Rice Research Institute: Los Ban˜os, 1991; pp 17–38.

16. Khush, G. S. Breaking the Yield Frontier of Rice. *GeoJournal.* **1995,** *35,* 329–332.

17. Peng, S.; Khush, G. S.; Cassman, K. G. Evaluation of a New Plant Ideotype for Increased Yield Potential, In *Breaking the Yield Barrier;* Cassman, K. G., Ed.; Proceedings of a Workshop on Rice Yield Potential in Favourable Environments, International Rice Research Institute: Los Baños, 1994, 5–20.

18. Huang, X.; Feng, Q.; Qian, Q.; Zhao, Q.; Wang, L.; Wang, A.; Guan, J.; Fan, D.; Weng Q.; Huang, T. High-Throughput Genotyping by Whole Genome Resequencing. *Genome Res.* **2009,** *19,* 1068–1076.

19. Wang, L.; Wang, A.; Huang, X.; Zhao, Q.; Dong, G.; Qian, Q.; Sang, T.; Han, B. Mapping 49 Quantitative Trait Loci at High Resolution Through Sequencing Based Genotyping of Rice Recombinant Inbred Lines. *Theor. Appl. Genet.* **2011,** *122,* 327–340.

20. Xie, W.; Feng, Q.; Yu, H.; Huang, X.; Zhao, Q.; Xing, Y.; Yu, S.; Han, B.; Zhang, Q. Parent-Independent Genotyping for Constructing an Ultrahigh-Density Linkage Map Based on Population Sequencing. *Proc. Natl. Acad. Sci. U. S. A.* **2010,** *107,* 10578–10583.

21. Yu, H.; Xie, W.; Wang, J.; Xing, Y.; Xu, C.; Li, X.; Xiao, J.; Zhang, Q. Gains in QTL Detection Using an Ultra-High Density SNP Map Based on Population Sequencing Relative to Traditional RFLP/SSR Markers. *PLoS ONE.* **2011,** *6* (3), 17595.

22. Huang, X.; Wei, X.; Sang, T.; Zhao, Q.; Feng, Q.; Zhao, Y.; Li, C.; Zhu, C.; Lu, T.; Zhang, Z. Genome-Wide Association Studies of 14 Agronomic Traits in Rice Landraces. *Nat. Genet.* **2010,** *42,* 961–967.

23. Huang, X.; Zhao, Y.; Wei, X.; Li, C.; Wang, A.; Zhao, Q.; Li, W.; Guo, Y.; Deng, L.; Zhu, C. Genome-Wide Association Study of Flowering Time and Grain Yield Traits in a Worldwide Collection of Rice Germplasm. *Nat. Genet.* **2012,** *44,* 32–39.

24. Huang, X.; Kurata, N.; Wei, X.; Wang, Z. X.; Wang, A.; Zhao, Q.; Zhao, Y.; Liu.; K. Lu, H.; Li, W. A Map of Rice Genome Variation Reveals the Origin of Cultivated Rice. *Nature.* **2012,** *490,* 497–501.

25. Abe, A.; Kosugi, S.; Yoshida, K.; Natsume, S.; Takagi, H.; Kanzaki, H.; Matsumura, H.; Mitsuoka, C.; Tamiru, M.; Innan, H. Genome Sequencing Reveals Agronomically Important Loci in Rice Using MutMap. *Nat. Biotechnol.* **2012,** *30,* 174–178.

26. Collard, B. C.; Mackill, D. J. Marker Assisted Selection: an Approach for Precision Plant Breeding in the Twenty-First Century. *Philos. Trans. R. Soc. Lond. B Biol. Sci.* **2008,** *363,* 557–572.

27. Peleman, J. D.; van der Voort, J. R. Breeding by Design. *Trends Plant Sci.* **2003,** *8,* 330–334.

28. Chen, S.; Lin, X. H.; Xu, C. G.; Zhang, Q. F. "Improvement of Bacterial Blight Resistance of 'Minghui 63', an Elite restorer Line of Hybrid Rice, by Molecular Marker-Assisted Selection." *Crop Sci.* **2000,** *40,* 239–244.

29. Neeraja, C. N.; Maghirang-Rodriguez, R.; Pamplona, A.; Heuer, S.; Collard, C. Y.; Septiningsih, E. M.; Vergara, Sanchez, G.; Xu, D. K.; Ismail, A. M.; Mackill, D. J. A Marker-Assisted Backcross Approach for Developing Submergence-Tolerant Rice Cultivars. *Theor. Appl. Genet.* **2007,** *115,* 767–776.

30. Septiningsih, E. M.; Pamplona, A. M.; Sanchez, D. L.; Neeraja, C. N.; Vergara, G. V.; Heuer, S. A.; Ismail, M.; Mackill, D. J. Development of Submergence Tolerant Rice Cultivars: The SUB1 Locus and Beyond. *Ann. Bot.* **2009,** *103,* 151–160.

31. Tanksley, S. D.; Nelson, J. C. Advanced Backcross QTL Analysis: a Method for Simultaneous Discovery and Transfer of Valuable QTL from Unadapted Germplasm into Elite Breeding Lines. *Theor. Appl. Genet.* **1996,** *92,* 191–203.

32. GUO, L.; Guo-you, Y. E. Use of Major Quantitative Trait Loci to Improve Grain Yield of Rice. *Rice sci.* **2014,** *21* (2), 65–82.

33. McCouch, S. R.; Sweeney, M.; Li, J. M.; Jiang, H.; Thomson, M.; Septiningsih, E.; Edwards, J.; Moncada, P.; Xiao, J. H.; Garris, A.; Tai, T.; Martinez, C.; Tohme, J.; Sugiono, M.; McClung, A.; Yuan, L. P.; Ahn, S. N. Through the Genetic Bottleneck: *O. rufipogon* as a Source of Trait Enhancing Alleles for *O. sativa. Euphytica.* **2007,** *154* (3), 317–339.

34. Wickneswari, R.; Bhuiyan, M. A. R.; Kalluvettankuzhy, K. S.; Lim, L. S.; Thomson, M. J.; Narimah, M. K.; Abdullah, M. Z. Identification and Validation of Quantitative Trait Loci for Agronomic Traits in Advanced Backcross Breeding Lines Derived from *Oryza rufipogon × Oryza sativa* Cultivar MR219. *Plant Mol. Biol. Rep.* **2012,** *30* (4), 929–939.

35. Li, J. M.; Xiao, J. H.; Grandillo, S.; Jiang, L. Y.; Wan, Y. Z.; Deng, Q. Y.; Yuan, L. P.; McCouch, S. R. QTL Detection for Rice Grain Quality Traits Using an Inter Specific Backcross Population Derived from Cultivated Asian (*O. sativa L.*) and African (*O. glaberrima S.*) Rice. *Genome.* **2004,** *47* (4), 697–704.

36. Brondani, C.; Rangel, P. H. N.; Brondani, R. P. V.; Ferreira, M. E. QTL Mapping and Introgression of Yield-Related Traits from *Oryza glumaepatula* to Cultivated Rice *Oryza sativa* Using Microsatellite Markers. *Theor. Appl. Genet.* **2002,** *104,* 1192–1203.

37. Eizenga, G. C.; Prasad, B.; Jackson, A. K.; Jia, M. H. Identification of Rice Sheath Blight and Blast Quantitative Trait Loci in Two Different *O. sativa/O. nivara* Advanced Backcross Populations. *Mol. Breed.* **2013,** *31,* 889–907.

38. Meuwissen, T. H. E.; Hayes, B. J.; Goddard, M. E. Prediction of Total Genetic Value Using Genome-Wide Dense Marker Maps. *Genetics.* **2001,** *157,* 1819–1829.

39. Heffner, E. L.; Sorrells, M. E.; Jannink, J. L. Genomic Selection for Crop Improvement. *Crop Sci.* **2009,** *49,* 1–12.

INDEX

T

Printed and bound by CPI Group (UK) Ltd, Croydon, CR0 4YY

23/10/2024

01777705-0013